nurses' handbook of fluid balance

NORMA MILLIGAN METHENY, R.N., M.S.N., Ph.D.

Associate Professor of Nursing,
Graduate Medical–Surgical Nursing Program,
Saint Louis University, Saint Louis, Missouri.

W. D. SNIVELY, JR., M.D., F.A.C.P.

Professor Emeritus of Life Sciences,
University of Evansville, Evansville, Indiana;
Clinical Professor of Pediatrics,
Indiana University School of Medicine,
Evansville Medical Center, Evansville, Indiana.

WITH 7 ADDITIONAL CONTRIBUTORS

J. B. LIPPINCOTT COMPANY *Philadelphia*
London   Mexico City   New York
Saint Louis   São Paulo   Sydney

nurses' handbook of

fluid balance

FOURTH EDITION

Acquisitions Editor: David T. Miller
Sponsoring Editor: David T. Miller
Manuscript Editor: Leslie E. Hoeltzel
Indexer: Julia Schwager
Art Director: Tracy Baldwin
Designer: Adrianne Onderdonk Dudden
Production Assistant: Susan A. Caldwell
Compositor: The Clarinda Company
Printer/Binder: The Murray Printing Company

 A David T. Miller Book

Library of Congress Catalog Card Number 78-24364
Printed in the United States of America

6 5 4 3 2 1

Library of Congress Cataloging in Publication Data

Metheny, Norma Milligan.
  Nurses' handbook of fluid balance.

  Includes bibliographies and index.
  1. Body fluid disorders—Nursing. 2. Body fluids.
I. Snively, William Daniel, 1911– . II. Title.
[DNLM: 1. Body fluids—Nursing texts. 2. Nursing care.
3. Water-electrolyte imbalance—Nursing texts. WY 150 M592n]
RB144.M47 1983   616   82–17148
ISBN 0-397-54381-6

The authors and publisher have exerted every effort to ensure that drug selection and dosage set forth in this text are in accord with current recommendations and practice at the time of publication. However, in view of ongoing research, changes in government regulations, and the constant flow of information relating to drug therapy and drug reactions, the reader is urged to check the package insert for each drug for any change in indications and dosage and for added warnings and precautions. This is particularly important when the recommended agent is a new or infrequently employed drug.

Contributors

Donna R. Helmer, M.A.
Medical Writer and Researcher
Evansville, Indiana

Anne G. Perry, R.N., M.S.N.
Assistant Professor of Nursing
Coordinator, Respiratory Option
Medical–Surgical Nursing Graduate Program
Saint Louis University
Saint Louis, Missouri

Darnell T. Roth, R.N.
Intravenous Therapy Consultant
Alexian Brothers Hospital
Saint Louis, Missouri

Roberta P. Scofield, R.N., M.S.N.
Oncology Clinical Nurse Specialist
Veterans Administration Medical Center
Saint Louis, Missouri

Catherine A. Smith, R.N., M.S.N.
Cardiovascular Clinical Nurse Specialist
Saint Louis, Missouri

Martha A. Spies, R.N., M.S.N.
Assistant Professor of Nursing
Saint Louis University
Saint Louis, Missouri

Ebert Westfall
Medical Artist
Evansville, Indiana

Preface

Accumulation of knowledge of practical import in the field of body fluid disturbances is at least keeping pace with advances in other fields of health science. Hence the authors and publisher believe that the time is here for a new, thoroughly revised edition of the *Nurses' Handbook of Fluid Balance*, a textbook whose acceptance has been a source of great gratification to those involved in its preparation and production!

In this Fourth Edition, particular emphasis is focused on the following:

1. Several of the basic body fluid disturbances have been added to the "basic" list.
2. An explanation of the complex syndrome of inappropriate antidiuretic hormone secretion has been added.
3. The entire section of acid–base disturbances has been rewritten and greatly enlarged with new approaches incorporated.
4. Phosphate, chloride, ammonium, and lithium and their divergent roles have been discussed.
5. New chapters have been added on special fluid and electrolyte problems in the geriatric patient and in the oncologic patient.
6. The chapter on endocrinology has been completely rewritten and enlarged with an emphasis on nursing assessment based on physiologic alterations.
7. The chapter on heat disorders has been completely rewritten with the most recent classification and advanced methods of therapy discussed.
8. A section on nutritional assessment has been added.
9. Tube feedings have been explained in greater up-to-date detail.
10. Sections on immobilization, low-sodium diets, and kidney stones have been updated and enlarged considerably.
11. Total parenteral nutrition has been discussed in greater detail.
12. Treatment of burns, ever changing, has been reviewed and updated.
13. Toxemia of pregnancy has been reviewed and updated, as has water intoxication associated with induction of labor with oxytocin.
14. The section on recognition and management of shock has been enlarged greatly.
15. Nursing assessment of fluid and electrolyte disorders has been updated and enlarged.
16. Chapters on management of fluid balance problems related to patients with cardiovascular and respiratory problems have been revised and updated thoroughly.

As before, this working handbook is divided into basic sections and sections on practical applications, targeted at nurses and other members of the allied health professions. Continuously, from the first edition in 1967, the authors have enjoyed the wholehearted support and cooperation of David T. Miller, Vice President and Editor, J. B. Lippincott Company, and his staff.

Norma Milligan Metheny, R.N., M.S.N., Ph.D.

W. D. Snively, Jr., M.D., F.A.C.P.

Contents

nurses' handbook of fluid balance

1

Foundation Concepts

Humans are 60% to 80% fluid by weight, depending upon age, sex, and body fat content. In general, the younger the person, the higher the percentage of body fluid. Women have less body fluid than do men, and a fat person contains less fluid than does a thin person, since fat has little water associated with it.

The body fluid is divided into two major compartments. The first compartment, the *cellular fluid*, comprises the fluid contained within the billions of body cells and accounts for about three fourths of the total body fluid. We might think of it as a vast multitude of tiny, encapsulated droplets suspended in the second compartment, the *extracellular fluid*. Extracellular fluid (ECF) constitutes about one fourth of the total body fluid (Figs. 1-1 and 1-2). The great French physiologist Claude Bernard designated cellular fluid as the "environment of life" and ECF as the "internal environment." It was he who first envisioned the two types of body fluid as separate entities.

Body fluid, which has proved to be remarkably fit for the external environment in which we exist, might properly be regarded as our own tiny portion of the sea. Why? Scientists explain it thus: about 2 billion years ago, life began in the ocean. Just what that first life form was is uncertain, but it may have been the protozoan. It appears reasonable that human cellular fluid had its origin in that simple single-celled creature of the pre-Cambrian sea. The ocean water surrounding that first microorganism and its infinite progeny contained everything necessary to maintain life: its physical characteristics were suitable, and dissolved in it were oxygen and other gases plus various nutrients. As the tiny organism floated about in its watery home, there was a constant exchange of materials between the sea water within the cell and the sea water surrounding it. The composition of our cellular fluid bears a striking resemblance to that of the pre-Cambrian sea, the sea incorporated into the single-celled body of the first life form.

Over the course of millions of years, the single-celled organisms joined together, forming many-celled metazoa. These creatures too were dependent on the ocean for their physicochemical needs. Eons later, certain sea creatures enclosed within their bodies additional sea water. This ECF, which surrounded the cells and filled the interstitial spaces between them, provided the creatures with an internal environment for their body cells and thus freed them of dependence on their external environment, the sea. This gigantic step in the evolutionary process made it possible for our

1

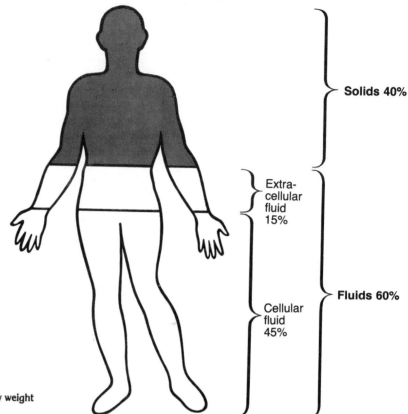

FIG. 1-1. Fluid and solid components of body weight in the adult.

FIG. 1-2. Fluid and solid components of body weight in the newborn.

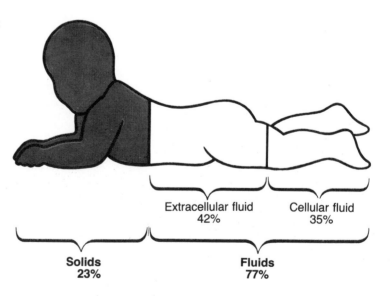

venturesome aquatic ancestor (related to the Coelacanth) to leave the Cambrian sea some 300 million years ago. In the same manner as cellular fluid resembles the pre-Cambrian sea, so does our ECF resemble the Cambrian sea in its chemical composition. The composition of sea water changes over the years as the continental land masses gradually dissolve and wash into it. So too, the two types of body fluid vary greatly in composition, with sodium predominant in ECF and potassium predominant in cellular fluid, since they were enclosed within the body during the two different geologic periods.

During the evolutionary process, the ECF underwent two further subdivisions. Once the sea water became enclosed within the body, it had to be circulated to carry nutrients and other substances to the cells and waste materials away from them. A heart and a circulatory system evolved, and part of the ECF—about a fourth of it—was appropriated for circulation through the system. This intravascular ECF is the *plasma*—the liquid fraction of the blood. The other three fourths of the ECF lies outside the blood vessels in the interstitial spaces between the body cells. This extravascular portion of the ECF is called *interstitial fluid*. Lymph and cerebrospinal fluid, while they have highly specialized and unique functions, are usually regarded as interstitial fluid. In some respects, lymph appears to be a sort of physiologic afterthought, since it serves many purposes: it combats infection, traps tumor cells, serves important circulatory functions, and even transports nutrients—chiefly fatty acids—from the intestine to the venous circulation. Cerebrospinal fluid is the interstitial fluid of the central nervous system.

Consider what might be termed *microscopic anatomy of the body fluids* (Fig. 1-3). If one were to take a cross section of a solid tissue, such as muscle, one might see something like this: ECF (including plasma, located within the blood vessels, and interstitial fluid, located outside the blood vessels) surrounding and bathing the cells. Cellular fluid would be seen inside body cells, including both the cells inside and those outside the blood vessels.

In addition to the two major divisions of body fluid, the cellular fluid and ECF, there are other essential fluids—the *secretions* and *excretions*. The secretions include the juices manufactured in the stomach, pancreas, liver, and intestine. Urine and feces are excretions. Perspiration is also regarded as an excretion, although its primary purpose is not to rid the body of wastes but rather to aid in body heat regulation. Even

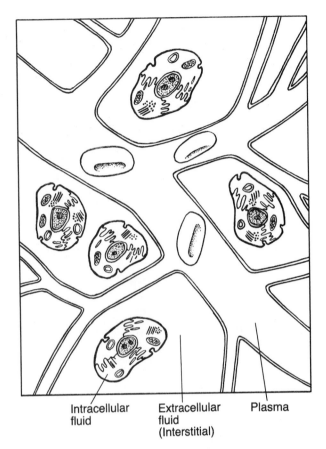

Intracellular fluid   Extracellular fluid (Interstitial)   Plasma

FIG. 1-3. Microscopic anatomy of body fluids.

though secretions and excretions are derived from ECF, they are not actually ECF.

A *topographic analogy of body fluids* (Fig. 1-4) might be helpful in understanding the relationships between the various body fluids. Cellular fluid can be regarded as a vast ocean that holds three fourths of the body fluid. It is fed by the ECF, which is pictured as a great river. The secretions and excretions represent streams and rivulets flowing from the ECF, which supplies them with all needed materials.

COMPOSITION OF BODY FLUIDS

Body fluid consists chiefly of *water* and certain dissolved substances sometimes referred to as salts, minerals, or crystalloids, but more correctly called *electrolytes*.

Water is the most essential nutrient to life. No

FIG. 1-4. Topographic analogy of body fluid.

known form of plant or animal life can exist for very long without it. Indeed, humans can live a long time without other nutrients, but only for a few days without water. Water possesses unique chemical and physical characteristics, and there is no substitute for it in the living cycle. Water's boiling point (100° C) and its freezing point (0° C) enable the human body to withstand all but the most drastic changes in heat or cold. Water is the closest substance known to a universal solvent. Since the adult body is from 60% to 70% water, each one of us is, in a real sense, a bag of more or less solid materials dissolved in water. Water is required for the countless chemical reactions of the body; no major physiologic function can proceed without it.

Electrolytes are so named because they ionize—develop electrical charges—when they are dissolved in water. Some electrolytes, including sodium, potassium, calcium, and magnesium, develop positive charges; some, including chloride, bicarbonate, sulfate, phosphate, proteinate, carbonic acid, and other organic acids, develop negative charges.

One can determine which electrolytes carry positive charges and which carry negative charges by placing the electrolyte in question in a wet electric cell through which an electric current is conducted (Fig. 1-5). Such a cell has a negative pole called a *cathode* and a positive pole called an *anode*. Since unlikes attract, positively charged particles travel to the negative pole, or cathode. Such particles, therefore, are designated as

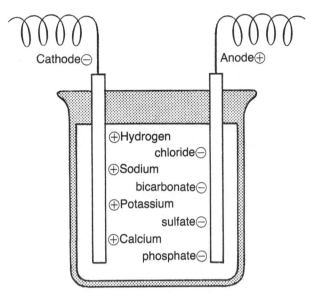

FIG. 1-5. In a wet electric cell, cations go to the negative pole, or cathode, and anions go to the positive pole, or anode.

cations. Negatively charged particles, called *anions*, migrate to the positive pole, or anode. The term *ion* refers to both cation and anion. In all electrical systems, positive charges and negative charges balance each other. Cations and anions exist in equal strength and numbers when measured according to their chem-

ical activity or ability to unite with other ions to form molecules (Fig. 1-6).

Each body fluid has its own normal composition of water and electrolytes. Cellular fluid contains large amounts of potassium, magnesium, and phosphate, and only small amounts of other electrolytes. One should not think for a moment that these differences in composition between cellular and ECF are unimportant. Indeed, so important are they that the body uses up one fifth of its energy stored in adenosine triphosphate (ATP) in maintaining them.

The chief difference between the plasma and the interstitial fluid is that plasma contains a much greater amount of proteinate, which acts as a sponge to prevent excess plasma from seeping into the interstitial fluid.

The composition of secretions and excretions varies considerably, particularly in illness. Therefore, the quantities of electrolyte per unit volume must be related to a constant, which consists of the levels of sodium, potassium, chloride, and bicarbonate in plasma. On this basis, we find that gastric secretions have about half the sodium concentration of plasma, about the same potassium, about half again as much chloride, and a fraction of the bicarbonate. Pancreatic secretions have about the same concentration of sodium and potassium, about a third the chloride, and about

FIG. 1-6. Cations and anions of the extracellular fluid (plasma) balance each other when expressed in milliequivalents.

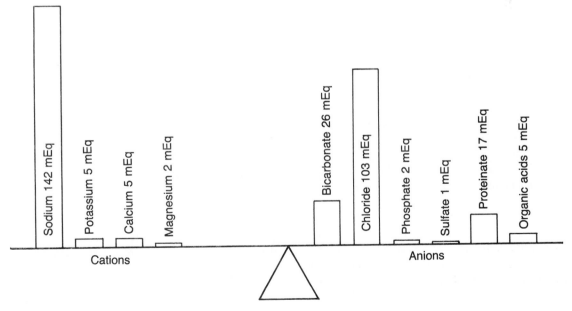

three times the bicarbonate. Perspiration, a dilute secretion, has more than half the sodium and chloride and about the same potassium. Diarrheal stools in children have a fraction of the sodium, several times as much potassium, a fraction of the chloride, and probably much more bicarbonate.

FUNCTIONS OF BODY FLUIDS

The main function of body fluids is to maintain healthy living conditions for the body cells. Although we may regard ourselves as highly complex, highly civilized organisms, the cells remain the basic units of life. A living body could no more exist without cells than could a brick building exist without bricks. Unlike bricks, however, our cells are live, dynamic, working units that must be nourished. The cellular fluid, which normally contains all the nutrients the cells need, serves as the supply source to replenish nutrients as they are used up. In addition, the cellular fluid must be cleared of wastes, such as carbon dioxide and breakdown products of protein; the ECF performs these services. Nutrients and other materials seep from the plasma into the interstitial fluid at the arterial end of the capillary beds, which exist in every part of the body, and are carried to the cells by means of the interstitial fluid. Waste materials pass from the cellular fluid into the interstitial fluid and back to the plasma through the venous capillaries. The plasma then sorts the waste products for storage or excretion and carries them to their proper destinations.

In addition to transporting nutrients and wastes, ECF transmits enzymes and hormones, plus many additional substances. It carries red blood cells through the body and white blood cells to attack bacteria.

ROUTES OF TRANSPORT

Materials are transported between cellular and extracellular fluid by several routes, including osmosis, diffusion, filtration, active transport, pinocytosis, and phagocytosis, of which *osmosis* is the most important.

OSMOSIS A quick way of describing osmosis is to say, "Water goes where salt is." This means that if there are two solutions separated by a semipermeable membrane, the solution with the greatest concentration of electrolyte draws water from the solution with the lesser concentration of electrolyte (Fig. 1-7). It is almost as if there were an effort on the part of each electrolyte particle to surround itself with its fair share of the available water.

FIG. 1-7. The principle of osmosis: "Water goes where the salt is."

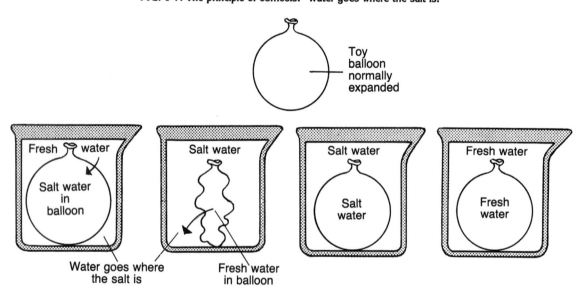

Osmosis is highly important in the body. If pure water not containing electrolytes were injected directly into the bloodstream, the red blood cells would absorb water and would swell and burst. If an extremely salty solution were injected into a vein, the red blood cells would lose water to their salty environment and would shrink, just as your fingers do if they have been in water too long. Osmosis occurs when the ECF develops an electrolyte content lower or higher than normal owing to disease or an accident such as drowning.

Osmotic pressure refers to the drawing power for water and depends on the number of molecules in solution. Electrolytes and substances such as mannitol with their low molecular weights exert ordinary osmotic pressure. Albumin and other substances with high molecular weights exert a special kind of osmotic pressure known as *colloid osmotic pressure* or *oncotic pressure*. Colloids possess special importance since they cannot pass through the capillary wall, the barrier between plasma and interstitial fluid. Mannitol, with its low molecular weight, raises the osmotic pressure of plasma when given intravenously, whereas human albumin, having an extremely high molecular weight, exerts an oncotic pressure.

DIFFUSION The physical process of transport, called diffusion, occurs through the random movement of ions and molecules, which tend to become equally concentrated in all parts of a vessel. The molecules move incessantly, bump into each other, and bounce away. They scatter from regions where their concentration is high and pass to regions where the concentration is low. Exchanges of oxygen and carbon dioxide by the lung alveoli and capillaries occur through diffusion.

ACTIVE TRANSPORT When it is necessary for ions to move from areas of lesser concentration to areas of greater concentration, active transport occurs, whereby ATP is released from a cell to enable certain substances to acquire the energy needed to pass through the cell membrane. This method of transfer, still not fully understood, may involve carrier systems located within cell membranes. Sodium ions, potassium ions, and amino acids are probably carried through all cell membranes by active transport, which also occurs in the transfer of chloride ions, hydrogen ions, phosphate ions, calcium ions, magnesium ions, creatinine, and uric acid through some cell membranes.

FILTRATION The transfer of water and dissolved substances through a permeable membrane from a region of high pressure to a region of low pressure is called filtration. The force behind filtration is *hydrostatic pressure*, produced by the pumping action of the heart. Examples of filtration include the passage of water and electrolytes from the arterial end of the capillary beds to the interstitial fluid and the passage of water and small molecules from the glomerular capillaries of the kidneys into the tubules. Opposing the hydrostatic pressure, which tends to force water and electrolytes out of the capillaries, is the oncotic pressure of the plasma proteins, which tends to hold them back.

PINOCYTOSIS AND PHAGOCYTOSIS A form of movement by which large molecular weight substances, such as protein, enter body cells is pinocytosis (protein molecules are taken into the cell by invagination of the cell membrane). In phagocytosis, the microorganisms, cells, or foreign particles are engulfed or digested by phagocytes. Leukocytes employ phagocytosis in attacking bacteria.

It is through these methods of exchange that the body cells are nourished and excrete wastes, which greatly alters the composition of ECF. The effects would be catastrophic were it not for chemical regulatory activities that are carried out by the body. This enables the pond within the skin to maintain its compositional integrity and, at the same time, to keep the body cells just as healthy as if they were still floating in the wide blue Cambrian sea.

ORGANS OF HOMEOSTASIS

For the maintenance of health, the body must maintain homeostasis—the normal state of the body fluids. The volume and electrolyte composition of ECF must be held very close to normal, in view of the many possible abnormalities involving water or electrolytes. Body cells constantly pour the results of chemical reactions into the ECF; they constantly withdraw from it substances needed for specific organ or cell activities. We eat and drink a wide variety of materials not needed by our body, and the digestive tract indiscriminately absorbs most of the substances offered it.

The volume of the liquids we drink would overcome us if there were no mechanism for maintaining a constant ECF volume. These problems in health are multiplied in disease, and this adds to the already gigantic task of maintaining homeostasis.

Fortunately, nature has provided each of us with a system of automation, which demonstrates the wisdom of the body. In its handling of a complex task, this system, called the *homeostatic mechanism*, puts man-made automation to shame. Claude Bernard, with his amazing foresight, sensed the presence of these mechanisms even though he had no precise knowledge concerning them, for, in 1857, he said, "All the vital mechanisms, however varied they may be, have always but one end, that of preserving the constancy of the conditions of life in the internal environment." What he clearly meant by the internal environment was the body ECF. The homeostatic mechanism utilizes every body system and every body organ, with the possible exception of the reproductive organs. The lungs, kidneys, heart, adrenal glands, pituitary gland, and parathyroid glands are particularly involved, and because these organs are so important, we call them the *organs of homeostasis* (Fig. 1-8).

The lungs serve to regulate the oxygen level and carbon dioxide level of blood. Since carbon dioxide stems from the carbonic acid in the blood, the lungs help maintain the extremely important balance between acids and alkalies in ECF. As far as fluid balance is concerned, perhaps the chief role of the lungs lies in excreting CO_2 (an acid material) when there is an excessive concentration of hydrogen ions in the ECF or in retaining CO_2 when a deficit of hydrogen ions exists.

FIG. 1-8. Key body homeostatic mechanisms.

Pituitary mechanism
- ADH from posterior pituitary

Parathyroid mechanism
- Parathormone

Renocardiovascular mechanism
- At least 8 cardinal functions

Pulmonary mechanism
- Hyperventilation/hypoventilation

Adrenal mechanism
- Cortex: aldosterone
- Medulla: epinephrine

Next, we have the great renocardiovascular mechanism. It includes the kidneys, the blood vessels, which bring blood to the kidneys, and the heart, which pumps the blood. How important the kidneys are! They might be called the master chemists of our body fluids, for it is not what we eat and drink, but how the kidneys function that determines the volume and chemical composition of ECF. They excrete chemical wastes and remove from the ECF a great variety of foreign substances indiscriminately absorbed by the digestive tract. The kidneys further sort out electrolytes as they pass from the blood through their filtering beds, excreting all but those that the body needs. Consider: the kidneys have an *excretory* function. They excrete the products of protein catabolism; they excrete powerful acids, such as phosphoric and sulfuric acids; they excrete certain drugs and toxins. They have a *regulatory* function, adjusting the concentration of electrolytes and the quantity of water in the ECF. In addition, they play an important role in regulating blood pressure through the secretion of renin from the juxtaglomerular apparatus of the kidneys. The kidneys also *regenerate*, producing bicarbonate when it is needed. They *convert* one form of vitamin D, which the body cannot use, to a form that the body can use. The kidneys even have a *manufacturing* function, synthesizing erythropoietin needed for normal red blood corpuscle function. We can exist for hours, days, or even longer without the use of our bones, muscles, digestive organs, nerves, endocrine glands, and even our cerebrum, but if the kidneys totally cease their chemical regulation of the ECF for an hour, physiologic deterioration promptly commences.

The kidneys are completely dependent on the heart, which pumps about 1700 liters of blood to the kidneys for cleansing every day. Some 180 liters of this amount is filtered through the kidneys, and approximately 1.5 liters of this daily filtrate pass out as urine—the rest is reabsorbed.

The adrenal glands, located above the kidneys, secrete numerous hormones that influence the body in many ways, both in health and in disease. These glands function in the retention and excretion of water and electrolytes, especially sodium, chloride, and potassium. As far as body fluid disturbances are concerned, perhaps the key adrenal hormone is aldosterone. Secreted in the zona glomerulosa of the adrenal cortex, it is the great sodium saver. While conserving sodium, aldosterone saves chloride and water as well and causes the excretion of potassium. Epinephrine, a hormone of the adrenal medulla, imitates the sympathetic nervous system in times of emergency. It increases the blood pressure, enhancing pulmonary ventilation, dilating blood vessels needed for meeting emergencies (e.g., those of the skeletal muscles), constricting those not needed (e.g., those of the gastrointestinal tract), and inhibiting the emptying of the contents of the gastrointestinal tract and bladder.

The pituitary gland, located within the brain, directly affects the body's conservation of water through its secretion of antidiuretic hormone (ADH). This water-conserving hormone is manufactured in the hypothalamus and then stored in the posterior pituitary gland, from where it is released when needed. ADH is the great water-controlling hormone, acting on the pores of the collecting ducts of the nephrons to determine the amount of water reabsorbed from the tubular urine.

The parathyroid glands are the last of the major organs of homeostasis. Located near or embedded in the thyroid gland, these pea-sized glands regulate the level of calcium in the ECF by their secretion of the important hormone parathormone, which acts to elevate the calcium level in the ECF. Without parathormone, one would die of calcium deficit within a few days' time.

The organs of body automation usually function efficiently; nevertheless, when any of the organs of homeostasis does malfunction, those persons caring for the patient face a formidable task.

UNITS OF MEASURE

Many years ago, a man named John Selden said,

. . . if they should make the standard for the measure we call a "foot" a Chancellor's foot; what an uncertain measure would this be! One Chancellor has a long foot, another a short foot, a third an indifferent foot . . .

Master Selden was pointing out, quite correctly, that, without standard or invariable units of measure, we are severely limited. In the same light, we could never understand, diagnose, and treat body fluid disturbances without simple and accurate units of measure.

Because of the constant exchanges between cellular and extracellular fluid, changes in either are reflected in the other. Therefore, imbalances in the ECF ultimately cause imbalances in the cellular fluid, and vice versa.

Because of its ready availability, we focus our chief attention upon the extracellular portion of the body fluid when we study imbalances of water and electrolytes.

MEASUREMENT OF VOLUME Among the most essential units of measure are the measurements of volume. The metric system of weights and measures is used universally in science. In the metric system, the *liter* is used for volume measurement. The liter is broken down into 1000 parts or milliliters (ml), each milliliter representing 1/1000 of a liter. A milliliter is virtually identical to the cubic centimeter (cc), but the milliliter is preferred in determining volume because the centimeter is a linear rather than a volumetric unit of measure. Expressed in terms of weight rather than volume, a liter of water weighs 1000 grams (g), or about 2.2 pounds.

MEASUREMENT OF CHEMICAL ACTIVITY The electrolytes of the body fluid are dynamic, active chemicals. Since we are interested in their activity, we must have a unit that expresses *chemical activity*, or *chemical combining ability*, or the power of cations to unite with anions to form molecules. Virtually any cation can unite with any anion. The cation sodium, for example, combines with the anion chloride to form the molecule sodium chloride. Sodium united with proteinate forms the molecule sodium proteinate. Or, it may combine with nitrate, forming sodium nitrate. Similarly, the cations potassium, magnesium, and calcium can unite with any of the anions. Therefore, why could we not merely use as our unit of measure the weight of the ions in which we are interested? Unfortunately, this does not solve our problem, since the *weight* of a chemical bears no relation to its *chemical activity*. One milligram of sodium unites chemically with *180* mg of proteinate to form the compound sodium proteinate, for example.

In searching for a unit of chemical activity, chemists have discovered the *milliequivalent* (mEq). A traditional measurement of physical power in our civilization is the power of an imaginary average horse; our unit of chemical power, the milliequivalent, is *equivalent* to the activity of 1 mg of hydrogen. In other words our *chemical horse is 1 mg of hydrogen.* One milligram of hydrogen exerts 1 mEq of chemical activity; so do 23 mg of sodium, 39 mg of potassium, 20 mg of calcium, 35 mg of chloride, or 4140 mg of protein-ate. Each of these weights represents 1 mEq of the ion in question, whether it is cation or anion.

When electrolytes are measured in milliequiva-lents, cations and anions always balance each other, and a given number of milliequivalents of cation al-ways reacts with exactly the same number of milli-equivalents of an anion. Here is something useful to memorize:

> One Milliequivalent Of Any Cation
> Is Equivalent Chemically
> To One Milliequivalent Of Any Anion

It matters not a bit whether the cation is sodium, potassium, calcium, or magnesium, or whether the an-ion is chloride, bicarbonate, phosphate, sulfate, or pro-teinate.

The milliequivalent value of an element or mole-cule is determined by taking the *millimole* (mM) (the atomic or molecular weight of the element or com-pound in milligrams) and dividing by the *valence* (the numerical measure of combining power for one atom of a chemical element). Valence also reflects the num-ber of hydrogen atoms that can be held in combination or displaced in a reaction by one atom of an element. If a substance is *univalent* (e.g., chloride), *1 mM equals 1 mEq*. If a substance is *bivalent* (e.g., calcium), *1 mM equals 2 mEq*. Therefore, *2 mM (2 mEq) of a univalent* substance reacts chemically with only *1 mM (2 mEq) of a bivalent* substance.

MEASUREMENT OF OSMOTIC PRESSURE The mil-liequivalent also is a *rough* unit of measure for the os-motic pressure, or drawing power, of a solution. How-ever, since not all substances that exert osmotic pressure can be measured in milliequivalents, a more accurate measure of osmotic pressure is the *millios-mole* (mOsm). To determine the number of millios-moles in a solution, we again refer to the millimole. *One millimole of a substance that does not dissociate into ions (e.g., glucose) equals 1 mOsm.* However, *a millimole of a compound that does dissociate into ions equals two or more milliosmoles,* depending on how many ions it dissociates into. Sodium chloride (NaCl), for example, dissociates into one sodium ion and one chloride ion, so 1 mM of this salt equals 2 mOsm. A millimole of a more complex salt such as disodium phosphate (Na_2HPO_4) dissociates into two sodium ions and one phosphate ion, so it equals 3

mOsm. The osmotic pressure of a solution, therefore, is calculated by adding up all the ions, or millios-moles, it contains.

The milliequivalents per liter and the milliosmoles per liter of plasma, interstitial fluid, and cellular fluid are as follows:

	mEq/liter	*mOsm/liter*
Plasma	308	296.50
Interstitial fluid	311	300.75
Cellular fluid	364	305.00

MEASUREMENT OF CHEMICAL REACTION The final unit of measure with which we shall concern our-selves is *p*H, which tells us whether the chemical re-action of a fluid is acid, neutral, or alkaline. Basically, the reaction of a fluid is determined by the number of hydrogen ions it contains. Hydrogen is present in the body fluid in only tiny amounts, between 0.0000001 g and 0.00000001 g/liter of ECF. As a convenient way of expressing such minute hydrogen ion concentrations without resorting to decimals, the symbol *p*H was de-vised.

*p*H represents the reciprocal of the logarithm of the hydrogen ion concentration. Put more simple, *p*H is the power of 1/10, and it tells us the grams of hydrogen per liter of ECF. For example, *p*H 3 = $1/10^3$. As you probably recall from elementary arithmetic, power rep-resents the product arising from the continued multi-plication of a number by itself. Therefore, *p*H 3 = 1/10 × 1/10 × 1/10, or 1/1000 g hydrogen per liter of ECF. Similarly *p*H 7 = $1/10^7$, which, when multiplied, equals 1/10,000,000 g hydrogen per liter of ECF.

It is not really necessary to do all that multiplying, however. All one has to do is put down 1/1 and add zeroes equal to the number of the *p*H. For example, if the *p*H is 3, you simply write down 1/1 and add three zeroes, and you see immediately that *p*H 3 = 1/1000 g hydrogen per liter of ECF.

The more zeros we have, of course, the smaller the amount of hydrogen. Since it is the amount of hydro-gen in a fluid that determines its acidity, as *p*H *goes up the fluid becomes less acid and as pH goes down the fluid becomes more acid.* *p*H 7, for example, is ten times more acid than *p*H 8; *p*H 6 is ten times more acid than *p*H 7, and 100 times more acid than *p*H 8. The normal *p*H of ECF is between 7.35 and 7.45.

This unit of measure is extremely important in un-derstanding acid–base (or acid–alkali) balance, which we shall discuss in Chapter 3.

Another value used for expressing the hydrogen ion concentration in body fluids is the nanomole (nM), which is 1/1,000,000,000 g, or 1/1,000,000 mg of hydrogen. (Recall that in the case of hydrogen—and only in the case of hydrogen— 1 mg = 1 mEq.) Now, nanomoles are not difficult to understand. They are *linear*, which means that when the number of nanomoles of H^+/liter rises, so does the acidity; when it goes down, so does the acidity. Nanomoles are an *arithmetic* rather than a *logarithmic* measurement. Nanomoles are measured by using, first, a pH meter and, second, a simple conversion table (Table 1-1).

The extreme ranges for nanomoles of H^+/liter of human body fluid start with a high of 100nM/liter—extremely severe acidosis, as acidic as one can become and still live. The other end of the range of 10 nM/liter is about as alkalotic (or alkaline) as one can become and survive. The normal value is about 40 nM H^+/liter; 50 represents acidemia, 30 represents alkalemia—but neither is really extreme.

GAINS AND LOSSES OF WATER AND ELECTROLYTES

Whether they be primary or secondary to other conditions, all body fluid disturbances are caused by abnormal differences between gains and losses of water and electrolytes.

TABLE 1-1 *Nanomoles of hydrogen for varying pH values*

pH		Nanomoles
7.0	=	100.0
7.1	=	79.4
7.2	=	63.0
7.3	=	50.1
7.4	=	39.8
7.5	=	31.6
7.6	=	25.1
7.7	=	20.0
7.8	=	15.8
7.9	=	12.5
8.0	=	10.0

A comparison of the routes by which we gain water and electrolytes and the routes by which we lose them is in order. The body gains water and electrolytes in various ways: water alone is gained by drinking distilled water and by oxidation of foodstuffs and body tissues; softened water, well water, mineral water, and most city water supplies provide both water and electrolytes. Food also supplies both, for although it consists largely of water, it is rich in electrolytes and other nutrients, such as protein, fat, carbohydrate, and vitamins. Hospitalized patients frequently gain water and electrolytes, as well as other materials, by means of nasogastric tube, intravenous needle, or, infrequently, rectal tube.

Normal losses of both water and electrolytes occur through the lungs in breath, through the eyes in tears, through the kidneys in urine, through the skin in perspiration, and through the intestines in feces. In addition, water alone is lost through the skin in insensible perspiration, which goes on constantly. Abnormal losses can occur during illness or injury, as in burn or wound exudate, hemorrhage, vomiting, and diarrhea. Rapid breathing, suction by gastric or intestinal tube, enterostomy, colostomy, and cecostomy also cause great losses of water and electrolytes. Drainage from sites of surgical operations or from abscesses contains both water and electrolytes, as does fluid extracted by paracentesis. Fluids surrounding the brain and spinal cord can be lost if there is an abnormal opening to the outside. Indeed, fluids may be lost even inside the body; when abnormal closed collections of fluid develop, as in intestinal obstruction, *these fluids are just as useless to the body economy as if they were outside the body* (Fig. 1-9).

Now, what is important about these gains and losses is this: when one becomes ill, his gains decrease (may cease altogether) and his losses almost invariably increase. So it is no wonder that every seriously ill person is a logical candidate for a body fluid disturbance. In the healthy adult, the volume of urine excreted is approximately equal to the volume of fluid ingested as fluid, and water derived from solid food and from chemical oxidation in the body approximately equals the normal losses of water through the lungs and skin and in the stool (Fig. 1-10). Thus, in health, gains approximately equal losses, whereas during illness, gains and losses are not always equal. Intake of food and fluids may cease or diminish, while the normal losses continue; the losses may outweigh the gains by as much as one half liter or more per day.

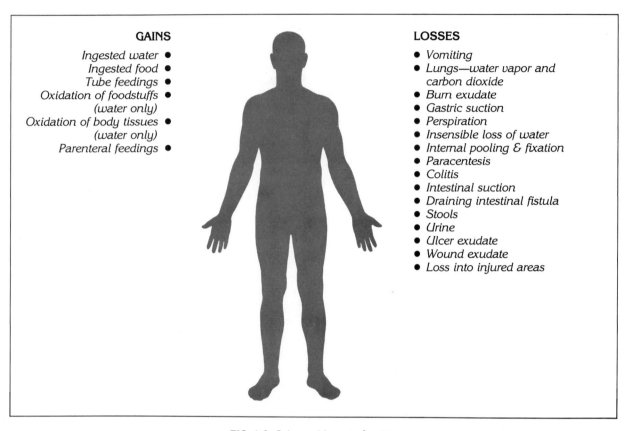

GAINS

Ingested water ●
Ingested food ●
Tube feedings ●
Oxidation of foodstuffs ●
(water only)
Oxidation of body tissues ●
(water only)
Parenteral feedings ●

LOSSES

● Vomiting
● Lungs—water vapor and carbon dioxide
● Burn exudate
● Gastric suction
● Perspiration
● Insensible loss of water
● Internal pooling & fixation
● Paracentesis
● Colitis
● Intestinal suction
● Draining intestinal fistula
● Stools
● Urine
● Ulcer exudate
● Wound exudate
● Loss into injured areas

FIG. 1-9. Gains and losses of water.

FIG. 1-10.
The balance portrays water balance in health.

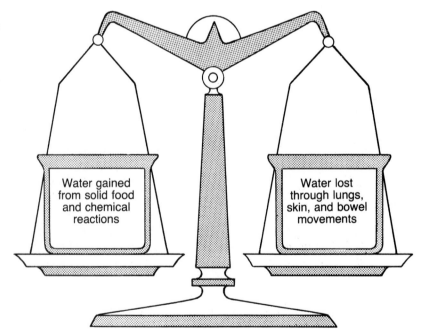

Water gained from solid food and chemical reactions

Water lost through lungs, skin, and bowel movements

The daily losses are cumulative; hence, a serious deficit can develop in a short time. If abnormal losses are occurring in addition, as in vomiting or diarrhea, the patient may become gravely ill within a matter of hours. The type of imbalance caused depends upon the kind of fluid lost, since body secretions and excretions vary greatly in electrolyte compositions and concentrations (Fig. 1-11).

Serious imbalances also occur when the gains are greater than the losses, as when the kidneys are not functioning properly. Excesses are just as dangerous as deficits and can prove fatal in a short time.

Thus, abnormal differences between gains and losses cause all of the eighteen basic imbalances in our diagnostic classification; these imbalances involve changes in volume, composition, and position of the ECF.

HELP FROM THE LABORATORY

The role of laboratory tests varies greatly from one body fluid disturbance to another: for example, laboratory tests contribute little to diagnosis of changes in position of ECF; on the other hand, in that form of sodium deficit known as syndrome of inappropriate secretion of antidiuretic hormone (SIADH), the serum

FIG. 1-11. Electrolyte composition of various body secretions or excretions.

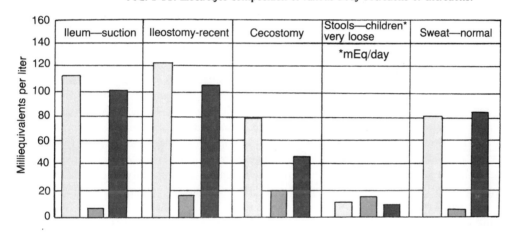

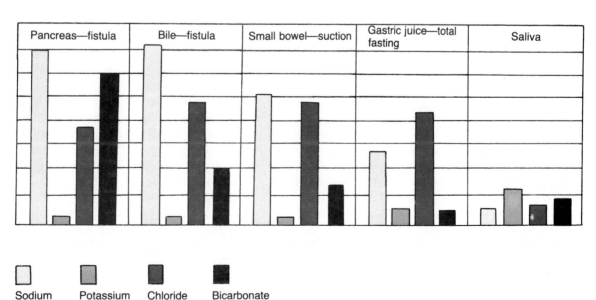

sodium level determination is *crucial*. A careful history and physical examination often indicate the correct diagnosis; almost invariably, the laboratory represents an invaluable ally. (The reader is encouraged to see Chap. 4 for a discussion of nursing assessment of the patient with a fluid balance disturbance.)

LIMITATIONS OF THE LABORATORY

What are some of the limitations of the laboratory? First of all, many of the usual test procedures have difficulties inherent in them that make consistently accurate results difficult to attain, even though the tests are performed by highly skilled technicians. Moreover, physicians do not always know the "normal" laboratory values for ill persons, since our table of values are based on those found in healthy persons. It may sometimes be undesirable—even dangerous—to attempt to restore the "abnormal" value of a patient to "normal," since what we regard as an abnormal value may be the result of a defensive action on the part of the body. For example, in primary carbonic acid deficit (respiratory alkalosis), the body lowers the level of base bicarbonate in order to maintain the carbonic acid–base bicarbonate balance. The plasma level of base bicarbonate is, therefore, lower than normal, but the normal referred to is the normal for health, not for this particular imbalance. The normal level of base bicarbonate for carbonic acid deficit is much lower than the level for health, and so, to bring this properly low base bicarbonate level up to the normal level for health by the injection of base bicarbonate into the blood would make the disturbance worse and might well kill the patient.

We can cite many other examples showing why we must take the laboratory reading with the proverbial grain of salt. For example, the level of potassium in plasma is but one indication of the cellular stores. For the long-term survival of the patient, these are vastly important, but, for the acute situation, the plasma level of potassium determines whether there will be an ECF deficit or excess of the cation. And, it is the ECF status of potassium that largely determines symptoms and findings. It also determines whether the patient requires measures to increase or decrease the plasma potassium level, which will be reflected both by the serum potassium and by the electrocardiogram (EKG).

Another example is the state of urine, ordinarily acid in acidosis, but sometimes alkaline even though acidosis is present. This fact is demonstrated in several clinical conditions, such as chronic renal disease, in which the infecting organism converts urea in newly formed urine to ammonia. And, although the urine is usually alkaline in alkalosis, we may find an acid urine when alkalosis is accompanied by a severe potassium deficit. Nevertheless, one can almost always improve recognition and management of body fluid disturbances when laboratory aids are available. In short, one must learn to use them intelligently.

LABORATORY AIDS AVAILABLE

In addition to the chemical methods for determining the plasma levels of calcium, chloride, phosphorus, magnesium, proteinate, albumin, bicarbonate, nonprotein nitrogen, and urea, the *flame photometer* has been immensely useful in determining the level of sodium and potassium in body fluids. The flame photometer reading depends upon the color of the flame given out when the electrolyte in question is burned.

The *electrocardiograph* is useful in diagnosing potassium deficit, potassium excess, calcium deficit, and calcium excess, although, like other laboratory tests, it is confirmatory rather than primarily diagnostic. *Roentgenography*, too, is a useful diagnostic aid, helping to diagnose certain imbalances in the calcium level. An important development in laboratory techniques is the number of tests that can be performed on a small quantity—for example, a few drops of blood. These *microchemical determinations* are particularly useful in the case of the small baby, who does not have much blood to spare for laboratory tests. The use of *radioisotopes* is becoming more and more important in medicine and may soon be of value to the practicing physician as he diagnoses abnormalities of the body fluid. In addition to laboratory tests that require complex equipment, there are many tests that physicians can do in their offices, including tests to determine the acidity or specific gravity of the urine, the hemoglobin level of the blood, the red blood cell count, and the packed cell volume.

Essential for the diagnosis of acid–base disturbances are the blood gases, covered in some detail in Chapter 3. The blood gas in which we are primarily interested from the standpoint of acid–base balance is carbon dioxide, measured as the P_{CO_2}, or carbon dioxide tension, expressed as millimeters of mercury (mm/Hg). Bear in mind that the P_{CO_2} is in equilibrium with the carbonic acid of the ECF ($0.03 \times H_2CO_3 = P_{CO_2}$). We therefore use the P_{CO_2} to represent the acid portion of the carbonic acid : bicarbonate ratio. P_{CO_2} is always performed on arterial or arterialized blood. (For a description of arterialized blood, see Chapter 3.)

pH, the reciprocal of the logarithm of the hydrogen ion concentration, is often regarded as a blood gas. The bicarbonate, although certainly not a gas, is also often regarded as one of the blood gases, since it is essential in evaluating the patient's acid–base status. On the other hand, the oxygen tension, P_{O_2}, although invaluable in evaluating pulmonary problems, is not ordinarily useful in acid–base disturbances. (However, an extremely low P_{O_2} predisposes to lactic acidosis and thus alerts one for this condition.)

There are other measurements that are not strictly laboratory tests but that are extremely valuable, including the determination of body weight and measurements of fluid intake and fluid output. Indeed, a careful measurement of both the kind and quantity of fluids going into and passing from the patient is of utmost importance for the proper treatment of body fluid disturbances. Since the nursing staff has responsibility for the 24-hour supervision of the patient, the important duty of measuring intake and output falls to them. Basic knowledge of body fluid disturbances and their treatment makes it apparent to nurses that accurate intake–output records are invaluable for the welfare of the patient. There is usually no substitute for them, since they are often the chief basis for accurate diagnosis and effective treatment. Knowledge of intake–output figures is vital to the physician treating a patient with an imbalance of the body fluids.

EVALUATION OF TEST RESULTS

Variations in laboratory values for persons of different ages are important considerations when evaluating the results of laboratory tests. Normal values for the red cell count, hemoglobin, packed cell volume, and plasma bicarbonate differ in infants, children, and adults. This is true not only for *average values* but *normal ranges* as well. Plasma levels are best reported as milliequivalents per liter (mEq/liter), but when they are reported as milligrams (mg), they can be quickly converted by reference to a conversion table.

A reasonable minimal daily program for laboratory surveillance of the patient should include, if facilities are available, plasma sodium, plasma potassium, plasma magnesium, plasma bicarbonate, and plasma chloride. Initially, the hemoglobin, urinary pH, and urinary specific gravity should be tested, and measurement of the plasma pH is always helpful. Blood gases, especially the P_{CO_2} are essential in evaluating body fluid disturbances. When tests are abnormal, they should be repeated daily until they return to normal. Although body weight and fluid intake–output measurements are not strictly laboratory procedures, they do provide important objective information, which helps to guard against gross overprovision or underprovision of water and electrolytes. If possible, these measurements should be performed daily until fluid balance is achieved.

The ancient Hebrews must have been interested in something akin to modern laboratory diagnosis, since they had a proverb that states: "The causes of all diseases are to be found in the blood." Today, the examination of blood, particularly of plasma, and of other body fluids is invaluable in the management of the patient with body fluid disturbances. Table 1-2 shows laboratory values useful in body fluid disturbances. Note that the left column names the lab test. The right column shows the normal range of values.

TREATMENT OF BODY FLUID DISTURBANCES: GENERAL PRINCIPLES

Rational therapy of patients with disturbances in their body fluids requires a working clinical diagnosis which is founded on a mental image of the nature, and a reasonably accurate estimate of the magnitude of the disturbances present.
—Michael James Sweeney

PRELIMINARY TESTING

In planning therapy for the patient with an actual or potential disturbance of the body fluids, the first step is to determine the status of the kidneys. Various types of solutions, particularly those containing potassium, can be hazardous if renal function is not adequate. The presence of any one of the following criteria indicates suppressed kidney function:

Specific gravity of urine above 1.030

Less than three voidings in 24 hours

No urine in the bladder

TABLE 1-2 *Laboratory values important in body fluid disturbances*

Lab test	Normal range
Albumin (serum)	3.5–5.0 g/dl
Alkaline phosphatase	Adults: 2.0–4.5 Bodanski units: Adolescents: 2.0–5.0 Infants: 2.0–14.0
Bicarbonate (serum)	Adults: 25–29 mEq/liter Children: 20–25 mEq/liter
BUN	8–25 mg/dl
Calcium (total serum)	8.5–10.5 mg/dl or 4.5–5.8 mEq/liter
Calcium (ionized)	56% of total calcium
Calcium (urinary)	150 mg/day or less
Chloride (serum)	98–106 mEq/liter
Creatinine	0.7–1.5 mg/dl
Delta	8–16 mEq/liter
Erythrocyte count	Women: 4.6 million/mm^3 Men: 5.4 million/mm^3
Glucose (fasting)	70–100 mg/dl
Hematocrit	Women: 39%–47% volume RBC/dl Men: 44%–52% volume RBC/dl
Hemoglobin	Women: 13–16 gm/dl Men: 15–18 gm/dl
Magnesium (serum)	1.4–2.3 mEq/liter
Osmolality (serum)	280–295 mOsm/kg water
Osmolality (urine)	300–900 mOsm/kg (average fluid intake)
PCO_2	38–42 mm Hg arterial or arterialized blood
pH (serum)	7.35–7.45
pH (urine)	4.5–8.2
Phosphorus	1.7–2.6 mEq/liter
PO_2	95–100 mm Hg arterial or arterialized blood
Potassium (serum)	4.0–5.6 mEq/liter
PTH	Less than 15 μl equivalents/150 μl
Sodium (serum)	137–147 mEq/liter
Sodium (urinary)	87–173 mEq/liter (diet-dependent, no real "normal")
Specific gravity of urine	1.010–1.030

Renal depression almost invariably occurs in patients who have suffered massive acute losses of ECF; hence, many patients with fluid imbalances have oliguria or even anuria. Before administering fluid therapy, it is necessary to determine if the suppression of urination is due simply to an ECF volume deficit or whether it is due to serious renal impairment. If urinary suppression is caused simply by volume deficit, the therapeutic test will reveal that fact by reestablishing urinary flow, making it safe to proceed with the fluid therapy program, including the administration of potassium-containing solutions.

The therapeutic test consists of administration of a special solution, frequently called an *initial hydrating solution* or a *pump-priming solution*. Such a solution often provides sodium, 51 mEq/liter; chloride, 51 mEq/liter; and glucose, 50 g/liter. It actually represents a solution containing one part of isotonic solution of sodium chloride in 5% glucose, and two parts of 5% glucose in water. The solution is administered at the rate of 8 ml/m^2 of body surface/minute for 45 minutes. When the kidneys begin to function, as shown by the restoration of the flow of urine, the initial hydrating solution is discontinued and therapy is started with other types of solutions. If the urinary flow is not restored, the infusion rate is reduced to 2 ml/m^2 body surface/minute and continued for another hour. If urination has not occurred at the end of this period, the physician assumes the problem is renal impairment rather than functional depression, thus demanding a battery of laboratory tests plus careful management by nephrologists or similarly qualified specialists.

THE GOALS OF FLUID THERAPY

Three specific goals should be envisioned in planning fluid and electrolyte therapy:

1. Any preexisting deficits of water and electrolytes must be repaired.
2. Water and electrolytes must be provided to meet the patient's maintenance requirements.
3. Continuing abnormal losses of water and electrolytes through such routes as vomiting, diarrhea, tubal drainage, wound drainage, burn drainage, diuresis, and the like must be replaced.

There are various, widely different methods for achieving these goals. Rather than describe several methods, we will present one simple method that has worked out well in practice and will help give the nurse a basic understanding of the principles involved in fluid therapy. From the standpoint of clinical results, this method is no better than other methods of fluid therapy and certainly has its drawbacks and limitations—we have chosen it for pedagogical reasons. The method that we shall describe briefly is that developed by Butler and his co-workers at the Massachusetts General Hospital, and that has, since that time, been used with great success in many parts of the world.

ADMINISTERING FLUID THERAPY

In perhaps 90% of patients with fluid imbalances, a single solution of the type such as that devised by Butler can be used to provide water and electrolytes for maintenance and repair. Such a solution is so designed that *when used to meet the patient's fluid volume requirement, it supplies electrolytes in quantities balanced between the minimal needs and the maximal tolerances of the patient.* The Butler-type solution is actually one third to one half as concentrated as plasma, providing free water for urinary formation and metabolic activities and, in addition, both cellular and extracellular electrolytes. It incorporates 5% to 10% of carbohydrate to minimize tissue destruction, reduce ketosis, and spare protein. The Butler-type solution utilizes the body homeostatic mechanisms, which select the electrolytes that are required and reject those that are not needed. Butler-type solutions, when used properly, have a great margin of safety and are enormously useful in managing body fluid disturbances that result from acute differences between intake and output of water and electrolytes not controllable by the body homeostatic mechanisms.

Body fluid disturbances that can usually be managed by the system of therapy just described include the following:

ECF volume deficit

Sodium excess of ECF

Potassium deficit of ECF

Primary base bicarbonate deficit of ECF

Primary base bicarbonate excess of ECF

For details concerning management of the above imbalances, as well as of the other imbalances in our classification, see Chapters 2 and 3, and later chapters involving alterations in specific body systems.

In all the imbalances listed above, balanced solutions can be used both for repair and for maintenance. Dosage is relatively simple and is based primarily on the patient's fluid volume, which is assessed mainly by acute changes in body weight. A moderate fluid volume deficit is indicated by weight loss up to 5% in a child or adult, up to 10% in an infant, and by symptoms moderate in degree, whereas a severe fluid volume deficit is indicated by an acute weight loss of over 5% in a child or adult and over 10% in an infant, with symptoms and laboratory findings severe in degree.

In the patient whose fluid volume is normal at the time therapy is started, his maintenance requirements can be met by the administration of 1500 ml of a Butler-type solution/m² body surface/day. In the patient who has a moderate preexisting deficit, 2400 ml/m² body surface/day will both correct this deficit and meet maintenance needs. If the patient has a severe preexisting deficit, one can correct the deficit and provide maintenance by giving 3000 ml of a Butler-type solution/m² body surface/day.

Various modifications of the Butler-type solution are made available by manufacturers of parenteral solutions. The usual solution for older children and adults contains 88 mEq of total cation (and hence of total anion), and that used for infants and small children contains 48 mEq of total cation (and hence of total anion) (See Chap. 6, Table 6-6). Manufacturers gladly provide information concerning available solutions and their proper use.

For replacing continuing abnormal losses (concurrent losses) from vomiting, drainage from an intestional tube or fistula, severe diarrhea, and so on, replacement solutions with a composition resembling the body fluid lost are added to the daily fluid ration. Thus, to replace gastric juice, one administers a gastric replacement solution intravenously and to replace intestinal or duodenal juices, an intestinal replacement solution is administered intravenously. (Gastric and duodenal replacement solutions are described in Chapter 6, Table 6-6.)

CALCULATING DOSAGE

Among the most beautiful passages in the medical literature are these from the clinical reports of early physicians who used parenteral fluid therapy. Latta, of Leith, Scotland, thus described the effect of an intravenous solution administered to a patient suffering from cholera in 1832:

. . . Like the effects of magic, instead of the pallid aspect of one whom death had sealed as his own, the vital tide was restored and life and vivacity returned . . .

Cantini, who practiced in Naples, Italy, and died in 1893, wrote as follows:

The cold, cyanotic, dehydrated, comatose patients, lying pulseless and almost lifeless, became animated after the subcutaneous infusion of warm salt water. Remarkably their pulse and voice often returned in a few minutes and they are even able to sit alone in bed . . .

Neither Dr. Latta nor Dr. Cantini felt much need for precise knowledge of parenteral fluid dosage, nor was a great need experienced for many decades to come, for as long as it remained an emergency measure, to be performed only as a last resort, the clinical response of the patient—if respond he did—was regarded as gauge enough for proper dosage. Now, with striking advances in the knowledge of clinical fluid disturbances during the past 4 decades, parenteral fluids have been administered much more extensively. The medical profession has, therefore, felt a mounting need for simple, widely applicable dosage gauges suitable for assisting the physician in determining the volume and the rate of administration of parenteral fluids.

The most frequently used gauges for dosage have been body surface area and body weight related to age. Some authorities feel that body weight related to age has an important shortcoming: requirements for water and electrolytes determined on the basis of body weight are different for patients of different ages. However, when using body surface area, expressed in terms of square meters of body surface per day, requirements for water and electrolytes are approximately the same for patients of all age-groups, as shown in Figure 1-12 for approximate water requirement and approximate sodium requirement.

Body surface area is proportional to many essential physiologic processes, including heat loss, blood volume, glomerular filtration rate, organ size, respiration, blood pressure, and nitrogen requirements. Butler and his co-workers at the Massachusetts General Hospital were the first to emphasize that the basic water and electrolyte requirements are also proportional to body surface area, regardless of the age or size of the patient. They pointed out that since body surface area provides a quantitative index of our total metabolic activity, it also provides us with a remarkably useful gauge for determining doses and proper rate for parenteral administration of water and electrolytes in fluid therapy. In addition, body surface area can provide nurses with a simple, rapid method of checking the correctness of parenteral fluid orders, especially in regard to volume and rate of administration, and in particular to determine if orders for sodium and potassium in parenteral solutions are within the realm of safety.

FIG. 1-12. The same values for water and sodium requirements can be used for persons of all ages if body surface area, rather than weight, is used as the dosage criterion.

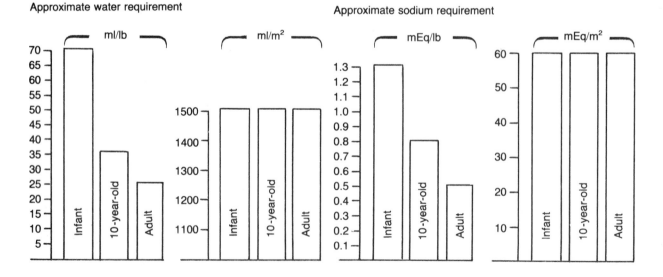

There are various methods of measuring body surface area, including the covering method, geometric method, skinning method, and other investigational methods. Clinicians employ simple nomograms that enable one to rapidly estimate body surface area from height and body weight. Actually, one can quickly obtain the approximate surface area from weight alone, as shown in Table 1-3.

The table is simple to use: an infant weighing 10 lb would have an approximate surface area of 0.27 m²; a child weighing 50 lb would have an approximate surface area of 0.87 m²; a male adult weighing 150 lb would have an approximate surface area of 1.75 m².

The body surface areas arrived at by use of the weight chart are *approximate* only and apply to persons of average body build. In general, obese or stocky persons have less surface area than tall lanky persons of the same weight (just as a long, low one-story ranch house has more external surface than a two- or three-story house of equal floor space). For example, our 10-lb infant would have a surface area of 0.2 m² if he were only 16 inches tall, but he would have 0.3 m² of surface area if he were 28 inches in height. Our 50-lb child would have 0.68 m² of surface area if he were 2 feet, 10 inches tall and 0.84 m² if his height were 3 feet, 10 inches. Our 150-lb adult would have 1.47 m² of surface area if he were only 4 feet, 4 inches tall and 1.86 m² if her had a height of 6 feet.

The body surface area gauge for dosage is especially useful for checking the correctness of water and electrolytes ordered for patients with body fluid disturbances. Remember the following rules:

1. For maintenance, administer 1500 ml/m² body surface/day.
2. For correction of a moderate deficit in ECF volume plus maintenance, administer 2400 ml/m² body surface/day.
3. For correction of a severe deficit in ECF volume plus maintenance, administer 3000 ml/m² body surface/day.
4. Concurrent losses are replaced with an appropriate replacement solution on a volume for volume basis.
5. Except in unusual situations, the daily fluid ration is given evenly over a 24-hour period.

By applying these rules, the nurse can quickly check the correctness of an order for fluid therapy. Let us suppose that the physician has written an order for a liter of solution to be given over a 24-hour period to an infant weighing 20 lb with a moderate fluid volume deficit. A quick check of the conversion chart shows that a 20-lb baby would have 0.45 m² of body surface; the volume of fluid for correction of a moderate fluid volume deficit in such a baby should be 0.45 times 2400 ml, or about 1080 ml/24 hours (45 ml/h).

Let us suppose, on the other hand, that a large man with a severe fluid volume deficit is to receive 3 liters of solution for the first 24 hours. Reference to the conversion chart shows that a 175-lb man would have a body surface area of 2.0 m²; 2.0 times 3000 ml, the requirement per m² of body surface per day for a severe fluid volume deficit, is 6 liters, or 6000 ml; 3 liters therefore is grossly inadequate.

Suppose that a 12-year-old boy with no fluid imbalance is to be given water and electrolytes for main-

TABLE 1-3 *Chart for converting weight to surface area*

Pounds	Surface area in square meters
(Figures are approximate and apply only to individuals of average build.)	
4	0.15
6	0.20
10	0.27
15	0.36
20	0.45
30	0.60
40	0.72
50	0.87
60	0.97
70	1.10
80	1.21
90	1.33
100	1.4
125	1.6
150	1.75
175	2.0
200	2.2
250	2.7

tenance for a few days following abdominal surgery. The physician orders 3 liters of his favorite maintenance solution over a 24-hour period. The boy's body surface area is 1.33 m^2 and maintenance should call for 1500 ml times 1.33, or 1995 ml, approximately 2 liters of fluid. The extra liter ordered might embarrass the boy's homeostatic mechanisms, and the physician should appreciate having the order questioned. Of course, in the presence of fever or heavy sweating, the extra fluid might well be entirely proper.

COMPLEX COMBINED IMBALANCES

The severe burn is an excellent example of a complex combination of body fluid imbalances. Therapy of this condition requires knowledge of the hazardous aftermaths of the burn, which include losses of body fluid, shock, renal depression, and remobilization of edema fluid. (See Chapter 8 for a discussion of burn therapy.) Other combined imbalances are discussed in the chapters dealing with alterations in specific body systems. Carefully controlled therapy must be directed at the correction of mixed imbalances.

TYPES OF SOLUTIONS

The nurse is urged to become familiar with the various types of parenteral solutions commonly in use. Besides being contained in the nursing literature, this information can be gained by talking with the hospital pharmacist, the intravenous therapist, or knowledgeable physicians and by carefully reading the literature provided by the pharmaceutical company that supplies the parenteral solutions. Certainly among the valuable resource personnel are the local pharmaceutical company representative and the medical departments of their respective companies. The reader is referred to Chapter 6 for an in-depth discussion of parenteral fluids and nursing responsibilities associated with their administration.

BIBLIOGRAPHY

Bernard C: Lecons sur les Proprietes Physiologiques en le Alterations des Liquides de l'Organisme, vols. 1 and 2. Paris, Balliere, 1859

Condon R, Nyhus L: Manual of Surgical Therapeutics, 4th ed. Boston, Little, Brown & Co, 1978

Factors Influencing SMAC Test Results. National Health Laboratories, Central Region Reference Laboratory, Louisville, 1981

Gamble J: Chemical Anatomy, Physiology, and Pathology of Extracellular Fluid, 6th ed. Cambridge, Harvard University Press, 1954

Goldberger E: A Primer of Water, Electrolyte and Acid–Base Syndromes, 6th ed. Philadelphia, Lea & Febiger, 1980

Goodman L, Gilman A (eds): The Pharmacological Basis of Therapeutics, 6th ed. New York, MacMillan, 1980

Henderson L: The Fitness of the Environment. Boston, Beacon Press, 1958

Kempe C, Silver H, O'Brien D: Current Pediatric Diagnosis & Treatment, 6th ed. Los Altos, Lange Medical Publications, 1980

Maxwell M, Kleeman C (eds): Clinical Disorders of Fluid & Electrolyte Metabolism, 3rd ed. New York, McGraw-Hill, 1980

Water, Trace Elements, and Major Body Fluid Disturbances

Life was born in water and is carrying on in water. Water is life's mater and matrix, mother and medium. There is no life without water. Life could leave the ocean when it learned to grow a skin, a bag in which to take the water with it. We are still living in water, having the water now inside.
                                                    —*Albert Szent-Gyorgyi*

WATER

Were you to ask a group of people what they regarded as the most essential nutrient, few of them would answer, "Water." Yet, if one were deprived of all liquid and solid nutrients, he would die of water deficit far sooner than he would die of deficit of any other nutrient. Doctor Thomas A. Dooley told of his horrible experience of being forced to watch the sufferings of a boy imprisoned in a bamboo cage without food or water; the boy died—of water deficit—after 8 days of agony. One need lose only some 10% of the body weight in water to be placed in danger of collapse and death.

Although we take water for granted, it is one of the most remarkable chemicals on earth. No known plant or animal can live without water in one form or another; it has no substitute. Fortunately, water is one of the most abundant and widely distributed chemicals on earth. (Of all the planets in the solar system, apparently only Earth has enough water to support human life.) Countless chemical compounds have water as part of the molecule, and few chemical reactions can occur unless water is present.

Living organisms contain more water than do most other substances. Water, therefore, exerts an enormous impact on our body's physiologic processes. Thus, we must examine both the distribution and regulation of body water, as well as the many substances dissolved in it, if we are to have any real understanding of the body's complex physicochemical mechanisms.

The percentage of fluid varies considerably from person to person, depending upon age, sex, and the amount of fat. A newborn infant is about 77% fluid; the average adult is about 60%. While there is no important difference between the proportion of body fluid in the two sexes until about 16 years of age, after that the male accumulates more fluid, until he has some 17% more. With advancing years, the total body fluid for both sexes decreases, but the sex difference remains. The average body fluid percentage of an elderly man is about 5% more than that of an elderly woman.

The percent of body fluid varies also with the fat content of the individual. Adipose, or fat, tissue is fairly free of water; hence, about 50% of the obese adult's body weight is fluid; in the extremely lean adult, it is about 75%. (Since females have more body fat than males, this explains why they have less body fluid in terms of percentage of body weight.) In disease states characterized by retention of excessive amounts of liquid (e.g., congestive heart failure or edema), the fluid percentage of the adult may approach that of the newborn.

As a solvent, water contains dissolved substances. Most of these substances are electrolytes (i.e., minerals that generate electrical charges when placed in water). The different locations of the body in which fluids are found are referred to as *compartments*. But this term may be misleading, for the compartments are by no means contiguous. The cellular compartment of body fluid, for example, is divided into trillions of tiny compartments consisting of the body cells. Yet, we refer to it as the *cellular compartment*. It is the largest compartment of our body fluid, about three times the volume of the *extracellular compartment*. This compartment is divided into the fluid inside the blood vessels (the plasma) and the fluid outside the blood vessels (the interstitial fluid). This fluid has the missions of carrying nutrients to the body cells from the arterial capillary bed and of carrying waste materials from the body cells to the venous capillary bed and to the lymphatic capillary bed (which assists the venous capillary system in returning waste materials to the venous circulation). Lymph is usually regarded as a subdivision of the interstitial fluid. So is the cerebrospinal fluid; it is really interstitial fluid that has been specialized to meet the needs of the brain and spinal cord.

In addition to these major body fluids—the cellular and the extracellular fluid (ECF)—there are other liquids in the body that usually are not classified as body fluids. There are the secretions and excretions of organs and glands—saliva, gastric juice, intestinal juice, bile, perspiration, and urine. Then there are the fluids of the body cavities—the pleural cavity, which surrounds the lungs; the pericardial cavitiy, which surrounds the heart; and the peritoneal cavity, which surrounds the digestive organs. Since these "cavities" are actually tightly packed with important organs, they are not really cavities at all. But they do contain small quantities of lubricating fluid, which can become excessive in disease, resulting in "pleural effusion" or "pericardial effusion" or "ascites," depending upon which cavity is involved.

Throughout the body, water serves a lubricating function, helping to protect the fragile body cells from injury. Water also acts as a temperature regulator, helping to keep the body temperature constant and within safe bounds.

How do we gain water? We obtain it when we drink liquids, we obtain it from solid food (which, after all, is mostly water), and we gain it from chemical oxidation of foodstuffs and body tissues. We can also gain it—in combination with electrolytes and dextrose—from parenteral infusions. Water can also be given by rectal infusion.

How do we lose water? In a host of ways. We lose it through the skin both as insensible and sensible perspiration. While insensible perspiration consists only of water, sensible perspiration contains water plus sodium, potassium, chloride, magnesium. and even some urea. We also lose important quantities of water in the moistened air we exhale; we lose it in the stools, and we lose a great amount in the urine. These are the normal losses that occur in health. When we become ill, we lose water in additional ways: as vomitus, in diarrheal stools, through excessive sweating if we have fever.

Considering all that water does, it appears logical that drinking generous quantities of it should be important for the maintenance of health. A generous intake of water is also indicated in most illnesses. For example, a temperature elevation between 101° and 103° F (38.3° C and 39.4° C) increases the 24-hour fluid requirement by at least 500 ml, and a temperature elevation above 103° F (39.4° C) increases it by at least 1000 ml. Often the first order written for a newly admitted patient is "force fluids" or "fluids freely." Physicians, nurses, nutritionists, and all members of the health team universally recognize the value of generous quantities of water.

TRACE ELEMENTS

The trace elements of the human body appear to represent an important portion of our heritage from the sea. Fabun, in the *Dynamics of Change*, presents (with tongue in cheek, we trust) the fantastic notion that the ocean is a great animal, so enormous that we earthlings cannot envision it as living. Fabun describes rivers as great tentacles of the ocean, the turbulent border between surf and shore as the sea animal's skin, the waves as the pulsations of a giant heart, and man himself as an emissary from the ocean who has invaded the land.[1] However bizarre this fantasy, it is not fantasy that the sea that covers most of earth's surface would be as lifeless as the land-locked Dead Sea were it not for its endless stores of those dynamic active chemicals, the trace elements. It is not fantasy that the ancestors of man carried small portions of these elements ashore when they left their ancient ancestral home. Even today, human body fluid contains that same varied cluster of trace elements and electrolytes possessed by the primeval oceans, despite the fact that some of the elements are not only useless, they are toxic. The stamp of the sea is an enduring one, and man bears it on every one of his trillions of cells.

One must bear in mind that the elements of the sea had their origin in the dissolving of the continents; they reflect, therefore, the composition of the planet earth. So, man's elements are earth elements and man is indeed an *earthling*. Had he developed on another planet, his trace-element legacy would have been different, and so would his basic physiology. Thus, the trace element profile of man's body carries a clarion message: *man cannot exist elsewhere in the universe*.

CHROMIUM People in the United States have far lower chromium levels than do Orientals, Africans, and persons from the Middle East. This may be explained by the fact that while raw sugars provide chromium in various quantities, refined sugars contain little or no chromium.

Chromium is required to maintain normal glucose metabolism in experimental animals, probably acting as a cofactor for insulin. Cases of disturbances in glucose metabolism in humans that responded favorably to administration of chromium have been described.

Chromium levels in tissues decline with age. The absorption and metabolism of chromium depend on the form in which the element is present. About 1% of chromium in simple salts is absorbed, while the availability from some food sources appears to be between 10% and 25% of a given amount.

We can obtain some idea of the human chromium requirement from the average daily loss, which is about 0.4μg to 1.8μg and which occurs through the urine. Nevertheless, a meaningful recommendation for daily chromium intake is not currently possible, owing chiefly to the great variation in availability of chromium in different foods. Chromium intake from typical Western diets is estimated between 50μg and 100μg/day. Patients on total parenteral nutrition (TPN) have been maintained by daily administration of 20μg chromium. Chromium deficiency, associated with impaired carbohydrate tolerance, has been reported during the long-term use of TPN.

Good sources of available chromium include most animal proteins, excluding fish; whole grain products; and brewers' yeast.

COPPER Copper is an essential component of various proteins and is required for healthy red blood corpuscles. It participates in the action of several essential enzymes; particularly important is copper's role in enzyme systems concerned with phospholipid synthesis.

Since milk is low in copper and copper is required for proper utilization of iron, some clinicians believe that nutritional anemia in infants who have been restricted to a milk diet improves more rapidly when both iron and copper are given.

In the adult, 1.24 mg of copper daily appears sufficient to maintain balance; infants and children require between 0.05 mg and 0.1 mg/kg body weight. Even generally poor diets appear to contain enough copper for the average person. Foods rich in copper include liver, kidney, shellfish, nuts, raisins, and dried legumes.

Copper deficiency syndrome (characterized by anemia, neutropenia, and hypoproteinemia) may occur with hyperalimentation. Oral copper supplements may be used when the patient receiving TPN has sufficient gastrointestinal (GI) function for their absorption; a copper solution can also be prepared by the pharmacist for intravenous administration.

FLUORINE The trace element fluorine, discovered in 1771, constitutes 0.027% of the earth's crust. It is widely, though sporadically, distributed; hence, it is a constituent of all normal diets. Traces of fluorine normally found in water can increase greatly when water passes through rocks and soils rich in fluorine.

Fluorine content of blood is about 0.2 parts per million (ppm), and of saliva, about 0.1 ppm. The largest amount of fluorine in the body is found in the teeth and bones. It is a normal component of tooth structure. Its presence is required for maximal resistance to dental caries, especially in infancy and early childhood; however, it is of benefit later in life also. Although the precise mechanism by which fluorine inhibits the development of dental caries remains unknown, a theory is that during the formative years, fluorine reduces the solubility of tooth enamel in acids produced by bacteria, thus reducing dental caries. At any rate, addition of fluorine to drinking water at a level of slightly less than 1 ppm gives remarkable protection against dental caries. Fluorine can also be given in a vitamin preparation, applied directly to the teeth by a dentist, or incorporated in toothpaste. Dentists properly emphasize that use of fluorine to prevent tooth decay is but one part of a broad preventive program designed to reduce dental caries.

Administration of sodium fluoride, in conjunction with vitamin D and calcium, has been suggested for the treatment of osteoporosis and Paget's disease of bone. Fluorides have also been suggested as one means of protecting the skeletal system from decalcification during space travel. However use of fluorides in therapy of bone disorders must be regarded as experimental at best, since they may well be harmful.

Intake of fluorine in concentrations of more than 2 ppm in water leads to mottling of the enamel of teeth. Endemic dental fluorosis occurs in communities in Texas and Colorado where the fluorine content of the water is above 2 ppm. Miners exposed to fluoride-containing dust for years develop hard, dense bones but no other physical abnormalities. A total intake from food and drinking water of 1.5 mg to 4.0 mg/day is tentatively recommended as safe and adequate for adults. For younger age-groups, the maximal level is 2.5 mg/day. During infancy, ranges of 0.1 mg to 1.0 mg/day appear adequate. When sodium fluoride is taken by mouth the dose is usually 2.2 mg, which provides 1 mg fluorine. Human beings excrete almost all ingested fluorine, up to 3 mg/day, in the urine. The degree of intestinal absorption depends upon the solubility of the ingested fluorine compound.

Sources of fluorine in addition to drinking water and pharmaceutical preparations are seafoods and seawater.

IODINE The essential trace element iodine is best known for its role as a basic component of thyroid hormone. Often classified among the rarer elements, iodine constitutes 0.00000006% of the earth's crust. A versatile element, it can be applied topically as a counterirritant, bactericide, and fungicide; it is also useful for disinfecting drinking water.

Iodine's prime function is as an essential component of the thyroid hormones thyroxin (T^4) and triiodothyronine (T^3). It is, therefore, required for normal growth and development and for maintenance of the metabolic rate.

The small intestine absorbs ingested iodine; the blood contains both inorganic iodide and protein-bound iodine. The iodine concentration of whole blood is about 1 part in 25 million; in the thyroid gland it is 1 part in 25 hundred. Drinking water contains varying quantities of iodine; estimates of the quantities of iodine in water provide a basis for determining the iodine concentration of nearby soil, as well as the iodine content of fruits, grains, grasses, and vegetables grown in the vicinity.

In the absence of sufficient iodine, the thyroid gland increases its secretory activity in a vain attempt to compensate for the deficit. The gland enlarges, causing an unsightly, uncomfortable mass on the anterior surface of the neck called *goiter*. In certain inland regions, including the Great Lakes area and the Alpine regions of Europe, water and soil provide inadequate iodine. The consequent high incidence of goiter can be diminished or eliminated by an iodine supplement, usually given as iodized table salt.

The drug propylthiouracil blocks the oxidation of iodide to iodine by the thyroid gland. Cabbage, rutabaga, and other members of the Bassica family contain goiter-causing substances.

Ingestion of iodine in therapeutic doses may cause skin lesions resembling acne in sensitive persons. Large quantities of iodine produce abdominal pain, nausea, vomiting, and diarrhea.

The recommended daily allowance (RDA) for iodine ranges from 5.83 µg/day/kg body weight for the infant during the first 6 months of life to 2.95 µg/day/kg for the adolescent male and lactating woman. The present iodine intake of most US citizens is adequate and safe.

Outside of iodized salt, the best sources of iodine are seafoods, vegetables grown in iodine-rich soils, and iodine-rich drinking water. The iodine content of dairy products and eggs depends upon the composition of the animal feed.

IRON Iron, a trace element for which an RDA has been established, constitutes 4.7% of the earth's crust. It is an essential constituent of hemoglobin, myoglobin, and various enzymes. It is required for healthy red blood corpuscles. Iron's role in oxygen transport and cellular respiration makes it indispensable for man.

The average adult stores only 1 g of iron, chiefly in the liver and spleen. Reservoir iron exists in the cells as a protein complex, either as a hemosiderin or as ferritin. Although the average person receives about 15 mg of iron in his daily diet, usually only 1.5 mg to 2 mg are absorbed. Ferrous iron is absorbed more efficiently than is ferric iron. Certain dietary and medicinal substances decrease iron absorption. These include calcium and phosphate salts, phytates, tannic acid in tea, and antacids.

The nursing infant receives no iron from mother's milk; iron stores in his liver decrease progressively during the rapid growth of the first few months of life. Use of iron-supplemented feeding formulas and vitamin preparations helps correct the iron deficit and forestall the development of iron deficiency anemia. Many pregnant women, especially multiparas, have iron deficiency. Women who experience heavy menstrual flow are also in danger of developing iron deficit. The adult male, on the other hand, rarely needs iron supplementation unless he is losing blood. Should a man develop iron deficiency anemia, a careful search should be carried out to determine the source of blood loss.

Approximately 2000 people—mainly young children—are poisoned by iron medications every year. The lethal dose of ferrous sulfate for 2-year-olds is about 3 g; for adults, the lethal does is between 200 mg and 250 mg/kg body weight. Excessive accumulation of iron can result from idiopathic hemochromatosis, transfusion, hemosiderosis, or prolonged excessive iron therapy, especially if the parenteral route of administration has been employed.

The RDA for iron varies from 10 mg/day for the first 6 months of life to a high of 18 mg or more/day for the pregnant woman—an amount that requires supplemental iron. For adolescents and women—except for the 51 + age group—18 mg/day is recommended.

Excellent sources of food iron include beef liver, lean meats, beans, oatmeal, spinach, egg yolk, peas, whole wheat bread, prunes, and dark molasses.

LITHIUM Despite the fact that lithium resembles sodium and potassium in many respects, it has no known biologic role. Imitating sodium more closely than does any other ion, it can replace sodium in isolated preparations of nerve, muscle, and erythrocytes. Once inside the body cells, however, lithium is not effectively removed by the sodium pump. Lithium is promptly and almost completely absorbed from the GI tract. Initially, the ion is distributed in the ECF, then is gradually localized in different tissues to varying degrees. Although passage through the blood–brain barrier is slow, the final concentration of lithium in the cerebrospinal fluid is about 40% of the plasma concentration. About 95% of a single dose of lithium is eliminated in the urine. Renal excretion is promoted by aminophylline, acetazolamide, and osmotic diuretics. It is excreted in feces, sweat, and milk. Women being treated with lithium should avoid breastfeeding to prevent poisoning the child.

During the past 2 decades, lithium has been found useful in controlling severe mania in manic-depressive psychosis. In some patients, it reduces the frequency of hypomanic relapse. Used in therapy, lithium offsets mood changes without producing sedation. Three to five days of treatment are usually needed for the lithium level in tissues to exert its effect. Antipsychotic agents may be given with lithium in the initial therapy of highly agitated, hyperactive manics. Lithium should be continued only in those patients in whom it reduces the frequency or the intensity (or both) of recurrent manic and depressive episodes. Should breakthrough depression occur, a tricyclic drug may be required. Similarly, should mania be observed during administration of tricyclic compounds for patients with recurrent depression, combination therapy with lithium may help.

When Medine and his co-workers first employed lithium carbonate to treat cluster headaches, they found that patients with 3 to 20 years of almost daily headaches found relief within a few days. The authors then studied the mechanism of action of lithium. It appeared that cluster headaches were directly related to the levels of serum serotonin and histamine. Lithium modified the course of the headaches and, by so doing, reduced serotonin and histamine levels. Thus, the changes in the amines appear to be a result, not a cause, of the headache relief.

Administration of lithium requires close clinical observation, careful dosage adjustment, and continuous monitoring of blood levels. Serum levels of lithium should be measured two or three times weekly during the acute phase of treatment, and monthly after the maintenance dose has been determined. A morning plasma lithium concentration between 0.6 mEq and 1.3 mEq/liter is desirable. Lithium can cause acute or chronic toxicity. Symptoms include nausea, vomiting, diarrhea, and severe nervous system effects, including coma and death. Since there is no specific antidote for lithium intoxication, the importance of monitoring the blood level is obvious. Prolonged therapy should also be accompanied by renal function studies.

Contraindications for lithium include patients with heart disease or with impaired lithium excretion, including those with decreased renal blood flow. In the presence of sodium deficit, lithium is selectively reabsorbed by the tubules; toxic levels are possible. For this reason, lithium should be given with caution to patients receiving diuretics or whose sodium intake is restricted. Special precautions are also indicated when lithium is used for the elderly. Reports of an increased incidence of congenital abnormalities in infants of mothers receiving lithium suggest its avoidance during the first trimester of pregnancy. Because of reports of synergistic antithyroid effects, lithium should not be employed with iodine.

ZINC With zinc comprising 0.2% of the earth's crust, it is amazing that there is so little in sea water as to make analysis virtually impossible. Only in the recent past has zinc been shown to be essential for humans. Yet, its presence in various tissues suggests that it must be essential: its concentration in the human body ranges from 16 ppm to 50 ppm, with the highest values in the pancreas, liver, kidney, pituitary, adrenal, prostate, epididymis, seminal fluid, spermatozoa, bone leukocytes, hair, and the choroid of the eye.

Most—perhaps all—of the enzymes primarily concerned with deoxyribonucleic acid (DNA) replication, repair, and transcription are zinc metalloenzymes. Blood zinc assists in carbon dioxide exchange. The relatively large stores of zinc in bone are not readily available to the body. Clinically manifest zinc deficiency appears in humans only in those areas where nutrition is at an extremely low level. In 1961, Prasad and associates studied young Iranian dwarfs apparently suffering from iron deficiency anemia. Later they found a similar syndrome in Egyptian dwarfs.[2] In both groups, zinc supplementation promoted body growth and sexual maturation. Further studies in Egypt by other workers supported Prasad's findings.

Other types of patients have been found to be deficient in zinc, including severely burned patients in whom zinc levels in neck hair were found to be depressed. Freeland–Graves and co-workers investigated the status of the zinc nutrition of 79 vegetarians and 41 nonvegetarians. Their findings indicated that available zinc in vegetarian diets may be limited and that female vegetarians may be more likely to develop marginal zinc deficiencies than male vegetarians. In contrast, Anderson and co-workers found that long-term vegetarians did not suffer from suboptimal iron and zinc nutrition, based on hemoglobin, serum iron, serum zinc, and hair zinc concentrations. They hypothesized that long-term vegetarians may adapt to their vegetarian diets by increased adsorption of dietary iron and zinc.

Of particular relevance to body fluid disturbances is the report of Allen and colleagues, who found that two patients developed severe zinc deficiency with acrodermatitis during parenteral hyperalimen-

tation. They assessed the response of circulating T-lymphocytes to phytohemagglutin during zinc deficiency and after IV zinc supplementation. Following 20 days of such supplementation, T-lymphocytes increased dramatically. Their findings agree with animal studies that showed the detrimental effect of zinc deficiency on cellular immunity.

The RDA for infants during the first 6 months is 3 mg/day. For persons over 11 years of age, it is 15 mg/day, except for pregnant women, for whom it is 20 mg/day, and for the lactating, for whom it is 25 mg/day. Foods especially high in zinc include grain products, fish, and maple syrup. Two workers have emphasized the possibility of an interaction between zinc and vitamin A. Hence, indications of a deficiency of one should prompt an elevation of the adequacy of the other.

The causes of zinc deficiency are dietary inadequacy; malabsorption, such as results from pancreatic insufficiency or chronic inflammatory intestinal disease; increased body losses, as from chronic blood loss, dialysis, diuretics, or burns; and prolonged intravenous feedings with inadequate zinc content. If zinc combines with carbohydrates, amino acids, or both, excessive quantities may be lost in the urine, especially during anabolic phases, when there is an abrupt increase in the body requirement for zinc. (See Chapter 6 for a discussion of zinc deficiency in TPN.)

Assessment of the status of zinc nutrition is difficult: there is no one reliable test for clinical situations. Plasma zinc levels are subject to acute variations; a single low value should be interpreted cautiously. Hair zinc levels correlate poorly with plasma levels but may be of value in epidemiologic studies. The most reliable sign of zinc deficiency is a positive clinical response to zinc administration.

Symptoms of zinc deficiency include anorexia, impairment of taste and smell, acrodermatitis, alopecia, impaired wound healing, impaired immunity, and depressed serum albumin (3 g/dl or below). Burnet suggests that dementia may result from lack of zinc in subjects genetically at risk and recommends zinc administration for such persons to delay the onset of dementia.

In treating zinc deficiency, one should recognize that zinc salts have a wide therapeutic index; doses should be adjusted according to the clinical response. Among the salts are zinc sulfate, elemental zinc, zinc acetate, zinc oxide, and zinc gluconate.

Zinc toxicity, which develops with ingestion of 2 g

or more daily, occurs only occasionally. It has been associated with oral zinc ingestion and with intravenous overdosage of zinc salts. It can also follow eating of acidic foods that have been stored in zinc-coated containers. Symptoms include anorexia, nausea and vomiting, lethargy, dizziness, diarrhea, bleeding gastric erosions, and, in the case of intravenous overdosage, acute renal failure and death. Excessive intakes of zinc may aggravate marginal copper deficiency.

OTHER TRACE ELEMENTS It is difficult to prove with direct evidence that the following trace elements are essential for humans. However, their essential nature can be reliably assumed because they are essential for other mammals and also because they possess elements that are components of human enzyme systems.

Cobalt Cobalt, discovered in 1739, is thought to speed red blood corpuscle regeneration after destruction by radiation. Sheep grazing in pastures low in cobalt have developed weakness. The element is an integral part of vitamin B_{12}. This role appears to be its only contribution to normal nutrition.

A daily requirement of 15µg has been suggested; this need is amply met by an ordinary diet. The best source of cobalt is organ meats; muscle meats contain less generous quantities.

Manganese Manganese was first isolated in 1774. It is needed for normal bone structure, reproduction, and normal function of the central nervous system. It is part of essential human enzyme systems. Like magnesium, it is an activator of enzymes. Manganese is concentrated in most organs and tissues, varying with different species.

The average human dietary intake of 2 mg to 9 mg/day appears to meet the requirement. Some authorities recommend the administration of 2 mg of manganese intravenously each day to patients receiving TPN.

Toxicity of ingested manganese is low; more than 1000µg/g of diet must be present to produce it. However, when the element is injected or inhaled as dust, adverse effects on the central nervous system occur with much smaller doses.

Excellent sources of the element include nuts and whole grains; vegetables and fruits also provide manganese.

Molybdenum Molybdenum, discovered in 1778, is contained in highest concentration in the liver, skele-

ton, and kidneys. It contributes to bone formation, body growth, and normal metabolism. Excessive quantities interfere with copper metabolism.

The estimated daily intake in the US is 45µg to 500µg, which probably meets the human need. Molybdenum can be toxic, however, especially because of its antagonism to copper. Hence, daily intake should not habitually exceed 0.5 mg.

Beef kidneys, some cereals, and some legumes appear to be good sources of molybdenum. Dark green, leafy vegetables and animal organs also contain this trace element.

Selenium The selenium content of crops varies from one region to another; this fact is reflected in differences in the selenium content of pooled human blood from different areas. Although this element has been shown to be essential for several species, little is known about the human need. Data from experimental animals suggest a maximal intake of 200µg/day for adults. Selenium is found in the liver, heart, kidney, and spleen. It prevents various ailments in a variety of animals, including the rat, pig, lamb, and calf. It appears to prevent breakdown of polyunsaturated fatty acids in mammals.

A concentration of 0.1µg/g in the diet prevents selenium deficit in animals; this concentration is present in the average mixed American diet.

Organ meats, such as liver, heart, kidney, and spleen, constitute a dietary source of selenium.

TRACE ELEMENTS OF UNKNOWN IMPORT FOR HUMAN NUTRITION Deficiencies of *arsenic, cadmium, nickle, tin, vanadium,* and *silicon* have been produced in experimental animals under rigidly controlled—and, hence, artificial—conditions. While the findings suggest that the elements are essential, their exact relationships to human nutrition are unknown.

FOURTEEN MAJOR BODY FLUID DISTURBANCES

If the volume or chemical composition of the body fluid deviates even slightly from the safe bounds of normal, disease results. Diseases involving body fluids are called *body fluid disturbances* or *body fluid imbalances.* Such disturbances may be primary or they may occur secondarily to other conditions. Indeed, every patient with a serious illness is a potential candidate for body fluid disturbances, and even the patient who is only moderately or mildly ill may be stricken with one imbalance or a combination of two or more.

Despite the frequency and the importance of body fluid disturbances, probably no group of clinical problems has been so poorly understood. Much of the early knowledge concerning imbalances of the body fluids originated with teachers who were biochemists first and clinicians second—if, indeed, they were clinicians at all. Quite naturally, they presented body fluid disturbances in the light of their own research. As a result, much of the early teaching about body fluid disturbances was concerned with detailed descriptions of all the possible imbalances that could occur in the one or two diseases on which the teacher had concentrated. Body fluid disturbances began to be regarded as *biochemical appendages* of disease states rather than as a broad group of problems representing the common denominator of many ailments. No overall view of these disturbances was presented. The subject as taught was so complex, so specialized, and so sophisticated that most members of the medical and nursing professions came to regard the subject as difficult, if not impossible. One student in the class of a famous pediatrician-biochemist described the subject: "as clear as if written backwards by Gertrude Stein in Sanskrit."

A completely different and enlightened approach to the study of body fluid disturbances might be termed the *clinical picture approach*. First introduced by Carl Moyer, it was enlarged upon considerably by Snively and Sweeney. Its basic breakthrough involved understanding that disturbances of water and electrolytes are produced by a fairly small number of mechanisms, most of which are simple and readily comprehensible.

Essential to the application of the clinical picture approach is a simple diagnostic classification that divides body fluid disturbances into some eighteen basic imbalances or clinical pictures. Each has its own set of causative mechanisms; its own symptoms, subjective and objective; and its own laboratory findings.

Since we do not have direct access to the cellular fluid except in highly sophisticated procedures, it is not practical to attempt to assess the state of the cellular fluid directly. However, we can determine the condition of the ECF directly (and, hence, of the cellular fluid indirectly) by examining plasma and other body liquids, such as sweat and urine. Thus, our diagnostic classification of body fluid disturbances is based on the ECF. When we discover an imbalance, we correct it by correcting the composition of the ECF, knowing that the problems within the body's trillions of cells will in all probabiblity also be corrected.

Two of the imbalances in our diagnositc classification involve changes in the *volume* of the ECF—either a deficit or an excess. The next two imbalances in our classification involve *shifts* of water and electrolytes from one ECF compartment to another. The final 14 imbalances involve alterations in the *concentrations* of electrolytes in the ECF in units of electrolytes per unit volume of body fluid. It is important to bear in mind that what is important from the standpoint of a disturbance of body fluid is usually the concentration of the electrolyte per unit volume and *not* the total quantity of the electrolyte. The volume imbalances, the shifts, and ten of the concentration imbalances are discussed in this chapter. The remaining four concentration imbalances, the acid–base disturbances, are discussed in Chapter 3.

There is nothing new to this pedagogic technique for studying disease. When ailments such as rheumatic fever, appendicitis, lobar pneumonia, or the contagious diseases are presented to students, they are presented as clinical pictures, and the students analyze them as clinical pictures. Each picture includes the history, the symptoms—both subjective and objective—and the laboratory data. The student who studies body fluid disturbances by this method analyzes the underlying mechanisms responsible for the disturbances through clinical pictures.

Sometimes one of these disturbances exists by itself; at other times, it occurs in combination with one or more additional imbalances. Frequently, body fluid disturbances are associated with other disease states and, indeed, interact intimately with them. Sometimes a succession of body fluid disturbances occurs, one after another. Clearly, one must understand single imbalances if one is to understand the combinations.

VOLUME CHANGES IN ECF

Both ECF volume deficit and ECF volume excess are fascinating to study. One of them (volume deficit) has caused more death than all the wars in history. Even today, it is a massive problem for countries without modern medical facilities.

Volume disturbances of the ECF represent either deficits or excesses of both water and electrolytes—not water alone—in approximately the same proportions as they are found in the normal state. Thus, although the ECF changes in volume, the percentages of water and electrolytes remain about the same. It is also important to note that when volume changes occur in the ECF, there are corresponding changes in the cellular fluid. An uncorrected volume deficit of ECF, for example, will ultimately cause a volume deficit of cellular fluid as well.

ECF VOLUME DEFICIT Among the terms frequently used to describe ECF volume deficit are *fluid deficit, hypovolemia,* and *dehydration.* The term *dehydration* is incorrect, however, since it involves the loss of water only.

Volume deficit results from an abrupt decrease in fluid intake, from an acute loss of secretions and excretions, or from a combination of decreased intake and increased loss (Fig. 2-1). It usually begins with one of the following: a loss of secretions and excretions, as occurs in vomiting, diarrhea, or fistulous drainage; a systemic infection, with its attendant fever and increased utilization of water and electrolytes; or intestinal obstruction. As secretions and excretions are depleted, they are replenished by water and electrolytes of the ECF, thus reducing the volume of the ECF. With continued depletion of the ECF, water and electrolytes are drawn from the cells, thus causing a deficit in the cellular fluid as well, although not immediately.

A volume deficit can develop slowly or with great rapidity, in which case it may cause death within hours after onset. In epidemics of Asiatic cholera, thousands upon thousands died of an ECF volume deficit produced by the severe vomiting and purging associated with the disease. Fortunately, increased knowledge concerning body fluids has virtually eliminated deaths caused by cholera, and today's cholera victim usually survives with no aftereffects if given prompt treatment for the ECF volume deficit that occurs during the initial period.

FIG. 2-1. Extracellular fluid volume deficit.

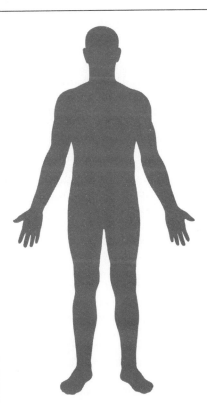

PRECEDING EVENTS
- *Loss of water and electrolytes*
- *Decreased intake of water and electrolytes*

CLINICAL OBSERVATIONS
- *Depressed fontanel in infant*
- *Longitudinal furrows in tongue (most reliable)*
- *Inelastic skin (in youth and middle-aged)*
- *Decreased tearing and salivation*
- *Slow-filling hand veins when arm is lowered*
- *Absence of moisture in axilla and groin*
- *Systolic blood pressure 10 mm Hg less standing than supine*
- *Weight loss: 2% = mild deficit, 5% = moderate deficit; 8% = severe deficit*
- *Decreased central venous pressure*
- *Temperature subnormal (unless infection is present)*

- *Urine flow rate under 40 ml/ hr (in adult)*
- *Pulse and respiration increased*
- *Flat neck veins in supine position*
- *Increased urinary specific gravity*
- *Pinched facial expression*

LABORATORY FINDINGS
- *Little help*

RELATED PROBLEMS
- *Kidney tubules deteriorate rapidly*
- *Quickly leads to other deficits*

Diagnosis ECF volume deficit frequently is difficult to diagnose because so many of its clinical symptoms appear in any seriously ill patient. In the infant, depression of the anterior fontanel is diagnostic. In others, observations that are generally useful include longitudinal wrinkles in the tongue (one of the most valuable); dry skin and mucous membranes; fatigue; a urine flow rate of less than 20 to 40 ml/hr in an adult, and materially lower in children, in proportion to body surface area. A systolic blood pressure of 10 mm Hg less in the standing position than in the supine is also indicative of the imbalance in the adult. Characteristically, the pulse is rapid, the temperature is normal or subnormal (unless infection is present, as it often is), the respiratory rate is elevated, and the venous pressure (which can be measured either centrally or peripherally) is decreased. Acute body weight loss is enormously helpful in diagnosis. For a mild deficit, weight loss might be 2%; for a pronounced deficit, 5%; and for a severe deficit, 8% or more, with the loss considerably more alarming if it is acute. Unfortunately, the patient often does not know his pre-illness weight.

The laboratory offers us little, if any, help in diagnosing volume deficit, although findings of hemoconcentration may be present in a severe deficit. Thus, diagnosis must be based on a careful history and searching physical examination.

One of the great hazards of a severe ECF volume deficit is inadequate perfusion of the kidney, since there is not enough ECF to bring the requisite amount of plasma to the glomeruli. When this imbalance is permitted to exist for more than a short period of time (minutes, if the imbalance is severe), the kidney tubules deteriorate and may soon become permanently damaged. The reason for this deterioration is not clear at present.

ECF volume deficit frequently is followed by other body fluid disturbances, including bicarbonate deficit and potassium deficit. If excessive water is lost in watery stools or with rapid breathing, sodium excess also may develop.

The goal of therapy in this imbalance is to provide both cellular and extracellular electrolytes to repair losses without altering the composition of the ECF. Although a treatment solution containing water and electrolytes in the proportions present in ECF (an isotonic solution) may appear logical, many prefer a hypotonic electrolyte solution because it provides the additional water needed for urinary excretion and for replacement of plain water lost through the skin (in insensible perspiration) and through the lungs in respiration.

CASE HISTORIES Eight-year-old R. S. was admitted to the hospital following 2 days of severe vomiting and diarrhea, which started after a school picnic (where the sanitation had been poorly supervised). He had not been able to keep anything down for 36 hours and had not voided for 24 hours. His weight before he became ill was about 60 lb (27 kg). Physical examination revealed lethargy, a rectal temperature of 101° F (38.3° C), pulse 105, respirations deep and regular at 24/min, blood pressure 110/80, dry skin and mucous membranes, an acetone odor to his breath, increased intestinal peristaltic sounds, weight 55 lb (25 kg), and a height of 50 in. Laboratory tests showed hemoglobin 15 g, plasma sodium 148 mEq/liter, plasma potassium 5 mEq/liter, plasma bicarbonate 12 mEq/liter, plasma chloride 108 mEq/liter, P_{CO_2} 28 mm Hg, and plasma pH 7.25. A catheterized urine specimen revealed a bladder urine volume of 10 ml, pH 5, specific gravity 1.032, a trace of albumin, and the presence of acetone. Nausea, vomiting, and diarrhea persisted throughout the first day. Estimated fluid lost from vomiting and diarrhea was 500 ml in the first 12 hours and 400 ml during the second 12 hours.

Commentary The increased respirations and the depressed P_{CO_2} (indicating hyperventilation) in this clear-cut case of ECF volume deficit are revealed as compensatory for HCO_3 deficit, since the P_{CO_2} was depressed 1 mm Hg from the normal of 40 for each 1 mEq/liter depression of the bicarbonate from the normal 24 mEq/liter. (If the P_{CO_2} has been depressed more than 1 mm Hg for each mEq/liter depression of the bicarbonate from the normal of 24 mEq/liter, one must suspect a complicating carbonic acid deficit, since such a change would indicate that something more than simple compensatory decrease of P_{CO_2} was occurring. If, on the other hand, the P_{CO_2} were depressed less than 1 mm Hg for each mEq/liter depression of the bicarbonate from the normal 24 mEq/liter, then carbonic acid excess must be superimposed on the compensatory decrease of P_{CO_2}, since such depression of P_{CO_2} would be less than normally occurs as compensation for bicarbonate deficit. For detailed explanation, refer to the section on acid–base disturbances in Chapter 3.)

E. D., age 57, was admitted to the hospital. Her husband stated that she had become nauseated 2 days before after a day or two of just "not feeling up to par," with loss of her usual excellent appetite. The day before admission, she complained of intermittent pain centering around the umbilicus. She vomited occasionally, usually when she was having pain. On physical examination, her skin felt dry, her tongue showed definite longitudinal wrinkes. Her blood pressure was found to be 130/90 in the supine position and 115/75 when she was held upright. Her abdominal wall was tight but not rigid. There was slight distention. Peristaltic sounds could be heard with the stethoscope, louder during periods when there was pain. No gas or feces were passed. On admission to the hospital, a Foley catheter was inserted; her urine flow rate in the ensuing 2 hours was 20 ml/hr. Urine specific gravity was 1.020 (concentrated), but the specimen was otherwise normal. All electrolytes and blood gases (Pco_2, Po_2, pH) were within limits of normal. Red and white blood counts and hemoglobin were normal. Her weight was 130 lb (59 kg), which represented no change from recent weights (patient had been on a reducing diet and had been weighing regularly). A diagnosis of complete intestinal obstruction was made, later found to be caused by obstruction of the small intestine by a tumor. The body fluid imbalance diagnosis was ECF volume deficit. Of key importance in making this diagnosis was the physician's recognition of the fact that as much as 8 liters of fluid that closely resembles ECF can accumulate daily within an obstructed intestine without any weight loss. Yet, this accumulated fluid, still within the confines of the body, is just as much lost to the body as if it were outside. The "lost" fluid was replaced immediately, before surgical intervention was carried out.

Commentary The various causes of intestinal obstruction can be divided under headings of *mechanical* (e.g., pressure on the intestine from adjacent tumors), *vascular* (e.g., thrombosis of a splanchnic artery or vein), or *neurogenic* (e.g., as occurs in pneumonia). Essential to remember in ECF volume deficit caused by intestinal obstruction is the fact that one cannot rely on acute weight changes to give a clue to the diagnosis. This truth holds, regardless of what portion of the intestine is obstructed. Important clues to the diagnosis in this type of ECF volume deficit are orthostatic hypotension, decreased urine flow rate (under 40 ml/hr in an adult), increased urinary specific gravity, and longitudinal wrinkles in the tongue.

ECF VOLUME EXCESS ECF volume excess is frequently called *fluid excess* or *overhydration*, but the latter term is incorrect because it represents an excess of water only.

A volume excess develops when the kidneys are unable to rid the body of unneeded water and electrolytes. This inability may result from simple overloading of the body by oral or parenteral administration of excessive quantities of an isotonic solution of sodium chloride. Somewhat paradoxically, it is the quantity of sodium in the body that determines the volume of the ECF; therefore, excess sodium from any cause may well precede ECF volume excess. The imbalance may occur in diseases that affect the function of the homeostatic mechanisms, such as chronic kidney dis-

ease, chronic liver disease with portal hypertension, congestive heart failure, and malnutrition. In all of these conditions, there is abnormal retention of water and sodium and, whether the volume excess is caused by simple overloading or by diminished function of the homeostatic mechanisms, the result is the same: the ECF becomes excessively salty. Therefore, in an attempt to maintain its normal composition, the ECF draws water from the cells.

Diagnosis Expanding volume of ECF produces numerous clinical symptoms that contribute far more in diagnosing this imbalance than they do in the case of ECF volume deficit (Fig. 2-2). Clinical observations include puffy eyelids; peripheral edema, which may actually be pitting; ascites (accumulation of fluid in the abdominal cavity); pleural effusion; pulmonary edema (often visible by roentgenography when undetectable by stethoscope); elevated central or peripheral venous pressure; acute weight gain, sometimes as much as 10% of the normal body weight; and moist rales in the lungs (late in the imbalance). In many patients, this imbalance occurs without frank edema because of the mysterious third factor, which causes diuresis before frank edema supervenes; in a severe excess, the patient may succumb from the pulmonary edema.

The laboratory gives little help in diagnosing ECF volume excess, except that the formed elements of the blood may be decreased because of plasma dilution. If renal impairment has occurred, both the blood urea nitrogen (BUN) and the plasma potassium may be elevated. An x-ray picture of the lungs will detect fluid accumulation long before it can be detected by auscultation.

Considering related problems, we should recall that the patient may succumb from pulmonary edema. The imbalance can occur with remobilization of edema fluid on the third day following a severe burn if the patient has been given, or permitted, excessive fluids during the first 2 days of the burn. If the cause of the excess is kidney failure, the BUN and potassium (K) would be elevated, thus providing a valuable clue to the cause of the accumulation.

Fluid therapy usually is not indicated in this imbalance; rather, therapy is directed toward the causative factors. Symptomatic treatment consists of restriction of fluids, administration of diuretics, or both. When the excess has resulted from excessive administration of isotonic solutions, discontinuing the infusion may be all that is required in the patient with functional homeostatic mechanisms. If the excess has

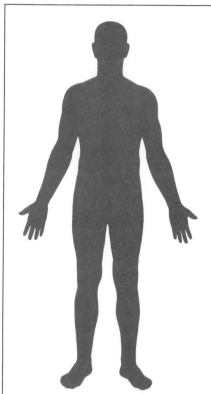

PRECEDING EVENTS

- *Any cause of retention of excessive sodium and water (e.g., chronic heart failure, chronic renal failure, excessive adrenocortical hormones, and overzealous administration of IV fluids—especially sodium-containing solutions)*

CLINICAL OBSERVATIONS

- *Puffy eyelids*
- *Peripheral edema*
- *Ascites*
- *Effusions into "third spaces" (e.g., pleural and abdominal cavities)*
- *Moist rales in lungs (can indicate an excess of 1500 ml or more)*
- *Acute weight gain*
- *Full, bounding pulse*
- *Slow emptying of hand veins when arm is raised*
- *Elevated central venous pressure*
- *Distended neck veins*
- *Increased urinary volume if kidneys normal (BUN and potassium elevated if kidney disease present)*

LABORATORY FINDINGS

- *Little help*

RELATED PROBLEMS

- *Patient may succumb from pulmonary edema*
- *Imbalance may occur with remobilization of edema fluid third postburn day*
- *If BUN and potassium are elevated, kidneys may be failing*

FIG. 2-2. **Extracellular fluid volume excess.**

developed over a long period of time, as may occur in the patient with chronic kidney disease, chronic liver disease, or congestive heart failure, withholding of fluids is not indicated. The homeostatic mechanisms do not function normally in patients with these conditions, and treatment is directed toward the underlying condition. The volume excess may also be corrected by use of a low-sodium diet. Finally, the patient with chronic malnutrition may develop a volume excess, largely because of plasma protein deficit; in this case, treatment is directed toward improving the patient's nutritional status. In some patients, one of the potent diuretics may prove useful. In extreme situations, hemodialysis or peritoneal dialysis can be lifesaving.

CASE HISTORIES Eight-year-old J. D. was unable to retain liquids by mouth following appendectomy and, therefore, was provided needed water and electrolytes by the intravenous route. For 2 days, he was given a continuous intravenous infusion of 5% dextrose in isotonic solution of sodium chloride in the amount of approximately 4 liters/day. Early on the morning of the third day, he complained of shortness of breath; his eyelids were puffy, his cheeks appeared full, and moist rales were heard in the lungs. He was found to have gained about 6% over his admission weight. His (RBC) was 4,000,000 and his hemoglobin was 10 g.

Commentary Isotonic solution of sodium chloride is sometimes called *normal saline* or *physiological solution of sodium chloride.* It is not *normal* from the chemical standpoint, and it is certainly not physiologic. It provides 154 mEq of Na/liter, somewhat in excess of the normal 142 mEq/liter. It also contains 154 mEq of Cl, far in excess of the normal 105 mEq/liter. Thus, it forces a considerable excess of chloride on the kidneys. Moreover, it contains no potassium, calcium, magnesium, or phosphate, all required for a solution to meet both the extracellular and cellular needs of the body. Finally, since it is isotonic with plasma, it provides no free water for renal function and metabolic processes. Because of its chloride excess, isotonic solution of sodium chloride tends to favor bicarbonate deficit (metabolic acidosis).

H. F., a 45-year-old male, visited his physician because of episodic weakness, paresthesias, tetany, polyuria, and polydipsia. In the course of a complete workup for hypertension, it was found that the patient's aldosterone excretion on a sodium intake of more than 10,000 mg/day exceeded 200µg daily. The patient was given spironolactone, an aldosterone antagonist, and all manifestations of the disease were reversed in 6 weeks. Adrenal adenoma or adenomata were suspected. At surgical operation, the patient was found to be suffering from bilateral adrenal hyperplasia, rather than from adrenal adenoma. No corrective surgery was carried out; instead, the patient was placed on mild sodium restriction, low doses of chlorthalidone, and supplementary potassium.

Commentary Adrenal hyperplasia, as does an adrenal adenoma, causes the production of excessive aldosterone, which is not responsive to the body's need for sodium conservation, volume conservation, and potassium excretion. Even when a large quantity of sodium is given, as in the above example, the patient continues to secrete large quantities of aldosterone. The symptoms of excessive aldosterone are due chiefly to the excessive retention of sodium, chloride, and water, and the excessive excretion of potassium. While spironolactone reverses the effects of aldosterone, its side-effects are often undesirable. The patient with an adrenal adenoma is fortunate, since removal of the tumor can be expected to cure the hypertension.

SHIFTS OF WATER AND ELECTROLYTES OF ECF

Normally, about one fourth of ECF exists as plasma, and about three fourths as interstitial fluid. These two fluids are quite similar in composition, except for the amount of protein they contain: plasma has about 18 mEq/liter and interstitial fluid only 1 mEq/liter. It is the oncotic pressure exerted by the plasma protein that prevents large amounts of the plasma's water and electrolytes from being forced into the interstitial fluid by the hydrostatic pressure produced by the pumping action of the heart. However, under certain circumstances, water and electrolytes do shift into the interstitial fluid, and, in other cases, water and electrolytes of interstitial fluid shift into the plasma.

We can readily understand why such shifts occur when the plasma protein content is altered. However, shifts also take place when plasma protein is normal, although the mechanisms for these shifts are obscure and appear to be of little value.

Before we study these shifts, we would do well to review what we learned about the delivery of water, electrolytes, glucose, amino acids, fatty acids, vitamins, and so on to the cells. This all occurs in the area of the arterial capillary bed. Indeed, the capillary bed is the whole reason for the existence of the cardiovascular system.

In the arterial capillaries, there is some 22 mm Hg force exerted by the plasma albumin, holding liquids in the capillaries; however, there is 32 mm Hg of force (hydrostatic pressure transmitted from the heart pump) pushing liquids from the capillaries. The net force propelling liquids from the capillaries, then, is $32 - 22 = 10$ mm Hg. This force is sufficient for water, electrolytes, and other nutrients to leave the capillaries, pass into the interstitial fluid, and, ultimately, to enter the cell. There are forces opposing and favoring this movement within the interstitial fluid, but their net force is negligible (Fig. 2-3).

Now let's turn our attention to the venous capillar-

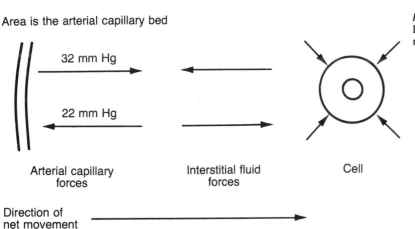

Area is the arterial capillary bed

32 mm Hg

22 mm Hg

Arterial capillary forces

Interstitial fluid forces

Cell

Direction of net movement

FIG. 2-3.
Delivery of water, electrolytes, and other nutrients to cells.

ies. They possess that same 22 mm Hg exerted by plasma albumin in the arterial capillaries, again acting as a sort of sponge, pulling materials into the capillaries. But the venous capillaries have only some 12 mm Hg hydrostatic pressure pushing liquids from the capillaries, and the force of the cardiac contraction has greatly dissipated itself by the time it reaches the venous capillaries. There is, therefore, a net movement from the cells into the venous capillaries. Urea, CO_2, and other wastes leave the cells and enter the venous capillaries to return to the large veins. In this return of wastes from the cells, the venous capillaries are assisted by the lymph capillaries, blind capillaries that help carry the cellular wastes through lymph channels to the thoracic duct, where they are poured into the left subclavian vein (Fig. 2-4).

With this brief review of the capillary bed and recalling the essential role played by the plasma albumin, let us consider the great shifts.

PLASMA-TO-INTERSTITIAL FLUID SHIFT In plasma-to-interstitial fluid shift, sometimes called *hypovolemia*, abnormal quantities of water and electrolytes move from plasma into the interstitial fluid. This decreases the volume of plasma and increases the volume of interstitial fluid; the normal 3:1 ratio is upset. The imbalance is closely related to shock and to edema and is almost invariably seen on the first or second day following a severe burn. The shift is also often seen following a massive crushing injury, severe trauma, or perforation of a peptic ulcer and may be observed with intestinal obstruction or following the acute occlusion of a major artery. Plasma-to-interstitial fluid shift may occur in the patient undergoing surgery, producing the condition known as *surgical shock*. The shift can re-

sult simply from depression of the plasma albumin, such as occurs in chronic illness, liver disease, hemorrhage, adult or childhood nephrosis, or in starvation ("starvation edema"). With the normal for the serum albumin 4 g/dl, a serious deficit has occurred when the level reaches 3 g/dl.

The clinical symptoms of plasma-to-interstitial fluid shift include pallor, tachycardia, low blood pressure, weak pulse—perhaps undetectable—cold extremities, disorientation, and, finally, coma. The clinical picture usually suffices for diagnosis of this imbalance, but laboratory tests are helpful, as the red blood cell count, hemoglobin, and packed cell volume will be elevated because of the decrease in plasma volume.

When plasma-to-interstitial fluid shift develops slowly over a long period of time, as in starvation, the patient displays a huge belly, as well as swollen ankles, legs, wrists, and arms—a condition known as *starvation edema*.

Therapy of this imbalance is directed toward limiting the shift, if possible; maintaining plasma volume; and treating vascular collapse and heart failure if they occur. To help limit the severity of the secondary interstitial fluid-to-plasma shift that occurs with remobilization of edema fluid, unnecessary administration of fluids during the plasma-to-interstitial fluid shift should be avoided. Should the serum albumin be depressed to 3 g/dl or below, salt-poor albumin should be given intravenously (Fig. 2-5).

CASE HISTORIES Eleven-year-old D. G. was admitted to the hospital with two thirds of her body covered with second- and third-degree burns suffered when a flamma-

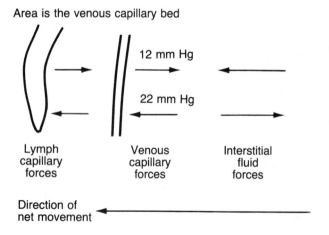

Area is the venous capillary bed

Lymph capillary forces

12 mm Hg

22 mm Hg

Venous capillary forces

Interstitial fluid forces

Direction of net movement

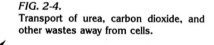

Cell

FIG. 2-4.
Transport of urea, carbon dioxide, and other wastes away from cells.

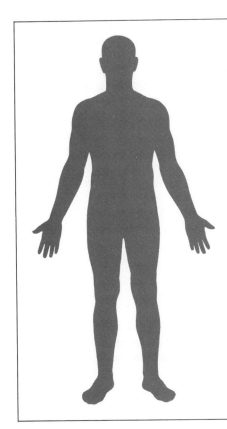

PRECEDING EVENTS

- *Severe trauma*
- *Critical abdominal event*
- *Depression of plasma albumin*

CLINICAL OBSERVATIONS

- *Pallor*
- *Rapid pulse*
- *Low blood pressure*
- *Cold extremities*
- *Apprehension*
- *Disorientation*
- *Coma*

LABORATORY FINDINGS

- *Formed elements of blood elevated*

RELATED PROBLEMS

- *Invariable occurrence during first 2 days of severe burn*

FIG. 2-5. Plasma-to-interstitial fluid shift.

ble party dress caught fire as she danced near the open fire in the fireplace. The burn team of the hospital immediately took charge and carried out their customary excellent routine for severe burns. Nevertheless, during her second 24 hours in the hospital, her hemoglobin and RBC rose rapidly as her blood pressure fell to 90/70 and her pulse increased to 125. She complained of cold hands and feet. The burned area appeared boggy, as if filled with fluid, as indeed it was, from a plasma-to-interstitial fluid shift. As with all severely burned patients, she was permitted no fluids—not even ice chips—by mouth for the first 48 hours of her hospital stay. All fluids permitted were monitored carefully and given by the intravenous route.

Commentary Just why a plasma-to-interstitial fluid shift should occur during the first 48 hours of a severe burn remains a mystery. If it is a defensive mechanism by the body, one is hard put to see what it accomplishes. It makes the burned patient thirsty—desperately thirsty. If burned patients are permitted to drink water (they used to be), they are prone to develop sodium deficit and plasma fluid overload when the interstitial fluid-to-plasma shift of the third day begins. Many severely burned patients used to die of sodium deficit because they were permitted water

ad libitum, but their deaths were mistakenly attributed to "burn poisoning."

Eight-year-old T. B. was suffering from childhood nephrosis, an ailment of unknown cause and largely unknown pathophysiology, which involves the loss of enormous quantities of plasma albumin in the urine. Because of the loss of the albumin, the oncotic pressure of the plasma is decreased, and water and electrolytes are allowed to escape from the plasma and accumulate in the interstitial space, causing edema so severe that sometimes the child's eyes are swollen shut. On one occasion when T. B. was given a checkup, her plasma albumin was 2.6 g/dl. The urine was, of course, loaded with albumin, since this was the route of loss.

Commentary Fortunately, most children recover from childhood nephrosis and may never have a recurrence. Treatments vary widely, but several appear effective. This ailment provides a striking example of the function of albumin in maintaining the oncotic pressure of the plasma.

INTERSTITIAL FLUID-TO-PLASMA SHIFT In interstitial fluid-to-plasma shift, or *hypervolemia*, water and electrolytes from interstitial fluid flow into the plasma. This imbalance always occurs during the recovery phase of a condition that previously caused a plasma-to-interstitial fluid shift, such as burn or fracture. In this case, the shift is called *remobilization of edema fluid*. If too much fluid is administered during the initial shift, the secondary shift may endanger life. Following hemorrhage, water and electrolytes move from interstitial space to plasma in an effort to replace those that were lost in blood. The shift also can occur when the oncotic pressure of the plasma has been increased by administration of excessive amounts of blood, plasma, dextran, or hypertonic solutions.

In many respects, the clinical observations of interstitial fluid-to-plasma shift are just the opposite of those seen in plasma-to-interstitial fluid shift. They include a bounding pulse, engorgement of the peripheral veins, moist rales in the lungs, pallor, weakness, and air hunger. Cardiac dilatation and ventricular failure may occur as well. Laboratory findings helpful in diagnosis include a decrease in the RBC, hemoglobin, and packed cell volume.

The rapid interstitial fluid-to-plasma shift that occurs with remobilization of edema fluid following the reverse shift or following excessive intravenous administration of hypertonic solutions is unpreventable and may be so severe as to overload the renocardiovascular system. When this occurs, the goal of therapy is to reduce the amount of fluid returning to the heart, accomplished by placing tourniquets around the extremities in such a way that they block venous return but do not interfere with arterial circulation. Phlebotomy may be necessary. If the shift occurs as a compensatory measure following internal or external loss of whole blood, it can be halted by transfusion of whole blood (Fig. 2-6).

CASE HISTORIES Fourteen-year-old K. B. was admitted to the hospital suffering from second- and third-degree burns over 60% of his body. His initial water and electrolyte

FIG. 2-6. Interstitial fluid-to-plasma shift.

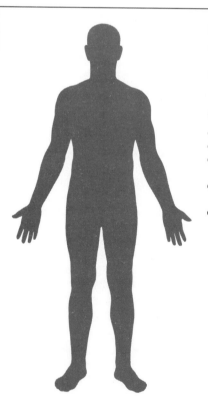

PRECEDING EVENTS

- *Occurs during recovery from plasma-to-interstitial fluid shift*
- *Compensatory shift following hemorrhage*

CLINICAL OBSERVATIONS

- *Bounding pulse*
- *Hypertension*
- *Abnormally large urinary volume*
- *Pulmonary edema (radiograph and auscultatory evidence)*
- *Cardiac failure*

LABORATORY FINDINGS

- *Formed elements of blood depressed*

RELATED PROBLEMS

- *Begins on third day after severe burn; can be fatal if too much fluid given first 2 days* therapy was overzealous. On his third postburn day, he complained of being unable to get his breath, and physical examination at that time revealed a bounding pulse, with engorgement of the peripheral veins. Moist rales were heard in both lungs, and the RBC was found to be 3,500,000, with a hemoglobin of 9 g/dl. He was clearly suffering from the remobilization of edema fluid that occurs during the third to fifth day following a severe burn. The shift can be hazardous, even fatal, if administration of water and electrolytes during the first 2 days postburn has not been carefully controlled, which usually means given entirely by the the intravenous route.

Commentary While this shift and the plasma-to-interstitial fluid shift that preceded it may well have a purpose, to date it has eluded discovery. In order to prevent such dangerous remobilization of shifted fluid, several useful formulas for administration of fluid to burned patients have been developed. All are excellent if used in accordance with the patient's ever-changing condition. Perhaps the most useful gauge for fluid therapy is the urine output per hour, as measured from an indwelling catheter. This is ideally 40 ml/hr in an adult, proportionately less in children and in infants.

D. L., a 24-year-old medical student, contributed 850 ml of his blood to a desperately injured friend, both of whom had a relatively rare blood type. For several days following the donation of blood, the student experienced a bounding pulse, noticed some engorgement of his peripheral veins, and was short of breath. But within a week, and with no treatment, he felt his normal self. The symptoms were due to a shift of water and electrolytes from interstitial fluid to plasma, compensatory effort on the part of the body to restore the normal viscosity of the circulating blood, rendered unduly viscous by the release of large quantities of red blood cells into the circulation by the spleen.

Commentary This case history illustrates at least one potential function of the spleen, which is certainly not necessary for life. It also demonstrates a useful shift of water and electrolytes from one ECF compartment to another. The other shifts described in this section serve no obvious useful purpose and may, indeed, prove fatal.

CONCENTRATION IMBALANCES OF ECF

Each electrolyte has special bodily functions and, although some play larger roles than others, all are necessary for the maintenance of life. Their normal concentrations are precisely geared to the body's needs, so they must be maintained at proper levels if the body is to remain healthy. Also vitally important are the concentrations of carbonic acid and base bicarbonate.

Discussed in this chapter are disturbances involving alterations in the normal concentration of the ECF's electrolytes, as expressed in mEq/liter of ECF. A change in the concentration of an electrolyte can be produced either by a change in total quantity of the electrolyte or in the total volume of water in ECF. For example, one could develop a sodium deficit either because of decreased intake or increased loss of sodium, or because of excessive intake or decreased loss of water. Similarly, one could develop an excess of sodium in ECF either because of increased intake or decreased loss of sodium, or because of decreased intake or increased loss of water. These mechanisms apply not only to changes in sodium concentration but also to all the other extracellular electrolytes.

SODIUM Sodium is the most abundant cation of ECF and the second most important cation of cellular fluid. While most of the body's sodium is to be found in the body fluids, some is stored in bone. Chemically, sodium is a metallic element, and its chemical symbol, Na, stems from the Latin word *natrium*, meaning sodium. In pure form, the element is extremely unstable, readily combining with oxygen either in air or in water. In nature and in body fluids, it exists only in combination with various anions, usually chloride; the compound sodium chloride is more commonly called table salt.

Sodium affects many vital functions of the human body. It is largely responsible for the osmotic pressure of ECF. With potassium, it moves continuously across the cell membrane, propelled by a mysterious pump still only partially understood. Inside cells, sodium is a factor in numerous vital chemical reactions. The kidneys regulate body water and electrolytes, based in part on sodium concentration in the ECF. Therefore, when sodium concentration rises, the kidneys retain water in an effort to maintain the normal composition of the ECF. When the amount of water increases, the kidneys retain sodium. Other functions of the element are as follows: probably through some type of chemical-electrical action, sodium stimulates reactions within nerve and muscle tissues; as a component of sodium bicarbonate, it is highly important in maintaining the delicate balance between the acids and alkalies of the body fluid.

Since sodium plays such a vital role, it is not surprising that the body possesses a highly efficient mechanism for regulating the electrolyte so that its normal concentration of 142 mEq/liters of ECF can be maintained. During periods of decreased intake or increased loss, the mechanism acts to conserve sodium. Important in sodium conservation is the hormone aldosterone, secreted by the adrenal cortex, which promotes potassium loss and sodium retention by means of a complex and little understood feedback mechanism involving the afferent arterioles, the juxtaglomerular apparatus, and the macula densa of the distal tubule. The action of the mechanism is so formidable that destruction of the adrenal glands or their surgical removal will result in fatal sodium deficit in a few days unless replacement hormones are administered. When fluid volume deficit occurs, a resultant decrease in glomerular filtration aids in sodium conservation simply because there is less sodium in the tubular urine to be exchanged for potassium and hydrogen in the peritubular capillaries of the distal tubules. Body sodium conservation is usually so effective that patients being maintained on a sodium–restricted diet for control of certain diseases, such as essential hypertension, are seldom in danger of developing sodium deficit as a result of the decreased intake. Should one be placed on a sodium-*free* diet (obviously impossible), sodium would be excreted in his urine for only a few days. After that, the body would be completely stingy with the precious substance, conserving almost all. Consider the ranges of sodium concentration in the ECF compatible with life: probably the lowest value is 100 mEq/liter, the highest, 200 mEq/liter. These, mind you, are extremes; few of us would survive if we reached them. The ratio between minimum and maximum is 1:2. Now consider two other electrolytes of crucial importance, hydrogen and potassium. The ratio between minimum and maximum for each of these is roughly 1:10. Sodium, therefore, enjoys a sort of primacy that is not only unequaled but that is not even remotely approached. Its primacy may be due—in part, at least—to the fact that the body endeavors to maintain ECF volume above all else. Volume consists primarily of water, and sodium is the "sponge" that holds the water upon which ECF volume so heavily depends.

When sodium intake increases, the regulating mechanism acts to rid the body of excesses, and there is a prompt increase both in sodium excretion and in the glomerular filtration rate. For many years, it was believed that the increased filtration of sodium through the glomeruli was responsible for the in-creased urinary excretion of the ion. Recent studies, however, indicate that the increase in sodium excretion is related to decreased reabsorption of sodium from the proximal tubules of the kidneys. A similar sodium loss occurs when salt-retaining hormones, such as aldosterone, are administered over a prolonged period to normal human subjects. It is believed that tubular excretion of sodium is controlled by a third factor, perhaps a hormone, the chemical structure and site of origin of which are unknown. This mysterious third factor may explain why patients with primary hyperaldosteronism do not develop marked edema.

Although the sodium-regulating mechanism is quite effective, an *acute* decrease or increase in the sodium level of ECF produces severe reactions.

Persuasive evidence indicates that persons who salt their food excessively or who are members of a culture that eats extremely salty food may stand a greater than average chance of developing hypertension. Heredity appears to play a determining role in who will and who will not develop high blood pressure from excessive salt intake. Most Americans eat from 2000 mg to 5000 mg or 10,000 mg of sodium/day. For some, the figure probably tops 15,000 mg. In one area of Japan, where the average intake of sodium is 26,000 mg daily, the number of persons who suffer from strokes and hypertension is startlingly high. Because of the possibility of harm from high-sodium intakes, some nutritionists are convinced that the salt content of the infant's diet should be kept low so as not to accustom the baby to salty foods.

A diet restricted in sodium has proved effective in the treatment of various ailments, including hypertension, congestive heart failure, edema, excessive weight gain by otherwise healthy middle-aged women, and toxemia of pregnancy. At least a third, perhaps more, of patients with hypertension have their blood pressure reduced by a diet containing not more than 500 mg of sodium/day. (Sodium-restricted diets are discussed in Chapter 5.)

The requirement for sodium has been stated as from 2.38 g to 7.14 g sodium, corresponding to 6 g to 18 g sodium chloride. The actual requirement, especially in persons who have become accustomed to a low-sodium intake, is far less. Persons have been maintained in good health for indefinite periods on diets containing from 250 mg to 500 mg of sodium daily. Certain primitive peoples have apparently done well on less than 100 mg/day.

Barring drug therapy or a stringent sodium-restricted diet, sodium deficit appears virtually impossible. In fact, most so-called low-sodium diets probably do not achieve effectiveness since it is so difficult to eliminate salt-containing foods in our society.

Sodium deficit of ECF Sodium deficit of ECF occurs when the concentration of sodium in ECF (really, in extracellular water—see below) falls below normal. The imbalance is also referred to as *hyponatremia, low-sodium syndrome, electrolyte concentration deficit*, and *hypotonic dehydration*. The imbalance is quantitated in terms of the sodium deficit in mEq/liter. There are four types of hyponatremia, based on etiology:

Sodium deficit caused by the loss principally of sodium, as from vomiting or diarrhea, in bronchial secretions, from overzealous administration of diuretics (e.g., the thiazides or furosemide), through losses from serous cavities (as in draining ascites), from losses of blood, or by sequestration of sodium within the body (as in intestinal obstruction). One example of this type of deficit is heat exhaustion that develops in unacclimatized persons. One might, for example, take a long walk on a hot day, sweat profusely, and drink plain water. Having replaced lost water, sodium, and chloride with plain water only, the person, therefore, develops sodium deficit. The low-sodium syndrome can occur when diuresis is excessive. (One patient we know of took four times the dose of a potent diuretic, appeared in the hospital emergency room with a plasma sodium of 90 mEq/liter, and could not be saved even by energetic treatment.)

Sodium deficit resulting from excessive intake of water that dilutes the serum sodium (often called "water intoxication"). Such dilution can result from drinking plain water after perspiring profusely; depressed renal blood flow; repeated tap-water enemas; excessive water retention because of congestive heart failure, cirrhosis of the liver, or the nephrotic syndrome; permitting severely burned patients or those undergoing gastric or intestinal suction to drink water freely; parenteral administration of electrolyte-free solutions; absorption of irrigating fluid during transurethral electro-resection of the prostate; compulsive drinking of water, as in certain psychic disturbances; hypodermoclysis of dextrose in plain water (the fluid pools beneath the skin and draws sodium into it); inhalation of fresh water in drowning; or habitual excessive beer drinking (more than 4 liters daily).

A combination of factors 1 and 2 discussed above occurs in inappropriate secretion of antidiuretic hormone (SIADH), covered below as a separate imbalance.

Osmotic hyponatremia is really a false hyponatremia caused by the accumulation in the plasma of glucose, lipids, or protein (as in multiple myeloma). It does not require therapy from the standpoint of the sodium concentration, which is normal. Its nature can be clarified by the following discussion: the total concentration of solute, measured in mOsm/kg fluid, is termed *osmolality*. Closely related to osmolality is *osmolarity*, referring to mOsm/liter fluid. These are not the same, since neither cellular fluid nor ECF has a specific gravity of 1.000, as does plain water. The osmolalities of both cellular fluid and ECF are apparently the same. Measurement of the solute concentration of plasma by determining the depression of the freezing point provides an excellent and available index of osmolality of these two major body fluids. Also useful, although indirect, is the determination of the serum sodium concentration in mEq/liter; however, allowance must be made for elevated urea concentration and for hyperglycemia, both of which increase the osmolality of plasma, and for lipemia and hyperproteinemia, which add to the plasma volume. Abnormal elevations of any of these four constituents may cause the sodium level to appear low when it actually is normal when considered in terms of its concentration in actual plasma water rather than in terms of its concentration in the abnormal expanded plasma sample analyzed. Such factitious hyponatremia, or pseudohyponatremia, does not require treatment.

To determine if the sodium concentration of extracellular water is normal, regardless of the measured serum sodium concentration, one must first measure the serum osmolality and then determine the corrected osmolality, using the method of Edelman, cited by Goldberger:

$$\text{Measured serum osmolality (mOsm/kg)} - \frac{\text{glucose (mg/dl)}}{18} - \frac{\text{BUN (mg/dl)}}{2.8}$$
$$= \text{corrected serum osmolality}$$

If the *corrected* serum osmolality falls within the limits of normal (260 mOsm to 275 mOsm/kg), the sodium content of the extracellular water is normal, regardless of the value for the measured serum sodium concentration. If the *corrected* serum osmolality is low, the sodium concentration of extracellular water is low and the hyponatremia should be corrected. If the *measured* serum osmolality is high, the reason for the elevation is usually obvious—diabetes, azotemia, perhaps accumulation of salicylates. Attention should be focused on one of these rather than on the apparently low serum sodium.

The clinical findings in sodium deficit vary greatly from patient to patient and are dependent on the cause, magnitude, and acuteness of the condition. One person may suffer severe symptoms from a deficit that would not phase another. As a rule, symptoms of hyponatremia due to a combination of sodium loss and water ingestion *differ greatly* from those that occur with acute water intoxication (SIADH) (Fig. 2-7 and 2-8). A typical patient with a moderately severe deficit not due to SIADH might experience the following:

The patient first suffers apprehension or anxiety, followed by a bizarre, undefinable feeling of impending doom. He is weak, confused, perhaps even stuporous, and may suffer abdominal cramps or muscle twitching, diarrhea, and, in severe deficit, convulsions. In an effort to counterbalance the deficit, the adrenal glands secrete increased amounts of aldosterone, which stimulate the kidneys to conserve water, sodium, and chloride. This action results in depressed urinary output (oliguria) or complete absence of urinary output (anuria). If the deficit is severe, there is vasomotor collapse, with the following symptoms: hypotension; rapid, thready pulse; cold, clammy skin; and cyanosis. An interesting finding that may be observed is fingerprinting over the sternum, consisting of a visible fingerprint apparent after pressure is applied with a finger or thumb on the skin overlying the sternum. Fingerprinting is due to increased plasticity of the tissues, which results from the transfer of water from the abnormally dilute ECF into the cells. As a result of this osmotic transfer of water, an abnormally large portion of the body fluid lies within the confines of the cells, and the fluid volume of the ECF—both plasma and interstitial fluid—is decreased. Consequently, tissues become more plastic than normal and tend to retain any shape attained by pressure deformation.

Laboratory findings are helpful in diagnosis of sodium deficit, since the plasma sodium is below 137 mEq/liter—it may be as low as 110, 100, or even 90—and plasma chloride is below 98 mEq/liter. The specific gravity of the urine is below 1.010.

FIG. 2-7. **Sodium deficit.**

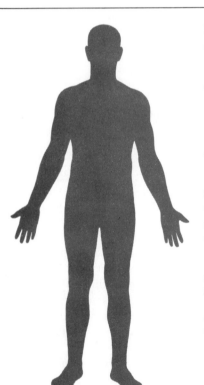

PRECEDING EVENTS

- *Loss of sodium-containing secretions without adequate replacement*
- *Use of potent diuretics (particularly if dietary sodium restriction is severe or another route of sodium loss is present)*
- *Excessive administration of sodium-free parenteral fluids*

CLINICAL OBSERVATIONS

Vary greatly, but often include:
- *Apprehension*
- *Excessive gastrointestinal stimulation (abdominal cramps, diarrhea, and nausea)*
- *Personality change*
- *Dizziness with position change*
- *Postural hypotension*
- *Cold, clammy skin*

LABORATORY FINDINGS

- *Plasma sodium below 137 mEq/liter*
- *Specific gravity of urine below 1.010*

RELATED PROBLEMS

- *Heat exhaustion in unacclimatized persons may result in sodium deficit*
- *Sodium deficit may occur with excessive use of diuretics*

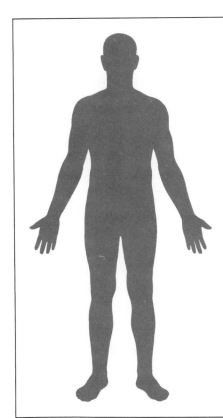

- *Trauma*
- *CNS disorders*
- *Certain malignancies*
- *Various pharmacologic agents*

CLINICAL OBSERVATIONS

- *Lethargy*
- *Anorexia*
- *Nausea*
- *Vomiting*
- *Abdominal cramps*
- *Weight gain*
- *Withdrawal*
- *Hostility*
- *Convulsions (onset of cerebral edema)*
- *Coma*

LABORATORY FINDINGS

- *Urine osmolality exceeds that of plasma*
- *Plasma sodium below 120 mEq/liter*
- *Urine specific gravity above 1.012*

RELATED PROBLEMS

- *Elevated osmolality of urine differs from other forms of hyponatremia*
- *First indication of presence is often depressed plasma sodium*
- *Signs and symptoms often appear only after patient is beyond help*
- *Great danger is cerebral edema, with herniation of brain*

FIG. 2-8. Sodium deficit (SIADH).

The goal of therapy in sodium deficit of the ECF is to restore the sodium concentration to normal as quickly as possible without producing a fluid volume excess. Treatment varies with the pathogenesis and the type of hyponatremia involved. If there is no emergency, salt and water can be replaced orally. In severe sodium deficits, small amounts of 3% or 5% sodium chloride in water may be given slowly by the IV route. In dilutional hyponatremia, diuresis plus restriction of water should be employed, although sometimes simple restriction of water suffices. Should the total sodium of the body and the body water be excessive even though a concentration deficit of sodium exists, both sodium and water should be restricted.

CASE HISTORY The day following the removal of her gallbladder, a 35-year-old woman, L. B., complained repeatedly of gas and abdominal discomfort. A gastric tube was inserted through the nose and suction drainage was started. Since the woman complained of thrist, she was given ice chips to suck. It seemed to the student nurse that she was consuming an inordinate quantity of these. During the following 2 days, a little more than 2 liters of fluid were removed from her stomach, replaced only by the plain water of ice chips. On the morning of the third day, the woman complained of being "nervous" and of having abdominal cramps, considerably more painful than the discomfort she had complained of on the first postoperative day. Examination of the intake–output records showed that the patient had passed only some 400 ml of urine during the previous 24 hours. Physical examination revealed an apprehensive, jittery patient with definite fingerprinting over the sternum. The serum sodium was determined to be 128 mEq/liter, and the chloride 86 mEq/liter. Other electrolytes were within limits of normal, except for the potassium, which bordered on a deficit. A catheterized specimen of urine revealed a specific gravity of 1.002, with no other abnormal findings.

Commentary All extracellular electrolytes were being removed through the gastric tube with no replacement, since only plain water was consumed. Since sodium is the chief cation of the ECF, and since the body tolerates relatively slight variations in sodium, the clinical syndrome that developed was that of sodium deficit. This sensitivity of the body to relatively modest deficits or excesses of sodium has been referred to as *the primacy of sodium* (discussed above).

Apprehensiveness and the more severe and even fatal neurologic symptoms of severe sodium deficit appear due to the absorption of water by the cerebrospinal fluid from the relatively hypotonic ECF by osmosis. Recall that a lag exists in the response of the cerebrospinal fluid to changes in the ECF. This lag is the reason why abnormalities of the ECF should be corrected slowly, to allow the cerebrospinal fluid time to "catch up."

Syndrome of inappropriate secretion of ADH The neurohypophyseal system maintains plasma osmolality within narrow limits by stimulating or inhibiting secretion of antidiuretic hormone (ADH). ADH is correctly regarded as our chief water-conserving hormone. Although appropriate release of ADH is essential for health, a bewildering variety of factors can stimulate inappropriate release of the hormone or synthesis of ectopic ADH-like compounds exerting the physiologic effects of true ADH.

Divergent factors stimulate ADH release, including pain, trauma (including surgical procedures), stress, and a long list of therapeutic agents. The syndrome of inappropriate secretion of *ADH* is abbreviated *SIADH*. Three facts contribute to its peculiar significance as a body fluid disturbance:

- It stems from a cluster of seemingly unrelated causes.
- Its onset is signaled by a dramatic drop in the serum sodium rather than by premonitory symptoms, as with most other body fluid disturbances.
- By the time symptoms do appear, it may be too late to reverse the pathophysiologic process—that is, the point of no return may have been passed.

Not without reason has SIADH been referred to as "Nemesis for the unwary."

In SIADH, aldosterone fails to perform its sodium-conserving mission. With normal ADH secretion, sodium deprivation stimulates secretion of aldosterone because of the reduction in ECF volume caused by sodium deficit. With excessive ADH secretion, however, water is retained. Since ECF volume is thus at least normal, there is no increase in aldosterone secretion,

hence no retention of sodium. SIADH then produces this chain of events: hyponatremia occurs, followed by ECF volume expansion and additional loss of precious sodium. The usual symptoms of hyponatremia occur. If the syndrome is not interrupted, cerebral edema develops as a terminal event. Why? Because of the relative isolation of the cerebrospinal fluid from the general ECF, there is no free exchange of solids. Water, however, passes the barrier readily. Drawn by osmosis into the cerebrospinal fluid, it causes edema of the brain. Should 5% dextrose in water be given parenterally, the dextrose would be used by the brain cells, leaving water in the cerebrospinal fluid and further promoting edema. If a head or facial injury causes the SIADH initially, the injury contributes to the edema. Should edema become so massive as to cause herniation of the brain stem through the foramen magnum, death promptly occurs. Agents causing SIADH are listed in Table 2-1.

In most cases, signs and symptoms of SIADH do not appear until the serum sodium descends to 115 mEq to 120 mEq/liter. The first symptoms are lethargy, loss of appetite, nausea, vomiting, headache, and abdominal cramps; later, personality changes—including withdrawal, confusion, hostility, and even violence—appear. Convulsions or coma may occur. Weight gain due to water retention is seen unless the patient is severely anorexic or suffering from advanced malignancy. Rapidity of the onset of symptoms has been said to parallel the speed of the development of the hyponatremia. The appearance of any combination of the symptoms described in a patient whose history indicates the possibility of SIADH should prompt the nurse to insist on serial serum sodium determinations, with prompt reporting of the results (if the order has not been previously given).

Diagnosis of SIADH rests on characteristic etiology, clinical findings, and laboratory findings. The latter, crucial for a firm diagnosis, include a serum sodium below 125 mEq/liter; urine is concentrated, with a specific gravity greater than 1.012 and a sodium above 20 mEq/liter, provided the patient is not on a sodium-restricted diet; urine osmolality is usually higher than the plasma osmolality; radioimmunoassay for ADH in the urine usually reveals increased excretion. Since sodium ions continue to be excreted in the urine despite the low serum sodium, the sodium excretion will equal or exceed the sodium intake (Table 2-2). In differential diagnosis, other causes of hyponatremia

TABLE 2-1 *Causes of SIADH*

Trauma or Therapeutic Procedures
- Head injury, including facial injury
- Prolonged mechanical pulmonary ventilation
- Heart valve operation
- Major surgical operation

Tumors
- Oat cell tumor of lung
- Adenocarcinoma of lung
- Carcinoma of pancreas
- Carcinoma of duodenum
- Thymoma

Central Nervous System Disorders
- Tuberculous meningitis
- Aneurysm
- Herpes simplex encephalitis
- Brain abscess
- Cerebral hemorrhage
- Guillain-Barré syndrome
- Limbic stimulation (pain, fear, major trauma)

Pulmonary Disorders
- Tuberculosis
- Pneumonia
- Chronic lung infection
- Status asthmaticus
- Aspergillosis with cavitation

Endocrine Disturbances
- Myxedema
- Adrenal insufficiency
- Hypopituitarism
- Porphyria

Pharmacologic Agents
- Hypoglycemic agents: chlorpropamide (Diabinese), tolbutamide (Orinase), phenformin (DBI, Meltrol), metformin
- Antitumor agents: vincristine (Oncovin), cyclophosphamide (Cytoxan)
- Diuretics: chlorothiazide (Diuril, Diupres)
- Analgesics: acetaminophen (Tempra, Tylenol, Datril)
- Tranquilizers: amitriptyline (Evavil, Amitril, Endep, Etrafon tablets), thioridazine (Mellaril), fluphenazine (Prolixin), thiothixene (Navane), carbamazepine (Tegretol)
- Antihypercholesterolemic agents: clofibrate (Atromid-S)
- Bronchodilators: isoproterenol (Isuprel)

TABLE 2-2 *Criteria for recognizing SIADH on the basis of lab findings*

- Hyponatremia with low plasma osmolality
- Continued renal excretion of sodium in excess of intake
- Absence of evidence of fluid volume depletion
- Urine osmolality inappropriately high for the plasma osmolality (*i.e.,* less than maximally dilute)
- Normal renal function
- Normal adrenal function
- Improvement with fluid restriction

must be ruled out. These include osmotic or pseudo-hyponatremia, dilutional hyponatremia, hyponatremia of sodium depletion, and psychogenic polydipsia (Table 2-3).

Treatment Treatment can be divided into two categories: treatment of the underlying clinical cause and alleviation of excessive water retention.

If the cause of excessive water retention can be readily corrected (such as by discontinuance of a drug), clinical and laboratory parameters will return to normal. If the cause is a malignancy, chemotherapy or radiation may be employed; later, long-term pharmacologic treatment may be indicated.

One of the primary treatments of SIADH is *fluid restriction*. Fluid is restricted to the extent that urinary and insensible losses induce a *negative water balance*. This treatment may be all that is needed in mild cases. Fluids are often restricted to approximately 500 ml to 700 ml/24 h.

In more severe cases, in addition to fluid restriction, the administration of a small volume of hypertonic saline (such as 3% sodium chloride solution) and furosemide (Lasix) is often indicated. Furosemide is administered intravenously; the simultaneous use of hypertonic saline and furosemide serves to replace sodium and aid in water excretion. (Urinary losses of sodium and potassium must be measured frequently and replaced.) Use of hypertonic saline alone only transiently elevates the serum sodium level, since most of the administered sodium is rapidly lost in the urine; in addition, there is a significant risk of pulmonary edema when hypertonic saline is not used in conjunction with a diuretic. For patients whose underlying cause of SIADH is chronic, other therapeutic measures

TABLE 2-3   *Typical findings in various hyponatremias*

	ECF volume	Edema	Urine specific gravity	Urine osmolality greater than serum osmolality	Sodium concentration of urine
SIADH	Normal or expanded	Absent*	Above 1.012	Yes	Above 20 mEq/liter
Dilution hyponatremia	Expanded	Present	Below 1.012	No	Below 20 mEq/liter
Sodium-depletion hyponatremia	Contracted	Absent	Below 1.012	No	Below 20 mEq/liter
Psychogenic polydipsia	Normal	Absent	1.000–1.004	No	Below 20 mEq/liter
Osmotic hyponatremia	Normal	Absent	Normal	No	Below 20 mEq/liter

*Except in terminal cerebral edema.

must be employed. (Fluid restriction is not tolerated well by most patients over a prolonged period.) The most prevalent measure for such patients is the administration of demeclocycline (Declomycin). Lithium carbonate has also been used for the same purpose. Both of these drugs probably act by interfering with the effects of ADH on the renal tubule; water excretion is increased, and thus, the serum sodium level is elevated. Demeclocycline is thought by most authorities to be superior; however, since its effects take several days to begin, it is not used in acute phases. Lithium carbonate is not as effective and has a greater potential for toxicity (particularly in patients with liver or heart disease). Phenytoin (Dilantin) has also been used to treat SIADH; it acts by inhibiting the release of ADH from the hypophysis.

Nursing responsibilities during therapy If fluid restriction is to be effective, no more fluid must be given than is indicated by the urinary output. Intake–output records are extremely important and provide the basis for regulation of fluid intake. Accuracy of these records is essential. If possible, the patient's cooperation should be gained; the need for fluid restriction should be explained to the rational patient. Great effort is required to prevent the mobile confused patient from drinking more than the prescribed quantity of fluid. "Fluid restriction" signs should be placed at the bedside. The water pitcher should be removed; visitors should be instructed not to give fluids to the patient. To minimize patient discomfort, fluids should be evenly spaced over the 24-hour period.

Since fluid restriction refers to the *total* amount of fluid intake, all fluid (oral and intravenous) should be considered. To minimize the risk of accidental over-administration of intravenous fluid, volume-controlled devices with microdrip sets should be used.

In addition to maintaining an accurate intake–output record, the nurse should study its pattern over a period of time. Increased urinary volume when fluid restriction is maintained indicates a positive response to therapy (actually, a negative water balance). One must remember that the excessive water load must be excreted before significant improvement can occur.

The nurse should analyze the weight chart; with proper therapy, one would expect to see an acute decline in body weight (due to excretion of excess water). One should recall that a liter of fluid is equivalent to 2.2 lb; for example: a weight loss of 6.6 lb over a period of a day or two would indicate a loss of approximately 3 liters of fluid.

Assessment of neurologic status is a *major* aspect of nursing of patients with SIADH. The nurse should observe for improvement in the overall neurologic status following appropriate therapy; increased level of consciousness, increased deep tendon reflexes, and increased muscle strength are all indicative of improvement. Unfortunately, neurologic damage induced by SIADH is not always reversible, particularly if treatment is delayed.

If the patient is comatose or immobile, particular care must be taken to prevent pneumonia, atelectasis, and pressure sores. Constipation may be a problem due to inactivity and fluid restriction; if it occurs, tap water enemas are contraindicated, since some of the water may be absorbed by the bowel (contributing to the already excessive water load). Other measures, such as suppositories, low-volume hyperosmolar enemas, or oil-retention enemas, may be required. Safety precautions (such as elevated bed rails) are indicated for the patient with a decreased level of consciousness. The nurse should be prepared to deal with seizures, since the patient with SIADH is a likely candidate for seizure activity.

Because SIADH is a potentially fatal condition, it is far better to prevent its occurrence than to deal with a full-blown case once it has developed. Recovery is usually rapid if the dilutional state and its causes are recognized early and appropriate measures are instituted. The nurse plays a vital role in this area.

CASE HISTORY On a rainy morning, a 30-year-old housewife was driving down a slippery highway to a nearby shopping center. Her car skidded off the road as she rounded a curve and careened into a tree. Passersby observing the mishap called an ambulance. The ambulance log indicated that the woman, C. T., was freed from the wreckage and taken to a nearby hospital. In the emergency room of the hospital, the patient was found to be conscious, rational, and oriented as to time, place, and person. She was found to have sustained fractures of her mandible, maxilla, and left zygomatic arch. She also had extensive facial contusions. Blood was drawn for baseline electrolyte values, and a urine sample was obtained. She was prepared for operation and, under general anesthesia, underwent surgical procedures in which her fractures were aligned and immobilized. Initial electrolyte values were found to be normal, and urinalysis revealed no abnormalities. Intake–output measurements were started because of the tendency of patients with head injuries to develop body fluid disturbances. The following morning, the patient appeared to be in good spirits. Serum electrolytes revealed a sodium of 130 mEq/liter and a chloride of 90 mEq/liter. Urine specific gravity was 1.010. The next morning, after a restless night, the patient's blood and urine were again tested. On ward rounds, the patient impressed her physician as apprehensive and jittery. The serum sodium was found to be 114 mEq/liter and the chloride 79 mEq/liter. Specific gravity of urine was 1.016. A repeat sodium was 115 mEq/liter and chloride 80 mEq/liter. Intake–output records indicated that the patient was not excreting water normally; moreover, she weighed 4 pounds more than on admission. The pronounced depression of the serum sodium and chloride plus the unduly high specific gravity of the urine indicated that the patient was suffering from sodium deficit (hyponatremia) due to

inappropriate secretion of ADH (SIADH) caused by the trauma to the skull and its contents. On the basis of this diagnosis, fluids were greatly restricted, and small quantities of a 5% solution of sodium chloride in water were slowly administered.

Commentary Note that this patient's urinary specific gravity was 1.016. Hyponatremia not caused by SIADH is characterized by a dilute urine with a specific gravity of 1.001 to 1.003. The excessively secreted ADH, while conserving water inappropriately, also appears to inhibit secretion of aldosterone, the chief sodium-conserving hormone. This probably explains the relatively high output of sodium in the urine, despite the severe deficit of sodium in the ECF. SIADH occurs not only in various forms of trauma to the central nervous system but also has been reported to occur with several malignant tumors, nontraumatic disorders of the central nervous system, pulmonary disease, the postoperative state, and with most anesthetics and a wide variety of drugs. It has been seen in some patients in whom no cause could be discovered.

Sodium excess of ECF Sodium excess of ECF, sometimes called *hypernatremia, salt excess, oversalting,* or *hypertonic dehydration,* is one of the most dangerous of all body fluid disturbances. It occurs when the sodium concentration of ECF rises above the normal level. The imbalance, which may be acute or chronic, results from decreased intake or increased output of water or from increased intake or decreased output of sodium.

Acute sodium excess may follow excessive administration of concentrated oral electrolyte mixtures. A tragic example of this occurred when salt instead of sugar was mistakenly used in preparing infant formulas. The infants fed the mixture developed sodium excess so severe that even the most vigorous therapy failed to save several of the babies. The imbalance also may occur in any condition in which more water is lost than electrolytes, as in tracheobronchitis—in which excessive water losses occur through the lungs as a result of fever and deep, rapid breathing, leaving an abnormally high concentration of sodium in the ECF—or in profuse watery diarrhea when treatment has been inadequate. Some sodium is lost in diarrheal stools, but the concentration of the remaining sodium is so high as to produce hypernatremia, emphasizing the fact that it is the *concentration* of sodium in mEq/liter—not the total quantity of sodium in the body—that is all important. Sodium excess can also develop because of a hypothalamic tumor. In this condition, ADH is no longer produced, and the person loses excessive water (having lost the water-conserving action of ADH), leaving an excessive sodium concentration. Unconscious patients may develop sodium excess simply because they cannot drink water. The person who has inhaled ocean water frequently develops a sodium excess, which must be corrected if the patient is to recover; indeed, sodium excess is often the actual cause of death in salt water drowning. Infants are particularly prone to develop acute sodium excess, and the mortality rate is quite high—about 50% in hospital-treated infants with diarrhea and sodium excess.

Clinical findings in sodium excess include dry, sticky mucous membranes, flushed skin, intense thirst, and rough, dry tongue. Body tissues are firm and rubbery because water from the cellular fluid, following the law of osmosis, flows into the more concentrated ECF. Oliguria or anuria is present, and the patient may have fever. He appears agitated and restless, may develop mania or convulsions, and his reflexes are decreased.

Laboratory findings are helpful in diagnosis, since the plasma sodium is usually above 147 mEq/liter, but may range far higher. Plasma chloride is above 106 mEq/liter, and the specific gravity of urine is above 1.030.

Long-term or chronic ingestion of large amounts of sodium, as when one salts his food excessively or eats extremely salty foods, also can cause sodium excess and may lead to hypertension. Although the daily requirement for sodium is probably below 100 mg in acclimatized persons, most persons ingest from 3000 mg to 5000 mg daily, and inhabitants in certain areas of Japan have an average daily intake in excess of 20,000 mg. Heredity factors largely determine whether this type of sodium excess will cause hypertension in a given individual, but researchers have induced uninherited but life-long hypertension in experimental animals by feeding them highly salted foods. Fairly recently, researchers questioned a group of Americans about normal use of salt and learned that only 1% of those who never salted their food at the table suffered from hypertension, as compared to 8% of those who salted to taste, and 10% of those who salted before tasting. Another study revealed that a large segment of the population in areas of Japan where the usual diet is extremely salty suffers from hypertension. In sodium excess, which is often accompanied by volume deficit, the goal of therapy is to provide water to dilute the concentration of electrolytes both in the ECF and in the cellular fluid while supplying maintenance amounts of electrolytes to reestablish the normal composition of the body fluid. The free water in balanced solutions dilutes the electrolyte concentration excess, and the body homeostatic mechanisms are also supplied with the electrolytes they may need to reestablish the normal water-to-electrolyte ratio.

In early sodium excess, merely giving plain water by mouth as desired is usually adequate. In advanced imbalances, intravenous infusion of a hypotonic solution of water and electrolytes is mandatory. Tonicity of the solution might range from 50 mEq to 75 mEq/liter. Simple dextrose in water should not be given, since some electrolyte appears necessary. Another mode of therapy is use of a diuretic, with plain water given by mouth. Spironolactone (Aldactone) may prove useful, since it combats the conserving action of aldosterone (Fig. 2-9).

CASE HISTORIES Six-year-old S. W. was admitted to the hospital because of extreme agitation and inability to sleep. The mother asserted that the little girl had "not slept a wink" for the past 48 hours. The history revealed that for the past six days, the child had been suffering from a mild, nonbloody diarrhea. Her mother had been treating her with a home remedy consisting of a solution of salt and soda in water. She had added a tablespoon of salt and a tablespoon of baking soda to each pint of water and had forced the child to drink about a pint of this mixture every day. On physical examination, the child was found to have dry, sticky mucous membranes and a rough, dry tongue. Her skin was flushed, and she complained of thirst. Her tissues had a firm, rubbery feel to them. The mother stated that the child had not urinated for the past 24 hours. The plasma sodium was 165 mEq/liter and the chloride 117 mEq/liter. The potassium was 3 mEq/liter. The child could not urinate.

Commentary This patient represents a simple example of sodium excess of the ECF caused by administration of an excessive quantity of sodium, both as NaCl and NaHCO$_3$. It demonstrates that the body does not tolerate sodium excess well. Discontinuance of the salt and soda mixture and oral administration of water could probably remedy the imbalance.

Four-year-old J. M. was admitted to the emergency room of the hospital, obviously extremely ill. His mother stated that he had had profuse watery diarrhea for the past week.

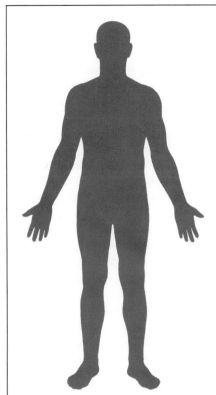

PRECEDING EVENTS

- *Decreased water intake (common in unconscious patients or others unable to perceive thirst)*
- *Increased urinary water losses when water intake is inadequate:*
  - *High-protein tube feedings without adequate water supplements*
  - *Osmotic diuresis in patients with hyperglycemia*
- *Neurosurgical patients with diabetes insipidus*
- *Excessive administration of sodium-containing parenteral solutions (e.g., 3% NaCl, or even isotonic saline; has occurred following use of $NaHCO_3$ in patients with cardiac arrest)*
- *Drowning in salt water*

CLINICAL OBSERVATIONS

- *Intense thirst*
- *Tongue rough, red, and dry*
- *Skin flushed*
- *Difficulty in speaking without first moistening lips*
- *Temperature elevated if room temperature above 18°C (65°F). (Temperature elevation is usually low-grade and disappears after rehydration)*
- *Firm, rubbery tissue turgor*
- *Restlessness and excitement*
- *Agitation (may progress to mania and convulsions)*
- *Oliguria or anuria*

LABORATORY FINDINGS

- *Plasma sodium above 147 mEq/liter*
- *Specific gravity of urine above 1.030 (if water loss is nonrenal)*

RELATED PROBLEMS

- *Often occurs following diarrhea of 5 or 6 days' duration*

FIG. 2-9. Sodium excess.

He had been given sips of water and a soft drink but had vomited part of this. The child appeared surprisingly alert. His cheeks were flushed, mucous membranes dry and sticky, and his tongue rough. His subcutaneous tissues felt firm and rubbery. While being examined, he had a watery stool not containing blood. His serum electrolytes were sodium 180 mEq/liter, chloride 137 mEq/liter, bicarbonate 16 mEq/liter, potassium 3.8 mEq/liter, and P_{CO_2} 33 mm Hg. A urine specimen could not be obtained.

Commentary This patient demonstrates that what is important in concentration disturbances of ECF is not the total quantity of an electrolyte in the body but its concentration in units per liter. This child had a life-threatening excess concentration of sodium, yet his total body sodium was partially depleted since some sodium had passed out in the watery stools. In treatment of such a patient, therefore, it is important to give a hypotonic solution of water and electrolytes plus dextrose so as to resupply not only the badly depleted water but also the sodium. Obviously, close observation of the body electrolytes is essential as the deficits are repaired.

CHLORIDE As the major anion of the ECF, chloride—combined with sodium—is the chief contributor to serum osmolality. It is essential for formation of the hydrochloric acid of gastric juice. Loss of chloride usually parallels loss of sodium, with which it is normally paired. But chloride loss caused by vomiting may result in a chloride deficit apart from loss of sodium. In such an instance, the bicarbonate portion of the anion column is increased (since total anions must always equal total cations), and alkalosis results. Persons whose sodium chloride intake is severely restricted because of disease of heart, kidney, or liver may need an alternative source of chloride, such as potassium chloride (provided the kidneys are relatively normal).

Determination of the urinary chloride, a simple test, is useful in detecting ECF volume deficit and bicarbonate excess maintained by ECF volume deficit.

The requirement for chloride usually parallels that for sodium: 3.62 g to 10.86 g daily. Foods high in sodium are usually also high in chloride.

POTASSIUM The symbol for potassium is *K*, after the first letter of its German name, *Kalium*. A naturally radioactive element, potassium played a major role in the geologic development of the earth. Although potassium is closely related to sodium from the chemical standpoint, it differs strikingly from sodium in its physiologic behavior. Indeed, in many situations, potassium and sodium appear to be paired off against each other. Potassium also is a close chemical relative of lithium and rubidium. Potassium is a soft, bluish silver metal that exists outside the test tube only when combined with other elements.

The primary function of potassium is its involvement with cellular enzyme activities. It plays a leading role in the intricate chemical reactions required to transform carbohydrate into energy and to convert amino acids into proteins. The difference in the concentrations of sodium and potassium across cell walls determines the electrical potentials of cell membranes, hence cell excitability and nerve impulse conduction. The cells require potassium to maintain their normal water content. It is essential for transmission of electrical impulses within the heart. Muscles cannot function normally without potassium.

The ECF contains about 70 mEq of potassium, contained in a little more than a level teaspoonful of potassium chloride. But the body cells contain 4000 mEq of potassium—the equivalent of more than 6 lb of potassium chloride. Thus, although one cannot live long without potassium, each of us has enough in our body cells to kill dozens of persons if it were injected quickly into the bloodstream.

Body secretions and excretions are rich in potassium. Especially plentiful in this element are sweat (especially after one has become acclimatized to prolonged heat), saliva, stomach and intestinal secretions, and stools. The potassium content of urine varies with the intake, although it never approaches zero, even with depletion of body potassium. In the kidneys, along with water and other electroytes and small molecules, potassium is completely filtered at the glomerulus and is then completely reabsorbed in the proximal convoluted tubules. Excretion occurs only in the distal convoluted tubules, and then in exchange for sodium. Scientists are baffled by the body's extremely negligent conservation of this essential element, which stands in striking contrast to the body's parsimonious husbanding of sodium. For even though the potassium intake is dangerously low, the urine may carry out as much as 40 mEq to 50 mEq daily. Why? Could it be that the body's conservation of sodium and lack of

conservation of potassium reflects the theory that man's ancient ancestor was an herbivore, an exclusive plant eater? (Recall that meats are rich in sodium, while vegetables and fruits abound in potassium.) Physiologic "habits," even of one's extremely remote ancestors, have a way of perpetuating themselves. Is it possible that the sodium-conserving, potassium-wasting pattern is due for a change as man evolves? Certainly the present physiologic pattern provides us with a stunning example of atavistic physiology.

While no RDA has been established for potassium, the daily requirement can be regarded as about 2.5 g (64 mEq), provided unusual stress, including prolonged high environmental temperature, is not present.

A well-balanced diet usually assures adequate potassium, provided no abnormal losses are occurring from vomiting, diarrhea, sweating, or diuresis. Leading food sources of potassium include apricots, bananas, dates, figs, oranges, peaches, prunes, raisins, tomato juice, orange juice, and vegetables in general. Meat and dairy products provide appreciable but lesser quantities of potassium. Pharmaceutical potassium supplements appear advisable if the person is to receive a potent potassium-losing diuretic for a prolonged period. (Diuretics and potassium supplements are discussed in Chapter 5.)

Potassium deficit of ECF Potassium deficit, or *hypokalemia*, can develop quickly when the patient's intake is inadequate, as in starvation. This imbalance occurs frequently, not only as a result of prolonged inadequate intake of potassium, but also as a result of excessive losses of potassium-rich secretions or excretions, as occurs in prolonged diarrhea.

Probably the leading cause of potassium deficit is administration of powerful diuretics, particularly the thiazides or furosemide, without adequate potassium supplementation. Surgical operations often cause losses of potassium-rich fluids, especially if the procedure involves the digestive tract. The imbalance also is often associated with GI suction, diseases involving the intestinal tract, familial periodic paralysis, pyelonephritis, thyroid storm, aldosterone-secreting tumor of the adrenal cortex, crushing injuries, broken bones, extensive bruising, and wound healing. Even emotional or physical stress can cause the imbalance. Excessive sweating, fever, and high environmental temperatures are further causes. Indeed, potassium loss plays a major role in the development of heat stress disease, which affects people of all ages who are subjected to high environmental temperatures, especially if they are exercising. Factory workers, athletes, and others who are exposed to high environmental temperatures often use salt tablets to replace the sodium lost in sweat, but, since the body conserves sodium and continues to lose potassium, potassium supplementation, at least as important as sodium supplementation, is often neglected.

Chronic excessive licorice ingestion can bring about hypokalemia and, along with it, sodium retention, metabolic alkalosis, renal potassium wasting, depressed renin, and hypertension. This is because natural licorice, an extract of *Glycyrrhiza glabra* root, contains a compound known to have mineralocorticoid activity. Artificial licorice flavoring is employed in commercial products in the United States, so pseudo primary hyperaldosteronism is rarely reported here. However, certain chewing tobaccos contain large amounts of true licorice and can produce the syndrome. In one such patient, the admission serum potassium was 1.8 mEq/liter.

Potassium deficit is frequently associated with metabolic alkalosis, as discussed in Chapter 3.

The clinical findings in potassium deficit can be remembered easily if one recalls that potassium is essential for the normal functioning of muscle cells; thus, symptoms of potassium deficit are caused largely by muscle weakness. Early symptoms are nonspecific; the patient has malaise or is simply not feeling well. In some patients, potassium deficit damages renal tubules and thus impairs the concentrating ability of the kidney, the result of which is polyuria and thirst. Later, the skeletal muscles become weak, and the reflexes are decreased or absent; eventually, muscle weakness leads to flabbiness, with the patient lying flat, like a cadaver. Cardiac disturbances accompanying potassium deficit may include atrial and ventricular arrhythmia, diminution of the intensity of heart sounds, weak pulse, falling blood pressure, and heart block. Degeneration of the myocardium with loss of cellular striations may follow prolonged potassium deficit, and the intestinal muscles, too, are affected. The patient suffers anorexia, vomiting, gaseous intestinal distention, and paralytic ileus. Weakness of the respiratory muscles produces shallow respiration, and death in potassium deficit apparently results from apnea and respiratory arrest rather than from cardiac standstill.

Laboratory findings helpful in diagnosis include repeated serum potassium determinations of below 4 mEq/liter (a single determination can be seriously misleading). The plasma chloride is often below 98 mEq/ liter, and plasma bicarbonate is above 29 mEq/liter. Potassium deficit also induces specific EKG findings, including low, flattened T wave, depressed ST segment, and prominent U wave (Fig. 2-10).

There is a pertinent note: after the potassium of the blood coursing through the glomeruli of the kidneys is filtered, all of the potassium (or almost all) is reabsorbed into the peritubular arterioles. Then, in the distal convoluted tubules, it is exchanged for the sodium of the tubular urine, under the influence of aldosterone. Recall, there are 4000 mEq of potassium in the cellular fluid and 70 mEq in the ECF; yet, it is the extracellular potassium that is all-important in causing symptoms and findings. Metabolic alkalosis is frequently associated with potassium deficit. The reasons for this are complex, and even leading authorities are not in full agreement on just what these reasons are. *Bear in mind: in considering potassium deficit, one of the most frequent of all body fluid disturbances, that the body conserves potassium poorly, in striking contrast to its parsimonious hoarding of sodium.*

Potassium deficit can usually be prevented by giving supplemental potassium to patients in a situation likely to favor such a deficit. Although many vegetables and fruits (notably oranges and bananas) contain generous quantities of the electrolyte, it would usually be necessary to ingest enormous quantities to forestall potassium deficit. (As a somewhat facetious example, consider: there is about 1 mEq of potassium in 1 inch of banana. To supplement the diet with 40 mEq of potassium, one would have to eat 40 inches of banana.) Fortunately, pharmaceutical companies market several pleasant-tasting potassium supplements. Most salt substitutes consist largely of potassium chloride; if the patient is using one of these, he has an economical and convenient source. It might be useful to recall that a level teaspoon of salt substitute contains about 75 mEq of potassium. (See Table 5-6.)

Therapy of potassium deficit is designed to repair the deficit while providing adequate amounts of potassium for maintenance needs. Potassium should be given orally if tolerated, parenterally otherwise. Potassium chloride should be used if alkalosis is present. Nursing considerations related to IV potassium administration are discussed in Chapter 6 (Fig. 2-10).

CASE HISTORIES J. K., a 55-year-old businessman, was found to have an elevated blood pressure on his first annual physical examination, with his reading 190/120. A

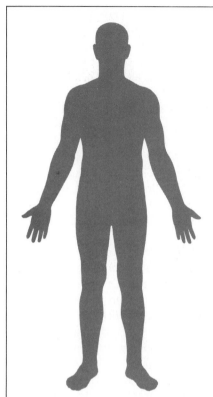

PRECEDING EVENTS

- *Prolonged inadequate intake*
- *Use of potassium-losing diuretics without adequate potassium replacement*
- *Excessive loss of gastrointestinal fluids*
- *Prolonged heat stress*
- *Hyperaldosteronism*

CLINICAL OBSERVATIONS

- *Chronic fatigue*
- *Muscle weakness (neuromuscular symptoms not usually seen until serum potassium has decreased to approximately 2.5 mEq/liter)*
- *Soft, flabby skeletal muscles*
- *Gaseous intestinal distention (ileus)*
- *Decreased bowel sounds*
- *Paresthesia*
- *Weak, irregular pulse*

- *Increased sensitivity to digitalis*
- *With severe deficit, paralysis and heart block*
- *Cause of death: apnea or heart block*

LABORATORY FINDINGS

- *Repeated plasma potassium below 4 mEq/liter*
- *Specific EKG findings: low voltage, flattening of T waves, depression of ST segment (EKG changes likely when serum potassium is less than 3 mEq/liter)*

RELATED PROBLEMS

- *Metabolic alkalosis frequently associated*

FIG. 2-10. **Potassium deficit.**

thorough workup ruled out an obvious cause for his hypertension, and a diagnosis of essential hypertension was made. He was given a thiazide diuretic, and his sodium intake was restricted to about 1000 mg/day. His pressure responded well and was maintained at about 140/88. He checked with his internist every 2 months. Five months after starting therapy, he began to notice that he was short of breath on climbing stairs, he was exceedingly fatigued, he was constipated yet troubled with gas, and he felt that his heart was beating irregularly. On reporting the symptoms to his physician, blood was drawn for electrolyte studies. His sodium was 142 mEq/liter, potassium 2.6 mEq/liter, bicarbonate 36 mEq/liter, and serum pH 7.6. A Pco_2 on arterial blood was 55 mm Hg. The EKG showed low, broad T waves with a double summit caused by superimposition of the U wave on the T wave. The P-R interval was prolonged. Treatment of the hypokalemia consisted of a potassium supplement, 100 mEq/day, plus the addition of generous quantities of orange juice and bananas to the diet. In 5 days, the sodium was 142 mEq/liter, potassium 4.5 mEq/liter, bicarbonate 24 mEq/liter, serum pH 7.4, and the Pco_2 40 mm Hg. The patient was advised to discontinue the potassium supplement and instead to use a KCl salt substitute in palatable quantities on his food.

Commentary Although the EKG is useful in making a diagnosis of potassium deficit, it may be impossible to determine the exact degree of hypopotassemia from the tracing. Potassium deficit and alkalosis, both nonrespiratory (bicarbonate excess) and respiratory (carbonic acid deficit), are closely interrelated. Either tends to cause the other, probably because of the twin facts that potassium and hydrogen are exchanged for sodium in the distal convoluted tubule of the nephron and, similarly, across the cell membrane. Suppose potassium deficit of the ECF exists. Then hydrogen will preponderantly exchange for sodium in the distal convoluted tubules and across the cell membrane. A relative deficit of hydrogen will develop in the ECF, resulting in alkalosis. On the other hand, suppose that alkalosis exists. Then there will be a dearth of hydrogen in the ECF, and potassium will preponderantly exchange for sodium in the two locations. Potassium deficit will develop.

M. J., 23 years old, told her physician of periodic attacks of complete paralysis, usually brought on by rest after exercise, exposure to cold, overeating in the evening, or on waking. The attacks were sometimes preceded by thirst, sweating, or muscular cramps. Not all attacks proceeded to complete paralysis; some consisted only of great weakness. Family history revealed similar attacks in her mother's family. On physical examination during an attack, no sensory changes were found, but the deep reflexes were lost and electrical excitability of muscle was absent. Examination was performed during a prolonged attack (attacks lasted from a few minutes to several days). Two serum potassium readings were 2.6 mEq/liter and 2.7 mEq/liter.

Commentary This unusual form of potassium deficit of ECF appears to be caused by passage of extracellular potassium into the body cells. It remains one of the many diseases of mystery. Strangely enough, some patients have similar attacks with *hyper*kalemia rather than *hypo*kalemia. Immediate ingestion of a potassium salt appears to help in hypokalemic periodic paralysis.

Potassium excess of ECF Because the kidneys so efficiently rid the body of potassium, potassium excess or *hyperkalemia* does not develop as often as potassium deficit. However, this does not mean that excesses are any less dangerous than deficits; indeed, they can be extremely hazardous. In most cases, potassium excess is caused by leakage of the electrolyte from the body cells; thus, following a severe burn or crushing injury, the ECF may be flooded with potassium from the damaged cells. If the kidneys are not functioning properly, the imbalance can result from excessive oral ingestion of potassium. Intentional excessive oral ingestion, as in a suicide attempt, is another cause. The imbalance also may result from overzealous parenteral administration of potassium-containing solutions. Mercuric bichloride poisoning, which damages the kidneys, can lead to potassium excess, since the imbalance is inevitable following renal failure (end-stage renal disease) and may be the cause of death. Severe excesses also result from cellular hypoxia, uremia, adrenal insufficiency—as in Addison's disease, in which aldosterone is lacking—and administration of spironolactone, which opposes aldosterone (recall that aldosterone promotes excretion of potassium). Metabolic acidosis may be associated with potassium excess, as discussed later in this chapter.

The clinical findings in potassium excess include irritability, nausea, diarrhea—a result, not a cause, of the imbalance—and intestinal colic. If the condition becomes severe, there is weakness and flaccid paralysis, perhaps difficulty in phonation and respiration, and there is oliguria, which may progress to anuria. Because transmission of stimuli through the heart muscle is slowed or prevented, intraventricular conduction disturbance occurs, with or without atrioventricular dissociation, and finally, ventricular fibrillation and cardiac arrest develop. Death in potassium excess results from poisoning of the heart muscle.

A repeated plasma potassium above 5.6 mEq/liter indicates potassium excess, and a test for renal function usually will show renal impairment. The imbalance also presents specific EKG findings: early, T waves are peaked and elevated; later, P waves disappear; finally, there are biphasic deflections resulting from fusion of the QRS complex, RST segment, and the T wave (Fig. 2-11).

Treatment Potassium intake by any route is contraindicated in persons with hyperkalemia. Serious hyperkalemia can be managed by measures that antagonize the effects of potassium, force potassium into the cells, or actually remove potassium from the body. The intravenous administration of calcium is an emergency measure that acts rapidly to provide temporary myocardial protection from the toxic effects of hyperkalemia without actually lowering the plasma potassium level. (Recall that calcium antagonizes the cardiac effects of potassium.) The intravenous injection of $NaHCO_3$ solution (such as 44 mEq over a 5-min period) causes rapid movement of potassium into the cells, temporarily lowering the plasma potassium level (1 to 2 h). In addition, $NaHCO_3$ provides sodium ion for antagonizing the cardiac effects of potassium. The intravenous infusion of hypertonic dextrose and regular insulin temporarily reduces the plasma potassium level by forcing potassium into the cells. Cation-exchange resins and either peritoneal dialysis or hemodialysis are used to actually *remove* potassium from the body. (See Chaps. 10 and 11 for a more thorough discussion of therapy for hyperkalemia.)

CASE HISTORIES P. T. was admitted to the hospital with a history of chronic renal disease resulting from acute glomerulonephritis. Recently, the symptoms had become more severe, with dyspnea, pallor, weakness, and fatigue from the slightest exertion. The physical examination

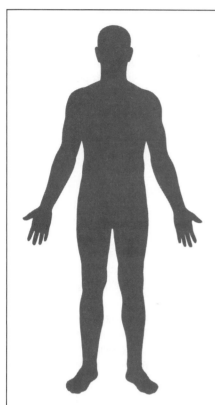

PRECEDING EVENTS

- *Kidney failure*
- *Excessive ingestion of potassium*
- *Excessive parenteral administration of potassium*
- *Leakage of potassium from cells*
- *Adrenal insufficiency*
- *Excessive use of potassium-conserving diuretics*

CLINICAL OBSERVATIONS

- *Irritability combined with anxiety*
- *Gastrointestinal hyperactivity (nausea, colic, and diarrhea)*
- *Cardiac arrhythmias (bradycardia can occur at 7 mEq/liter, heart block at 9 mEq/liter)*
- *Weakness*

- *Paresthesia (can occur at a level of 6 mEq/liter)*
- *Levels greater than 8.5 mEq/liter are often fatal (due to cardiac standstill or arrhythmia)*

LABORATORY FINDINGS

- *Repeated plasma potassium above 5.6 mEq/liter*
- *Specific EKG findings*

RELATED PROBLEMS

- *Follows kidney failure*
- *Metabolic acidosis frequently associated*

FIG. 2-11. **Potassium excess.**

revealed hypertension in the supine position, hypotension when upright. There was generalized muscle weakness, a pericardial friction rub, and mild flank tenderness on both sides. The patient vomited during the examination. Slight edema of the ankles was detected. Laboratory findings included serum creatinine 4 mg/dl (normal, 0.7–1.5), sodium 130 mEq/liter, potassium 7 mEq/liter, bicarbonate 16 mEq/liter, P_{CO_2} 32 mm Hg, Delta 16 mEq/liter, calcium 7 mg/dl (3.5 mEq/liter), phosphate 8 mg/dl (4.8 mEq/liter). Proteinuria was calculated at 15 g/day; no red blood cells were seen, and finely and coarsely granular tubular cell and hyaline casts were present in moderate numbers in the urinary sediment.

Commentary This case is presented as an example of potassium excess of the ECF, although other findings of end-stage renal disease are present. Note the elevated creatinine, the depressed sodium, the evidence of bicarbonate deficit (metabolic acidosis), with a Delta of 16. This indicates the acidosis is due to accumulation of organic acids titrating bicarbonate. One would expect the calcium to be depressed and the phosphate to be elevated in end-stage renal disease, and this is the case. The elevation of

the potassium is only moderate. It would be more severe with oliguria and less severe—even normal or subnormal—with copious urine flow. Note the urine findings, indicating chronic nephritis. The loss of protein accounts for the edema observed. (For details concerning the values for calcium and phosphate refer to the section on calcium below. For details on the acid–base disturbance, refer to the chapter on acid–base disturbances.)

E. H., a 44-year-old business executive, was admitted to the hospital with a history of blood pressure averaging about 190/120, which had failed to respond to chlorthalidone, 50 mg b.i.d., plus propranolol, 40 mg/day. He had been on a sodium-restricted diet, estimated at 1000 mg (43 mEq) daily and had been using a salt substitute consisting largely of KCl in quantities estimated at 2925 mg (75 mEq) daily. He brought the salt substitute with him to the hospital and, unknown to the nurses, began using it on his food. The physician ordered a new vasodilator plus a potassium-sparing diuretic for the patient. Inadvertently, 40 mEq of supplemental potassium was administered daily to E. H. On the fourth hospital day, the patient began complaining of tingling about the mouth and in the fingers and toes. He stated he felt extremely weak. Upon examination, his muscles proved to be hyperirritable and his heart rate irregular. A battery of electrolyte determinations revealed serum potassium 8 mEq/liter, bicarbonate 16 mEq/liter, and P_{CO_2} 32 mm Hg. Other serum electroytes were normal. Urine *p*H was 5, but otherwise the specimen was normal. An EKG revealed no P waves. The physician made a firm diagnosis of potassium excess and discontinued the salt substitute and the potassium supplement. The potassium dropped to normal within 4 days, and all symptoms disappeared.

Commentary Totaling the patient's dietary intake of potassium (about 75 mEq), that from the salt substitute (about 75 mEq), and that from the potassium supplement (40 mEq) showed that he had been ingesting about 190 mEq of potassium daily. A person with normal kidneys can tolerate about 135 mEq without developing potassium excess. Both the serum potassium and the EKG indicated potassium excess. Potassium excess is usually associated with acidosis (bicarbonate deficit), and this was the case in this instance, with the bicarbonate and the P_{CO_2} both depressed. Depression of the P_{CO_2} was of such extent as to indicate that is was compensatory in nature. (See Chapter on acid–base disturbances for details.) The error cited above has been reported from numerous institutions.

CALCIUM The cation calcium *(Ca)* is the most abundant electrolyte in the body; about 99% is concentrated in bones and teeth, the remainder throughout the plasma and body cells. The normal concentration of calcium in the ECF is 5 mEq/liter. Calcium is closely associated with phosphate; together, they make bones and teeth rigid, strong, and durable. The plasma levels of calcium and phosphate are regulated by the parathyroid glands; normally, an increase or decrease in the serum phosphate concentration is associated with the opposite decrease or increase in the plasma calcium level, and vice versa. Vitamin D promotes intestinal absorption of calcium and increases kidney excretion of phosphate.

Calcium is also a necessary ingredient of cell cement, which holds the body cells together, and, in addition, determines the strength and thickness of cell membranes. Calcium exerts a sedative effect on nerve cells and, thus, is important in maintaining normal transmission of nerve impulses. The electrolyte also aids in the transfer of energy and in the absorption and utilization of vitamin B_{12}. It activates enzymes that stimulate many essential chemical reactions in the body and plays a role in blood coagulation.

About half of the calcium in blood is ionized and the remainder is bound to albumin; it is the *ionized* form of calcium that is physiologically active. A larger percentage of serum calcium becomes ionized as the serum protein level decreases. Unfortunately, determination of the ionized portion is exceedingly difficult (although recently more centers are gaining this capability).

Clinically then, it is important to evaluate the serum calcium concentration in light of the serum albumin level. *Each rise (or fall) of serum albumin by 1 g/dl (beyond the normal of 4 g/dl) is associated with a rise (or fall) of serum calcium concentration of approximately 0.8 mg/dl.* For example: in hypoalbuminemia, the measured total calcium gives a misleadingly low estimate of the ionized calcium.

The effect of acidosis is to decrease the amount of calcium bound to albumin, thereby raising the ionic calcium concentration. Thus, in acidosis, there will seldom be signs of hypocalcemia even if a total body calcium deficit exists. Alkalosis does the opposite, resulting in decreased calcium ionization; signs of hypocalcemia can occur even though the total body calcium is normal.

Usually, more than one calcium determination is advisable to confirm a diagnosis. Spurious elevations of calcium can occur after prolonged venous stasis or exposure of the sample to cork stoppers. A false depression in the value results when serum is allowed to stand without separating the erythrocytes, since they absorb calcium.

Frequently, the human diet is deficient in calcium, especially in countries where milk is scarce and where other food sources of calcium are lacking. Calcium deficit is especially serious during the rapid growth of early infancy, during puberty, and in pregnancy and lactation. Only from 20% to 30% of the calcium in the diet is absorbed. If vitamin D intake is inadequate, even this percent of absorption is reduced. The parathyroid glands, through their hormone parathormone, help maintain the plasma level of calcium: they stimulate bone breakdown and absorption when calcium is needed to maintain the plasma calcium level.

The RDA for calcium for formula-fed infants is 360 mg/day up to 6 months of age and 540 mg/day from 6 months to 1 year. Breastfed infants' needs are fully met by breast milk. Other RDAs are as follows: children 1 to 10, 800 mg/day; males and females 11 to 18, 1200 mg/day; and males and females 19 to 51, 800 mg/day. Pregnant or lactating women require an additional 400 mg/day.

The best food sources of calcium are milk and cheese. Less rich sources include dried beans, kale, brazil nuts, and bone. (While bone is made up largely of calcium, the absorption of calcium from bone is low.)

Calcium deficit of ECF Calcium deficit of ECF, also called *hypocalcemia*, can result from abnormalities in body metabolism, from inadequate dietary intake of calcium, or from excessive losses of calcium in diarrheal stools or wound exudate. The imbalance is often associated with sprue, acute pancreatitis, hypoparathyroidism, surgical removal of the parathyroids, massive subcutaneous infections, burns, and generalized peritonitis. Excessive infusion of citrated blood, as in an exchange transfusion in an infant, may bring about calcium deficit, and the imbalance can also result from rapid correction or overcorrection of acidosis, wherein calcium ionization is increased. When the plasma pH returns to normal, or above normal, calcium ionization decreases; unless adequate calcium is provided, the decreased ionization may cause hypocalcemia. Magnesium excess also tends to cause calcium deficit, possibly because magnesium plays a role in parathyroid function. Renal failure also looms large as a cause of calcium deficit for this reason: phosphate is reciprocal with calcium; in early renal failure, the kidneys aren't able to dispose of phosphate normally. Hence, the phosphate rises and the calcium falls. Subsequently, added secretion of parathormone stimulates the calcium to rise. But because of kidney malfunction, the body is unable to utilize this calcium normally. Finally, in the late stages of renal failure, the calcium falls again in a final decline.

The clinical findings in calcium deficit are as follows: tingling of the ends of the fingers and of the circumoral region; muscle cramps, affecting both abdominal and skeletal muscles; carpopedal spasms; tetany; and convulsions. If calcium deficit is prolonged, calcium will be drawn from the bones to replenish the ECF, after which the bones become porous and break at the slightest provocation.

Laboratory findings helpful in diagnosis include a plasma calcium determination below 4.5 mEq/liter, or below 10 mg/dl. However, plasma calcium determinations can be misleading since only the ionized calcium is physiologically active. The ionized calcium can be measured by determining calcium levels before and after treatment with dextran gel or by ultrafiltration. In the simple and sometimes useful urinary Sulkowitch test, Sulkowitch reagent is added to urine and urinary calcium is precipitated as calcium oxalate. If no precipitate forms, hypocalcemia is present, whereas a fine white precipitate indicates normal plasma calcium, and a dense precipitate indicates calcium excess. The Sulkowitch test results do not correlate well with urinary calcium excretion, but the test is valuable in tetany, in hyperparathyroidism, and for regulation of vitamin D intake in a patient with hypoparathyroidism. Finally, a specific abnormal EKG finding in calcium deficit is a prolonged Q-T interval.

An interesting manifestation of prolonged calcium deficit is a condition called *osteomalacia*. In this condition, bones lose their calcium, phosphate, and other electrolytes, becoming soft and pliable, and the patient actually shrinks in height. In the first recorded case of osteomalacia, the physician reported that his patient had shrunk some 17 inches. Fortunately, osteomalacia is rare, yet it is seen during famines and among pregnant and lactating women in countries where calcium intake is grossly inadequate.

A not so rare manifestation of calcium deficit is osteoporosis, which has a high incidence among women over 45 and men over 55; 25% of women and 20% of men over the age of 70 have osteoporosis. In this condition, the bones maintain their chemical composition but become thinner, lighter, and more porous. The ailment reveals itself in back pains, decreased height, and frequent and painful fractures of the vertebrae, ribs, and the bones of the arms, legs, hands, and feet. (See Chapter 18.)

Problems related to calcium deficit include the elevated phosphate in end-stage renal disease. In health, the kidneys convert a nonusable form of vitamin D to one that the body can use. But in serious renal disease, this conversion does not take place: calcium is not utilized and may precipitate out in tissues as harmful concretions. Recently, a usable form of vitamin D that can be administered to patients with end-stage renal disease was developed. Dietary deficiency of vitamin D causes inadequate absorption and utilization of calcium. Insufficient exposure to ultraviolet light from the sun, in the absence of oral vitamin D, causes calcium deficit.

Calcium deficit requires administration of vitamin D or, when a renal problem exists, of activated vitamin D. Should the hypocalcemia be caused by magnesium

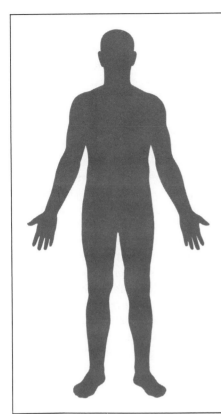

PRECEDING EVENTS

- *Loss of calcium-rich intestinal secretions*
- *Immobilization of calcium*
- *Parathormone deficit*
- *Phosphate reciprocity*

CLINICAL OBSERVATIONS

- *Numbness with tingling of fingers and circumoral region*
- *Hyperactive reflexes*
- *Muscle cramps*
- *Positive Trousseau's sign (carpopedal spasm of hand when blood supply decreased or nerve stimulated by pressure)*
- *Chvostek's sign positive (tapping facial nerve causes spasm of lip and cheek)*
- *Tetany*
- *Laryngeal stridor*
- *Convulsions*
- *Fractures due to bone porosity (seen in chronic hypocalcemia)*

LABORATORY FINDINGS

- *Plasma calcium below 4.5 mEq/liter or 10 mg/100 ml*
- *Specific EKG findings*

RELATED PROBLEMS

- *Elevated phosphate*
- *Vitamin D deficiency*
- *Inadequate ultraviolet exposure*

FIG. 2-12. Calcium deficit.

depletion, magnesium should be given by mouth or parenterally. Calcium can be replenished by the oral route, using the lactate, chloride, gluconate, levulinate, or carbonate salt. Acute calcium deficit is treated by intravenous administration of a 10% solution of calcium gluconate, crucially important if tetany or convulsions have occurred. No treatment of osteoporosis has proved entirely successful. Physical therapy helps keep the bones metabolically active. A high-protein diet plus generous supplements of vitamin D and calcium are employed. Use of fluorides is generally regarded as experimental and potentially hazardous (Fig. 2–12).

CASE HISTORIES A. G., 70-years old, was admitted to the hospital because of sprue of some 2-months' duration. The attack had followed 10 years spent in the tropics, in the Amazon Valley of Brazil. During his first 2 weeks in the hospital, the patient had numerous large, foul-smelling stools daily. Because of the frequent bowel movements, he was given a potassium supplement in addition to the regular diet. During the third week in the hospital, he complained of tingling of the ends of his fingers and of abdominal cramps. Physical examination revealed hyperactive deep reflexes and bilateral carpopedal spasms. A Sulkowitch test on the urine revealed no precipitation, and the serum calcium was found to be 7 mg/dl (3.5 mEq/liter).

Commentary The impaired fat digestion of sprue causes large quantities of calcium to be excreted in the stools as calcium soaps. Calcium deficit of ECF can readily develop under such circumstances. The clinician showed good judgment in this case in giving potassium, but he should have added a calcium and vitamin D preparation. (Vitamin D promotes the absorption of calcium from the intestine.)

D. A., an energetic sales executive, had been in robust health as far back as he could remember. In his thirtieth year, he was visiting the home office to have a conference with his superior. Sitting in the latter's office, he suddenly fell from his chair to the floor and immediately began convulsing violently. So severe was the convulsion that he became cyanotic, as observed by a company physician with an office nearby. Steps were taken to prevent him from swallowing his tongue. By the time an ambulance arrived, his convulsion had subsided. He was taken to a local hospital and was admitted. His previous history was noncontributory. The patient had used alcohol only moderately. Upon admission to the hospital, he complained of severe abdominal pain—continuous, boring, and partially relieved by sitting up. He was nauseated and vomited. His temperature was 38.9° C (102° F). He had epigastric tenderness and moderate abdominal distention. After 2 days in the hospital, ecchymoses appeared about the umbilicus. Laboratory tests revealed a white blood cell count of 16,000 and serum amylase 700 Somogyi units (normal 50 to 200). Serum calcium was 5 mg/dl (2.5 mEq/liter). There were no positive radiographic findings. The diagnosis was acute hemorrhagic pancreatitis with calcium deficit.

Commentary Despite the serious illness this patient experienced, he made an uneventful recovery and 10 years later was still in good health. Such a result is unusual in acute hemorrhagic pancreatitis. How does pancreatic infection cause calcium deficit? The current theory, which has been recently reviewed and corroborated, is that somehow, calcium is removed from the ECF in considerable quantities by the exudate surrounding the inflamed pancreas. The same phenomenon appears to occur in massive infections in other areas.

Calcium excess of ECF Calcium excess of ECF, or *hypercalcemia*, can develop from drinking too much milk—as an ulcer patient might do to soothe his pain—or too much hard water with a high calcium content. The imbalance may be caused by tumor or overactivity of the parathyroid glands, multiple myeloma, or excessive administration of vitamin D (over 50,000 units daily) in the treatment of arthritis. Multiple fractures, bone tumors, and prolonged immobilization also produce symptoms of calcium excess when calcium stores released from bone flood the ECF. Osteomalacia and osteoporosis, although manifestations of calcium deficit, cause symptoms of hypercalcemia during the early stages when calcium is moving out of bones and into the ECF. Thus, while the bones may be dangerously deficient in calcium, there still exists calcium excess of the ECF. The cause of idiopathic hypercalcemia is unknown.

The clinical observations in calcium excess include relaxed skeletal muscles, deep bony pain (caused by honeycombing of bones), flank pain (caused by kidney stones, which form from the excess calcium presented to the kidneys for excretion), and pathologic fractures (caused by weakening of bone by leeching away of calcium; this symptom was formerly the first inkling that the patient had calcium excess. Recall that the "excess" refers to the ECF, not to the bony tissues from which calcium has been removed.)

The plasma calcium concentration is above 5.8 mEq/liter or above 11 mg/dl; the urinary Sulkowitch test shows dense precipitation, and calcium excess is indicated by radiographic examination revealing generalized osteoporosis, widespread bone cavitation, or radiopaque urinary stones. (Radiographic examination early reveals rarefaction, or thinning, of the bone just under the periosteum of long bones, a most valuable early test that becomes positive before serious and perhaps irreparable damage has been done to the kidneys.) An elevated BUN indicates that the urinary stones have damaged the kidneys (Fig. 2-13).

Hypercalcemic crisis is the most important syndrome of calcium excess, for it represents an emergency situation requiring immediate attention to prevent cardiac arrest. The symptoms include intractable nausea and vomiting, dehydration, stupor, coma, and azotemia.

Treatment Treatment is directed at restriction of calcium intake and the promotion of calcium excretion. Intravenous NaCl solutions, 0.9% or 0.5%, are frequently used to increase calcium excretion. (Saline infusion enhances urinary calcium excretion because sodium inhibits tubular reabsorption of calcium.) Saline is infused rapidly to induce calcium diuresis; to prevent fluid overload, furosemide (Lasix) is given intravenously in frequent small doses. Furosemide also promotes calcium excretion and thus serves a dual purpose. It is important that urinary water, sodium, and potassium losses be measured and replaced. Failure to replace the fluid loss promoted by furosemide will lead to volume depletion and rise in the serum calcium level. Phosphates are sometimes used to cause a reciprocal drop in the serum calcium level. Glucocorticoids decrease bone turnover and tubular reabsorption of calcium and thus are sometimes used in the treatment of hypercalcemia. Calcitonin will temporarily lower the serum calcium level and may be used if saline therapy is ineffective. (Therapy for hypercalcemia is discussed further in Chapters 12 and 19).

Medical management of hypercalcemic crisis has

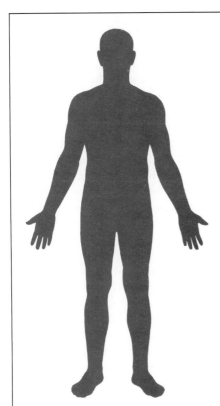

PRECEDING EVENTS

- *Hyperparathyroidism*
- *Widespread bony metastases*
- *Prolonged immobilization*
- *Excessive vitamin D*
- *Multiple myeloma*
- *Paget's disease*

CLINICAL OBSERVATIONS

- *Hypotonic skeletal muscles (recall that calcium acts as a sedative at the myoneural junction)*
- *Anorexia*
- *Nausea and vomiting*
- *Lethargy*
- *Polydipsia and polyuria*
- *Deep bony pain*
- *Pain in flanks (related to kidney stones)*
- *Stupor*
- *Coma*
- *Cardiac arrest*

LABORATORY FINDINGS

- *Plasma calcium above 5.8 mEq/liter or 10 mg/100 ml*
- *Radiographic examination shows generalized osteoporosis, widespread bone cavitation, radiopaque urinary stones*
- *Elevated BUN (due to fluid volume deficit or renal damage)*

RELATED PROBLEMS

- *Kidney damage caused by stones*

FIG. 2-13. Calcium excess.

generally been unsatisfactory, since, owing to slow action or inherent toxicity, none of the many regimens tried has been consistently successful. Saline solutions and inorganic phosphate solutions are among the most frequently used. If the patient is also being treated for other fluid imbalances, only calcium-free solutions should be used.

CASE HISTORIES A. W., 35 years old, had commented to her friends how "poorly" she had been feeling for the past several weeks. On their urging, she consulted her family physician. Her current symptoms included weakness, anorexia, nausea, polyuria, and thirst. Her past and family histories made no contribution. The physical examination revealed only that the patient looked ill and had lost some 22 lb (10 kg). But radiographic and laboratory findings were more revealing: the radiographic study showed subperiosteal resorption of the cortex of the phalanges of the fingers but no evidence of renal calcinosis. The serum calcium was 16 mg/dl (8 mEq/liter), and

the phosphate was 2 mg/dl (1.18 mEq/liter). At operation, a moderate sized parathyroid adenoma was discovered.

Commentary A parathyroid adenoma elaborates excessive quantities of parathormone, which mobilizes calcium from the bony skeleton. If the process has gone on long enough, fractures can occur, or stones can form in the renal pelves, in time destroying the substance of the kidney. Early roentgenography of phalanges should be carried out in anyone even remotely suspected of suffering from a parathyroid tumor because it reveals subperiosteal resorption of the cortex. Diagnosis based on the presence of kidney stones is tragically late. In about half the patients with calcium excess of the ECF, the serum phosphate is normal. Hypercalcemia may be masked by the effect of high phosphate intake but can be made evident by phosphate deprivation.

W. T., 54 years old, reported to his family physician because of a train of symptoms that had developed within the past 3 weeks. His first symptom was that of excessive dryness of the mouth, accompanied by thirst and excessive urination. After about 10 days, he began to experience nausea and occasional vomiting. Then he began to become drowsy and lethargic. His wife was particularly disturbed by his brief periods of disorientation and confusion. Prior to the present illness, the patient had enjoyed robust good health. The physical examination revealed exaggerated deep reflexes, some questionable muscle weakness, and definite tenderness in the right upper quadrant of the abdomen and somewhat less tenderness in the right lower quadrant of the abdomen and the right lumbar region.

The patient was admitted to the hospital for a thorough workup. Examination of the serum electrolytes revealed a serum calcium of 7.5 mEq/liter (15 mg/dl), bicarbonate 26 mEq/liter, parathyroid hormone normal, alkaline phosphatase normal, and other serum electrolytes normal. The urine revealed numerous red cells but was otherwise normal. Radiographic examination revealed a mass in the right kidney about the size of a small potato. An intravenous pyelogram confirmed this finding. Radiographs of the long bones, skull, and lungs revealed no significant findings. A diagnosis of carcinoma of the right kidney, probably without metastasis, was made, with secondary hypercalcemia, or calcium excess, responsible for the symptoms. Appropriate therapy of the hypercalcemia was instituted and surgical removal of the carcinoma contemplated.

Commentary About 20% of patients with malignant disease develop hypercalcemia. Among the tumors known to cause it are carcinoma of the lung, carcinoma of the kidney, ovarian tumor, bony metastases, and sarcoidosis. Unlike hypercalcemia caused by hyperparathyroidism, the parathyroid hormone and alkaline phosphatase are not elevated. Other causes of hypercalcemia include diseases involving immobilization and ingestion of an antacid. In arriving at a diagnosis of hypercalcemia, it is wise to obtain two serum calciums to avoid the chance of laboratory error. One or more of three mechanisms operate to produce calcium excess of the ECF: increased bone resorption, increased intestinal resorption of calcium, and decreased excretion of calcium by the kidneys. In the case of W. T., the most likely mechanism would be the third, although the first is a possibility.

PHOSPHATE Phosphate teams up with calcium in contributing to the supportive and dynamic functions of bones and teeth, which contain about 75% of the body's phosphate; most of the rest resides in the cells, with a much smaller portion located in the plasma. As the chief anion of the cells, phosphate participates in many important chemical reactions. Many of the B vitamins are effective only when combined with phosphate. The element helps energy transfers within the body cells, promotes normal nerve and muscle action, helps maintain the acid–base balance of the body fluids, and participates in carbohydrate metabolism. Moreover, this busy element is required for cell division and for transmission of hereditary traits from parent to offspring.

Seventy percent of dietary phosphate is absorbed, in contrast to only 20% to 30% of the calcium of the diet. Adequate intake of vitamin D promotes absorption of both calcium and phosphate; hence, when vitamin D consumption is inadequate, intestinal absorption of both elements drops precipitously. In renal failure, phosphates are abnormally retained; as a result, the calcium concentration of the plasma falls because of the reciprocal relationship between calcium and phosphate. The parathyroid glands then secrete increased amounts of parathormone, causing excessive removal of calcium from bone and abnormal deposition of calcium in body tissues.

Because so many foods contain generous quantities of phosphate, the human diet seldom is deficient in it. The RDAs for phosphate closely resemble those for calcium, differing only in the RDAs for the first year of life. The RDA for formula-fed infants is 240 mg/day for the first 6 months and 360 mg/day for the second 6 months. Breastfed infants' needs are fully met by breast milk. RDAs for all other age-groups are the same as those for calcium.

Foods especially high in phosphate include beef, pork, dried beans, and dried mature peas.

When the serum phosphate falls below 3 mg/dl, phosphate deficit is present. Should the level drop below 1 mg/dl, a profound deficit exists. Preceding events for phosphate deficit include alcoholism, severe burns in the healing phase, diabetic ketosis, bicarbonate excess, carbonic acid deficit, and liver disease. Specific etiologic factors include decreased intake of phosphate, impaired absorption accompanying use of carbonate antacids, hyperparathyroidism, acidosis, and hyperalimentation.

Acute deficits are characterized by hemolytic anemia, rhabdomyolysis, defective functioning of leukocytes, defective platelets, and encephalopathy. In chronic deficits, one sees anorexia, weakness, bone pain, and pathologic fractures. Urinary concentration of calcium, magnesium, and bicarbonate is increased. Absorption of calcium from the intestine is increased. Serum calcium levels are normal. Plasma parathyroid hormone levels are normal or low.

Phosphate excess exists when the serum phosphate level is above 4.5 mg/dl. Such increase occurs chiefly in acute or chronic renal disease and is worsened by catabolic stress. Phosphate excess can occur in lactic acidosis. Large doses of phosphate orally, parenterally, or rectally cause temporary elevation of serum phosphate, even with normal kidneys. Phosphate excess and calcium deficit may occur in association after administration of large amounts of phosphate in kidney failure and with chemotherapy for neoplasms.

MAGNESIUM Magnesium makes up 1.2% of the earth's crust and 3.69% of the ocean's solids. The element's physiological action was first recognized by the villagers of Epsom, England, who discovered in 1618 that the bitter-tasting water from a village pool produced catharsis. The water from the pool gained a reputation—deserved or not—for promoting health, and people from all over Europe flocked to Epsom to drink it. Later, the active ingredient of the water was found to be magnesium sulfate and was named Epsom salts. Known chemically as Mg, magnesium is the fourth most abundant cation in the human body. The average adult contains about 20 g, about half of which is stored in bone cells; 49% is distributed throughout the specialized cells of the liver, heart, and skeletal muscles. The ECF contains only 1%, most of which is in the cerebrospinal fluid, and the normal serum level is 1.67 mEq/liter.

The minimum daily requirement for magnesium is a matter of controversy, but most medical authorities agree on 250 mg for the average adult, 150 mg for the infant, and 400 mg for the pregnant or lactating woman. The usual American diet provides from 180 to 300 mg daily.

Calcium and magnesium, both of which are regulated by the parathyroid glands, share a common route of absorption in the intestinal tract and appear to have a mutually suppressive effect on each other. If calcium intake is unusually high, calcium will be absorbed in preference to magnesium, and, conversely, if magnesium intake is high, more of it will be absorbed and calcium will be excluded. A normal intake of both allows for adequate absorption of both.

The body's magnesium content also directly affects the potassium concentration because, if magnesium is deficient, the kidneys tend to excrete more potassium. Consequently, potassium deficit may also develop.

Magnesium's importance was not understood until its recent recognition as the activator of numerous vital reactions related to enzyme systems. Among the systems that magnesium activates are those that enable the B vitamins to function and those associated with utilization of potassium, calcium, and protein. Magnesium is particularly important in nervous tissue, in skeletal muscle, and in the heart, having considerable therapeutic value in correcting arrhythmias and in counteracting the toxic side-effects of certain powerful drugs used in the treatment of heart disease. When used in conjunction with hypotensive agents, magnesium is a useful therapeutic agent in treatment of hypertension. And lastly, toxemia of pregnancy, which formerly had a high mortality rate, responds to magnesium therapy.

Although nuts, legumes, fish, and whole grains provide much magnesium, the usual American diet, with its emphasis on meat, eggs, and dairy products, makes magnesium deficit a possibility. In contrast, the diet of most Orientals abounds in magnesium-rich cereals and vegetables.

Magnesium deficit of ECF Magnesium deficit of ECF, also referred to as *hypomagnesemia*, develops when the magnesium concentration of that fluid decreases.

Alcoholism looms high on the list of factors causing magnesium deficit and, indeed, ingestion of alcohol appears to promote the imbalance, even in the face of what normally would be an adequate intake. The combination of liver disease and sustained losses of GI secretions is almost certain to produce magnesium deficit. Diabetes mellitus appears to predispose to magnesium deficit, and so does a high intake of calcium, since absorption of magnesium from the intestinal tract varies inversely with calcium absorption. Other causes are primary hyperaldosteronism, severe renal disease, toxemia of pregnancy, diseases of the small intestine that impair GI absorption, vigorous drug-induced diuresis, or prolonged administration of magnesium-free solutions. Severe malnutrition, such as kwashiorkor or pluricarencial syndrome (perhaps better designated protein/calorie malnutrition), is also a cause of magnesium deficit. This condition affects countless infants and young children in developing countries in both hemispheres.

Although magnesium deficit is relatively simple to diagnose, a review of the literature suggests that many persons die of undiagnosed magnesium deficit. Perhaps the reason the imbalance is so frequently overlooked is that it is easily mistaken for potassium deficit, with which it is often associated. Magnesium deficit should always be considered a possible imbalance when a patient being treated for potassium deficit

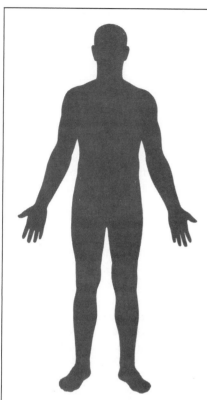

PRECEDING EVENTS

- *Usually a combination of liver malfunction and loss of intestinal contents*
- *Alcoholism*
- *Severe malnutrition*

CLINICAL OBSERVATIONS

- *Neuromuscular irritability with tremor and hyperactive deep reflexes*
- *Disorientation, confusion, and visual or auditory hallucinations*
- *Leg and foot cramps*
- *Tachycardia*
- *Hypertension*
- *Convulsions*
- *Improvement with magnesium sulfate*

LABORATORY FINDINGS

- *Plasma magnesium below 1.4 mEq/liter*

RELATED PROBLEMS

- *This an easy imbalance to miss; think of it when preceding events are suggestive*
- *If potassium deficit is present and improvement fails to occur with its correction, suspect magnesium deficit*
- *Alcoholics may develop magnesium deficit on a well-balanced diet*

FIG. 2-14. Magnesium deficit.

does not respond to appropriate therapy, or in the alcoholic with a body fluid disturbance.

While the clinical picture of magnesium deficit varies from patient to patient, certain symptoms are frequently seen. These include tremor, athetoid or choreiform movements, tetany, a positive Chvostek or Trousseau sign, excessive neuromuscular irritability, painful paresthesia, and—an ominous symptom—convulsions. The patient usually is confused and may hallucinate. Another symptom is tachycardia; when it occurs, the blood pressure rises. A therapeutic test consisting of administration of a 1% solution of magnesium sulfate or magnesium citrate by mouth may prove of great value in diagnosing magnesium deficit. If the imbalance is present, there will be an immediate response; all symptoms, however, may not disappear for some 60 to 80 hours.

The most useful laboratory test in magnesium deficit is the plasma magnesium determination, since, if it is below 1.4 mEq/liter, magnesium deficit is probably present.

Magnesium deficit is repaired by oral, intramuscu-

lar, or intravenous administration of magnesium salts. Most frequently used is hydrated magnesium sulfate. Tetany caused by magnesium deficit cannot be corrected by administration of calcium (Fig.2-14).

CASE HISTORIES R. S., 27 years of age, was admitted to the hospital with regional enteritis. Diarrhea was severe. He was given the usual hospital diet plus a potassium supplement. After a month in the hospital, the patient began to have periods of disorientation. Physical examination revealed a tremor of the fingers, hyperactive deep reflexes, and a positive Chvostek sign (twitching of facial muscles when zygoma is tapped). On one occasion, the patient had a brief convulsion. When serum electrolytes were determined, the plasma potassium was 5 mEq/liter, but the magnesium was found to be 0.9 mEq/liter. When magnesium sulfate was administered orally as a therapeutic test, improvement was immediate.

Commentary Deficit of potassium and of magnesium have many of the same causes. Therefore, when a patient with potassium deficit does not improve when the potassium is raised to normal, magnesium deficit should be suspected. Magnesium deficit is especially important in the alcoholic, since persons with this problem appear to have a tendency to develop magnesium deficit even when receiving a diet containing generous quantities of magnesium.

J. G., a 20-month-old child, was brought into a clinic because of emaciation and failure to thrive. He had been fed only cola drinks, corn syrup, and palatable (empty) snacks. He had never been breastfed, nor had he had a milk formula. Examination revealed generalized edema; scaly, irritated skin; thin, reddish hair; an enlarged fatty liver; and apathy so extreme that is was pitiful to see him. The laboratory findings were as follows: potassium 2 mEq/liter, magnesium 0.8 mEq/liter, hemoglobin 6 g/dl, RBC 3,000,000, serum albumin 2.7 g/dl. He was given a high-protein diet (it was no easy task devising one that he would eat), an iron preparation, a potassium supplement, and a vitamin supplement containing all essential vitamins. He failed to improve. After 3 weeks, a magnesium supplement was added to the therapeutic program, after which he improved rapidly.

Commentary The child described was suffering from kwashiorkor, also called pluricarencial syndrome in Latin America and protein/calorie malnutrition in the United States. It was discovered in Africa in 1957 that infants with this severe form of malnutrition frequently do not improve until magnesium is added to their therapeutic program. Magnesium is extremely important for its promotion of cellular enzymatic reactions. It deserves its name of the *activator.*

Magnesium excess of ECF Magnesium excess, or *hypermagnesemia,* represents an imbalance of increasing importance; hence, we have added it to our diagnostic classification. It can occur when magnesium is not being excreted normally, as in chronic renal disease or untreated diabetic acidosis. Administration of excessive quantities of magnesium, as in a child with congenital megacolon who is given magnesium sulfate rectally, can cause an excess, and the imbalance has also been reported when perfusion fluid used in hemodialysis contained excessive magnesium. Deficiency of aldosterone, as in Addison's disease, can cause it, as can hyperparathyroidism.

Symptoms of magnesium excess, including lethargy, coma, and impaired respiration, can terminate in death and appear when the plasma magnesium exceeds 6 mEq/liter (Fig. 2-15).

Important in prevention of magnesium excess is the rule that magnesium salts should never be given to patients with acute or chronic renal disease. Should magnesium excess develop, administration of magnesium-containing compounds should be *immediately* discontinued. ECF volume deficit, if present, should be corrected. Calcium gluconate, useful as a magnesium antagonist in emergency situations, is often administered intravenously. If respiratory failure occurs, artificial respiration should, of course, be instantly instituted.

CASE HISTORIES B. N., a 26-year-old male, was admitted to the hospital with hopelessly infected kidneys. Following bilateral nephrectomy, he was put on hemodialysis pending availability of a transplant. Following the third hemodialyzing session, he became extremely lethargic and appeared to have some difficulty in breathing. Complete serum electrolyte studies revealed a magnesium concentration of 3 mEq/liter. The source of the excess was sought without success until one of the nurses recalled that water from a new supplier had been used for the last perfusion bath. When analyzed, the water was found to be unduly high in several electrolytes, including magnesium. The supplier of water for the perfusing equipment was immediately changed, and the patient's next hemodialyzing session was uneventful.

W. B., 52-year-old female, was suffering from advanced chronic glomerulonephritis. (Such patients are unable to excrete magnesium normally.) Because of gastric discomfort, the patient was given a magnesium-containing antacid (*e.g.,* Maalox). She immediately developed EKG abnormalities, including ventricular premature contractions, and central nervous system symptoms and signs, including lethargy, which progressed to coma. Her serum magnesium was 7.5 mEq/liter. Since dialysis is far from efficient in removing magnesium, patients on dialysis should not be given magnesium-containing medications.

PROTEIN Protein has been called the *keystone* of the nutritional arch. Because all living matter is composed largely of proteins, there would be no life without protein. Through its drawing power for liquids—oncotic pressure—protein helps prevent leaking of plasma from the blood vessels. In addition, the amino acids of which proteins are composed are the building blocks for the growth of new tissue. Protein is also required for the elaboration of enzymes, for the fabrication of

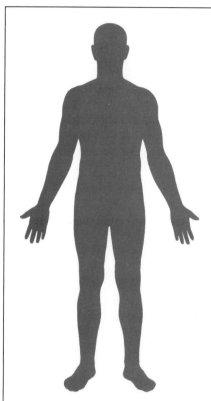

PRECEDING EVENTS

- Chronic renal disease, including uremia
- Parenteral or oral administration of magnesium to patients with renal insufficiency
- Administration of antacids containing magnesium salts (e.g., Maalox) to patients with renal impairment

CLINICAL OBSERVATIONS

- Flushing and sweating
- Hypotension (can occur at a level of 3 to 5 mEq/liter)
- Drowsiness
- Weak to absent deep tendon reflexes (may be depressed when plasma level exceeds 4 mEq/liter; patellar reflex disappears when level reaches 8 to 10 mEq/liter)
- Lethargy (can occur at a level between 5 to 7 mEq/liter)

- Impaired respiration (respiratory paralysis can occur when the plasma concentration reaches 10 to 12 mEq/liter)
- Bradycardia
- Coma (can occur at a plasma level of 12 to 15 mEq/liter)

LABORATORY FINDINGS

- Plasma magnesium above 2.3 mEq/liter
- EKG shows prolonged P-R interval, prolonged QRS interval, tall T waves, A-V block, evidence of ventricular premature contractions

RELATED PROBLEMS

- Hypermagnesemia is rarely seen except in patients who have renal impairment

FIG. 2-15. **Magnesium excess.**

those blood-borne messengers we call hormones, for the manufacture of some vitamins, and for proper function of immune mechanisms, since many antibodies are proteins. Because it is practically impossible to develop protein excess, this imbalance is not included in our diagnostic classification and will not be discussed.

Protein deficit of ECF Since protein is an anion (e.g., Ca^+ caseinate$^-$), we have included protein deficit of ECF in our diagnostic classification. The imbalance is also known as *hypoproteinemia* and *protein malnutrition;* Kwashiorkor, or the pluricarencial syndrome, and hypoproteinosis are closely related clinical states.

Protein deficit, which develops slowly, frequently does not attract the attention of the attending physician and can be deadly. The imbalance can result from decreased intake, increased loss, or impaired utilization of protein. Naturally, the person on an inadequate diet—for whatever reason—will develop protein deficit, most extreme in starvation, which causes wasting

and internal cannibalization of the body tissues. Bleeding, whether severe, repeated, or of long duration, drains the body's protein stores, sooner or later causing a deficit, as does infection. Burns, fractures, and surgical procedures help deplete protein stores through the destruction and apparently purposeless wastage of tissues; this is called *toxic destruction of protein.* Diseases that affect the digestive tract interfere with protein intake and cause protein poverty. Protein deficit is often associated with potassium deficit, since adequate potassium is essential for protein synthesis. Indeed, every patient with any disturbance of the body fluids is a candidate for protein deficit, particularly if he has been ill for a long time, since the sort of conditions that cause body fluid disturbances can also cause protein deficit.

The clinical findings of protein deficit include weight loss and wasting of the muscles; body tissues become flabby and soft. The patient complains that he is always tired and becomes chronically depressed. His poor appetite becomes still poorer and he may vomit repeatedly. If he is injured, his wounds will not heal. His recuperative powers are greatly decreased; he may experience one infection after another. Lastly, the protein-deficient patient frequently suffers from anemia, which may be either microcytic or macrocytic.

Laboratory aids in diagnosis of hypoproteinemia are limited; they include depressed hemoglobin, depressed hematocrit, and low RBC—these findings are significant only when iron intake is adequate. In severe imbalances, the plasma albumin may drop below 4 g/100 ml. However, there is no accurate test for mild or moderate deficits. Bear in mind that it is the albumin that is significant; the total protein can actually be elevated in the face of a protein deficit, provided the globulin has risen in response to an infection or allergic disturbance. When the hemoglobin is depressed, it probably indicates a protein deficit *provided* the patient has been receiving adequate iron. If iron deficiency anemia is present, the depressed hemoglobin would be due—in part, at least—to that.

Under related problems, we should mention that clinicians often neglect to consider protein deficit since it is so slow to develop. Its onset can often be described as insidious. Protein deficit is likely to develop with fad diets. Several persons at a prominent western university developed protein deficit when they were following macrobiotic diets. A "pure" vegetarian diet (*i.e.*, a vegetarian diet without milk or eggs) can induce protein deficit if protein complementarity is not achieved. (Recall that protein complementarity refers to a wise choice of protein so as to achieve a satisfactory amino acid mixture.)

Protein deficit (Fig. 2-16) is slow to develop and cannot be rapidly repaired. Administration of a high-calorie, high-protein diet is the best method of treatment. For patients unable to take food by mouth, high-protein tube feedings can be employed. (Nursing responsibilities in the administration of tube feedings are discussed in Chapter 5.) Although anabolic hormones are sometimes used in protein deficit, their efficacy is questionable. Administration of proteins in peripheral veins is becoming increasingly popular.

FIG. 2-16. **Protein deficit.**

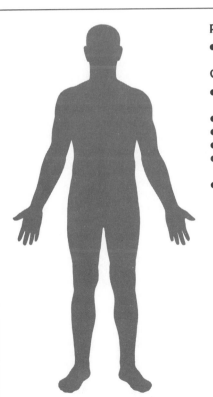

PRECEDING EVENTS
- *Prolonged loss of protein*

CLINICAL OBSERVATIONS
- *Mental and emotional depression*
- *Anorexia*
- *Loss of muscle mass and tone*
- *Weight loss*
- *Plasma-to-interstitial fluid shift (nutritional edema)*
- *Reduced resistance to infection*

LABORATORY FINDINGS
- *Plasma albumin below 4 g/100 ml*

RELATED PROBLEMS
- *Easy to fail to think of protein deficit because it develops so slowly*
- *May develop with fad diets; pure vegetarian diet may cause deficit if protein complementarity is not achieved.*

(See Chapter 6.) With the development of parenteral hyperalimentation, patients with nonfunctional GI tracts can have the state of their protein nutrition restored and then maintained. The requirements for proper hyperalimentation are demanding, for both the preparation and administration of the solutions demand high expertise. The hazards are many, but the results achieved are truly remarkable and unquestionably offset the disadvantages when other routes are contraindicated. (TPN is discussed in Chapter 6.)

CASE HISTORIES S. G. was a 28-year-old female admitted to the hospital following an automobile accident in which she was severely injured. She had sustained multiple fractures, one of which was compound, and had lost considerable blood. She received several blood transfusions during the first few days in the hospital. After this initial period, her only nutrition consisted of the usual hospital diet, at which she only nibbled. Both fractures and flesh wounds healed slowly. She was emotionally and mentally depressed, she was pale, she had almost no appetite, her muscles were soft and flabby, and she lost 22 lb (10 kg) during the first month in the hospital. Her serum potassium was 3.2 mEq/liter, albumin 3 g/dl, hemoglobin 8 g, and RBC 3,000,000. After the diagnosis of protein deficit was made, numerous steps were taken to restore her protein nutrition.

Commentary Protein has been called the *keystone nutrient* and, indeed, it is. It provides amino acids, the building blocks of growth and repair; it provides oncotic pressure for the cardiovascular system; it defends the body against infection; and it is required for the formation of enzymes and hormones. The severely injured patient not only is prone to eat far less protein than normal, but there is also an enormous loss of protein in the form of nitrogen in the urine—the so-called toxic destruction of protein. Protein deficit usually comes on insidiously and is often recognized quite late, as was the case with this patient.

R. H. was a 20-month-old boy brought to the doctor because he had almost no appetite, was underweight and underheight, was pale (his mother called him a "pastel child"), had poor posture, was irritable, and developed one upper respiratory infection after another. He was an only child, reared in a home in which the grandparents were living. According to the history, the first year of life was uneventful, except that there was some resistance to weaning. The infant, therefore, was subsisting largely on the bottle well past the end of the first year. Trouble began

about the fourteenth month. This is a time when growth slows drastically, and so does appetite. Many mothers do not understand this, hence they are greatly disturbed when the formerly voracious eater loses much of his interest in food. This happened with R. H. His mother reacted by trying to force him to eat, by one means or another. The child resisted, and the mother forced all the more. In this, she was enthusiastically supported by Grandma. The infant resorted to the formula bottle and rejected the introduction of new solid foods. By the time the child was taken to the physician, he was living primarily on milk—a splendid food when taken in moderation and in conjunction with other foods, but never intended as the sole food for the preschool child.

The physical examination showed that not all was well. It revealed all that the mother had complained about; in addition, the muscles were flabby and there was a moderately severe hypochromic microcytic anemia. Serum albumin was 3.8 g/dl. Electrolytes were within limits of normal. The diagnosis was moderate protein deficit. Drastic revision of the child's eating program was indicated if serious future trouble was to be avoided.

Commentary The hazard of an exclusive milk diet for children was highlighted by Dr. Joseph Brennemann (whom many regard as the father of modern pediatrics) when he protested:

Milk, the great 'protective food,' has been crammed down our and our children's throats, in season and out of season, although the observing, practical pediatrician has long known that even in the use of milk, children should be dealt with as individuals and that the slogan of 'a quart of milk or more every day' originated in the laboratory, and has a sweeter sound to the milk producer than to the pediatrician.
Personal communication

Meat, fish, eggs, and cheese are solid, concentrated sources of high-quality protein and should be included in the preschool child's diet, provided he does not have an allergy to one of them.

REFERENCES

1. Fabun D: Dynamics of Change. Englewood Cliffs, New Jersey, Prentice–Hall, 1967
2. Prasad A: Metabolism of zinc and its deficiency in human subjects. In Zinc Metabolism, pp 250–303. Springfield, Illinois, Charles C Thomas, 1966

BIBLIOGRAPHY

Aggett P, Harries J: Current status of zinc in health and disease states. Arch Dis Child 54:909, 1979

Allen J, Kay N, McClain C: Severe zinc deficiency in humans: Association with a reversible T-lymphocyte dysfunction. Ann Intern Med 95:154, 1981

Anderson B, Gibson R, Sabry J: The iron and zinc status of long-term vegetarian women. Am J Clin Nutri 34:1042, 1981

Bates J, McClain C: The effect of severe zinc deficiency on serum levels of albumin, transferrin, and prealbumin in man. Am J Clin Nutr 34:1655, 1981

Blachley J, Knochel J: Tobacco chewer's hypokalemia: Licorice revisited. N Engl J Med 302:784, 1980

Burnet F: A possible role of zinc in the pathology of dementia. Lancet Jan 24, 1981, p 186

Cooke C, Turin M, Walker W: The syndrome of inappropriate antidiuretic hormone secretion (SIADH): Pathophysiologic mechanisms in solute and volume regulation. Medicine 58:240, 1979

Decaux G, Unger J, Brimioulle S, Mockel J.: Hyponatremia in the syndrome of inappropriate secretion of antidiuretic hormone. JAMA 247:471, 1982

Dranov J: Question and answer: Kidney disease and sodium intake. JAMA 246:1463, 1981

Findling J, Beckstrom D, Rawsthorne L et al: Indomethacin-induced hyperkalemia in three patients with gouty arthritis. JAMA 244:1127, 1980

Freeland–Graves J, Bodzy P, Eppright M: Zinc status of vegetarians. J Am Diet Assoc 77:655, 1980

Freitag J, Miller L (eds): Manual of Medical Therapeutics, 23rd ed. Boston, Little, Brown & Co, 1980

Friedman H: Problem-Oriented Medical Diagnosis, 2nd ed, pp 227–233. Boston, Little, Brown & Co, 1979

Goldberger E: A Primer of Water, Electrolyte and Acid–Base Syndromes, 6th ed. Philadelphia: Lea & Febiger, 1980

Goodwin F (ed): Lithium ion: Impact on treatment and research. Arch Gen Psychiatry 36:833, 1979

Grim C, Luft F, Miller J et al: Effects of sodium loading and depletion in normotensive first-degree relatives of essential hypertensives. J Lab Clin Med 94:764, 1979

Hamburger S, Rush D: Syndrome of inappropriate antidiuretic hormone activity. Crit Care Q, Sept 1980

Luft F, Rankin L, Bloch R et al: Cardiovascular and humoral responses to extremes of sodium intake in normal black and white men. Circulation 60:697, 1979

McClain C, Souter C, Zieve L: Zinc deficiency: A complication of Crohn's disease. Gastroenterology 78:272, 1980

Maxwell M, Kleeman C (eds): Clinical Disorders of Fluid and Electrolyte Metabolism, 3rd ed. New York, McGraw–Hill, 1980

Medine J, Fareed J, Diamond S: Lithium carbonate therapy for cluster headache. Arch Neurol 37:559, 1980

Missri J, Alexander S: Hyperventilation syndrome. JAMA 240:2093, 1978

Newsome H: Vasopressin: Deficiency, excess and the syndrome of inappropriate antidiuretic hormone secretion. Nephron 23:125, 1979

Perks W, Walters E, Tams I et al: Demeclocycline in the treatment of the syndrome of inappropriate antidiuretic hormone secretion. Thorax 34:324, 1979

Recommended Dietary Allowances, 9th ed. Washington, DC, National Research Council, 1980

Rosenbaum A, Maruta T, Richelson E: Drugs that alter mood: I. Tricyclic agents and monoamine oxidase inhibitors: II. Lithium. Mayo Clin Proc 54:335, 401, 1979

Schwartz W, Bennett W, Curelop S: A syndrome of renal sodium loss and hyponatremia probably resulting from inappropriate secretion of antidiuretic hormone. Am J Med, Oct 1957, p 529

Shen R: The 'supply side' economics of sodium metabolism. JAMA 246:1311, 1981

Shike M, Harrison J, Sturtridge W et al: Metabolic bone disease in patients receiving long-term total parenteral nutrition. Ann Intern Med 92:343, 1980

Snively W, Helmer D: Syndrome of inappropriate secretion of antidiruetic hormone (SIADH): Nemesis for the unwary. J Indiana State Med Assoc 74:514, 1981

Solomons N, Russell R: The interaction of vitamin A and zinc: Implications for human nutrition. Am J Clin Nutr 33:2031, 1980

Spring G: Neurotoxicity with combined use of lithium and thioridazine. J Clin Psychiatry 40:135, 1979

Szent–Gyorgi A: Biology and pathology of water. Perspect Biol Med, Winter 1971

West K: Modern clinical nutrition, Part 2: The importance of trace element in total parenteral nutrition. Am J IV Ther Clin Nutr 8, No. 9 (Oct), 1981

White P, Crocco S (eds): Sodium and Potassium in Foods and Drugs. Chicago, American Medical Association, 1980

Wyant G, Ashenhurst E: Chronic pain syndromes and their treatment: I. Cluster headache. Can Anaesth Soc J 26:38, 1979

Acid–Base Disturbances

The normal composition of the body fluid depends not only upon the concentration of the various electrolytes but also upon the concentration of acids and alkalies. Acids are substances that contain hydrogen ions that they can release to other substances, whereas alkalies possess no hydrogen ions of their own but are able to accept them from acids. The strength of an acid is determined by the number of hydrogen ions it contains per unit of weight, and the power of an alkali is measured by the number of hydrogen ions it can accept per unit of weight.

It appears strange that hydrogen—by far the most plentiful chemical in the universe—should be present in the human body only in minuscule quantities. But even in these tiny amounts, its power is awesome. A highly reactive particle, hydrogen consists of a single proton, representing a hydrogen atom that has lost its only electron. Thus deprived, hydrogen is avid to combine with other molecules that carry a negative electric charge. The hydrogen ion probably never exists as an independent entity but, instead, attaches itself to water molecules to form hydronium ions, or H_3O^+. The hydronium ions, which we will refer to simply as hydrogen ions, occur in the extracellular fluid (ECF) in amounts extremely small in relation to other key solutes. As Gerhard Giebisch of the Yale University School of Medicine tells us, "For every hydrogen ion circulating freely in the ECF under normal physiologic circumstances, there are approximately 100,000 potassium ions and over 3 million sodium ions."[1]

Together, the acids and alkalies of the body fluids produce the chemical reactions necessary for life; in order for such reactions to proceed normally, the body must maintain a precarious balance between the burning acids on the one side and the corrosive alkalies on the other. This is called *acid–base balance*. When acid–base balance is maintained, the reaction of the body fluid is neutral, or normal; if the balance becomes upset, the body fluid becomes either acid or alkaline in its reaction. *It is the number of hydrogen ions present in the body fluid that determines whether its reaction is acid, neutral, or alkaline.*

Recall that we use the term pH to express the reaction of the body fluid (see Chapter 1, the section Units of Measure). A pH scale ranges from 1 to 14; 7 is neutral, below 7 is acid, and above 7 is alkaline. The normal pH of ECF ranges from 7.35 to 7.45 (slightly alkaline). If the plasma pH drops below 7.35, acidosis, or acidemia, is said to be present—even though in the technical sense, a true acid reaction must be below 7.

Rarely does plasma *pH* drop below 7, although it may be as low as 6.8 in extremely ill acidotic patients. If plasma *pH* rises above 7.45, alkalosis, or alkalemia, is said to be present. Death occurs when the plasma *pH* is below 6.8 or above 7.8. *pH* is an algebraic expression for the concentration of hydrogen ions, developed many years ago by general chemists. It represents the reciprocal of the logarithm of the hydrogen ion concentration. To meet the needs of physiologic chemistry, it would be far simpler to express hydrogen ions in nanomoles per liter (nm/liter), a nanomole being a billionth of a gram or a millionth of a milligram. Since the *pH* expression is still used in most medical facilities, we should know how it is derived. We start with the pK of carbonic acid, which is 6.1. The expression *pK* is defined as the negative logarithm of the ionization constant of an acid. It represents the *pH* at which equal concentrations of the acid and of the basic form or salt of the acid (in the case of carbonic acid, bicarbonate) are present. Now, the ratio of bicarbonate to carbonic acid at the normal *pH* of ECF is 20:1. The log for this ratio is 1.3. Adding the pK of carbonic acid of 6.1 to the log of 1.3 gives us a *pH* of 7.4, the approximate normal *pH* of ECF. This *pH* roughly corresponds to 40 nm of hydrogen/liter. It is important to remember two facts: (1) as the *pH* figure becomes larger, the number of H^+ ions becomes fewer; and (2) *pH* values move by tens; thus, *pH* 7 is ten times as acid as *pH* 8 and one tenth as acid as *pH* 6. If we were to convert *pH* 7 to nanomoles of H^+, we would find it equals 100 nm; if we convert *pH* 8 to nanomoles, we find it equals only 10 nm. Notice that as the nanomoles of H^+ figures goes up, the number of H^+ ions goes up; as the nanomoles of H^+ figures goes down, the number of H^+ ions goes down. So, *pH* 7 = 100 nm H^+; *pH* 8 = 10 nm H^+; and *pH* 7.4 (approximately normal) = 40 nm H^+. Unfortunately, the laboratories aren't ready for the simpler and more forthright nanomole measure. One day (perhaps soon) it will come to all hospitals, while at the present it is confined to a few teaching centers. In the meantime, we shall employ the time-honored *pH* method of expressing H^+ ion concentration.

BODY MECHANISMS THAT CONTROL ACID–BASE BALANCE

How does the body control H^+ ion concentration of the body fluids? It accomplishes this monumental task by the help of the body fluid buffers, the lungs, and the kidneys. Buffers tend to prevent changes in H^+ ion concentration when acid or alkali is added to the body from the environment or generated within the body. Buffers occur in pairs; each pair consists of a weak acid and the salt of that acid. There are four major buffer pairs: the *bicarbonate pair*, active in ECF; the *hemoglobin pair*, active in blood; the *plasma protein pair*, exerting its effect on plasma; and the *phosphate pair*, which does its work within the cells. The clinically important buffer is the bicarbonate–carbonic acid pair. Since it is in equilibrium with the other body buffers, we need be concerned only with it. Here is the formula for the bicarbonate–carbonic acid buffer:

$$\frac{HCO_3^- = 24 \text{ to } 28 \text{ mEq/liter}}{H_2CO_3 = 1.2 \text{ to } 1.4 \text{ mEq/liter}}$$

The ratio, obviously, is about 20:1. Remember, it is the ratio of bicarbonate to carbonic acid that determines the H^+ ion concentration of the ECF.

Carbon dioxide (CO_2) unites with water in the ECF to form carbonic acid (H_2CO_3). In the ECF, H_2CO_3, and CO_2 are in equilibrium. The chemical reaction involved is, of course, $H_2O + CO_2 = H_2CO_3$. If CO_2 is blown off through the lungs, then H_2O accompanies it, and the quantity of H_2CO_3 in the ECF is decreased. Conversely, if CO_2 is retained in the ECF, then the reaction is $CO_2 + H_2O = H_2CO_3$ and greater acidity results. Catalyzing the reaction between CO_2 and H_2O is carbonic anhydrase, universally present in the body. Now, H_2CO_3 ionizes to provide H^+ ions. Since CO_2 is the equivalent of H_2CO_3, it too must be regarded as an acid substance. Base bicarbonate is formed when the cations sodium, potassium, calcium, and magnesium unite with the anion bicarbonate.

The normal ratio of carbonic acid to base bicarbonate is 1:20; thus, as long as there is 1 mEq of carbonic acid for every 20 mEq of base bicarbonate in the ECF, the hydrogen ion concentration lies within normal limits. Normally, there are 1.2 mEq of carbonic acid to every 24 mEq of base bicarbonate.

It is important to note that *absolute* quantities of carbonic acid and base bicarbonate are not important in maintaining the balance; it is the *relative* quantities that are important. For example, acid–base balance will not be disturbed if base bicarbonate is increased, perhaps doubled, as long as the carbonic acid is also increased by the same factor; or, both can be decreased by the same factor without upsetting the balance. Imbalances result only when the normal 1:20 ratio is upset.

Think of acid–base balance as a teeter-totter, with carbonic acid on one end and base bicarbonate on the other. In health, the teeter-totter is level, but any condition that increases carbonic acid or decreases base bicarbonate tilts the teeter-totter toward the carbonic acid side and causes acidosis, or acidemia; any condition that increases base bicarbonate or decreases carbonic acid tilts the teeter-totter toward the base bicarbonate side and produces alkalosis, or alkalemia.

The balance can be tilted by two general types of body disturbances: one type adds or subtracts base bicarbonate and the other type adds or subtracts carbonic acid. Body metabolism affects the base bicarbonate side of the balance and, for this reason, imbalances caused by alterations in base bicarbonate concentration are called *metabolic disturbances* of acid–base balance. The kidneys largely regulate base bicarbonate concentration. When the H^+ ion concentration is excessively high (*low* pH), the kidneys increase the excretion of hydrogen by several ingenious chemical mechanisms, and they save and regenerate bicarbonate. When the H^+ ion concentration drops to an abnormally low level, the kidneys cease their excretion of hydrogen and permit excessive bicarbonate to pass in the urine.

The amount of carbon dioxide blown off by the lungs affects the carbonic acid side of the balance, since by speeding up or slowing down respiration, the lungs increase or decrease the level of carbonic acid in the ECF. When hydrogen accumulates excessively in the body, the lungs simultaneously speed up respiration and blow off carbon dioxide. Recall that CO_2, being the equivalent of H_2CO_3, is an acid material. When, on the other hand, the body lacks hydrogen, the lungs slow down respiration and retain CO_2. The lungs' action in blowing off CO_2 is more vigorous than its role in retaining CO_2. Neither may occur in newborns, whose homeostatic controls are imperfectly developed. Pulmonary ventilation has a rapid effect on the pH of the body fluids. If you hold your breath for one minute, your ECF will become acidotic. If, on the contrary, you hyperventilate actively for one minute, your ECF will become alkalotic. When lung function is abnormal, acid–base balance is disturbed, causing the resultant imbalances referred to as *respiratory disturbances* of acid–base balance.

Pulmonary loss of hydrogen ions by exhalation of CO_2 and H_2O is enormous: 14,000 mEq/day. The kidneys excrete only 60 mEq to 90 mEq of hydrogen each day, but they reabsorb 5100 mEq of bicarbonate ions daily by mechanisms we shall examine. These are the figures in health; the quantities excreted in acid–base disturbances vary widely from these. Of fundamental importance is this axiom: *the primary concern of the lungs is regulation of carbonic acid (i.e., CO_2 and H_2O); that of the kidneys is regulation of bicarbonate.*

Let us now examine in some detail the ingenious mechanisms by which the kidneys deal with disturbances in acid–base balance. Although the lungs' role in coping with these disturbances is relatively simple, the same cannot be said for the role of the kidneys! (One speaker, referring to the roles of lungs and kidneys in contending with acid–base disturbances, quipped, "The lungs are dumb but the kidneys are extremely smart.") The 60 mEq to 90 mEq of hydrogen that the kidneys excrete each day is derived from fixed acids, which cannot be broken down in the forthright manner in which carbonic acid is changed to carbon dioxide and water. The first of these fixed acids is phosphoric acid, originating in the body's phosphates. It reacts with sodium bicarbonate to form disodium hydrogen phosphate, carbon dioxide, and water. Later, we shall see the fate of disodium hydrogen phosphate. Sulfuric acid, on the other hand, comes from sulfur-containing amino acids. It reacts with sodium bicarbonate to form sodium sulfate, carbon dioxide, and water. How sodium sulfate is disposed of will be explained below.

Regulation of acid–base balance by the kidneys consists of three steps: bicarbonate stabilization; formation of titratable acid; and ammonia production. First let us consider renal stabilization of bicarbonate, which, in turn, can be divided into three separate processes: reabsorption of filtered bicarbonate; excretion of excess bicarbonate; and replenishment of depleted bicarbonate by excretion of titratable acid and by excretion of ammonia.

Bicarbonate reabsorption occurs in the following way: with a glomerular filtration rate of 200 liters/day, approximately 5200 mEq of bicarbonate are filtered through the kidneys. Only 2 mEq are excreted, leaving 5198 mEq to be reabsorbed. Bicarbonate in excess of the renal threshold is simply excreted in the urine. When the body's bicarbonate is partially depleted, it is replenished. This process is quite different from reabsorption of filtered bicarbonate, and it involves, as we shall see, excretion of hydrogen as titratable acid and as ammonia. Reabsorption of bicarbonate occurs in the proximal tubular cells. The process is shown in Figure 3-1.

In the distal convoluted tubules, quite a different

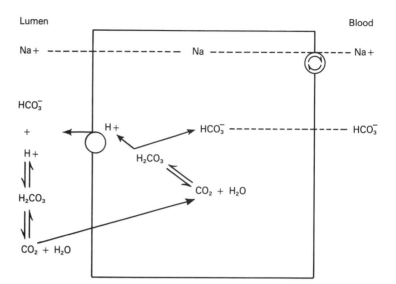

FIG. 3-1.
Proximal tubular cell.

series of events occurs. Here, disodium hydrogen phosphate is converted to sodium dihydrogen phosphate, contributing sodium to the tubule cell and removing hydrogen. Beginning with the raw materials of metabolite and oxygen, bicarbonate is regenerated, passing along with sodium into the peritubular capillaries. This is illustrated in Figure 3-2.

Figure 3-3 shows the breakdown of amino acids to ammonium, ammonia, and hydrogen within the tubule cells. When the urine is acid, hydrogen combines with ammonia, forming ammonium, which then passes out in the urine and provides an exit for excess hydrogen. This mechanism is particularly important when the kidneys' capacity to excrete titratable acid has been exceeded. When, on the other hand, the urine is alkaline, the reaction shifts to the right, with ammonium accumulating in the peritubular capillaries. Sulfur-containing amino acids are the source of urinary sulfates, accompanied by various cations, including sodium, potassium, ammonium, or hydrogen.

FIG. 3-2. Distal tubular cell.

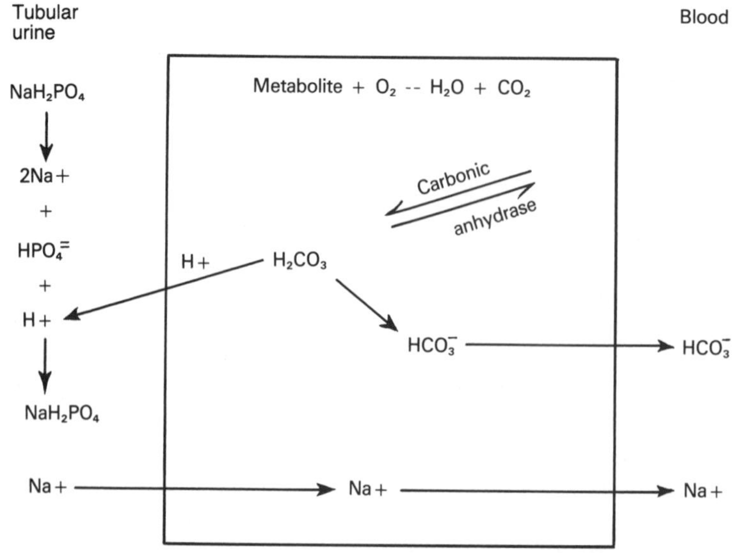

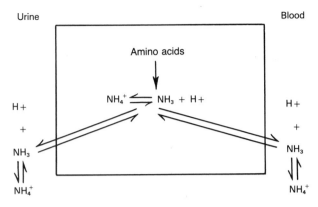

FIG. 3-3. Breakdown of amino acids.

MEASUREMENT OF ACID–BASE BALANCE

To determine the status of one's acid–base balance, it is necessary to study the blood gases. Usually when one speaks of blood gases, he is referring to oxygen and carbon dioxide. However, sometimes two additional laboratory values—pH and bicarbonate—are referred to as blood gases, since they are intimately involved in acid–base balance: pH, of course, represents the value for H^+, which is a gas and which deserves to be designated a blood gas. Bicarbonate, on the other hand, represents the chief alkaline, or basic, constituent of ECF and is a chemical, not a gas. Nevertheless, bicarbonate is so intimately concerned with the three blood gases already mentioned that it is convenient (if incorrect) to classify it among the blood gases.

The pH of the plasma, of course, tells us whether we have acidosis or alkalosis, although it does not indicate the nature of the imbalance, that is, whether it is metabolic or respiratory. pH can be measured directly by a pH meter or with a colorimeter, using either heparinized arterial blood or "arterialized" blood from a warmed ear lobe or fingertip. Blood from a vein can be used if it is promptly drawn and kept stoppered under oil.

Carbonic acid—really CO_2 + HOH—can be measured by a PCO_2 meter; from this, the amount of carbonic acid can be determined by use of a nomogram. Normal PCO_2 values range from 38 mm Hg to 42 mm Hg if arterial blood is used, and from 40 mm Hg to 41 mm Hg if venous blood is used.

Various laboratory tests can be used to measure the bicarbonate level of the plasma, including the carbon dioxide (CO_2) content, CO_2 capacity, and CO_2 combining power. The names of these tests can be misleading unless one recalls that what is being measured is not the carbon dioxide of plasma but, rather, the bicarbonate. In all these tests, bicarbonate is treated with sulfuric acid, which causes it to release carbon dioxide. The carbon dioxide is then measured, and after an adjustment has been made to allow for the amount of carbon dioxide dissolved in the plasma, the bicarbonate concentration is calculated on the basis of the carbon dioxide released from the bicarbonate as a result of the chemical treatment.

The test for CO_2 content measures the total carbon dioxide freed when plasma is acidified, representing not only the carbon dioxide derived from bicarbonate—which is a measure of plasma alkali—but also carbon dioxide in the form of dissolved CO_2 or carbonic acid—which measures the acid portion of the plasma. The normal CO_2 content ranges from 24 mEq/liter to 33 mEq/liter. If there is a respiratory acid–base disturbance, the CO_2 content will not represent an accurate measure of the bicarbonate. In carbonic acid excess (respiratory acidosis), because of the CO_2 retention by the lungs, the carbonic acid is elevated; in carbonic acid deficit (respiratory alkalosis), because of the blowing off of CO_2, it is depressed. There are tests, however, that give one a more accurate estimate of the plasma bicarbonate, even in the presence of a respiratory disturbance; these include the CO_2 capacity test and the CO_2 combining power test.

In these tests, the actual carbonic acid and carbon dioxide concentration of the plasma is adjusted to normal either by bubbling a 5.5% carbon dioxide gas mixture through the plasma (CO_2 capacity test) or by having the laboratory technician blow his breath through it (CO_2 combining power test). Thus, if the test plasma's CO_2 is elevated because of respiratory acidosis, bubbling the normal concentration of CO_2 through it will reduce it to normal, a process known as *equilibration*. If the plasma's CO_2 is depressed because of respiratory alkalosis, then equilibration will raise it to normal. In essence, equilibrating plasma amounts to recirculating the plasma through the lungs of a normal person so that it will achieve its normal CO_2-carbonic acid level. The result is that either the CO_2 capacity test or the CO_2 combining power test gives a fairly close approximation of the test specimen's bicarbonate. Normally, CO_2 capacity varies from 24 mEq to 33 mEq/liter, and CO_2 combining power varies from 24 mEq to 35 mEq/liter (depending on reference source).

The CO_2 content, bicarbonate, pH, P_{CO_2}, and carbonic acid content can all be calculated by use of a nomogram if two of the values are known. A straight line is drawn between the two known points and the desired value read from the other three scales.

Considerably more complex and somewhat more accurate methods of measuring acid–base disturbances have been introduced by Astrup, by Siggard–Andersen, and by Kintner, all involving logarithmic relationships. They are neither easy for the nonmathematician to visualize nor necessary for the usual clinical purposes.

SINGLE ACID–BASE DISTURBANCES

PRIMARY BASE BICARBONATE DEFICIT OF ECF

In their severe form, all acid–base disturbances can threaten life. The first disturbance we shall discuss, usually referred to as *metabolic acidosis,* can be defined as decreased arterial pH resulting from a decrease in bicarbonate (hence our preference to refer to this imbalance as *bicarbonate deficit*). Usually accompanying the imbalance is a decreased arterial P_{CO_2}.

Metabolic acidosis occurs when bicarbonate is titrated by endogenous acids (such as lactic acid or ketones) or exogenous acids (such as methanol, ethylene glycol, or salicylates). It can also occur with decreased acid excretion in acute or chronic renal disease or from excessive loss of bicarbonate in the stools of diarrhea.

The body defends against the disorder by a triple compensatory mechanism consisting of chemical buffering, increased alveolar ventilation, and renal compensation. The first of these mechanisms, all of which minimize shifts in arterial pH, is chemical buffering. In acute bicarbonate deficit, about 40% of the chemical buffering involves titration of acid by the serum bicarbonate. Titration of the proteins of red blood cells, soft tissues, and bone contributes the other 60% of chemical buffering.

The buffering comes to the defense of the body immediately. Just as quickly, there is increased alveolar ventilation, decreasing the P_{CO_2}, excreting CO_2 and

water through the lungs, and softening the pH shift. But this crucial help is not maximal until 12 to 24 hours after the onset of acidosis. Renal ammonium excretion increases to some extent in acute bicarbonate deficit but is considerably more important in chronic acid loading.

The physiologic consequences of acidosis deserve attention. In severe acute bicarbonate deficit—pH 7.1 or less or a bicarbonate of 5 mEq/liter or less—myocardial contractility decreases. In less severe acidosis, pulmonary vascular resistance is increased, and peripheral vascular resistance is decreased. Kussmaul respiration, with an increase in both rapidity and depth of respiration, accomplishes increased alveolar ventilation. There are two additional physiologic consequences of import: first, there is a shift in the hemoglobin dissociation curve to the left, with increased release of oxygen. Second, there is a decrease in the 2, 3 diphosphoglycerate (DPG) of erythrocytes. This decrease causes the dissociation curve to reverse itself. Decreased unloading of oxygen results. For this reason, rapid correction of acidosis could be hazardous, especially in patients already suffering from anoxia. In addition, central nervous system (CNS) symptoms, such as stupor, seizures, and coma, are sometimes seen. These are usually due not to the acidosis, but to accompanying fluid balance disturbances. Lethargy and confusion do result from the acidosis, and a pH under 7.1 can bring hypotension and shock.

What help does the lab give us? Urine pH is not particularly helpful, but serum pH is depressed, depending upon the success of compensation, and the plasma bicarbonate falls below 24 mEq/liter. P_{CO_2}, in compensation, drops below 40 mm Hg; for uncomplicated metabolic acidosis, one can expect a reduction of 1.2 mm Hg to 1.5 mm Hg in the P_{CO_2} for each reduction of 1 mEq of bicarbonate. The plasma potassium deserves special interest because in acidosis, potassium tends to move from cells to ECF. Should the cause of the acidosis remain in doubt, such determinations as the blood urea nitrogen (BUN), creatinine, urinalysis, serum glucose, lactate, ketones, salicylates, ethylene glycol, and methanol should be considered.

Recently, much interest has focused on the unmeasured anions of the serum in bicarbonate deficit. Sometimes referred to as the *anion gap,* sometimes as the *Delta,* the value is derived by subtracting the sum of the chloride and bicarbonate in mEq/liter from the sodium in mEq/liter. A Delta of 8 mEq to 16 mEq/liter is regarded as normal. Acidosis can occur with these values, but it is an acidosis due to inorganic acids. If, on the other hand, the Delta is above 16, there is an excellent chance that the acidosis is caused by organic acids. A Delta of 20 or more *strongly* favors an organic acid acidosis. Normal Delta acidosis include bicarbonate deficit from severe diarrhea; ureteroenterostomy, if conditions are such that chloride is absorbed; renal tubular acidosis; and hyperalimentation, because of the high cation content of amino acids. Abnormal Delta, or anion gap, acidoses include certain types of renal failure; poisoning by such substances as methanol, ethylene glycol, and salicylates; diabetic ketoacidosis; and lactic acidosis caused by tissue hypoxia or hypotension.

Diagnosis rests, first of all, on a careful history and physical examination. Laboratory findings fitting the pattern for metabolic acidosis provide helpful confirmation. Delta over 20 is an almost sure sign of bicarbonate deficit caused by organic acids.

In chronic bicarbonate deficit, oral alkali therapy with bicarbonate may be all that is required. Certainly with a pH higher than 7.2, parenteral bicarbonate is not required. In administering bicarbonate parenterally, the serum bicarbonate should be elevated gradually and—initially at least—to not above 10 mEq/liter. Body weight determined in kilograms is useful in arriving at the dosage. Depending upon the severity of the acidosis, from 40% to 100% of the body weight in kilograms should be multiplied by the desired increment in bicarbonate per liter. Example: a patient has a serum bicarbonate of 5 mEq/liter. It is desired to raise it to 10. The person weighs 60 kg. 60×5 (the desired increment in the bicarbonate) = 300—that is, 300 mEq of bicarbonate. As the acid–base balance is gradually restored, the serum electrolytes, especially potassium, and blood gases should be frequently checked. If the patient's pH is 7 or less, maximal doses of bicarbonate should be given. Under such a circumstance, much of the administered bicarbonate is required to buffer cellular hydrogen ions. As the pH values change, one should closely watch the potassium. A shift of 0.1 of a pH will usually change the potassium in the opposite direction by about 0.5 mEq/liter. Some patients with bicarbonate deficit already have hypokalemia. Moreover, since potassium shifts from the cells into the ECF in bicarbonate deficit, the hypokalemia is undoubtedly more serious than the serum reading indicates. In giving bicarbonate, one should also give potassium, since the addition of bicarbonate and the elevation of the pH will further depress the already low potassium.

Respiratory paralysis due to potassium deficit has been reported in patients given bicarbonate without potassium. Some patients with severe acidosis, especially those with life-threatening poisoning, should be considered for hemodialysis or peritoneal dialysis (Fig. 3-4).

CASE HISTORIES S. T., 18 years old and suffering from diabetes, was admitted to the hospital because of uncontrolled diabetes mellitus. Two days before admission, she had developed a fever and sore throat. At the same time, she found glucose when she tested her urine. The day before admission, she had ingested little except water but kept up her usual insulin dose. During the 8 hours before admission, she complained of abdominal pain, vomited several times, and became increasingly lethargic. She had urinated at frequent intervals. Physical examination revealed a lethargic young woman with rectal temperature 103°F (39.4°C); pulse 110; respirations 30, of Kussmaul type (deep and labored); blood pressure 110/75; acetone odor to the breath; dry skin and mucous membranes; acute pharyngitis; weight 115 lb (52.5 kg), usual weight 120 lb. Laboratory tests revealed white blood cell count (WBC) 20,000, with 85% polymorphonuclear cells; blood sugar 300 mg-dl. Serum electrolytes were sodium 138 mEq/liter, potassium 5 mEq/liter, chloride 98 mEq/liter, bicarbonate 10 mEq/liter. P_{CO_2} was 26 mm Hg. Urine revealed pH 5.5, specific gravity 1.035, sugar 4+, acetone present, acetoacetic acid present. The diagnosis was bicarbonate deficit (metabolic acidosis or diabetic ketoacidosis) and ECF volume deficit. Calculating the Delta, we arrive at a figure of 30, indicating acidosis caused by organic acids.

Commentary As is so often the case, we have not just one, but two fluid imbalances—bicarbonate deficit and ECF volume deficit. In treatment, they are managed simultaneously. In this instance, insulin and antibiotics would be used in addition. We see in this case how extremely useful Delta can be in telling us whether the acidosis is due to inorganic or organic acids.

Two-and-one-half-year-old R. T. drank an undetermined quantity of oil of wintergreen (methyl salicylate). He vomited and screamed, holding his hands over his abdomen. In the emergency room of the hospital, he was immediately gavaged, then admitted to the pediatric ward, where he perspired profusely, held his head as if in pain, became restless and excited, then went into convulsions. His pulse was 120/min and feeble. He was pale and appeared

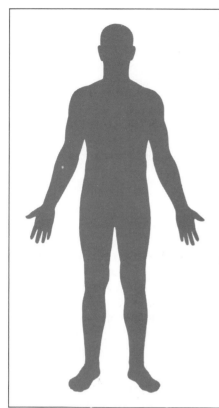

PRECEDING EVENTS

- *Flooding of ECF with acid, endogenous or exogenous*
- *Clinical situations: high delta (increased anion gap above 16 mEq/liter)*
  - *Diabetic ketoacidosis*
  - *Uremic acidosis*
  - *Ingestion of toxins or drugs (e.g., salicylates, ethanol [ethyl alcohol], methanol [methyl alcohol], formic acid, paraldehyde, boric acid, ethylene glycol, oxalic acid)*
  - *Lactic acidosis*
  - *Starvation ketoacidosis*
- *Clinical situations: normal delta (anion gap 16 mEq/liter or below)*
  - *Diarrhea*
  - *Renal tubular acidosis*
  - *Dilutional acidosis*
  - *Acidifying drugs*
  - *Hyperalimentation*
  - *Acetazolamide*

CLINICAL OBSERVATIONS

- *Hyperventilation*
- *Weakness*
- *Disorientation*
- *Coma*

LABORATORY FINDINGS

- *Urine pH often below 6*
- *Plasma pH below 7.35*
- *Plasma bicarbonate below 24 mEq/liter*
- *P_{CO_2} below 40 mm Hg*

RELATED CONSIDERATIONS

- *Body correction, both renal and pulmonary*
- *Often associated with potassium excess*

FIG. 3-4. **Primary base bicarbonate deficit (metabolic acidosis).**

short of breath. Blood pressure was 100/60. His serum HCO_3 was 12 mEq/liter, P_{CO_2} 26 mm Hg, pH 7.2. Urine pH was 5, and the urine tested positive for salicylates. The diagnosis was severe bicarbonate deficit caused by salicylate poisoning.

Commentary Salicylate poisoning is usually seen in infants and children, but it can be observed in adults. Salicylates produce two opposed acid–base disturbances. First, stimulation of the respiratory center of the brain produces pronounced hyperventilation and primary carbonic acid deficit (respiratory alkalosis). Within one hour or longer, the salicylate disrupts carbohydrate metabolism and causes depletion of liver glycogen. Lactic and pyruvic acids accumulate and cause primary base bicarbonate deficit (metabolic acidosis). Thus, there is a shift from respiratory alkalosis to metabolic acidosis, especially pronounced in children under 5 years of age. Treatment is directed at removing the salicylates from the stomach, if possible, and in correcting the body fluid disturbances. In severe salicylate intoxication, hemodialysis or peritoneal dialysis can prove lifesaving.

PRIMARY BASE BICARBONATE EXCESS OF ECF

A primary excess of base bicarbonate, usually referred to as *metabolic alkalosis* or *alkalemia*, can result from any clincial event that increases the amount of base bicarbonate in the ECF. The causes include loss of acid through the kidneys because of primary or secondary hyperaldosteronism or loss of acid and chloride from nasogastric suction or prolonged vomiting. Gain of HCO_3 from exogenous sources is an uncommon cause because excess HCO_3 is usually promptly excreted by the kidneys. Nevertheless, the milk–alkali syndrome can occur.

A trio of body compensatory mechanisms minimizes shifts in pH: chemical buffering, respiratory hypoventilation, and —most significant—renal correction of generated or administered alkali. Chemical buffering constitutes about one third of the buffering against bicarbonate excess. Bicarbonate reacts with the cellular hydrogen that moves into the plasma; the bicarbonate is titrated with production of HOH and CO_2.

In most instances, hypoventilation occurs in compensation, usually resulting in an increase of 0.5 mm Hg to 0.7 mm Hg in P_{CO_2} for each increase of 1 mEq/liter of HCO_3. Respiratory compensation through retention of CO_2 is probably as effective as the opposite compensation that we saw in bicarbonate deficit; yet, the arterial P_{CO_2} does not usually exceed 60 mm Hg. The explanation? Hypoventilation is limited by the fall in the P_{O_2}. Nevertheless, some patients' P_{CO_2} levels have risen to 70 mm Hg, with later return to normal. The compensation for bicarbonate excess relies primarily on the kidneys, however. First, we must understand why sometimes the kidney promptly excretes excessive HCO_3 and why sometimes it clings tenaciously to the HCO_3: the most important of several factors affecting the HCO_3 level are the ECF volume and the potassium concentration. The level of P_{CO_2} and the sodium concentration are also significant. If you are able to neutralize the factors that *maintain* the alkalosis, you will correct the alkalosis, although you do not remove the underlying cause.

Consider the patient with alkalosis who has primary aldosteronism plus mild volume expansion: his alkalosis is generated by his kidneys. But it is the decrease in his serum potassium that maintains the alkalosis. If the serum potassium is returned to normal, the alkalosis is corrected and further rise in the serum bicarbonate is prevented.

Now take the patient with secondary hyperaldosteronism caused by volume depletion, nasogastric suction, diuretic administration, or Bartter's syndrome (primary juxtaglomerular cell hyperplasia with secondary hyperaldosteronism). The generation of alkalosis stems either from gastric or renal acid losses, alkalosis being maintained primarily by volume depletion but secondarily by potassium deficit. When the volume depletion is corrected, the potassium deficit is repaired.

The posthypercapnic state deserves attention. Patients with this problem enter the hospital with chronic P_{CO_2} elevation and high HCO_3. Their primary lung condition is either corrected or they are given assisted ventilation; the ventilation relieves the hypercapnia, but the HCO_3 remains elevated. This high HCO_3 has, of course, been generated by renal compensation and is largely maintained by occult volume depletion. Administration either of sufficient sodium chloride or of acetazolamide increases the HCO_3 excretion and corrects this type of alkalosis.

You may see patients with edema caused by congestive heart failure, nephrotic syndrome, or cirrhosis who have received diuretics. Already such patients have some degree of secondary hyperaldosteronism with a decrease in circulating blood volume. Action of the diuretic now moves the tubular fluid to the distal exchange site, and sodium is exchanged for potassium: hypokalemia results. Next, sodium is exchanged for hydrogen, and alkalosis occurs. The alkalosis would not have occurred without the diuretic.

Clinical findings are largely nonspecific, although generalized hypertonicity has been cited. Probably most symptoms result from coexisting imbalances. When volume depletion is present, it may cause postural symptoms and weakness; the almost invariably present potassium depletion is responsible for polyuria, polydipsia, and muscle cramps.

Laboratory findings are helpful. The serum pH and bicarbonate are increased. The P_{CO_2} rises but seldom goes above 62 mm Hg. As a rule, one can expect a rise of 0.5 mm Hg to 0.7 mm Hg in the P_{CO_2} for each 1 mEq increase in the HCO_3. Serum potassium and serum chloride are almost always decreased. Perhaps the most useful laboratory finding in bicarbonate excess is the urinary chloride, with important implications both for diagnosis and treatment. Urinary chloride is an excellent indicator of the ECF volume, provided the patient has not been given a diuretic within the preceding 48 hours.

Patients who are volume depleted will have brisk sodium reabsorption to maintain circulating blood volume, the body's first priority. But when the cation sodium is reabsorbed, an anion must be reabsorbed with it to maintain electrical neutrality. Anions in the glomerular filtrate consist of chloride and bicarbonate. Bicarbonate is not as reabsorbable as chloride, hence chloride is reabsorbed, with a resultant decrease in urinary chloride excretion. The therapeutic correlary is that administration of sodium chloride will repair the ECF volume deficit and correct the alkalosis.

Consider the opposite situation: patients with an ECF volume excess, such as those with primary hyperaldosteronism, have an increased or normal urinary sodium excretion. Their chloride excretion is also increased or normal, paralleling the sodium. Such patients will not have their alkalosis corrected by saline administration since their alkalosis is not maintained by volume depletion. Rather, the alkalosis is maintained by potassium depletion and an excess of aldosterone. Thus we see that the urinary chloride tells us not only the cause of the alkalosis, but also what to do to correct it. We shall get into specifics later.

As always in acid–base disturbances, a thorough history and physical examination rate top priority in diagnosis. Laboratory findings consonant with metabolic alkalosis provide needed confirmation. Recall that the P_{CO_2} will usually be elevated, but only rarely above 62 mm Hg.

If the patient with metabolic alkalosis is volume depleted, reexpansion of ECF volume with normal saline allows the kidneys to excrete the retained bicarbonate and correct the acid–base imbalance. A useful guide: when the urinary chloride has risen to 20 mEq/liter, enough saline has probably been given.

There are some patients, however, whose alkalosis is maintained not by ECF volume deficit, but by mineralocorticoid (chiefly aldosterone) excess plus potassium deficit. For such patients, one must replace the potassium and, if primary hyperaldosteronism is present, give the competitive antagonist of aldosterone, spironolactone (Aldactone) while evaluating the patient for possible surgery (if he has a discrete tumor [Conn's Syndrome]) of the adrenal cortex. To correct the potassium deficit corrects the alkalosis, but this can be difficult simply because not enough potassium is given or because the potassium is excreted in the urine. (Recall that aldosterone promotes excretion of potassium.) Nursing responsibilities in potassium administration are discussed in Chapter 6. A minimum of sodium should be given: if potassium is administered in saline, then both potassium and sodium are delivered to the exchange site in the kidneys where aldosterone acts to foster exchange of sodium for potassium, with the latter lost in the urine.

There are special situations: patients with edema from congestive heart failure, cirrhosis, the nephrotic syndrome, or chronic obstructive pulmonary disease with cor pulmonale should not be given saline because they will retain it and become even more edematous. Moreover, the saline will not improve the renal capacity to excrete HCO_3. In such cases, one should try to correct the cause of the edema and administer potassium.

One drug that is often extremely useful is acetazolamide (Diamox), a carbonic anhydrase inhibitor that enhances renal excretion of HCO_3 and, in effect, causes renal tubular acidosis; 250 mg two or three times daily is often effective.

Another special problem is presented by the patient with acute postoperative renal insufficiency that sometimes occurs following gastric operation or after constant nasogastric suction necessitated by prolonged ileus. The disabled kidneys cannot excrete bicarbonate, so its serum level mounts rapidly, causing severe alkalosis with depressed ventilation, hypocapnia, and hypoxia. The chances of postoperative complications increase greatly. To meet this emergency, one can give cimetidine (Tagamet), which decreases the amount of gastric acid production. This, of course, decreases the hydrogen and chloride lost through vomiting or nasogastric suction. Another expedient is to give dilute hydrochloric acid through a central venous catheter. In desperate situations, consideration should be given to hemodialysis, using a high-chloride, low-acetate bath. This measure corrects the metabolic alkalosis while countering the renal insufficiency (Fig. 3-5).

CASE HISTORIES L. K., 20 years old and 4 months pregnant, was admitted to the hospital because of severe vomiting, with a presumptive diagnosis of hyperemesis gravidarum. She was fed a general diet, most of which she vomited. About all she was able to keep down was a little water. She was given a vitamin-mineral supplement but no parenteral fluid therapy. On the fourth day in the hospital, she exhibited tetanic movements of the fingers. Physical examination revealed that her muscles were hypertonic, and the physician believed that her respirations were suppressed. Laboratory tests revealed normal urine and these serum findings: sodium 135 mEq/liter, potassium 3.3 mEq/liter, HCO_3 38 mEq/liter, P_{CO_2} 30 mm Hg, pH 7.65, and chloride 94 mEq/liter. Diagnosis was bicarbonate excess (metabolic alkalosis) with sodium deficit and potassium deficit. The bicarbonate excess can be explained on the basis of the loss of chloride and potassium in the vomitus.

Commentary Chloride loss promotes bicarbonate excess since the bicarbonate rises in compensation as chloride is lost. (Total cations must always equal total anions.) Potassium deficit always tends to favor alkalosis. Gastric juice, you will recall, contains generous quantities of potassium and bicarbonate.

J. S., age 45, felt he could cure his peptic ulcer by taking generous doses of bicarbonate of soda (baking soda) several times a day. After a few weeks on this program, he became so tense and jittery that he consulted his physician. The doctor found him to be a tense, jittery person with hyperactive deep reflexes, suppressed respiration, and radiographic evidence of a large gastric ulcer. Labo-

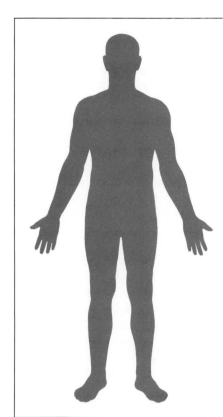

PRECEDING EVENTS

- *Loss of chloride, with compensatory bicarbonate rise*
- *Excessive intake of alkalies*

CLINICAL OBSERVATIONS

- *Numbness and tingling of extremities*
- *Hypertonicity of muscles (calcium ionization decreased)*
- *Slow, shallow respiration, with periods of apnea (compensatory)*
- *Bradycardia*
- *Tetany*

LABORATORY FINDINGS

- *Urine pH sometimes above 7*
- *Plasma pH above 7.45*
- *Plasma bicarbonate above 24 mEq/liter*
- *Pco$_2$ above 40 mm Hg*

RELATED PROBLEMS

- *Body correction, both renal and pulmonary*
- *Often associated with potassium deficit*

FIG. 3-5. Primary base bicarbonate excess (metabolic alkalosis).

ratory findings showed urine pH 7, otherwise normal. Serum electrolytes were: sodium 146 mEq/liter, chloride 93 mEq/liter, HCO$_3$ 36 mEq/liter, potassium 3.6 mEq/liter, Pco$_2$ 57 mm Hg, and pH 7.65. The HCO$_3$ was elevated because of the ingestion of baking soda, the Pco$_2$ was elevated as a pulmonary compensatory measure (which was why respirations were suppressed), chloride was lowered because the bicarbonate was elevated (total cations must always equal total anions), potassium was lowered because of the effect of alkalosis in causing potassium deficit, the the serum pH indicated alkalosis.

Commentary Baking soda now sees little use as an antacid, since it is absorbed and causes systemic alkalosis. Rather, nonabsorbable alkalies are employed for oral treatment of peptic ulcer.

COMMENT While the acid side of the acid–base balance goes up and down in the two imbalances just described, it is important to remember that base bicarbonate is the cause, and changes in the carbonic acid side occur only secondarily to alterations in the concentration of base bicarbonate. In the next two imbalances, the carbonic acid side becomes the cause, whereby changes in the base bicarbonate side become secondary to alterations in the concentration of carbonic acid.

PRIMARY CARBONIC ACID DEFICIT OF ECF

Carbonic acid deficit always results from hyperventilation, characterized by increased frequency of breathing, increased depth of breathing, or both. It can be voluntary, as in the case of the swimmer, unaware of the great danger, who desires to swim as far as possible underwater. It can also be caused by pain, bicarbonate deficit, carbonic acid excess, or from poisoning, as by salicylates or aspirin. The extensive list of causes includes fever, various CNS lesions, hepatic coma, and overly energetic assisted ventilation.

Perhaps the most important cause of hyperventilation, hence of carbonic acid deficit, is the so-called hyperventilation syndrome. This exceedingly common, frequently disabling ailment with no known organic cause is frequently ignored by physicians. Patients with the syndrome often go from one physician to another, ending up with a diagnosis of neurosis, with the hyperventilation unrecognized. One authority suspects that the hyperventilation syndrome may well be the most commonly missed diagnosis in internal medicine. The explanation may be that physicians do not think of it or that the textbook picture of overbreathing, paresthesia, and tetany is not always present.

Hypoxemia, such as occurs when sea-level dwellers climb to an altitude of over 10,000 feet, can stimulate hyperventilation. Cirrhosis, congestive heart failure, and pulmonary disorders such as pneumonitis can stimulate pulmonary stretch receptors and trigger hyperventilation.

As hyperventilation continues, both peripheral and cerebral arterial Pco_2 fall. The pH of both plasma and cerebrospinal fluid rises. Because of the Bohr effect, cerebral vasoconstriction takes place, bringing on cerebral hypoxia and the clinical findings of carbonic acid deficit.

Not all hyperventilating persons increase both the frequency and depth of breathing: perhaps half of those with the syndrome do not have an abnormally rapid respiratory rate. The patient feels light-headed or dizzy. Periodically, he may become breathless, feeling impelled to fill the lungs more fully. He may have blurred vision, tingling of the extremities, or numbness. Sometimes tetany occurs. The mouth may become dry and the patient may yawn, followed by epigastric distress and air swallowing. Many hyperventilators have chest wall pain, detected by palpation. The electrocardiogram (EKG) may reveal ST depression, for reasons unclear.

The presenting complaint may be chronic exhaustion. Other symptoms that may be elicited include tremors, stiffness, palpitation, tachycardia, precordial pain (imitating angina pectoris), psychological tension, sleeplessness, and nightmares. In some 75% of patients, deliberate overbreathing in the physician's presence will reproduce the clinical symptoms. If this occurs, a significant step has been taken in successful management of the condition. This procedure may not take more than a few deep breaths, or it may require rapid breathing for a minute or more. Strangely, hyperventilation by persons not suffering from the syndrome usually produces few or no symptoms.

The serum pH is elevated and the Pco_2 depressed. The kidneys endeavor to compensate for the imbalance by permitting bicarbonate to escape. In the case of acute respiratory alkalosis, for each 10-mm Hg decrease in Pco_2, we can expect a 2 mEq to 3 mEq/liter decrease in bicarbonate. If the condition is chronic, there will probably be a 5 mEq to 6 mEq/liter decrease in bicarbonate. The blood lactate and pyruvate levels increase, and the ionized calcium falls.

A careful history and physical examination usually provide the clues for the diagnosis. Laboratory findings indicated above are corroborative.

The chief goal of therapy is to correct the condition or situation causing the hyperventilation. Sedation or oxygen may prove useful in appropriate patients. For the patient whose hyperventilation is of psychic origin, convincing him of this fact is the cardinal step in treatment. He may have to be reminded from time to time, by both his physician and his friends. Follow-up visits will almost always be advisable (Fig. 3-6). Parenteral infusion of a solution containing chloride ions to neutralize bicarbonate may be helpful in restoring balance while the pulmonary problem is being corrected.

Related problems include voluntary hyperventilation before swimming underwater, an exceedingly dangerous practice often resorted to by swimmers trying to set distance records. Why is it so dangerous? When one hyperventilates, he blows off CO_2, the chief stimulus to respiration (it is not oxygen lack, as is widely believed). Therefore, he can go a long time (perhaps a minute or two or three) without receiving an impelling urge to breathe. Obviously, this makes long swims underwater possible. So, the hyperventilated underwater swimmer moves along blissfully, quite pleased with his ability to stay under so long. But in many instances (hundreds? thousands?), the swimmer's brain becomes anoxic before he has a strong urge to breathe; he becomes unconscious, yet still swims automatically along—until, that is, his CO_2 builds up and stimulates the medulla to cause inhalation. Then he inhales water, not air. If the swimmer is observed and rescued at this point, he has a 50% chance of being resuscitated; the chances are not greater because his lungs are full of water. (Some believe that the vigorous expulsion of water from the lungs by the justly famous "hug of life," or Heimlich maneuver, might improve the swimmer's chances, and this impresses us as reasonable.) The moral: *don't hyperventilate before swimming underwater!*

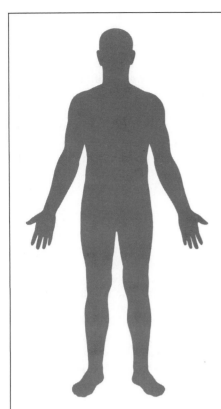

PRECEDING EVENTS

- *Hyperventilation from any cause*

CLINICAL OBSERVATIONS

- *Deep, rapid breathing (hyperventilation)*
- *Tetany (calcium ionization decreased)*
- *Paresthesia*
- *Tingling and numbness of extremities and circumoral region*
- *Inability to concentrate*
- *Tinnitus*
- *Blurred vision*
- *Sweating*
- *Dry mouth*
- *Coma*

LABORATORY FINDINGS

- *Urine pH above 7*
- *Plasma pH above 7.45*
- *Plasma bicarbonate below 24 mEq/liter*
- *P_{CO_2} below 40 mm Hg*

RELATED PROBLEMS

- *Body compensation must depend on kidneys alone; therefore, slow*
- *Voluntary hyperventilation before swimming underwater deadly*

FIG. 3-6. Primary carbonic acid deficit (respiratory alkalosis).

CASE HISTORIES Eighteen-year-old M. S. was admitted to the hospital for study because of tetanic spasms of the fingers and occasional convulsions. The medical history revealed that she had been extremely apprehensive because of her fear of failing in school. Her mother had noticed that she had been breathing deeply and rapidly for several weeks. Physical examination revealed a tense young woman with hyperactive reflexes. Laboratory tests showed a serum with pH 7.6, HCO_3 20 mEq/liter, P_{CO_2} 25 mm Hg, and urine normal. The serum pH indicated alkalosis, the HCO_3 indicated compensatory lowering of the HCO_3 by the kidneys, and P_{CO_2} was lowered by the hyperventilation.

Commentary By having a patient with suspected neurotic hyperventilation deliberately hyperventilate, one can usually reproduce the symptoms of which the patient has been complaining and, thus, corroborate the diagnosis.

J. N., age 16 years, was a victim of bulbar poliomyelitis and was placed on a respirator because he was unable to breathe unaided. About the second day on the respirator, he began complaining of a "tense" feeling and of spasmodic twitchings of his fingers. Serum P_{CO_2} was found to be 28 mm Hg, HCO_3 21 mEq/liter, and pH 7.6. The respiratory rate of the machine was slowed, and the symptoms disappeared. The conclusion was that the patient had been suffering from respirator-induced carbonic acid deficit (respiratory alkalosis).

Commentary Either carbonic acid excess or carbonic acid deficit can occur when the respiratory rate on a respirator is not adjusted to fit the needs of the patient.

PRIMARY CARBONIC ACID EXCESS OF ECF

Primary carbonic acid excess of ECF, commonly called *respiratory acidosis* or *acidemia*, is caused by any condition that obstructs respiratory exchange. Retention of carbon dioxide by the lungs can result from several mechanisms. Thus, simple alveolar hypoventilation can be caused by depression of the medullary respiratory center by drugs, anesthetics, neurologic disease, or the cardiopulmonary obesity syndrome (in which obesity simply straitjackets the lungs and adjacent structures). Abnormalities of the chest bellows, with musculoskeletal weakness, such as occur in polio, myasthenia gravis, Guillain-Barré syndrome, or crush injuries of the chest, can impair breathing, as can a pronounced reduction of the alveolar surface area, such as occurs in chronic obstructive lung disease (COLD), severe pneumonia, pulmonary edema, asthma, and pneumothorax. Laryngeal or tracheal obstruction also can lead to respiratory acidosis.

With the acidemia resulting from respiratory acidosis, both the cerebrospinal fluid and the brain cells may become acidic. This can induce neurologic changes. The hypoxemia and compensatory buildup of bicarbonate, which often go along with respiratory acidosis, accentuate the neurologic abnormalities. Headache and drowsiness may occur. The encephalopathy responsible for them may progress to stupor and coma. Multifocal myoclonus usually occurs. With encephalopathy, there may be dilation of retinal venules and papilledema.

The serum pH is depressed. PCO_2, the chemical cause of the imbalance, is elevated. The kidneys compensate by excreting hydrogen, reabsorbing bicarbonate, and regenerating bicarbonate. For each 10-mm Hg increase in PCO_2, one can expect a 1 mEq to 2 mEq/liter increase in bicarbonate in acute carbonic acid excess. In chronic carbonic acid excess, in which the kidneys have had more time to react, the increase may be 3 mEq to 4 mEq/liter. When diuretic therapy is employed, the compensatory increase in bicarbonate may have superimposed metabolic alkalosis.

Again, the history and physical examination provide invaluable clues to diagnosis. Radiographic studies may highlight the organic causes of the carbonic acid excess. Laboratory findings as outlined are supportive.

Therapy is directed primarily toward correcting the condition that is responsible for the impaired lung function, and administration of bicarbonate or lactate may also be necessary to restore balance while the lung ailment is being corrected. Mechanically assisted ventilation may be needed but should be carried out with restraint, particularly when the respiratory acidosis is chronic. Sedative drugs should invariably be avoided. In some patients, oxygen may be useful. Others react to oxygen with a fall in respiratory minute volume and further elevation of the PCO_2. Should the PCO_2 exceed 50 mm Hg to 55 mm Hg, mechanical ventilation should be considered. But the PCO_2 should be lowered slowly. Too rapid lowering of the PCO_2 causes a shift in the oxyhemoglobin dissociation curve to the left (less oxygen released from oxyhemoglobin), as well as cerebral vasoconstriction. The result can be seizures and death. Any electrolyte deficits (e.g., potassium or chloride) should be promptly repaired (Fig. 3-7).

CASE HISTORIES An intern, impatient with 84-year-old M. O., who had been upsetting his entire ward in a chronic disease hospital, foolishly ordered ½ grain of morphine for him. The intern then left for a date. The patient quieted down immediately but soon became so lethargic and difficult to arouse that the nurse on night duty became alarmed. She called the intern on duty, who found that the patient could be aroused only with considerable effort. The patient's muscles were relaxed, his deep reflexes were unobtainable, his skin was pale, he was perspiring profusely, and his respirations were down to 2 to 3/min. Even these respirations were shallow, irregular, and heavy. Resuscitative measures were immediately started without waiting to obtain a blood specimen for analysis. Had one been obtained, it would undoubtedly have revealed a carbonic acid excess (respiratory acidosis), with a PCO_2 in excess of 65 mm Hg. The HCO_3 would not have been elevated since the kidneys had not had time to swing into action. The serum *p*H would probably have been about 7.1 or 7.2.

Commentary Both the elderly and the very young have poor tolerance to narcotics. One-fourth grain of morphine would have been dangerous for this oldster; ½ grain was nearly lethal. Fortunately, the patient did survive and was again joyfully keeping the ward upset the next day.

P. K., age 40, came from a family afflicted with a cluster of allergies. His past history revealed that he had been shown to be allergic to pollens, molds, horse dander, ragweed, and various medications. On this admission to the hospital, he was suffering from status asthmaticus that had been preceded by a severe attack of bronchitis that had simply not let up. Instead, it had gradually worsened: for the past several days and nights, the patient had sat on the edge of his bed propped up with his arms in a desperate effort to expand his airway. On physical examination there was evident prolongation of expiration with

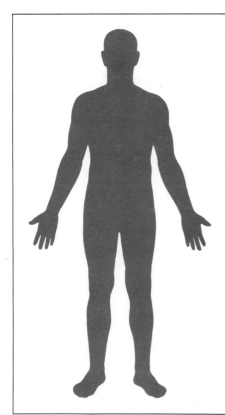

PRECEDING EVENTS

- *Respiratory obstruction*

CLINICAL OBSERVATIONS

- *Impaired respiration (may be detectable)*
- *Disorientation*
- *Weakness*
- *Headache*
- *Coma*

LABORATORY FINDINGS

- *Urine pH below 6*
- *Plasma pH below 7.35*
- *Plasma bicarbonate above 24 mEq/liter*
- *P_{CO_2} above 40 mm Hg*

RELATED PROBLEMS

- *Body compensation must depend on kidneys alone, therefore slow*

FIG. 3-7. **Primary carbonic acid excess (respiratory acidosis).**

rales throughout the chest. Respiration was labored, and the patient was slightly cyanotic. The laboratory findings were as follows: HCO_3 36 mEq/liter, P_{CO_2} 65 mm Hg, pH 7.2. The urine pH was 5.5; otherwise, urine was normal. The P_{CO_2} clearly indicated carbonic acid excess, supported by the serum pH of 7.2. The HCO_3 was elevated as a renal compensatory measure. Diagnosis: carbonic acid excess resulting from status asthmaticus.

Commentary In mild status asthmaticus, carbonic acid deficit (respiratory alkalosis) prevails. When the status worsens, the carbonic acid deficit changes to carbonic acid excess, or respiratory acidosis. This change represents an unfavorable prognostic sign. Urgent action is required, with ventilation, oxygen, and other measures judiciously applied.

Trends in uncompensated and compensated acid–base disturbances are shown in Tables 3-1 and 3-2.

TABLE 3-1 *Early uncompensated pH disturbances*

Imbalance	HCO_3	Arterial P_{CO_2}	Arterial plasma pH
Metabolic alkalosis	Above 25 mEq/liter	Normal	Above 7.42
Metabolic acidosis	Below 23 mEq/liter	Normal	Below 7.38
Respiratory alkalosis	Normal	Below 38 mm Hg	Above 7.42
Respiratory acidosis	Normal	Above 42 mm Hg	Below 7.38

Many clinicians regard the term *acidosis* as any clinical situation tending to produce acidemia, even though it has not produced it. Similarly, they regard the term *alkalosis* as any clinical situation tending to produce alkalemia, whether or not it has succeeded in so doing.

Using this terminology, one might encounter such a paradoxical situation as alkalemia (caused by carbonic acid deficit, or respiratory alkalosis) coexisting with base bicarbonate deficit (metabolic acidosis). In this instance, the respiratory alkalosis would, in effect,

TABLE 3-2 *Partially compensated pH disturbances*

Imbalance	HCO₃	Arterial Pco₂	Arterial plasma pH	Urine pH
Metabolic alkalosis	Above 25 mEq/liter	Above 42 mm Hg	Still above 7.42, but closer to normal	Above 7
Metabolic acidosis	Below 23 mEq/liter	Below 38 mm Hg	Still below 7.38, but closer to normal	Below 6
Respiratory alkalosis	Below 23 mEq/liter	Below 38 mm Hg	Still above 7.42, but closer to normal	Above 7
Respiratory acidosis	Above 25 mEq/liter	Above 42 mm Hg	Still below 7.38, but closer to normal	Below 6

overpower the effects of the metabolic acidosis. Similarly, one might conceivably find acidemia (caused by carbonic acid excess, or respiratory acidosis) coexisting with base bicarbonate excess (metabolic alkalosis). In this instance, the respiratory acidosis would overpower the metabolic alkalosis.

Some would refer to the former situation as a compensated metabolic acidosis coexisting with a respiratory alkalosis, and to the latter as a compensated metabolic alkalosis coexisting with a respiratory acidosis. Obviously, mixed metabolic and respiratory acid–base disturbances can be enormously complex.

One important clue to whether we are dealing with a mixed nonrespiratory and respiratory imbalance is the Pco₂. (We'll see shortly how this can help us.) Another help, which cannot be overemphasized, is the clinical history. Also important is knowing what the more frequent combination acid–base disturbances are. As we proceed, we'll use our "etiologic" classification of acid–base disturbances—namely, bicarbonate deficit, bicarbonate excess, carbonic acid deficit, and carbonic acid excess.

CARBON DIOXIDE ALTERATIONS: COMPENSATORY OR PRIMARY?

As we have seen, the lungs help correct metabolic, or nonrespiratory, acid–base disturbances by increasing or decreasing exhalation of CO_2 and H_2O, thereby increasing or decreasing hydrogen concentration in the ECF. The kidneys, on their part, assist in countering respiratory acid–base disturbances by elevating or repressing the level of bicarbonate in the ECF. It would be most helpful if we could determine whether the changes in Pco₂ or HCO₃ are merely compensatory or

whether they indicate secondary or mixed metabolic and respiratory acid–base disturbances. Fortunately, Cohen has provided us with useful rules of thumb for just such estimations.[2]

In the case of bicarbonate deficit, for each reduction of 1 mEq/liter of HCO₃, one can expect 1.2-mm Hg to 1.5-mm Hg reduction in Pco₂ *if the fall is purely compensatory.* Should the decrease be greater than that, a secondary carbonic acid deficit may exist. If the decrease in Pco₂ is less than expected by Cohen's rule, then carbonic acid excess may be superimposed on the bicarbonate deficit.

Consider bicarbonate excess: for each increase of 1 mEq/liter of HCO₃, one can expect 0.5-mm Hg to 0.7-mm Hg Pco₂ increase. If the rise is less than that, secondary carbonic acid deficit may be present; if more, a secondary carbonic acid excess may be superimposed.

In carbonic acid excess, for each 10 mm Hg Pco₂ increase, the normal compensation would consist of 1-mEq to 2-mEq/liter increase in HCO₃ for acute imbalances, 3-mEq to 4-mEq/liter increase in chronic disturbances. A level of HCO₃ higher than this would indicate a secondary bicarbonate excess, while lower levels would point to a secondary bicarbonate deficit.

In the case of carbonic acid deficit, one can expect for each 10-mm-Hg Pco₂ decrease, a 2-mEq/liter to 3-mEq/liter decrease in HCO₃ for an acute disturbance, a 5-mEq to 6-mEq/liter fall in a chronic imbalance. If the HCO₃ falls less than the rule, a secondary bicarbonate excess would be favored; a greater decrease would indicate a possible secondary bicarbonate deficit.

Compensatory changes in bicarbonate in respiratory acid–base disturbances tend to be greater in chronic than in acute disturbances because in the former, the kidneys have had more time to react. Recall that renal correction of acid–base disturbances is usually far slower than is the pulmonary reaction (Fig. 3-8 and Table 3-3).

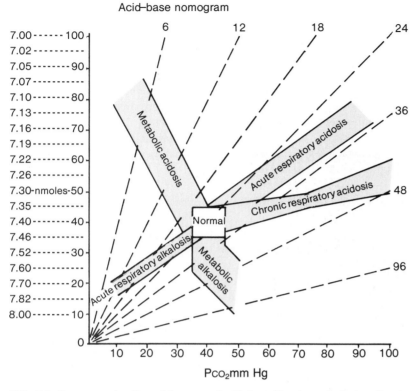

Acid–base nomogram

The confidence bands describe the normal (*i.e.*, with 95% confidence) steady-state response of humans to pure, uncomplicated respiratory or metabolic acid–base disorders. Because the respiratory adaptation to metabolic disorders occurs promptly, it is not necessary to segregate these disorders into acute or chronic components. Acid–base values (plotted as plasma pH on the ordinate and plasma P_{CO_2} on the abscissa) falling within a set of confidence bands are consistent with but not diagnostic of a pure disorder of that type. Values falling outside the confidence bands strongly suggest a mixed disorder of that type. The plasma bicarbonate isopleths are not required for analytic purposes. They simply reflect the HCO_3 concentration associated with a given set of pH and P_{CO_2} values.

FIG. 3-8. Nomogram for determining type of acid–base disorder and whether disorder is "pure" or "mixed." (Goldberg M, Green S, Moss et al: Computer-based instruction and diagnosis of acid–base disorders: A systematic approach. JAMA 223:269, 1973; text reproduced from Bunn P, Smith M: Fluid and electrolyte alterations. Audio-Digest Intern Med 27, No. 24, 1980)

TABLE 3-3 *Summary of pulmonary and renal compensatory action: rules of thumb in stable primary acid–base disturbances*

BASE BICARBONATE DEFICIT
- Decrease of HCO_3 by 1 mEq/liter = decrease of P_{CO_2} by 1.2–1.5 mm Hg

BASE BICARBONATE EXCESS
- Increase of HCO_3 by 1 mEq/liter = increase of P_{CO_2} by 0.5–0.7 mm Hg

CARBONIC ACID DEFICIT
- Decrease of P_{CO_2} by 10 mm Hg = decrease of HCO_3 by 2–3 mEq/liter if acute, 5–6 mEq/liter if chronic (kidney had time to react)

CARBONIC ACID EXCESS
- Increase of P_{CO_2} by 10 mm Hg = increase of HCO_3 by 1.2 mEq/liter if acute, 3–4 mEq/liter if chronic (kidney had time to react)

Now let us examine the clinical conditions that are especially likely to produce combination nonrespiratory and respiratory acid–base disturbances:

Cardiac arrest tends to produce carbonic acid excess because of impairment of breathing and bicarbonate deficit because of the production of lactic acid in anoxia. (This would be a high Delta acidosis.)

Septic shock, salicylate intoxication, and hepatorenal syndrome produce carbonic acid deficit because of hyperventilation induced by these states and bicarbonate deficit because of alkali-neutralizing toxins released into the circulation.

A patient with bronchial asthma who is on a sodium-restricted diet tends to develop carbonic acid excess because of inadequate respiration and bicarbonate excess because of deficient chloride intake with compensatory rise in bicarbonate generated by the kidneys.

Suppose we have a patient with carbonic acid excess who had already developed an elevated plasma HCO_3 because of renal compensation. We give the patient assisted ventilation, which converts the carbonic acid excess to a carbonic acid deficit. The already elevated bicarbonate would swing the patient over into a bicarbonate excess.

Here is a simple formula that can determine the P_{CO_2}, HCO_3, or pH if you know only two of the three;

it can enable one to find just how much good body compensation does in body fluid disturbances when functioning in the normal manner:

$$\text{nm/liter} = 24 \times \frac{\text{Pco}_2 \ (\text{in mm Hg})}{\text{HCO}_3 \ (\text{in mEq/liter})}$$

with nm/liter referring to hydrogen in nanomoles per liter. One nanomole of hydrogen is one billionth of a gram, one millionth of a milligram. See the conversion table below for converting nanomoles to pH, or vice versa:

nm/liter		pH
100.0	=	7.0
79.4	=	7.1
63.0	=	7.2
50.0	=	7.3
39.8	=	7.4
31.6	=	7.5
25.1	=	7.6
20.0	=	7.7
15.8	=	7.8
12.5	=	7.9
10.0	=	8.0

Using the formula and conversion table, try your hand at the following examples:

- Pco_2 is 60 mm Hg, HCO_3 is 34 mEq/liter. What is the nm/liter? (42, or pH 7.35)
- Pco_2 is 30 mm Hg, HCO_3 is 7 mEq/liter. What is the nm/liter? (103, or pH 7.0)
- nm/liter is 16, HCO_3 is 40 mEq/liter. What is the Pco_2? (27 mm Hg)
- nm/liter is 80, Pco_2 is 30 mm Hg. What is the HCO_3? (9 mEq/liter)

A SYSTEMATIC APPROACH TO DIAGNOSIS OF ACID–BASE DISTURBANCES

Donna McCurdy, M.D., of the University of Pennsylvania, has presented a logical and systematic approach to the diagnosis of acid–base disturbances. It makes such eminent good sense that we are presenting it here:

1. Examine the clinical history for processes that might lead to simple acid–base disturbances.

2. Note findings on physical examination that suggest an acid–base disturbance.
3. Study the laboratory reports on the electrolytes, the HCO_3, the Na, the K, the pH, the Pco_2, and the Cl. Calculate the Delta.
4. Examine other laboratory data for disease processes associated with acid–base disorders.
5. As you examine the Pco_2, determine whether any change stems from compensatory action or whether it signals a complicating primary respiratory acid–base disturbance.

COMMENT The clinical examples of body fluid disturbances in these chapters were chosen so as to represent relatively pure examples of single disturbances. Actually, these examples are not uncommon. Sometimes one sees combined imbalances, and in some patients a considerable number of imbalances can be present simultaneously. Usually, however, one or two imbalances dominate the clinical picture. The infant with an acute onset of vomiting and diarrhea, for example, early has volume deficit of ECF, and shortly thereafter, if not properly treated, will develop primary base bicarbonate deficit. If potassium is not given and the diarrhea continues, he will, after 3 or 4 days, develop potassium deficit. If he continues to have loose, watery stools, and, particularly, if he is also given salt mixtures by mouth, he may well develop a sodium excess.

The addition of each one of these fluid balance disturbances has the effect of increasing, perhaps doubling the mortality. Various disease states are characterized by a wide variety of body fluid disturbances, depending upon the severity of the disease. Among these conditions are burns, severe trauma, diabetes mellitus, gastrointestinal disease, and intestinal obstruction. (There conditions are discussed at length in later chapters.)

It cannot be overemphasized that if one is to understand either combined or complex body fluid disturbances, one must understand the single imbalances, the mechanism of their development, plus their relationships to one another and to disease.

TERMINOLOGY PROBLEMS

Understanding body fluid disturbances is made considerably more difficult by the fact that so many terms frequently apply to the same condition. In an attempt to resolve this problem, we are including in this chapter a table showing in parallel columns each chemical imbalance, then the conventional term applied to the imbalance, and, finally, the related clinical term, if any. A careful study of Table 3-4 will help to resolve apparent conflicts in terms.

REFERENCES

1. Giebisch G: Derangements of acid–base balance. In The Sea Within Us. New York, Science and Medicine, 1975
2. Cohen J: In Thier S, Cohen J, Relman A et al: Acid–base disturbances. Boston, ACP Annual Session, 1978

TABLE 3-4 *Terminology chart for physicochemical imbalances and related clinical states*

Physicochemical imbalance	Conventional terminology	Closely-related clinical states
ECF	Fluid deficit	
	Dehydration (incorrect term, since ECF volume deficit means loss of water *and* electrolytes)	
	Hypovolemia	
ECF	Fluid volume excess	
	Overhydration (incorrect term, since this imbalance means an excess of both water *and* electrolytes)	
Sodium deficit	Electrolyte concentration deficit	Heat exhaustion
	Hyponatremia	Sodium-losing kidney
	Low-sodium syndrome	Severe muscle cramps
	Hypotonic dehydration	
Sodium excess	Hypernatremia	
	Hypertonic dehydration	
	Salt excess	
	Oversalting	
Potassium deficit	Hypokalemia	Potassium-losing kidney
	Potassium deficiency	Muscle necrosis
Potassium excess	Hyperkalemia	
Calcium deficit	Hypocalcemia	Calcium deficiency leg cramps
Calcium excess	Hypercalcemia	Pathologic fractures
Primary base bicarbonate deficit	Metabolic acidosis	Diabetic ketosis
	Acidemia	Renal acidosis
Primary base bicarbonate excess	Metabolic alkalosis	Milk-alkali syndrome
	Alkalemia	
Primary carbonic acid deficit	Respiratory alkalosis	Hyperventilation
	Alkalemia	
Primary carbonic acid excess	Respiratory acidosis	Impaired ventilation
	Acidemia	
Protein deficit	Hypoproteinemia	Kwashiorkor
	Protein malnutrition	Pluricarencial syndrome
		Hypoproteinosis
Magnesium deficit	Hypomagnesemia	
Shift of water and electrolytes from plasma to interstitial space	Hypovolemia	Shock Edema
Shift of water and electrolytes from interstitial space to plasma	Hypervolemia	Remobilization of edema fluid

BIBLIOGRAPHY

Bunn P, Smith M: Fluid and electrolyte alterations. Audio-Digest Internal Medicine 27, No. 24, 1980

Fraley D, Adler S, Bruns F et al: Metabolic acidosis after hyperalimentation with case in hydrolysate. Ann Intern Med 88:352, 1978

Friedman H: Problem-Oriented Medical Diagnosis, 2nd ed, pp 233–243. Boston, Little, Brown & Co, 1979

Ganong W: Review of Medical Physiology, 10th ed, pp 579–581. Los Altos, Lange Medical Publications, 1981

Giebisch G: Derangements of acid–base balance. In The Sea Within Us, pp 36–46. New York, Science and Medicine, 1975

Goldberg, M, Green S, Moss M et al: Computer-based instruction and diagnosis of acid–base disorders: A systematic approach. JAMA 223:269, 1973

Goldberger E: A Primer of Water, Electrolyte and Acid–Base Syndromes, 6th ed. Philadelphia, Lea & Febiger, 1980

Snively W, Leitch G, Beshear D: Acid–base disturbances, a programmed text. Am J IV Ther 4:22 (Nov), 1977; 5:26 (Jan); 5:36 (Mar); 5:26 (May), 1978

Szwed J: Fluids, Electrolytes, and Acid–Base. Indianapolis, Indiana University School of Medicine, 1981

Thier S, Cohen J, Relman A et al: Acid–Base Disturbances. Boston, ACP Annual Session, 1978

Regardless of the multiplicity of laboratory determinations, proper interpretation and therapy depend nonetheless upon frequent and accurate clinical evaluation. —Robert E. Cooke

Nursing Assessment of the Patient with Body Fluid Disturbances

FORMULATION OF NURSING DIAGNOSIS

The nurse must possess enough knowledge of body fluid disturbances so that she can make an intelligent nursing diagnosis in order to locate pertinent nursing problems. Once the problems (or potential problems) are identified, the nurse can plan for meaningful observations, measures to prevent imbalances, intelligent execution of medical directives, and other effective nursing interventions.

One must be a careful observer in all areas of nursing, and fluid balance is no exception. Changes in the patient who is developing a fluid imbalance are often subtle and perceptible only to those familiar with him and his condition. Thus, observations made by the nurse are particularly valuable because she spends more time with the patient than does the physician. The observations must be planned and based on an understanding of basic physiologic processes; otherwise, they are of little or no value.

To assist in nursing diagnosis formulation, the nurse should answer the following:

1. Is there present any disease state that can disrupt body fluid balance? (For example: diabetes mellitus, emphysema, or fever) In what type of imbalance does this condition usually result? (Table 4-1.)
2. Is the patient receiving any medication or treatment that can disrupt body fluid balance? (For example: steroids or thiazide diuretics) If so, how might this therapy upset fluid balance? (Table 4-2.)
3. Is there an abnormal loss of body fluids and, if so, from what source? What type of imbalance is usually associated with the loss of the particular body fluid or fluids? (See Tables 4-3 and 4-4.)
4. Have any dietary restrictions been imposed? (For example: low-sodium diet) If so, how might fluid balance be affected or altered in this case?
5. Has the patient taken adequate amounts of water and other nutrients orally or by some other route? If no, how long?
6. How does the total intake of fluids compare with the total fluid output?

The nurse should remember that the symptoms seen in an imbalance, and their severity, depend on how long the imbalance has been present, its magnitude and rapidity of onset, and how efficiently the
(Text continues on p. 91)

TABLE 4-1 *Imbalances likely to occur in various clinical conditions*

Illnesses, Burns, Injuries		
Condition	*Imbalances likely to occur*	*Reason*
Uncontrolled severe diabetes mellitus	Metabolic acidosis	Increased utilization of fat for energy needs causes accumulation of ketone bodies (acids)
	Fluid volume deficit	Hyperglycemia causes osmotic diuresis
Adrenal insufficiency	Sodium deficit	Inadequate amounts of adrenal cortical hormones produced, causing the body to retain potassium and release too much sodium
	Potassium excess	
Acute pancreatitis	Calcium deficit	Fixation of calcium by fatty acids liberated from necrotic mesenteric fat deposits
Perforated viscus and chemical peritonitis	Plasma-to-interstitial fluid shift	Loss of fluid into peritoneal cavity as result of inflammatory process
	Calcium deficit	Fixation of calcium by fatty acids liberated from necrotic mesenteric fat deposits
Occlusion of breathing passages, bronchial pneumonia, pneumothorax, hemothorax	Respiratory acidosis	Inability to exchange carbon dioxide
Tracheobronchitis	Sodium excess	Excessive water vapor loss caused by very rapid breathing
Hysteria (psychogenic hyperventilation)	Respiratory alkalosis	Overbreathing, resulting in excessive loss of carbon dioxide
Oxygen-lack with hyperpnea	Respiratory alkalosis	Overbreathing results in excessive loss of carbon dioxide
Fever	Fluid volume deficit	Increased loss of water from lungs, and loss of water and electrolytes from kidneys and skin
	Respiratory alkalosis	Increased respirations resulting from overstimulation of respiratory center
Severe diarrhea	Metabolic acidosis	Excessive loss of potassium-rich, alkaline intestinal fluid (increased peristalsis shortens absorption period)
	Potassium deficit	
	Fluid volume deficit	
Emphysema, asthma, pulmonary edema	Respiratory acidosis	Inability to exchange carbon dioxide
Congestive heart failure	Fluid volume excess	Increased retention of sodium and water by kidneys caused by:
		• decreased renal perfusion secondary to failing pumping action of heart
		• increased aldosterone secretion
Renal disease	Metabolic acidosis	Inability of the kidneys to adequately excrete acid metabolites, potassium, and fluid
	Potassium excess	
	Fluid volume excess	
Hyperparathyroidism	Calcium excess	Increased secretion of parathyroid-hormone causes calcium to be released from bone matrix into extracellular fluid
Hypoparathyroidism	Calcium deficit	Decreased bone resorption causes depressed plasma calcium level
Hyperaldosteronism	Fluid volume excess	Excessive secretion of aldosterone by adrenal cortex causes the body to retain sodium, chloride, and water and to lose potassium
	Potassium deficit	
Meningitis and encephalitis	Respiratory alkalosis	Overstimulation of respiratory center in medulla produces overbreathing

(continued)

TABLE 4-1 *Imbalances likely to occur in various clinical conditions (continued)*

	Illnesses, Burns, Injuries	
Condition	*Imbalances likely to occur*	*Reason*
Gastric outlet obstruction with repeated loss of gastric juice through vomiting	Metabolic alkalosis	Loss of acid gastric juice
Vomiting with greater loss of alkaline intestinal juice than of gastric juice	Metabolic acidosis	Loss of alkaline fluid exceeds loss of acidic fluid
Chronic alcoholism	Magnesium deficit	Alcohol produces magnesium diuresis; also, dietary intake of magnesium is often low
Oat cell lung tumor	Sodium deficit (water intoxication)	Inappropriate secretion of an ADH-like substance from the tumor
Burn, early	Potassium excess	Increased cellular destruction results in release of potassium from cells into extracellular fluid
	Plasma-to-interstitial fluid shift	Plasma leaks out through the damaged capillaries at the burn site
Burn after third day	Potassium deficit	Extracellular potassium shifts back into cells
	Interstitial fluid-to-plasma shift	Edema fluid shifts back into intravascular compartment as capillaries heal
Massive crushing injury	Potassium excess	Cellular damage results in release of potassium from the cells into the extracellular fluid
	Plasma-to-interstitial fluid shift	Plasma leaks out through damaged capillaries at injury site
Head injury	Sodium deficit (water intoxication)	Inappropriate secretion of ADH occurs in some head injury patients

TABLE 4-2 *Imbalances caused by medical therapy*

Condition	*Imbalances likely to occur*	*Reason*
Excessive administration of adrenal glucocortical hormones	Fluid volume excess Potassium deficit	Adrenal cortical hormones cause the body to retain sodium and water and to lose potassium
	Metabolic alkalosis	Potassium deficit is frequently associated with metabolic alkalosis
	Negative nitrogen balance	Adrenal glucocorticoids are catabolic
Administration of potent potassium-losing diuretics (thiazides, furosemide, ethacrynic acid, mercurials)	Potassium deficit	These agents promote potassium loss
Administration of potassium-conserving diuretics (spironolactone, triamterene) to patients with oliguria	Potassium excess	These agents promote potassium retention
Morphine, meperedine or barbiturates in excessive doses	Respiratory acidosis	These agents depress respirations and thus decrease the elimination of carbon dioxide
Excessive or too rapid parenteral administration of potassium-containing fluids	Potassium excess	Kidneys unable to excrete potassium rapidly enough to prevent buildup in bloodstream
Early salicylate intoxication (children and adults)	Respiratory alkalosis	Toxic salicylate level stimulates respiratory center and causes overbreathing
Salicylate intoxication (not early) in young children	Metabolic acidosis	Toxic salicylate level results in inadequate utilization of carbohydrate and, thus, increased metabolism of body fats (ketosis)

(continued)

TABLE 4-2 *Imbalances caused by medical therapy* (continued)

Condition	Imbalances likely to occur	Reason
Excessive administration of vitamin D	Calcium excess	Increased resorption of bone induced by excessive doses of vitamin D
		Increased intestinal absorption of calcium
Excessive parenteral administration of calcium-free solutions	Calcium deficit	Dilution of plasma calcium level by calcium-free fluids
Excessive parenteral administration of magnesium-free solutions	Magnesium deficit	Dilution of plasma magnesium level by magnesium-free fluids
Excessive infusion of large molecular fluids	Interstitial fluid-to-plasma shift	Large molecular substances pull fluid into intravascular space
Excessive administration of citrated blood (particularly to patients with liver damage)	Calcium deficit	Citrate ions combine with ionized calcium in bloodstream
Excessive ingestion of sodium chloride	Sodium excess (if water intake is inadequate)	Sodium level increased as result of excessive intake
	Fluid volume excess (if water intake is adequate)	Large sodium intake causes body to retain water
Gastric suction plus drinking plain water	Sodium deficit	Washout of gastric electrolytes
	Metabolic alkalosis	
	Potassium deficit	
Recent correction of acidosis	Calcium deficit	Increased alkalinization of plasma causes calcium to ionize less freely than in acidosis
Mechanical respirator inaccurately regulated (causing too deep or too rapid breathing)	Respiratory alkalosis	Excessive loss of carbon dioxide
Mechanical respirator inaccurately regulated (causing too shallow or too slow breathing)	Respiratory acidosis	Retention of carbon dioxide
Prolonged immobilization	Calcium excess	Disuse osteoporosis (absence of weight bearing causes calcium to be resorbed from bone matrix into extracellular fluid)
	Respiratory acidosis	Decreased chest expansion and stasis of secretions
	Negative nitrogen balance	Nitrogen losses exceed intake (increased protein catabolism)
Inhalation anesthesia	Respiratory acidosis	Respiratory depression
Tight abdominal binders or dressings	Respiratory acidosis	Decreased respiratory excursions cause retention of carbon dioxide
Elemental diets and high-protein tube feedings (with inadequate water intake)	Sodium excess	Water drawn from tissues to supply the needed volume for urinary excretion of the increased solute load
	Elevated blood urea nitrogen level	Eventually inadequate fluid is available to excrete the high solute load and metabolites are retained
Excessive use of phosphate-binding antacids, especially when combined with maintenance hemodialysis	Phosphate depletion	Excessive loss of phosphate (phosphate depletion can occur under these circumstances although renal failure is more commonly associated with phosphate retention)

homeostatic mechanisms compensate for it. Also, remember that imbalances rarely occur alone; usually more than one imbalance is present, and this makes identification more difficult.

ANTICIPATION OF FLUID IMBALANCES ASSOCIATED WITH SPECIFIC BODY FLUID LOSSES

Because the anticipation of an imbalance makes its appearance easier to detect, the nurse should learn to anticipate imbalances. It is extremely difficult to recognize significant changes in the patient when one does not know what to look for, and prevention is easier to practice when one knows which imbalance is likely to occur.

The type of imbalance (or imbalances) that accompanies the loss of a specific body fluid varies with the content of the lost fluid. The nurse can learn to anticipate most specific imbalances when the chief constituents of the lost fluids and their functions are known. Table 4-3 lists the approximate sodium, chloride, and potassium concentration of many of the body fluids. A brief discussion of some of the body fluids and types of imbalances associated with their loss may help to clarify how the nurse can learn to anticipate fluid disturbances.

TABLE 4-3 *Approximate electrolyte content of body fluids*

Body fluid	Na^+	K^+	Cl^-
	← mEq/liter →		
Gastric secretions	60	10	100
Ileostomy (recent)	130	20	110
Ileostomy (adapted)	50	10	60
Bile (fistula)	140	10	100
Pancreatic juice (fistula)	140	10	75
Transverse colostomy	50	10	40
Diarrheal fluid	50	60	40
Perspiration (normal)	50	5	60
Plasma*	137–147	4.0–5.6	98–106
Transudates†	130–145	2.5–5.0	90–110

*Also contains protein, 6–8 g/dl.
†Protein content is similar to plasma.

GASTRIC JUICE

The usual daily volume of gastric juice is approximately 2500 ml, and the pH is usually 1 to 3, but the amount of gastric fluid can vary from 100 ml to 6000 ml in abnormal states. Gastric juice contains hydrogen (H^+), chloride (Cl^-), sodium (Na^+), and potassium (K^+) ions. Imbalances that may result from severe vomiting or prolonged gastric suction include:

Extracellular fluid (ECF) volume deficit. This imbalance is due to the loss of both water and electrolytes.

Metabolic alkalosis (primary base bicarbonate excess). Alkalosis develops because H^+ and Cl^- are lost from the body. The loss of H^+ from the body causes the pH to become more alkaline. The carbonic acid–base bicarbonate buffer system compensates for Cl^- loss by increasing the amount of HCO_3^-; thus, base bicarbonate excess develops. Because both Cl^- and HCO_3^- are anions, as the amount of Cl^- decreases, the body releases more HCO_3^- to keep the total number of anions equal to the total cations. Because HCO_3^- is basic, however, the pH becomes more alkaline.

Sodium deficit. Note in Table 4-3 that sodium is rather plentiful in gastric juice.

Potassium deficit. There is sufficient potassium in gastric juice to result in a potassium deficit if vomiting or gastric suction is prolonged. Potassium deficit is also associated with alkalosis.

Tetany (if metabolic alkalosis is present). Although there is no loss of calcium in gastric juice, the patient may develop tetany from a deficit of *ionized* calcium. Since calcium ionization is readily influenced by pH and calcium ionization is decreased in alkalosis, the tetany accompanying alkalosis is corrected when the pH is restored to normal, unless there is a severe calcium deficit.

Ketosis of starvation. Patients with gastric suction or with prolonged vomiting are usually allowed nothing by mouth. The parenteral route may supply only a fraction of the caloric need; hence the patient undergoes starvation. Ketosis results from the excessive catabolism of body fat during starvation. The accumulation of ketone bodies in the bloodstream may tend to counteract metabolic alkalosis caused by loss of gastric juice (the accumulation of sufficient ketones can convert the alkalosis into acidosis).

Magnesium deficit. Although this imbalance is rare, it can occur with the loss of gastric juice, particularly when prolonged nasogastric suction is used and magnesium-free parenteral fluids are given. There is 1 mEq/liter of magnesium in gastric juice.

INTESTINAL JUICE

The daily volume of intestinal juice is usually about 3000 ml, and its pH is alkaline (7.8–8.0). Diarrhea, intestinal suction, or fistulas can result in the loss of Na^+ and HCO_3^- in excess of Cl^-. Losses from these sources can lead to the following:

Extracellular fluid volume deficit. This imbalance is due to the loss of both water and electrolytes.

Metabolic acidosis (primary base bicarbonate deficit). The loss of HCO_3^- is compensated for by an increase in the number of Cl^-. (As mentioned earlier, this change occurs to keep the total number of anions equal to the total number of cations in the body). The loss of HCO_3^-, which is basic, results in acidosis.

Sodium deficit. The amount of sodium lost from the intestines in diarrhea or intestinal suction can be great.

Potassium deficit. This imbalance develops because relatively large amounts of potassium are lost in the secretions.

BILE

The normal daily secretion of bile is approximately 1500 ml, and the pH is alkaline. Abnormal losses of bile can occur from fistulas or from T-tube drainage following gallbladder surgery. Imbalances that can result from excessive loss of bile include the following:

Sodium deficit. Note in Table 4-3 that the sodium content of bile is quite high.

Metabolic acidosis. This imbalance develops because of the loss of HCO_3^- and the compensatory increase of Cl^-.

PANCREATIC JUICE

The normal daily secretion of pancreatic juice is approximately 1000 ml, and the pH is 8 (alkaline). Losses of pancreatic juice result in depletion of Na^+, HCO_3^-, and Cl^-. The loss of HCO_3^- exceeds the loss of Cl^- because it is more plentiful in pancreatic juice, which is an integral part of intestinal secretions; thus, loss of pancreatic juice is accompanied by losses of other intestinal secretions. Imbalances that can result from pancreatic fistulas include those listed below:

Metabolic acidosis. This imbalance occurs because the basic ion, HCO_3^-, is lost in excess of Cl^-.

Sodium deficit

Calcium deficit

Decrease in ECF Volume

SENSIBLE PERSPIRATION

Excessive sweating caused by fever or high environmental temperature can result in large losses of water, sodium, and chloride. Normally, sweat is a hypotonic fluid containing sodium, chloride, potassium, magnesium, ammonia, and urea. Severe perspiration can lead to the following imbalances:

Sodium deficit. This is especially likely to occur when plain water is ingested in large amounts after profuse sweating, replacing the lost water but not the electrolytes.

Sodium excess. This imbalance can occur if the heavily perspiring person has an inadequate water supply. The sodium concentration in sweat averages around 50 mEq to 80 mEq/liter (considerably less than the sodium content of plasma). Still, proportionately more water than sodium is lost in sweat and results in a relative plasma sodium increase. If water intake is deficient, plasma sodium elevates further.

INSENSIBLE WATER LOSS

The insensible loss of water without solute through the lungs and skin is normally about 600 to 1000 ml daily and is increased by anything that accelerates metabolism. For example, the increase in insensible water loss is roughly 50 ml to 75 ml per degree of Fahrenheit temperature elevation for a 24-hour period.

Increased respiratory activity causes an increased loss of water vapor by way of the lungs. Damage to the skin's surface also results in loss. Increased insensible water loss through either the lungs or skin can lead to the following:

Sodium excess. A loss of water alone results in an increased concentration of sodium in the ECF, that is, water deficit or dehydration.

NURSING OBSERVATIONS RELATED TO BODY FLUID DISTURBANCES

The following summary of familiar nursing routines will show how meaningful observations can reveal a wealth of information concerning the patient's fluid balance status. When significant symptoms appear, the nurse should relay them to the physician to facilitate early diagnosis and treatment. The ability to sort out observations demanding urgent action comes with a working understanding of fluid balance and experience in applying this knowledge. The tables presented in this section refer to symptoms and the diagnoses that they may possibly indicate.

BODY TEMPERATURE

Changes in body temperature can be symptomatic of fluid balance changes or may actually *cause* fluid balance problems.

Dehydration (sodium excess) can cause the body temperature to rise considerably. In part, the temperature elevation is probably related to lack of available fluid for sweating; in addition, dehydration almost certainly has a direct effect on the hypothalamic centers to cause fever.

Fluid volume deficit (when uncomplicated by infection) can cause a decrease in body temperature (Table 4-4). Changes in the temperature of the extremities may be noted by touch; external skin temperature gives some insight into the state of peripheral circulation. A profound decrease in circulating blood volume causes the extremities to be cold to touch.

Fever causes an increase in metabolism and, thus, in formed metabolic wastes, which require fluid to make a solution for renal excretion; in this way, fluid loss is increased. Fever also causes hyperpnea, an increase in breathing resulting in extra water vapor loss through the lungs. Because fever increases loss of body fluids, it is important that temperature elevations be detected early and appropriate interventions taken.

PULSE

The pulse should be evaluated in terms of rate, volume, regularity, and ease of obliteration. Pulse *rate* can be affected by imbalances of potassium, sodium, and

TABLE 4-4 *Significance of body temperature variations*

Symptom	Imbalance indicated by symptom
Depressed body temperature: 97–99° R 95–98° R	Moderate fluid volume deficit[1] Severe fluid volume deficit[1]
Elevated body temperature	Sodium excess (excessive water loss)
Extremities cold to touch	Plasma-to-interstitial fluid shift Profound fluid volume deficit

TABLE 4-5 *Significance of pulse variations*

Symptom	Imbalance indicated by symptom
Weak, irregular, rapid pulse	Severe potassium deficit
Weak, irregular, slowing pulse	Severe potassium excess
Increased pulse rate	Sodium excess Magnesium deficit Hypovolemia (due to plasma-to-interstitial fluid shift or fluid volume deficit)
Decreased pulse rate	Magnesium excess
Bounding pulse (not easily obliterated)	Fluid volume excess Interstitial fluid-to-plasma shift
Bounding, easily obliterated pulse	Impending circulatory collapse
Rapid, weak, thready pulse, easily obliterated	Circulatory collapse Severe fluid volume deficit Hemorrhage Plasma-to-interstitial fluid shift
Irregular pulse	Magnesium deficit

magnesium; other imbalances that can affect pulse rate include fluid volume depletion and plasma-to-interstitial fluid shift. Pulse *volume* is affected by changes in circulating blood volume. *Arrhythmias* can be associated with potassium and magnesium imbalances. Table 4-5 shows possible implications of pulse variations.

RESPIRATION

The nurse should become skilled in observing respiration for indications of body pH changes, since the lungs play a major role in regulating body pH by varying the amount of carbon dioxide retention. In order to detect variations from normal, the nurse must evaluate respiration in terms of rate, depth, and regularity and be familiar with the respiratory changes accompanying both metabolic alkalosis and metabolic acidosis.

Severe metabolic alkalosis affects all aspects of breathing: the rate is decreased; the depth is shallow; the respiratory pattern is disrupted by periods of apnea lasting from 5 to 30 seconds. Note that the lungs attempt to compensate for the alkalosis by retaining carbon dioxide, hence carbonic acid, and, of course, hydrogen ions. (Slow, shallow respiration favors carbon dioxide retention.) The decreased ventilatory rate in metabolic alkalosis is caused by inhibition of the respiratory center by alterations in the arterial blood gases. The hypoventilation that occurs in potassium deficit, potassium excess, and calcium excess is, at least in part, ascribable to significant muscle weakness.

Severe metabolic acidosis (as occurs in diabetic ketoacidosis, lactic acidosis, and alcoholic ketosis) affects primarily the rate and depth of breathing: the breathing rate is increased and may be as fast as 50 per minute; depth is greatly increased. The increased volume of lung ventilation is striking; all of the respiratory accessory muscles are used to increase the capacity of the thorax. Note that the lungs attempt to compensate for the acidosis by "blowing off" carbon dioxide. (Fast, deep respiration favors a loss of carbon dioxide from the lungs.)

Usually, respiration is observed for 30 seconds and multiplied by two for a full minute's count; when abnormalities are noted, respiration should be observed for 2 full minutes. More accurate results are obtained

if the patient is unaware that his breathing is being observed. Variable factors, such as increased activity or emotional upsets, may influence respiration.

Other factors related to fluid balance may also influence respiration (Table 4-6). Rales in the absence of pulmonary disease indicate accumulation of alveolar fluid and imply an increased plasma volume or heart failure or both. If rales are due to fluid volume excess, the acute increase in volume is at least 1500 ml.[2]

BLOOD PRESSURE

Normal adults have an average systolic pressure of 90 mm Hg to 140 mm Hg and an average diastolic pressure of 60 mm Hg to 90 mm Hg. The pulse pressure (difference between systolic and diastolic pressures) is usually between 30 mm Hg and 50 mm Hg. The systolic pressure indicates the pressure within the blood vessels when the heart is in systole, whereas the diastolic indicates the pressure when it is in diastole. Pulse pressure varies directly with cardiac output.

When blood pressure changes are due to blood loss, the systolic pressure falls more rapidly than the diastolic, resulting in diminished pulse pressure. A fall in systolic pressure exceeding 10 mm Hg from the lying to the standing or sitting position (postural hypotension) is a cause for concern and usually indicates fluid volume deficit. Symptoms such as dizziness while standing, sudden apprehension, or a weak pulse are indications to check the blood pressure.

Blood pressure variations are immensely helpful in evaluating body fluid disturbances and should be used often as a means of evaluation when there is a real or potential water and electrolyte balance problem. When an abnormal reading is obtained, it is wise to check the pressure in both arms. Variables that may influence blood pressure, such as increased activity, position change, and emotional upsets, should be considered. Table 4-7 lists possible implications of blood pressure variations.

NECK VEINS

The jugular veins provide a built-in manometer for following changes in central venous pressure. Examination of these veins requires no invasive technique and can be highly reliable when done correctly. Changes in fluid volume are reflected by changes in neck-vein filling, provided the patient is not in heart failure. Recall

TABLE 4-6 *Significance of variations in breathing*

Symptom	Imbalance indicated by symptom
Shallow, slightly irregular, slow breathing (compensatory attempt of lungs to correct alkalosis by retaining CO_2)	Metabolic alkalosis
Shortness of breath on mild exertion (in absence of cardiopulmonary disease)	Mild metabolic acidosis
Deep rapid breathing—close observation reveals an effort with expiration (compensatory attempt of lungs to correct acidosis by blowing off CO_2)	Metabolic acidosis
Shortness of breath (in absence of cardiopulmonary disease)	Fluid volume excess
Moist rales (at first, heard only with stethoscope)	Fluid volume excess
	Pulmonary edema
Shallow breathing (secondary to weakness or paralysis of respiratory muscles)	Potassium deficit (severe)
	Potassium excess (severe)
Respiratory stridor	Calcium deficit (severe)
Severe dyspnea	Acute pulmonary edema
Hyperventilation	Phosphate depletion
Decreased respirations	Magnesium excess (respiratory center paralyzed at a plasma level of 10–15 mEq/liter)

TABLE 4-7 *Significance of blood pressure variations*

Symptom	Imbalance indicated by symptom
Hypotension	Plasma-to-interstitial fluid shift
	Contracted plasma volume
	Severe potassium deficit or excess
	Magnesium excess (3 to 5 mEq/liter)
Hypertension	Early interstitial fluid-to-plasma shift
	Fluid volume excess
	Magnesium deficit
Normal blood pressure while patient is flat in bed— systolic pressure drops in excess of 10 mm Hg when head of bed elevated	Contracted plasma volume

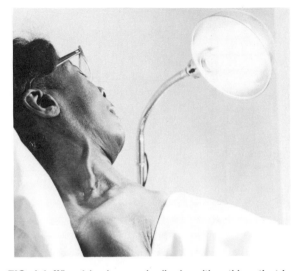

FIG. 4-1. When lying in a semireclined position, this patient has distended neck veins, which indicates that the heart is incapable of receiving and pumping adequately all the incoming venous blood. (Reproduced with permission of the American Heart Association)

that central venous pressure is elevated by an increased plasma volume or by heart failure and is lowered by a decreased plasma volume.

Normally, with the patient supine, the external jugular veins fill to the anterior border of the sternocleidomastoid muscle. Flat neck veins in a supine patient indicate a decreased plasma volume. In a healthy person positioned sitting at a 45 degree angle, the venous distentions should not extend higher than 2 cm above the sternal angle. High venous pressure is manifested by neck veins distended from the level of the manubrium (top portion of sternum) up to the angle of the jaw (Fig. 4-1). To estimate cervical venous pressure the nurse should follow the procedure below:

1. Position the patient in semi-Fowler's (head of bed elevated to a 30 to 45 degree angle), keeping the neck straight.
2. Remove any clothing that could constrict the neck or upper chest.
3. Provide adequate lighting to effectively visualize the external jugular veins on each side of the neck.
4. Measure the level to which the veins are distended up the neck above the level of the manubrium.

HAND VEINS

Observation of hand veins can be helpful in evaluating the patient's plasma volume. Usually, elevation of the hands causes the hand veins to empty in 3 to 5

seconds; placing the hands in a dependent position causes the veins to fill in 3 to 5 seconds (Figs. 4-2 and 4-3).

A decreased plasma volume causes the hand veins

FIG. 4-2. Appearance of hand veins when the hand is held in a dependent position.

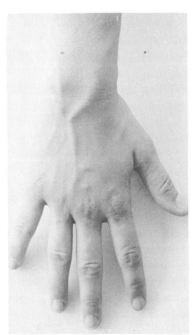

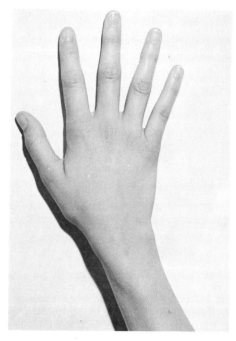

FIG. 4-3. Appearance of hand veins when the hand is held in an elevated position.

to take longer than 3 to 5 seconds to fill when the hands are in a dependent position. The decreased plasma volume may be secondary to an ECF volume deficit or to a shift of fluid from the plasma to the interstitial space. The veins are not readily apparent when plasma volume is reduced. The slow filling of hand veins often precedes hypotension when the patient is in the early stage of shock.

An increased plasma volume causes the hand veins to take longer than 3 to 5 seconds to empty when the hands are elevated. The increased plasma volume may be secondary to an increased ECF volume or to a shift of fluid from the interstitial space into the vascular compartment. When this is the case, the peripheral veins are engorged and clearly visible.

SKIN AND MUCOUS MEMBRANES

Changes in skin elasticity and in mucous membrane moisture are important for evaluation of changes in fluid volume and electrolyte concentration, as pointed out in previous chapters. (However, skin turgor is less reliable than tongue turgor in older persons because their skin is less elastic.)

A dry mouth may be due to a fluid volume deficit or may be due to mouth breathing. When in doubt, the nurse should run her finger along the oral cavity and feel the mucous membrane where the cheek and gum meet; dryness in this area indicates a true fluid volume deficit. Dry mouth is relieved by gargling a small amount of water, but thirst is not. (See Table 4-8 for possible implications of skin and mucous membrane variations.)

In a normal person, pinched skin will fall back to its normal position when released. In a person with fluid volume deficit, the skin may remain slightly raised for many seconds (Fig. 4-4). In part *A* of Figure 4-4, the skin of the forearm is picked up; in part *B*, 30 seconds later, the skin has not returned to its normal position. The patient in this instance is a young man with moderately severe ECF volume deficit. The skin over the sternum and forehead tends to be the most reliable for assessing fluid volume deficit.

Normally, there is some degree of moisture constantly present in the axilla and groin areas because of

TABLE 4-8 *Significance of skin and mucous membrane variations*

Symptom	Imbalance indicated by symptom
Poor skin turgor (best tested over forehead and sternum)	Fluid volume deficit
Dry mucous membranes with longitudinal furrows on tongue	Fluid volume deficit
Dry but otherwise normal axilla or groin (normally a small amount of moisture should be present in these areas)	Fluid volume deficit
Decrease in tearing and salivation	Fluid volume deficit
Warm flushed skin (peripheral vasodilatation)	Metabolic acidosis
Flushed dry skin	Sodium excess
Pale, cold, and clammy skin (peripheral vasoconstriction to compensate for hypovolemia)	Decreased plasma volume
Pitting edema	Fluid volume excess
Fingerprinting on sternum	Sodium deficit (sign of cellular fluid volume excess)
Coarse dry skin and alopecia	Chronic calcium deficit
Dry sticky mucous membranes	Sodium excess
Red swollen tongue	Sodium excess

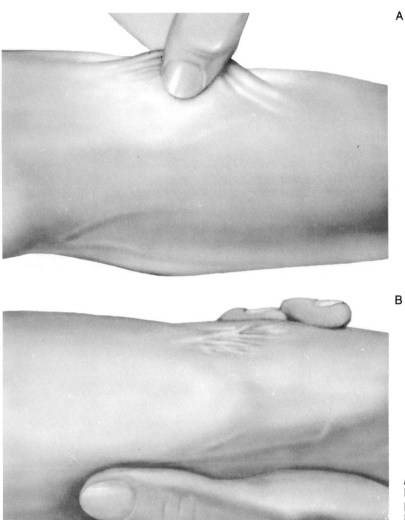

FIG. 4-4.
Poor skin turgor. *(A)* Skin of forearm is picked up, *(B)* It does not return to its normal position 30 seconds later. (Moyer CA: Fluid Balance, A Clinical Manual. Chicago, Year Book Medical Publishers, 1952)

action of the apocrine sweat glands; complete dryness may indicate a major fluid deficit (unless a skin disorder has caused malfunction of the sweat glands). Absence of moisture in the axilla or groin probably is indicative of a fluid volume deficit of at least 1500 ml.[3]

EDEMA

Edema is the presence of excess interstitial fluid in the tissues. It implies that the total body sodium is increased. (Edema is *not* produced by retention of water alone.) Clinically, edema is not usually apparent until 5 to 10 lb of excess fluid have been retained. It is sometimes gauged on a scale of plusses, ranging from +1 to +4 (+1 indicating barely perceptible edema, up to +4 indicating severe edema). Among the types of edema seen in clinical practice are pitting edema, dependent edema, and refractory edema.

Pitting edema is a phenomenon manifested by a small depression when one's finger is pressed over an edematous area (Fig. 4-5). Gradually, within 5 to 30 seconds after the pressure is removed, the "pit" disappears. Usually, pitting edema is not evident until a 10% increase in weight has occurred. It is often tested for over a bony prominence, such as the pretibial area. Dependent edema refers to the flow of excess fluid by gravity to the most dependent portion of the body (feet

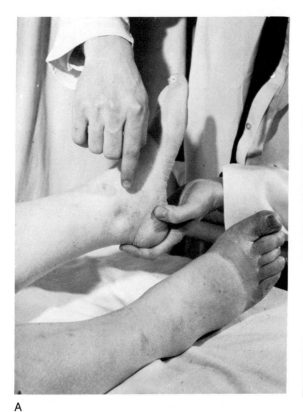
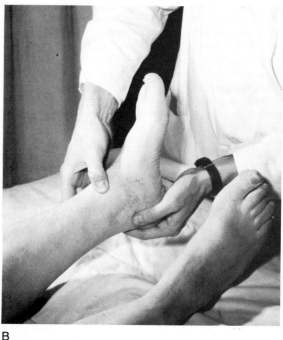

FIG. 4-5. (A) Pitting edema of feet and lower legs, (B) The same patient after edema has been relieved by treatment. (Courtesy of CIBA Pharmaceutical Co., Summit, New Jersey)

and ankles if standing, back and buttocks if lying down). Refractory edema refers to edema that persists despite treatment with low-sodium diet and diuretics.

Besides the subcutaneous tissues, excess body fluid may accumulate in the lung tissue (pulmonary edema), peritoneal cavity (ascites), pleural cavity (hydrothorax), and pericardial cavity (hydropericardium).

FATIGUE THRESHOLD

Many factors contribute to a low fatigue threshold, but those related to fluid balance may include deficits in ECF volume, potassium, sodium, and protein. The nurse should compare the patient's activities and fatigue level with that of previous days to detect significant changes. Episodes of muscular weakness, fatigability, and diminution of stamina and endurance should be noted—the latter symptoms are particularly descriptive of potassium deficit.

FACIAL APPEARANCE

A patient with a severe ECF volume deficit has a drawn facial expression; the eyes are sunken and feel much less firm than normal.

A patient with an excess of ECF may have puffy eyelids, and the cheeks may appear fuller than usual.

BEHAVIOR

Behavior changes may be indicative of water and electrolyte disturbances. (See Table 4-9 for possible implications of behavior changes.) These changes are subtle at first, and often the patient's family is the first to notice them. The attitude of the patient toward his illness is significant; a severely depleted patient is usually aware that he is seriously ill. Aged patients are particularly prone to develop personality changes and impaired mental function with fluid imbalances because

TABLE 4-9 *Significance of behavior changes*

Symptom	Imbalance indicated by symptom
Lassitude and indifference	Fluid volume deficit
Irritability and restlessness	Potassium excess
	Sodium excess in children
	Phosphate depletion
Carphologia (picking at bedclothes)	Potassium deficit
	Magnesium deficit
Apprehension and giddiness	Sodium deficit
	Plasma-to-interstitial fluid shift
Lethargy, confusion, delirium, psychosis, coma[5]	Sodium excess
	Sodium deficit
	Calcium excess
	Calcium deficit
	Magnesium deficit
	Magnesium excess
	Acidosis
	Alkalosis

their homeostatic mechanisms do not function as efficiently as those of younger persons.

Disturbances in plasma pH manifest themselves primarily as central nervous system (CNS) changes. In acidosis, the major problem is depression of the CNS; a sharp decrease in plasma pH toward 7.0 or below causes disorientation and finally coma. In alkalosis, the major problem is overexcitability of the CNS; symptoms may manifest themselves as nervousness or, in susceptible persons, as convulsions.

SKELETAL MUSCLES

Usually the condition of skeletal muscles is readily observable by the nurse. Subjective complaints from the patient, such as weakness or cramping, should be noted.

Muscle weakness is prominent in most electrolyte disturbances and, thus, by itself, is not diagnostic. Neuromuscular symptoms are prominent in severe hypokalemia (usually when the plasma potassium level is below 2.5 mEq/liter). Usually the weakness is most prominent in the quadricep muscles of the legs; muscles innervated by cranial nerves are almost never af-

fected. Fatigability is common in chronic hypokalemia and may be present for months. Muscle cramps, tenderness, and paresthesias may be pronounced in the patient with hypokalemia. With profound potassium deficit, the respiratory muscles become involved; "fishmouth" (pursed-lip) breathing may occur as respiratory failure ensues. Death may be caused by respiratory failure. Occasionally an extremely high plasma potassium level (greater than 7 mEq/liter and usually greater than 8.5) causes a rapidly ascending muscular weakness that leads to flaccid quadriplegia.[4]

Seizures may commonly occur in sodium imbalances, hypocalcemia, hypomagnesemia, and alkalosis. They less commonly occur in hypokalemia, hypercalcemia, and respiratory acidosis. Seizures do not occur in metabolic acidosis, hyperkalemia, or hypermagnesemia. (See Table 4-10 for possible implications of symptoms related to skeletal muscles.)

SENSATION

Patients with water and electrolyte imbalances frequently report changes in sensation. Some of the more common sensation changes are listed in Table 4-11.

TABLE 4-10 *Significance of skeletal muscle changes*

Symptom	Imbalance indicated by symptom
Muscle weakness (particularly in legs)	Chronic potassium deficit
	Phosphate depletion
Flabbiness (like half-filled hot water bottles)	Potassium deficit
Flaccid paralysis of respiratory muscles and extremities	Severe potassium deficit or excess
Hypertonus:	Calcium deficit
a. Chvostek's sign* may be positive	
b. Carpopedal spasm (Trousseau's sign†) (Fig. 12-2)	Alkalosis (decreased calcium ionization)
c. Tremors in mild deficit	Magnesium deficit
d. Convulsions in severe deficit	
Painful tonic muscle spasms	Calcium deficit
Muscle rigidity (particularly in limbs and abdominal wall)	Calcium deficit

*Chvostek's sign is a local spasm of the lip, nose, or side of the face following a tap just below the temple where the facial nerve crosses the jaw.
†Trousseau's sign refers to the hand assuming a position of palmar flexion after circulation to the hand is constricted for several minutes.

TABLE 4-11   *Significance of sensation changes*

Symptom	Imbalance indicated by symptom
Numbness and tingling of fingers and toes; circumoral paresthesia	Calcium deficit
	Alkalosis (decreased calcium ionization)
Light-headedness and tinnitus	Respiratory alkalosis
Abdominal cramps	Sodium deficit
	Potassium excess
Painful muscle cramps	Potassium deficit
	Calcium deficit
Numb, dead feeling, particularly in extremities (precedes flaccid paralysis)	Severe potassium deficit and excess
Nausea	Calcium excess
	Potassium excess
	Potassium deficit
Deep bony pain (decalcification of bones)	Calcium excess
Flank pain (calcium deposits in kidneys)	Calcium excess
Abnormal sensitivity to sound	Magnesium deficit
Paresthesia	Phosphate depletion
Warm sensation throughout body	Magnesium excess
Dizziness when turned quickly in bed	Sodium deficit
Headache (due to brain swelling)	Sodium deficit (water intoxication)

DESIRE FOR FOOD AND WATER

ANOREXIA The patient's interest in food and water is useful in evaluating his body fluid status. Anorexia is common in potassium deficit and in protein deficit. Many fluid imbalances are accompanied by nausea, vomiting, and anorexia.

THIRST Thirst is a subjective sensory symptom and has been defined as an awareness of the desire to drink. The sense of thirst is so strong a defender of the plasma sodium level in normal persons that hypernatremia never occurs unless thirst is impaired or rendered ineffective because the patient is unconscious or is denied access to water.

Gamble has estimated the daily water requirement of the adult at rest as 1500 ml. Butler and his associates cite a somewhat higher figure, 1500 ml/m² of body

surface/day. These figures are minimal, since the average active adult, not ill, requires from 2000 ml to 3000 ml of water per day, including 1000 ml to 1500 ml for insensible perspiration and 1000 ml to 1500 ml for urine excretion. Among the conditions that can increase the requirement for water are the following:

Fever

Excessive perspiration

Abnormal loss of fluids from vomiting, diarrhea, intestinal suction, and fistulas

Hyperthyroidism or any other cause of increased metabolic rate

Diminished renal concentrating ability, such as occurs in old age

The nurse should constantly remember how important it is to meet the daily water requirement of the patient by keeping accurate records of his intake and output and by carefully observing the patient for signs of water deficit.

Thirst is often caused by dryness of the mouth resulting from decreased salivary flow, which can be caused by ECF volume deficit. In this case, true thirst exists. Decreased salivary flow can also be caused by administration of atropine, in which case there is a desire to relieve the unpleasant dry sensation, but no true thirst.

Thirst is not always a reliable indicator of need and should not be the sole factor influencing fluid intake. The aged patient often does not recognize thirst and, even if he does, may be too weak to reach his water supply. Confusion or aphasia may stop him from making known his desire to drink. Patients with fluid volume excess caused by cardiac and renal damage are sometimes quite thirsty; a problem in this situation is to meet their need without adding to their fluid overload, and the nurse must use caution in allowing patients to ingest as much water as they desire. Seriously burned patients experience great thirst; if allowed to drink all the water they desire, serious sodium deficit will develop. Thirst in the burned patient should be met with specially prepared oral electrolyte solutions with the quantity carefully calculated or with intravenous infusion of fluids. (See Table 4-12 for implications of changes in desire for food and water.)

CHARACTER AND VOLUME OF URINE

SPECIFIC GRAVITY OF URINE To maintain fluid and osmolar balance, the kidneys must be able to dilute and concentrate urine. The specific gravity test is a

TABLE 4-12 *Significance of changes in desire for food and water*

Symptom	Imbalance indicated by symptom
Anorexia	Potassium deficit
	Protein deficit
	Calcium excess
	Fluid volume deficit (moderate and severe)
Thirst	Sodium excess (hypertonic ECF)
	Blood volume deficit due to hemorrhage
	Calcium excess (hypertonic ECF)
Absence of thirst	Sodium deficit (hypotonic ECF)

convenient and simple method for evaluation of the kidney's ability to perform this function.

The specific gravity of urine is its weight compared with the weight of an equal volume of distilled water; it indicates the amount of dissolved solids in the urine. The specific gravity of distilled water is 1.000; the specific gravity of urine, in health, ranges from 1.003 to 1.030. Random urine samples usually have a specific gravity of 1.012 to 1.025.[6] The higher the solute content of urine, the higher the specific gravity.

The chief urinary solutes are nitrogenous end products (urea), sodium, and chloride; others include potassium, phosphate, sulfate, and ammonia. Urinary solutes are mainly derived from ingested foods and from metabolism of endogenous protein and other substances. Diet, then, influences specific gravity of the urine. The usual diet supplies approximately 50 g of urinary solutes in 24 hours. Patients on low-sodium or low-protein diets cannot concentrate urine to high levels because they are ingesting an inadequate amount of solute. The inability of the patient eating a normal diet to concentrate urine is an indication of renal disease. Renal function is easily assessed by measuring the specific gravity of the first morning urine specimen. If the patient is able to concentrate a protein- and glucose-free urine to a level of 1.016 or higher, renal function probably is adequate.

The patient's state of hydration can be assessed by measuring specific gravity, provided the kidneys are healthy. A highly concentrated urine implies water deficit; a dilute urine implies adequate hydration or possibly overhydration.

Urine specific gravity measurement can help differentiate between the scanty urinary output of acute renal failure and that of water deficit. In acute renal failure, the specific gravity is fixed at a low level (1.010

to 1.012); in water deficit, the specific gravity is high. (Specific gravity persistently below 1.015 is a sign of significant renal disease.)

In the condition referred to as inappropriate secretion of antidiuretic hormone (SIADH), the urinary specific gravity is inappropriately high in relation to the plasma osmolality. (See the discussion of SIADH in Chapter 2)

The nurse may be asked to measure the urinary specific gravity of patients with burns, renal disease, cardiovascular disease, febrile conditions, and general surgical conditions. The following points should be kept in mind:

1. The nurse should be familiar with the equipment used to test the specific gravity of urine:
   a. The apparatus used for the test consists of two parts—the cylinder to contain the urine, and the urinometer (Fig. 4-6).
   b. Note that the urinometer is calibrated in units of .001, beginning with 1.000 at the top and progressing downward to 1.060. A urinometer is read from top to bottom (Fig. 4-7).
   c. New urinometers should be checked for accuracy against distilled water before use, and rechecked from time to time thereafter—even a slight discrepancy can be significant.
2. The urine sample must be fresh.
3. The urine sample must be well mixed—remember, the specific gravity test measures solute concentration, and a uniform solution must be used to yield an accurate reading.
4. The cylinder should be filled ¾ of the way with urine.
5. After the urinometer is placed in the cylinder, it is given a gentle spin with the thumb and forefinger to prevent it from adhering to the cylinder's sides. (See Fig. 4-6).
6. To read specific gravity, it is necessary that the urinometer be at eye level (Fig. 4-8). It is read by imagining a line where the lower portion of the meniscus crosses the scale on the urinometer.

The results of specific gravity tests are evaluated in relation to other clinical signs shown by the patient. (See Table 4-13 for conditions that may cause low or high specific gravity of urine.)

URINE OSMOLALITY The simultaneous measurement of plasma and urine osmolality is a more accurate way to measure the renal concentrating ability than is urinary specific gravity. Recall that the normal

Urinary pH tests measure the hydrogen ion concentration in urine and, because the excreted electrolytes vary according to the body's need, the urinary pH can range widely (from 4.5 to 8.0) and still be within normal limits. However, the pooled daily urine output averages around 6.0 (acidic) because more acid than alkali is formed in the body during metabolism, and these acids must be removed continually by the kidneys. Several factors may cause normal fluctuations in urinary pH.

FIG. 4-6. Urinometer.

FIG. 4-7. Urinometer scale.

plasma osmolality is 280 mOsm to 300 mOsm/liter. The extreme range of urine osmolality is 40 mOsm to 1600 mOsm/liter; typical normal urine has an osmolality of 500 mOsm to 800 mOsm/liter.[7]

*p***H OF URINE** The kidneys play a major role in maintaining the acid–base balance in the body by excreting electrolytes that are not required and by retaining those that are needed in the body. When an acid factor is in excess, the kidneys excrete hydrogen ions and conserve basic ions; when an alkaline factor is in excess, the kidneys excrete basic ions and retain hydrogen ions.

TABLE 4-13 *Conditions associated with persistently low or high urinary specific gravity*

Low specific gravity (1.010 or less)	Sodium deficit:
	    Drinking large quantities of water
	    Excessive parenteral administration of electrolyte-free solutions
	    Severely restricted dietary intake of sodium chloride
	    Diuresis from potent diuretics
	Diabetes insipidus (deficiency of antidiuretic hormone)
	Acute renal failure

High specific gravity (1.030 or higher)	Sodium excess:
	    Decreased water intake
	    Excessive loss of water (nonrenal)
	    Excessive ingestion of sodium chloride
	Glycosuria

Sleep causes urine to become highly acid, since respiration is depressed during sleep and a mild state of respiratory acidosis is induced. (Shallow respiration favors retention of carbon dioxide and an increase in the acid side of the carbonic acid/base bicarbonate ratio.) Note the low pH during the night hours in Figure 4-9. A rise in pH usually occurs upon awakening.

A rise in pH usually occurs following meals, since they stimulate the production hydrochloric acid, a process causing hydrogen ions to be extracted from the blood. Note the rise in pH following meals in Figure 4-9.

The type of food or fluid ingested affects urinary pH. (For example: a large intake of citrus juices or carbonated soft drinks tends to alkalinize urine; cranberry juice in large quantities tends to acidify urine.) For further discussion of urinary pH, see Urinary Calculi in Chapter 10.

Certain drugs significantly alter urinary pH. Drugs that cause urine to become more acid include ammonium chloride, methenamine mandelate, sodium acid phosphate, and ascorbic acid; those that cause urine to become alkaline include alkaline salts, such as sodium bicarbonate or potassium citrate.

In spite of the variables that influence urinary pH, most random urine samples show a pH of less than 6.6.

The nurse may be asked to perform urinary pH tests at the bedside for patients with a variety of metabolic disorders. Table 4-14 lists conditions in which

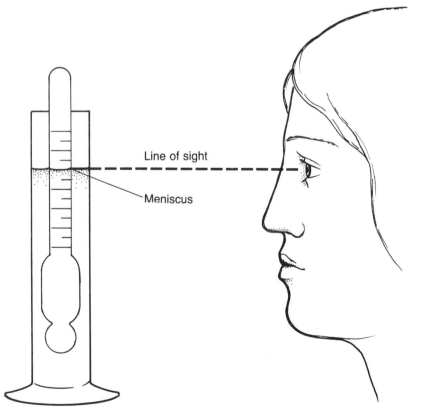

FIG. 4-8.
The urinometer must be at eye level for an accurate reading of specific gravity.

Line of sight

Meniscus

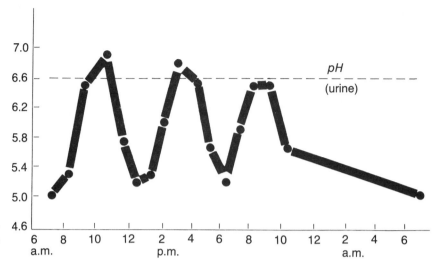

FIG. 4-9.
Daily fluctuations of urinary pH. (Picto-clinic 9:6, No. 6, Ames Co., Elkhart, Indiana, 1962)

urine is persistently acid or alkaline. Several simple methods may be used by the nurse to measure urinary pH. One such method is the use of Nitrazine Paper.

Squibb nitrazine paper When Nitrazine Paper is dipped into urine, a chemical change takes place that causes the paper to change color. The color can vary from yellow to blue and is matched against the scale on the paper dispenser. The color changes correspond to specific pH levels ranging from 4.5 to 7.5 in 0.5-unit increments. The nurse should keep the following points in mind while performing this test:

1. Only fresh urine should be used. (When urine is allowed to set for a time, urea breaks down into ammonia and the pH becomes more alkaline.)
2. The paper should be well moistened with urine but should not be left in the urine more than a few seconds—excessive fluid can wash away the chemicals from the paper.
3. After the paper has been dipped into the urine, the excess urine should be shaken off and the reading made immediately.
4. The color comparison between the Nitrazine Paper and the color scale on the dispenser should be made in a good light—avoid color comparison in pure fluorescent light. (Deduct 0.2 pH units for urine.)

CHANGES IN URINARY VOLUME Expected urinary volume can range from 1000 ml to 2000 ml/24 hr (40 ml–80 ml/hr) in patients in a basal state; it may decrease temporarily to 750 ml to 1200 ml/24 hr (30 ml–

50 ml/hr) following trauma or stress. A urine output of less than 500 ml/day constitutes oliguria. The expected urinary volume varies with each patient and is influenced by many factors. Urine volume is dependent upon the following factors:

1. The amount of fluid intake
2. Water needs of the lungs, skin, and gastrointestinal tract
3. The amount of waste products to be excreted by the kidneys
4. The ability of the kidneys to concentrate urine
5. Blood volume
6. Hormonal influences
7. Age

Fluid Intake

1. In health, urinary volume is approximately equal to the volume of liquids taken into the body—this rule does not hold true in illness. (See Table 4-15.)
2. A large intake of liquids causes a large urinary output; a small intake causes a small urinary output.
3. Adults usually pass between 1000 ml and 1500 ml of urine in 24 hours. (Most persons void five to ten times during the waking hours; each voiding is usually between 100 to 300 ml)

Water needs of the lungs, skin, and gastrointestinal tract

1. Water is available for urine formation only after the needs of the skin, the lungs, and the gastrointestinal tract have been met.

TABLE 4-14 *Clinical conditions associated with persistently acid or alkaline urine*

Symptom	Imbalance indicated by symptom
Acid urine	Metabolic acidosis: Diabetic acidosis Ketosis of starvation Severe diarrhea Respiratory acidosis: Emphysema Asthma Metabolic alkalosis accompanied by severe potassium deficit
Alkaline urine	Metabolic alkalosis: Excessive ingestion of alkalies Severe vomiting Respiratory alkalosis: Oxygen lack Fever with its hyperpnea Primary hyperaldosteronism Acidosis accompanied by the following clinical conditions: *Chronic renal infection,* in which the infecting organism converts urea in newly formed urine to ammonia *Persistent acidosis* secondary to renal tubular dysfunction in infants

2. Excessive sweating causes a decreased urinary volume—daily urinary volume is several hundred ml less in the summer.
3. Increased water loss from the lungs occurs with hyperpnea—this loss can cause a reduction in the urinary volume.

TABLE 4-15 *24-hour average intake and output of water in a healthy adult*

Intake		Output	
Oral liquids	1300 ml	Urine	1500 ml
Water in food	1000 ml	Stool	200 ml
Water of oxidation	300 ml	Insensible:	
		Lungs	300 ml
		Skin	600 ml
Total	2600 ml	Total	2600 ml

4. Excessive losses of fluid in vomiting and diarrhea can also cause a decreased urinary volume.

Amount of waste products to be excreted

1. Under most circumstances, urine volume varies directly with the urinary solute load—a solute excess causes an increased need for water excretion.
2. Excessive fluid loss may occur as the result of increased solute loads found in diabetes mellitus, thyrotoxicosis, fever, and response to stress.
3. A decreased solute load causes a decreased urinary volume.

Ability of the kidneys to concentrate urine

1. Kidneys able to concentrate urine normally have less need for water than damaged kidneys—a normal person with a urinary specific gravity of 1.029 to 1.032 requires 15 ml of water to excrete 1 g of solute, whereas a person with nephritis and a urinary specific gravity of 1.010 to 1.015 requires 40 ml of water to excrete 1 g of solute.
2. Low concentrating ability of the kidneys results in large urinary output.
3. Provided renal concentration is normal, the least amount of urine needed for excretion of daily metabolic wastes is 400 ml to 500 ml.
4. An acutely ill patient, or one with poor kidney function, may need to excrete as much as 3000 ml of urine daily to rid the body of waste products.

Blood volume

1. A decreased blood volume causes decreased urinary output due primarily to changes in arterial pressure and pressure in the glomeruli. (This phenomenon explains the oliguria or anuria of profound shock.)
2. An increased blood volume causes increased urinary output, again due primarily to changes in arterial pressure and pressure in the glomeruli.

Hormonal Influences

1. An increased secretion of antidiuretic hormone occurs when the blood volume is decreased—the kidneys increase water reabsorption, and the urinary volume is decreased.
2. An increased secretion of aldosterone occurs when the blood volume is decreased—the kidneys increase sodium reabsorpiton, and the urine volume is decreased.
3. A decreased secretion of antidiuretic hormone oc-

curs when the blood volume is increased—the kidneys decrease water reabsorption, and the urinary volume is increased.

4. A decreased secretion of aldosterone occurs when the blood volume is increased—the kidneys excrete more sodium and the urinary volume is increased. (See Table 4-16 for conditions associated with urinary volume changes.)

MEASURING AND RECORDING FLUID INTAKE AND OUTPUT

INTAKE–OUTPUT RECORDS A workable intake–output record has appropriate columns for all types of fluid gains and losses. The intake side of most records provides a column for oral intake and another column for parenteral fluids or other avenues of fluid gain. The output side usually provides a column for urine output and another column for gastrointestinal fluid losses or other routes of fluid loss. The record should be sufficiently simple so that the method of its use is self-evident, and necessary instructions should be incorporated in the record.

The type of intake–output record used often depends on the patient's condition. A patient with a severe fluid balance problem may require an hourly summary of his fluid gains and losses, so that the physician can plan treatment according to immediate needs. Many patients require only 8-hour summaries of their fluid gains and losses. Figure 4-10 depicts a bedside record suitable for this purpose.

TABLE 4-16 *Significance of urinary volume changes*

Symptom	Imbalance indicated by symptom
Oliguria	Fluid volume deficit
	Severe sodium excess
	Shock
	Renal insufficiency (late in disease)
Polyuria	Interstitial fluid-to-plasma shift
	Diabetes insipidus
	Increased renal solute load:
	Diabetes mellitus
	Infection
	Calcium excess
	Hyperthyroidism
	Renal insufficiency (early)

INDICATIONS FOR INTAKE–OUTPUT MEASUREMENT Physicians often indicate which patients are to have records kept of their daily fluid intake and output. However, the failure of the physician to request intake–output measurement is no reason to omit it when the patient has a real or potential water and electrolyte balance problem. Patients with the following conditions should automatically be placed on the fluid intake–output measurement list:

Following major surgery

Thermal burns or other serious injuries (especially head and facial bone injuries)

Suspected or known electrolyte imbalance

Acute renal failure

Oliguria

Congestive heart failure

Abnormal losses of body fluids

Diuretic therapy

Inadequate food and fluid intake

Coma

NEED FOR ACCURATE INTAKE–OUTPUT RECORDS Most discrepanices between gains and losses of body fluids can be detected when an accurate record is kept of the total fluid intake–output. The content of the gained and lost fluids is as important as their volume and, because the electrolyte content of the individual body fluids varies widely, the amount of each fluid lost should be designated on the intake–output record. The amounts and kinds of fluids taken into the body should also be recorded.

Ideally, the physician should be able to use the nursing intake–output records as a major tool in diagnosis as well as in the formulation of fluid replacement therapy. However, most physicians regard the usual intake–output record with the proverbial grain of salt. They are justified in complaining that an accurate account of a patient's intake–output is extremely difficult to obtain in the average hospital.

CAUSES OF INACCURATE INTAKE–OUTPUT RECORDS Although the principle of intake–output measurement is simple, neglect in keeping accurate nursing intake–output records is widespread. Causative factors are extremely difficult to pinpoint; however, the difficulty seems to be related to a combination of the following factors:

Failure to comprehend the value of accurate intake–output records

Liquid measurements

Coffee cup (¾ full)	150 ml	Paper soup bowl	210 ml
Coffee pot	240 ml	Paper cup (full)	180 ml
Drinking glass (¾ full) ..	210 ml	Paper cup (with ice)	120 ml
Ice glass (¾ full)	120 ml	Soup bowl (¾ full)	120 ml
Milk carton	240 ml	Soup cup (full)	120 ml
Paper coffee cup	180 ml	Squat glass (¾ full)	180 ml

When force fluids/restrict fluids:
Amount desired in 24 h _____ ml

I.V.		P.O.	
7-3 _____ ml		7-3 _____ ml	
3-11 _____ ml		3-11 _____ ml	
11-7 _____ ml		11-7 _____ ml	

Date: _____

Intake						Output						
						Urine		Other		Irrigation		
Oral	Time	Parenteral Fluids and Additives	Initial amount ml	Amount ml Absorbed	Time	Source	Amount ml	Source	Amount ml	Source	In/out	
		carryover										
8 h total			8 h total			8 h total		8 h total		8 h total		
		carryover										
8 h total			8 h total			8 h total		8 h total		8 h total		
		carryover										
8 h total			8 h total			8 h total		8 h total		8 h total		
24 h total			24 h total			24 h total		24 h total		24 h total		

Urine		Other				
U	= urine	NG	= nasogastric tube	Liq. S.	= liquid stool	
Cath.	= catheter	J	= jejunostomy	HV	= Hemovac	
U. Cath.	= ureteral catheter	TT	= T tube	ST	= Shirley tube	
Inc.	= incontinent	Vomitus	= vomitus	O	= other	
D	= diaper					

Total intake		Total output	
Oral		Urine	
Parenteral		Other	
Other			
Total		Total	

DECATUR MEMORIAL HOSPITAL
INTAKE AND OUTPUT RECORD

Form #9314   Revised: 2/81

FIG. 4-10. **Sample intake and output record. (Courtesy of Decatur Memorial Hospital, Decatur, Illinois)**

Understaffed patient-care units

Lethargic and improperly motivated personnel

Lax supervision

Unqualified personnel giving direct care to seriously ill patients

Inadequate in-service education programs for all levels of personnel

Failure to devise or implement a workable intake–output record

ACHIEVING ACCURATE INTAKE–OUTPUT RECORDS It is not technically difficult to measure fluid intake and output or to record the measurements. Yet, persistent effort is required if one is to achieve an accurate account of gains and losses of fluids.

There are innumerable possiblities for error in the measurement and recording of fluid gains and losses; however, some errors occur much more frequently than others. Common errors and suggestions on overcoming them are shown in Table 4-17.

Pflaum examined the accuracy of a routine intake–output technique in a general hospital and found it lacking; when compared with measurement of body weight, the mean daily error was 800 ml.[8] From her results she concluded that measurement of intake–output is of little value and that a more accurate indicator, body weight, should be adopted instead. While these authors agree with the inaccuracy of intake–output records in many situations, we generally feel that *both* indicators (intake–output measurement *plus* body

(Text continues on p. 110)

TABLE 4-17 *Overcoming common errors* (continued)

Common errors	Suggestions
Errors Involving Both Intake and Output:	
Failure to communicate to the entire staff which patients require intake–output measurement	a. A "measure intake–output" sign should be attached to the patient's bed to serve as a reminder.
(Body fluids are often discarded without being measured, and oral fluids are not recorded, merely because staff members are not aware of the patients on intake–output)	b. A list of all patients requiring intake–output measurement should be posted in a convenient work area for quick reference.
	c. Also for quick reference, the cardex should contain a list of all patients requiring intake–output measurement.
	d. An adequate patient report should be given to all personnel.
Failure to explain intake–output to the patient and his family	a. Both the patient and his family should receive a simple explanation of why intake–output measurement is necessary.
(Most patients will cooperate *if* they know what is expected of them)	b. Careful instructions are necessary to acquaint the patient and his family with their role in helping to achieve an accurate intake–output record.
Well-meaning intentions to record a drink of water or an emptied urinal at a later, more convenient time are often forgotten	Measurements should be recorded at the time they are obtained.
Failure to measure fluids that can be directly measured because it takes less time to guess at their amounts	Measure *all* fluids amenable to direct measurement—guesses should be reserved for fluids that cannot be measured directly.
Errors Related to Intake:	
Failure to designate the specific volume of glasses, cups, bowls, and other fluid containers used in the hospital	The bedside record should list the volumes of glasses, cups, bowls, and other fluid containers used in the hospital (see Fig. 4-10).
(Each person may ascribe a different volume to the same glass of water)	
Failure to obtain an adequate measuring device for small amounts of oral fluids	Small calibrated paper cups should be kept at the bedside for such a purpose.
(Patients frequently drink small quantities of fluids; the amounts must be estimated unless a calibrated cup is available—frequent estimates increase the margin of error)	
Failure to consider the volume of fluid displaced by ice in iced drinks frequently causes an overstatement of ingested oral fluids	Only small amounts of ice should be used for iced drinks so that the accurate amount of fluid ingested can be recorded.

(continued)

TABLE 4-17 *Overcoming common errors* (continued)

Common errors	Suggestions
Errors Related to Intake (continued):	
Assuming that the contents of empty containers were drunk by the patient	The patient should be asked what fluids he drank.
(Patients sometimes give their coffee or juice to a visiting relative or other patients in the room; they may forget to tell the person checking the tray)	
Failure to accurately record the amount of parenteral fluid administered on each shift	a. The actual amount of parenteral fluid run in on each shift should be recorded.
	b. The amount of solution left in the bottle at the end of a shift should be noted, in pencil, on the bedside record—this makes it easier for the next shift to determine the amount run in on their time.
Errors Related to Output:	
Failure to estimate fluid lost as perspiration	a. An attempt should be made to describe the amount of clothing and bed linen saturated with perspiration—it has been estimated that one necessary bed change represents at least 1 liter of lost fluid.
(Many nurses fail to recognize perspiration as a major source of fluid loss)	b. Some intake–output records require the nurse to estimate perspiration as +, + +, + + +, or + + + + (+ represents sweating that is just visible, and + + + + represents profuse sweating).
Failure to estimate "uncaught" vomitus	The amount of fluid lost as vomitus should be estimated, and recorded as an estimate—it is better to make a guess than to give no indication at all as to the amount.
(Frequently, "uncaught" emesis is recorded merely as a lost specimen)	
Failure to estimate the amount of incontinent urine	The amount of incontinent urine should be estimated—it is helpful to note the amount of clothing and bed linen saturated with urine.
(Intake–output records often indicate the number of incontinent voidings but give no indication of the amounts; obviously, such records are of little value)	
Failure to estimate fluid lost as liquid feces	a. The patient should be encouraged to use the bedpan rather than the toilet so that the fluid loss can be directly measured.
	b. The amount of fluid lost in incontinent liquid stools should be estimated.
Failure to estimate fluid lost as wound exudate	a. The amount of drainage on a dressing should be measured and charted—this can be done by measuring the width of the stained area and determining the thickness of the dressing.
	b. If extreme measures are necessary, the dressing can be weighed before application and again when removed.
Failure to check a urinary catheter for patency when there is decreased drainage of urine	Decreased drainage from a urinary catheter is an indication to irrigate it and check for patency before charting the absence of, or decrease in, urinary output.
(It is sometimes too quickly assumed that decreased drainage from a catheter is due to decreased urine formation)	
Failure to obtain an adequate measuring device for hourly or more frequent checks on urinary output	A collecting device calibrated to measure *small amounts* of urine should be utilized. (see Fig. 4-11.)
(An error of even 10 ml could be significant when dealing with small amounts of urine)	
Failure to record the amount of solution used to irrigate tubes and the amount of fluid withdrawn during the irrigation	a. One method for dealing with this problem is to add the amount of irrigating solution to the intake column, and to add the amount of fluid withdrawn to the output column.
	b. Another method is to compare the amount of irrigating solution used and the amount of fluid withdrawn during the irrigation—if more fluid was put in than was taken out, the excess is added to the intake column; if more fluid was taken out than was put in, the excess is added to the output column.

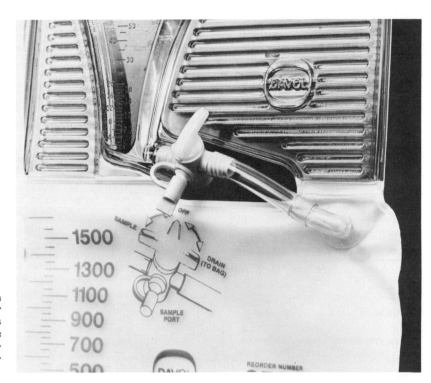

FIG. 4-11.
Davol #3590 Uri-Meter drainage bag, a collecting device that can be used for precise measurement of small volumes of urine. After hourly collection, the urine can be diverted into the large drainage bag. (Courtesy of Davol, Inc., Cranston, Rhode Island)

weight) are necessary to adequately monitor the patient with a fluid balance problem.

BLOOD UREA NITROGEN/CREATININE LEVELS

The nurse should be aware of implications of changes in blood urea nitrogen (BUN) and serum creatinine levels; both of these values are useful in evaluating fluid balance status and renal function.

Recall that urea nitrogen is the end product of protein metabolism; most of the urea formed daily is excreted in the urine. (The normal BUN ranges between 10 mg to 20 mg/dl.) Elevated BUN levels can be due to:

> Increased protein catabolism (as in excessive protein intake, starvation, infection, trauma, and bleeding into the intestine)
>
> Fluid volume depletion, as in excessive diuresis, vomiting, diarrhea, and low fluid intake (resulting in lowered glomerular filtration rate and thus in decreased excretion of nitrogenous wastes)
>
> Poor perfusion (as in diminished cardiac output or sudden hypotension in a hypertensive patient) resulting in lowered glomerular filtration rate

Creatinine, a product of muscle metabolism, is another nitrogenous waste; its normal serum level ranges between 0.7 mg to 1.5 mg/dl. Creatinine is a more reliable assessment tool for renal function than is BUN because it is largely unaffected by diet and fluid intake.

The normal BUN/creatinine ratio is approximately 10:1. When the ratio increases in favor of the BUN, conditions such as hypovolemia, low perfusion pressures to the kidney, or increased protein metabolism may be present. When both the BUN and creatinine levels rise (even though a 10:1 ratio is maintained) the problem is likely renal in nature.

BODY WEIGHT

The daily weighing of patients with potential or actual fluid balance problems is of great clinical value because accurate body weight measurements are much easier to obtain than accurate intake–output measurements and rapid variations in weight closely reflect changes in fluid volume. A loss of body weight will occur when the total fluid intake is less than the total fluid output; conversely, a gain in body weight will occur when the total fluid intake is greater than the total fluid output (Table 4-18). For instance, a gain or loss of 1 kg (2.2 lb) of body weight is approximately equivalent to the gain or loss of 1 liter of fluid. Or ex-

TABLE 4-18 *Signficance of weight changes*

Symptom	Imbalance indicated by symptom
Rapid loss of 2% total body weight	Mild fluid volume deficit
Rapid loss of 5% total body weight	Moderate fluid volume deficit
Rapid loss of 8% total body weight	Severe fluid volume deficit
Rapid gain of 2% total body weight	Mild fluid volume excess
Rapid gain of 5% total body weight	Moderate fluid volume excess
Rapid gain of 8% total body weight	Severe fluid volume excess
Chronic weight loss	Protein deficit

pressed another way, a gain or loss of 500 ml of fluid is equivalent to a gain or loss of 1 lb. The use of body weight as an accurate index of fluid balance is based on the assumption that the patient's dry weight has remained stable and that the change observed is due to changes in body water. This is a reasonably safe as-

sumption, since tissue catabolism and anabolism are not likely to result in significant differences in weight from one day to the next. (Recall that even under starvation conditions, a person loses only ⅓ lb to ½ lb per day.)

The nurse should recall that fluids can be lost to the body in the pooling that occurs, for example, with intestinal obstruction. Such losses, which can cause a serious plasma fluid volume deficit, are not reflected by weight changes.

Daily weight measurement may be indicated in the same conditions listed earlier as indications for fluid intake–output measurement. At the minimum, all patients should be weighed on admittance, so that a baseline can be established for later comparison.

It is important that the same scale be used for repeated measurements, since any two portable scales seldom give the same reading. The daily variations in weight should reflect true body weight changes rather than variations between scales.

Ambulatory patients may be weighed on small portable scales. Seriously ill patients confined to bed can be weighed with a bed scale (Fig. 4-12).

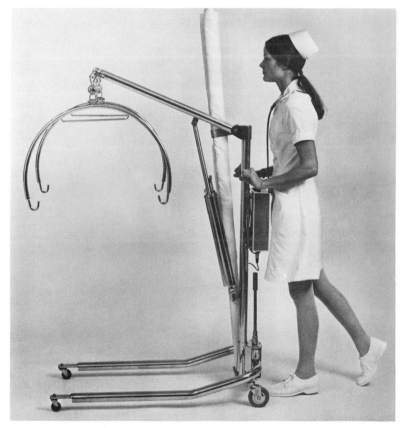

FIG. 4-12.
Scale-Tronic 2001. Seriously ill patients confined to bed can be weighed on a bed scale. (Courtesy of Scale-Tronic, White Plains, New York)

It is useless to weigh the patient daily if the procedure is not performed the same way each day. The nurse should strive for accurate weight measurements because the physician often bases the administration of fluids and diuretics on the recorded weight changes. The following practices should be followed:

The same scales should be used each time.

The weight should be measured in the morning before breakfast.

The patient should empty his bladder before each weight measurement.

The same or similar clothing should be worn each time (the clothing should be dry).

FREQUENCY AND CHARACTER OF STOOLS

The nurse should record bowel movements and any significant facts related to them on the chart. The consistency of the stool (solid or liquid) and the frequency of evacuation should be noted.

Abnormal fecal losses and their causal relationship to fluid imbalances are discussed earlier in this chapter. Table 4-19 lists some of the fluid imbalances accompanied by changes in the character and the frequency of bowel movements.

CONCLUSION

The preceding discussion of planned nursing observations has been quite general. The reader is encouraged to further explore and assimilate pertinent and detailed nursing observations (in context with appropriate nursing actions) in the chapters dealing with specific conditions.

TABLE 4-19 *Significance of stool changes*

Symptom	Imbalance indicated by symptom
Hard fecal mass	Fluid volume deficit
Intestinal colic with diarrhea	Sodium deficit
	Potassium excess
Ileus, abdominal distention (little or no stool or flatus passed)	Potassium deficit
Diminished bowel sounds	

REFERENCES

1. Schwartz S (ed): *Principles of Surgery, 3rd ed*, p 69. New York, McGraw–Hill, 1979
2. Condon R, Nyhus L: *Manual of Surgical Therapeutics, 4th ed*, p 216. Boston, Little, Brown & Company, 1978
3. *Ibid*, p 213
4. Maxwell M, Kleeman C (ed): *Clinical Disorders of Fluid & Electrolyte Metabolism, 3rd ed*, McGraw-Hill, 1980
5. *Ibid*, p 500
6. Condon and Nyhus, Surgical Therapeutics, p 165
7. *Ibid*, p 209
8. Pflaum S: Investigation of intake-output as a means of assessing body fluid balance. Heart Lung 8, No. 3:495, 1979

It is better to prevent disturbances of water, electrolytes, and other nutrients than to have to treat them after they have developed. Any patient can develop a serious nutritional disturbance if his needs for nutrients are not considered and met; consequently, the practical application of the principles of nutrition is as much a part of nursing as is the administration of medications. Examples of common clinical events that can lead to nutritional disturbances are listed below:

1. Inadequate intake
2. Excessive losses from such conditions as
   a. Fever
   b. Vomiting
   c. Gastric or intestinal suction
   d. Diarrhea

   Medical therapy, such as the prolonged use of diuretics, laxatives, enemas, or steroids
   Catabolic effects, such as those induced by
   a. Immobilization
   b. Draining decubitus ulcers
   c. Surgical operations (hemorrhage, toxic destruction of protein)
   d. Trauma

When one considers how frequently the nurse encounters situations such as these in her daily practice, the need for a carefully considered plan of action on her part becomes evident.

Compared to other healing disciplines, the profession of nursing is unique in that its ministrations are applied continuously to the hospitalized patient. Initiative is mandatory on the part of the nurse if the nutritional therapy is to produce optimal results. Hence, the nurse has an important role in contributing to the nutritional assessment of patients.

NUTRITIONAL ASSESSMENT

Clinical malnutrition exists in hospitals today to an extent that hardly seems possible, especially among the aged, and among surgical patients, cancer patients, and other chronically ill persons (such as those with arthritis and chronic obstructive pulmonary disease [COPD]). The aged are at particular risk because they typically have diminished incomes, frequently live alone, often wear (or need) dentures, and tend to lack interest in food.

In part, the rampant malnutrition in hospitals is

"He who does not prevent a crime when he can, encourages it."

—Seneca

Nutritional Interventions to Prevent or Minimize Imbalances of Water, Electrolytes, and Other Nutrients

caused by certain hospital practices; for example, the routine preparations for surgery and for many laboratory and radiographic tests are often nutritionally debilitating, particularly if they must be repeated. (Unfortunately, it is not uncommon for errors to occur, necessitating that tests be repeated or surgery cancelled and then rescheduled). Rarely is the amount of protein and calories omitted in those lost meals made up at a later time. Not only does hospital malnutrition exist, it often increases in *incidence* and *severity* with increased length of hospital stay!

Malnourished patients have difficulty withstanding medical and surgical therapies and in achieving wound healing and recovery from disease. To deal with the problem, some institutions have developed specialized nutritional support teams, ideally consisting of a physician–clinical nutritionist, a therapeutic dietitian, a nurse, and a pharmacist. Such teams conduct nutritional evaluation, prescribe preventive and therapeutic nutrition modalities, monitor the therapies, and review the results. Where such services do not exist, malnourished patients must be diagnosed and treated by the primary care physician, using the services of the clinical dietician, the staff nurse, and the pharmicist.

NUTRITIONAL HISTORY

The nurse can help identify malnourished patients by addressing the following pertinent areas during the nutritional segment of the nursing history:

Is the usual body weight 20% above or below normal?

Has there been recent loss or gain of 10% of usual body weight?

Are there ill-fitting dentures?

Is there any evidence that income is inadequate for food purchasing?

Does the patient live alone and prepare his own meals?

Are there any swallowing difficulties?

Is there any paralysis of extremities (making feeding of self difficult)?

Are there any mobility problems, interfering with food preparation and eating?

Is there any chronic illness (such as arthritis, COPD, renal or liver disease, malabsorption)?

Is there any recent major illness, surgery, or injury?

Has the patient been maintained more than 10 days on "routine" intravenous fluids?

Is there a history of excessive use of alcohol?

Is there frequent use of a monotonous diet?

Is the patient taking any medications interfering with appetite or metabolism or absorption?

Have abnormal losses of body fluids occurred such as vomiting, loose stools, or drainage from enterostomy, colostomy, fistulas, or wounds)?

Significant findings should be discussed with both the physician and clinical dietician and a plan of action determined.

PHYSICAL ASSESSMENT

Gross physical examination can help reveal nutritional deficiencies manifested by the conditions listed below:

Does the patient appear skinny or obese?

Is hair sparse, dry, and easily broken?

Can small clumps of hair be pulled out with moderate force and no pain, especially from the side of the head?

Are the nails brittle?

Do nails on more than one extremity have crosswise ridges or grooves?

Does the skin "break down" easily?

Do wounds fail to heal normally? (Examination of wounds will give a general idea of the patient's nitrogen status; healing may be markedly retarded if a prolonged protein deficit is present.)

Do muscles have a "wasted" appearance?

Does the skin display pallor, areas of hyperpigmentation or petechiae?

Is there hepatomegaly or swollen abdomen?

Is there paleness of bucchal mucosa and conjunctiva?

Is the patient listless or apathetic?

Is there swelling and redness of the mouth and lips?

Is edema present?

NURSING RESPONSIBILITIES During the hospital stay, the nurse has primary responsibility for the following areas:

Assisting the patient to eat a therapeutic diet

Describing actual food consumption, including any provided by nonhospital sources, and noting inadequate intake (sadly, most hospital charts contain little information about what the patient eats or does not eat)

Working with the physician and clinical dietician to promote optimal nutrition, particularly when a problem exists

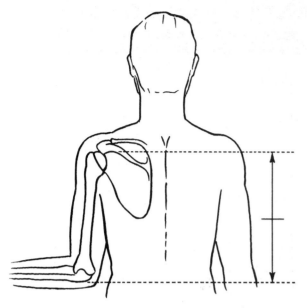

FIG. 5-1. Assessing midpoint of upper arm.

Determining height and weight on admission and measuring weight at regular intervals throughout hospitalization

Measuring all routes of fluid intake and output

Anthropometrics In addition to the above duties, the nurse may be responsible for measuring other parameters of nutritional status. As valuable as height and weight measurements are for indicating nutritional status, they have limitations; for example, edema induced by protein-calorie malnutrition or other disease states can give falsely high weight readings. Daily weight fluctuations tend to reflect body water changes rather than real body mass changes; weight trends over a period of time tend to be more helpful in measuring nutritional status. To accurately assess nutritional status in patients with probable malnutrition, it is sometimes recommended that skinfold thickness and mid-upper arm circumference be measured; from these two measurements the lean muscle mass can be estimated. (A skinfold consists of two layers of subcutaneous fat without any muscle or tendon included.)[1]

Triceps skinfold thickness measurements are good indicators of body fat and energy reserves; subcutaneous fat is estimated by measuring the triceps skinfold located at the back of the upper arm midway between the elbow and shoulder (i.e., halfway between the acromian process of the scapula and the olecranon process of the ulna; Fig. 5-1). The skinfold thickness is measured over the triceps muscle; a fold of the skin is gently grasped with the thumb and forefinger of one hand while the thickness is measured with a caliper (Fig. 5-2). Three readings are taken in succession, allowing the calipers to exert pressure for 3 seconds before each reading. An average of the 3 readings is recorded. (The measurement of skinfold thickness offers a rough estimate of total body fat, since it is known that approximately 50% of body fat is subcutaneous.)

Another parameter measured in nutritional assessment of patients with suspected malnutrition is the mid-upper arm circumference. Arm circumference is measured with a tape calibrated in millimeters. (Use of a steel tape avoids the errors introduced by stretching of a cloth or paper tape.) The examiner must be careful to exert the same amount of tension on the tape during each measurement (Fig. 5-3). Arm circumference and skinfold measurements should be taken on the pa-

FIG. 5-2.
Measurement of triceps skinfold with calipers. (Blackburn G: Techniques of nutritional assessment. In: Nutritional Assessment, p 8. Berkeley, Cutter Laboratories, Inc., 1980)

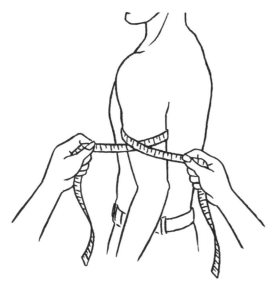

FIG. 5-3. Measurement of mid-upper-arm circumference.

tient's nondominant side when possible; the arm circumference measurement should be taken while the arms hang freely. *Muscle* circumference can be calculated once mid-upper arm circumference and skinfold thickness have been determined. (See footnote at bottom of Table 5-1.) Table 5-1 also lists normals for triceps skinfold thickness, mid-upper arm circumference, muscle circumference, and the level of depletion associated with each.

In summary, muscle circumference (lean muscle mass) is an important indicator of nutritional status,

TABLE 5-1 *Standard values for arm and muscle circumference and triceps skinfold*

	Men*	Women*	Depletion status
Mid-upper arm Circumference (cm)	>26 18–26 <18	>26 18–26 <18	None Moderate Severe
Muscle Circumference[†] (cm)	>23 16–23 <16	>20 14–20 <14	None Moderate Severe
Triceps skinfold (mm)	>11 7.5–11 <7.5	>15 10–15 <10	None Moderate Severe

*Adults only
[†]Calculation of muscle circumference: C = πD.
   C = circumference; π = 3.14; D = diameter.
Mid-upper arm circumference ÷ 3.14 = arm diameter
Arm diameter − triceps skinfold thickness = muscle diameter
Muscle diameter × 3.14 = muscle circumference
(Halpern S: Quick Reference to Clinical Nutrition, p 369. Philadelphia, JB Lippincott, 1979)

since it becomes diminished when protein intake has been inadequate for a prolonged period. Inadequate protein ingestion causes catabolism of muscle mass in order to maintain plasma protein.

LABORATORY TESTS

In patients with normal renal function, 24-hour urinary creatinine excretion is related to lean body mass. Estimation of the creatinine–height index (CHI) indicates the status of muscle stores and is a very sensitive indicator of protein depletion in cachectic states. It is established by comparing the patient's 24-hour creatinine excretion with normal creatinine clearance in a person of similar height and ideal weight. With protein malnutrition, creatinine excretion decreases.

The following laboratory tests are also useful in initial nutritional assessment and in the continuous assessment of the form of nutritional support used for the patient:

   Serum albumin
   Total white blood cell count
   White blood cell differential
   Total iron-binding capacity (TIBC)

The normal serum albumin level is approximately 3.4 g to 5.0 g/dl; the nutritionally depleted patient will have a below normal albumin level. (Remember, however, that the serum albumin level may be altered by the state of hydration and by the administration of blood products.)[2] For example, excessive administration of "routine" intravenous solutions (such as D_5W) causes depression of the serum albumin level.

Since albumin has a 16-day turnover rate, this protein is usually the last to be resynthesized after nutritional replacement therapy has begun; therefore, monitoring for improvement with therapy is best done by estimating the serum transferrin (a more sensitive gauge for acute protein-calorie malnutrition since it has a turnover rate of 7 days).[3] Serum transferrin can be estimated when the TIBC is known; a depressed transferrin level should be suspected when the TIBC is less than 250 μg/dl.[3]

The total white blood cell count and differential are significant in evaluating the immune response capabilities of patients with suspected malnutrition. A total lymphocyte count of less than 1500 may be suggestive of protein-calorie malnutrition.[4] In the presence of infection, failure of lymphocyte count to increase may well be an indication that cellular defenses have been impaired due to protein catabolism. When polymorphonuclear leukocytes are unable to respond to infection by leukocytosis, the patient is in danger of fatal septicemia.

Other laboratory tests may be helpful in identifying malnourished patients. For example, a blood urea nitrogen (BUN) concentration below the normal range suggests inadequate protein intake, and below normal serum levels of calcium and phosphate suggest vitamin D deficiency.[4] In semistarvation, 2-hour postprandial blood sugar levels may fall within the diabetic range.[5]

SKIN ANTIGEN TESTING

Hypersensitivity skin antigen testing may be used to further analyze the patient's immune system. Certain challenge antigens (such as *Candida* extract, mumps, tuberculin, and streptokinase/streptodornase) are administered intradermally and the patient's response observed; failure to respond positively (greater than 5 mm induration) may be regarded as a sign of impaired immune competence, since the majority of persons have been sensitized to most of these antigens in the course of normal living.[3] A negative response, when coupled with other abnormal nutrition indices, may, be due in part to protein-calorie malnutrition. Nutritional status can be monitored periodically be repeating skin antigen testing; the conversion from a negative to a positive skin response indicates improvement in the nutritional state. Variations exist in the rate of development of positive reactions and in their duration according to the antigen, the subject tested, and the number of prior skin tests administered.[6] For example, reactions to mumps antigen usually peak faster and subside more rapidly than those to PPD. Persons being tested for the first time tend to develop reactions more slowly than those who are tested repeatedly, and the reactions of those tested repeatedly tend to subside more rapidly than those tested for the first time.[6] Skin tests may be unreliable if the patient is receiving steroids.

MALNOURISHED STATES

There are essentially three types of protein-calorie deficiency states:

Calorie deficiency (marasmus)

Protein deficiency (kwashiorkor)
Protein-calorie malnutrition

Calorie deficiency is characterized by decreased body weight and decreased skinfold thickness; however, there is normal muscle mass, normal visceral protein (albumin and transferrin), and normal immunity. Persons in this category have primarily utilized fat stores rather than protein stores during their period of inadequate food intake.

Protein deficiency is characterized by decreased serum albumin, decreased transferrin, decreased immune competence, and decreased muscle mass. Body weight is normal, as in skinfold thickness. Protein deficiency occurs most frequently in stress states (such as trauma or sepsis) in which energy needs are met, at least in part, with amino acids.

Protein-calorie malnutrition is characterized by decreased body weight, muscle mass, and skinfold thickness as well as by decreased immune competence.

After the degree and type of nutritional deficit have been established, the physician will prescribe the appropriate nutrients. The route of replacement must also be determined, namely, oral feedings, tube feedings, or parenteral fluids. If oral feedings are selected, the type of diet must be designated. If tube feedings are selected, there are a variety of formulas from which to choose. Finally, if the intravenous route is selected, there are numerous types of fluids to consider. A patient with adequate reserves of fat and skeletal muscle may often be maintained on appropriate amounts of fluids, electrolytes, vitamins, minerals, and enough protein to offset obligatory nitrogen losses. However, the severely depleted patient needs prompt and vigorous nutritional support in the form of total parenteral nutrition (TPN). (Tube feedings and elemental diets are discussed later in this chapter; various forms of parenteral nutrition are discussed in Chapter 6).

NURSING RESPONSIBILITIES RELATED TO INADEQUATE INTAKE

ORAL ROUTE

With respect to the intake of water, electrolytes, vitamins, and other nutrients, the oral route has long been preferred for the prevention of deficits and the correction of mild deficits already present. It offers several advantages:

The oral route appeals to the patient as being more natural.

Because it allows gradual absorption, the oral route provides far greater efficiency in the utilization of administered nutrients.

Water, electrolytes, and other nutrients ingested orally are less apt to upset water and electrolyte balance than when taken by the parenteral route.

The oral route is far less expensive.

Naturally, the gastrointestinal tract must be functional if the oral route is to be used.

NURSING MEASURES TO PROMOTE EATING Illness or imposed dietary restrictions may greatly decrease the desire to eat, and offering an adequate diet does not guarantee that the patient will accept it. The real test of nursing skill is not to order and serve a diet; rather, it lies in getting the patient to eat it. The nurse should keep the following points in mind:

Review the patient's eating habits, his food likes and dislikes, and the cultural and religious beliefs that influence his eating habits.

Permit the patient to make dietary choices when possible, since this will increase the acceptability of the diet.

Eliminate unpleasant sights and odors at mealtime (e.g., change soiled dressings, empty ileostomy bags, remove from sight such items as bedpans, urinals, dead flowers).

Minimize strong emotions near mealtime, when possible, since they affect both appetite and digestive processes adversely. (Incidents that irritate the patient should be noted on the nursing care plan and avoided, if possible, in the future.)

Avoid unpleasant procedures in the vicinity at mealtime.

See that the patient is clean before eating (change soiled linens, wash face and hands).

Provide comfort measures as needed before meals (e.g., the patient with a distended bladder may need to be catheterized; the patient in pain may require a p.r.n. medication).

Position the patient in high Fowler's, if permitted, to facilitate ease of swallowing.

Situate the tray so that it is easy for the patient to reach.

Help the patient prepare the tray if this is necessary (open milk cartons, pour coffee, cut up food). Remember that a weak patient or one with an IV device in his hand cannot easily perform these activities.

Serve food portions appropriate to the patient's appetite (remember that a patient with a small appetite can be repelled by large portions of food).

Avoid unnecessary interruptions during meals (the not-too-eager appetite can be easily discouraged by the poor planning of nursing activities).

Be sure that patients with dentures have them in place before serving the meal. Patients with ill-fitting dentures or few teeth may require soft or liquid foods to provide adequate nourishment. Dietary departments will usually grind hard-to-chew foods for such patients if requested.

Serve trays last to patients who must be fed. (It is frustrating for patients to wait for someone to return to feed them.)

Feed patients in a friendly, unhurried manner, allowing sufficient time to chew and swallow comfortably. Offer liquids and solids alternately in manageable amounts.

Allow the patient's family to bring him favorite foods from home, if permitted, under nursing supervision.

Encourage the patient to be as physically active as allowed in order to stimulate his appetite.

Every effort should be made to prevent nausea and vomiting after meals; therefore, the patient should not be turned quickly or subjected to excessive physical activity. On the other hand, mild exercise should not be condemned, for it may aid digestion. If nausea should occur, appropriate medication may be given in accord with medical orders. In the absence of a p.r.n. order for nausea, the nurse might seek a medical order. The cause of the nausea should be sought and noted on the chart.

Although the points just enumerated are well-known to most nurses, *we are convinced that they cannot be overemphasized.* One does well to remember that any program of diet therapy is worthless if the patient does not eat! Conversely, any measure capable of promoting food and liquid intake contributes to the patient's welfare. The nurse should impress the importance of these facts on the auxiliary staff members, who are frequently charged with important responsibilities associated with feeding the patients.

ELEMENTAL DIETS Elemental diets are powdered mixtures of basic nutrients (eight essential amino acids plus a number of nonessential amino acids, simple sugars, electrolytes, minerals, trace elements, and essential fatty acids (Table 5-2). When mixed with water, the powdered mixture forms a stable solution, similar in content to those used for parenteral hyperalimentation. Elemental diets are frequently used to maintain caloric and nitrogen balance over long periods of time when the patient is unable to eat conventional foods. These diets may be administered orally or by nasogastric tube, gastrostomy, or jejunostomy. Elemental diets

TABLE 5-2 *Electrolyte content of two elemental diets*

	High Nitrogen Vivonex (Norwich–Eaton) 80-g packet	Standard Vivonex (Norwich–Eaton) 80-g packet
	mg	mg
Sodium	158.6	140.5
Potassium	351.9	351.5
Calcium	100.0	166.7
Magnesium	40.0	66.67
Manganese	0.281	0.468
Iron	1.8	3.0
Copper	0.2	0.333
Zinc	1.5	2.5
Selenium	0.015	0.025
Molybdenum	0.015	0.025
Chromium	0.005	0.008
Chloride	244.6	216.6
Phosphate	100.0	166.6
Acetate	459.9	331.8
Iodide	0.015	0.025

are low in bulk and require only minimal digestive effort before total absorption in the small intestine; they are appropriate for patients who have a normally functioning gastrointestinal tract and for those who have digestive or absorptive malfunction (such as short bowel syndrome, inflammatory bowel disease, pancreatic insufficiency, or enteric fistulas). Some elemental diets are available with low sodium content for patients requiring low sodium intake. Points to consider in the use of elemental diets include the following:

1. Elemental diets are hyperosmolar, causing them to draw fluid into the stomach if consumed too rapidly, thus resulting in gastric distention, nausea, vomiting, cramping, and diarrhea. Some diets have added flavorings to mask their metallic taste; these flavorings further increase the osmolarity of the feeding.
2. To prevent these untoward gastrointestinal symptoms, the patient should be instructed to sip a glass of the solution slowly over a period of an hour. The mixture should never be drunk all at once (although patients may try to do so to avoid prolonging the unpleasant sulphurlike flavor). If given by tube feeding, the solution should be administered *slowly*; bolus volumes of these solutions cause delayed gastric emptying and diarrhea.
3. Chilling improves the palatability of most flavors; however, patients with swallowing problems should take the liquid at room temperature.
4. *After the elemental diet is reconstituted, it becomes perishable and must be refrigerated.* It is best to prepare no more than is needed in a 24-hour period.
5. Intake and output records should be kept and extra water should be provided to maintain an adequate urinary output. Periodic measurement of urinary glucose, acetone, and pH may be ordered. Because of the diet's high carbohydrate content, some depleted patients may develop hyperglycemia, requiring insulin for regulation.
6. Because elemental diets are hypertonic, they should be used cautiously in patients sensitive to hyperosmolarity (such as diabetics).
7. Because of the hypertonicity of elemental diets, the solution should be delivered in a more diluted form during the initial days; if untoward effects do not occur, the concentration can be gradually increased.
8. Remember that stool bulk is greatly diminished when elemental diets are the sole source of intake. (These diets were originally designed for astronauts with the purpose of producing minimal fecal waste while providing adequate nutrition.
9. Daily weights should be recorded at first, then at least biweekly. After positive nitrogen balance has been achieved, the patient will begin gaining weight.
10. It is wise to monitor the BUN, glucose, serum electrolytes, and proteins at least weekly for patients receiving elemental diets.
11. Patient tolerance of elemental diets is judged by the presence or absence of nausea, vomiting, cramps, or diarrhea and by attention to laboratory values that indicate hyperosmolarity and dehydration.

TUBE FEEDINGS

INDICATIONS Oral intake might be impossible, even though adequate gastrointestinal function is present, in such situations as those listed on page 120.

Semiconsciousness or unconsciousness

Swallowing problems

Extreme anorexia

Head and neck surgery

Weakness caused by chronic debilitating conditions

Serious mental diseases, such as major psychoses

The nurse must be aware of these conditions and report inadequate intake when any of them are present. The patient should not be permitted to develop malnutrition before another route for intake of nutrients is adopted. One solution to the problems posed by such patients is the use of tube feedings. Tube feedings have many advantages, especially when the problems requiring them are likely to be of long duration.

When oral intake is inadequate, tube feedings are used to administer the protein and calories necessary for tissue healing in such conditions as fractures, decubitus ulcers, and burns. An intake of at least 1600 calories to 2000 calories must be provided if the protein is to be used for tissue repair; otherwise, it will be used to meet energy needs.

Disorders of the gastrointestinal tract, such as biliary or pancreatic fistulas or delayed emptying of the stomach, may necessitate passing a feeding tube beyond the affected area in order to administer nutrients. One of the few contraindications for tube feedings is the complete obstruction of the lower gastrointestinal tract.

Even if the underlying disease process cannot be corrected, the patient can live more comfortably during the weeks and months remaining if he is adequately nourished. The incidence of trophis ulcers, wasting, and debilitation is decreased by adequate nourishment.

Anorexia and malnutrition create a vicious cycle: anorexia leads to malnutrition; malnutrition promotes anorexia. The use of tube feedings for several weeks improves the patient's nutrition, as well as his appetite, thus interrupting the cycle.

ROUTES Routes for administration of tube feedings include the following:

Nasogastric intubation (or extension of the tube into the small intestine)

Gastrostomy (insertion of the tube through a stab wound)

Jejunostomy (insertion of tube into jejunum through a stab wound; a plastic catheter may be passed into the jejunum by means of a needle [needle-catheter jejunostomy])

Cervical esophagostomy (insertion of a tube into the hypopharynx through a stab wound in the side of the neck; the distal end of the tube is then advanced through the esophagus into the stomach)

A nasogastric tube is usually not difficult to insert but can be associated with a number of complications, including irritation of the nose, upper respiratory tract, and larynx (particularly when a large diameter, firm tube is used); reflex esophagitis and esophageal erosion with stricture (again, particularly when a firm, large diameter tube is used); and occasional otitis media, sinusitis, and pressure necrosis of the pharynx. Soft, small diameter feeding tubes are recommended to avoid these complications. Nasogastric tube feedings are best reserved for patients requiring short-term enteral nutritional support. The foremost contraindication for nasogastric tube feeding is unconsciousness or lack of protective laryngeal reflexes, which may result in tracheal aspiration of regurgitated gastric contents. (For such patients, intubation of the small intestine is less apt to result in aspiration, particularly if the feeding is delivered slowly over the 24-hour period.)

For long-term management, a gastrostomy or jejunostomy tube is sometimes used. These tubes must be surgically inserted and need time to seal; generally they are not safe for infusion until 3 days after insertion.[7] A newer method of access to the jejunum by needle-catheter jejunostomy is advocated by some.[8] With a gastrostomy or jejunostomy, the peritoneum is violated and the possibility of infection exists; in addition, leakage of secretions can cause severe skin excoriation.

Gastrostomy tube feedings eliminate the problem of irritation of the nose, throat, larynx, and esophagus. Because of the danger of tracheal aspiration, gastrostomy feedings, like nasogastric tube feedings, are contraindicated for mentally obtunded patients with inadequate laryngeal reflexes. Jejunostomy tube feedings are usually used for patients in whom nasogastric or gastrostomy tube feedings are contraindicated (such as comatose patients or those with high gastrointestinal fistulas or obstruction).[9]

Cervical esophagostomy is sometimes used for patients with swallowing problems requiring long-term feedings. It consists of introducing a feeding tube into the hypopharynx through a stab wound in the side of the neck; the distal end of the tube is then advanced through the esophagus into the stomach (Fig. 5-4). The

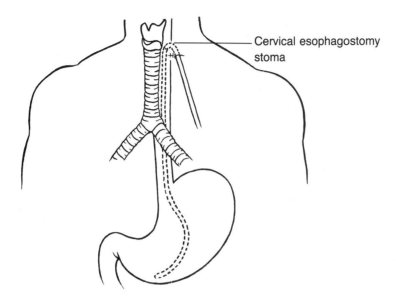

Cervical esophagostomy stoma

FIG. 5-4. **Cervical esophagostomy.**

feeding tube remains sutured in place for a week to 10 days, allowing the channel that was created for insertion of the tube to mature.[10] After healing takes place, the tube may be removed and reinserted at each feeding through the stoma. The stoma can usually be concealed beneath a blouse or shirt.

MIXTURES Any tube feeding mixture should provide all the required solid nutrients—protein, carbohydrate, fat, minerals, and vitamins—and should incorporate generous quantities of water. Additional water, however, must be supplied if the daily requirement is to be met. It is especially necessary to provide adequate water for the elderly, confused, lethargic, and comatose, since these patients frequently do not experience or express normal thirst. The amount of water that must be given should be gauged for each patient by assessing variance in total fluid intake and output, clinical signs, and laboratory data such as BUN and sodium levels. (See the section on prevention of tube feeding complications later in this chapter for further discussion of water requirements in tube feedings.)

Whole foods can be liquefied in a blender and administered by means of a tube; commercial strained baby foods can also be administered as tube feedings. Numerous convenient ready-to-use feeding mixtures are available commercially and are widely used; their major appeal is ease of preparation. It is necessary to review the contents of the various feeding mixtures

and to follow manufacturer's directions for their administration. Most commercial feeding preparations contain between 1 calorie and 2 calories/ml when mixed as directed.

Low bulk (elemental) diets, described earlier in this chapter, may also be used for tube feedings. Because they are hyperosmotic, they should be administered slowly by gravity drip or by a feeding pump.

Tube feedings should be initiated in a dilute solution at a slow rate; gradually the concentration and rate can be increased until maximum tolerance is attained. Isotonic solutions (osmolality of approximately 300 mOsm/Kg) may be initiated at full strength.

METHODS Tube feedings can be administered by gravity flow or by mechanical pump. Many variations have been devised for administering tube feedings by gravity flow.

An Asepto syringe can serve as a funnel; it is sometimes used to administer the feeding mixture (Fig. 5-5). The syringe attaches directly to the nasogastric tube, and flow rate is regulated by the height at which the nurse holds the syringe. Although the equipment is simple and easy to use, this method has several disadvantages. The tube feeding mixture is often administered too rapidly because the nurse lacks sufficient time to remain at the bedside for a slow delivery of the feeding. Also, the flow rate can be only crudely adjusted by raising or lowering the syringe. Highly concentrated (hyperosmolar) feedings should not be delivered by

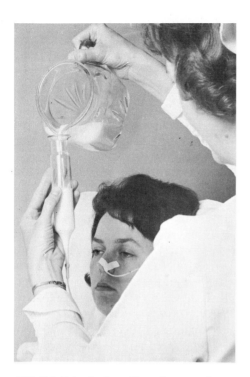

FIG. 5-5. Tube feeding with an Asepto syringe.

this route. Bolus feedings are complicated by gastric retention and pulmonary aspiration more often than are continuous feedings and thus are less desirable.

Disposable gravity-flow gastric feeding units are available commercially and are desirable because the flow rate of the feeding mixture can be adjusted in drops, and the constant attendance of the nurse is not required. One kind of disposable bag is the Davol Gastric Feeding Unit (Fig. 5-6). It consists of a graduated polyethylene bag, drip chamber, petcock shut-off for measurement of flow, wide plastic tubing, and a Sims connector for attachment to the feeding tube.

The *mechanical pump method* is more dependable than the gravity drip method for slow constant delivery of tube feedings. Factors that may cause an uneven flow rate in the gravity drip method include the following:

- A change in the patient's position after the flow rate has been adjusted; almost invariably the patient will desire to change his position during the feeding, especially when the mixture is given slowly.

- Failure to adjust the flow rate as frequently as necessary; frequent checks are indicated to ensure maintenance of the desired flow rate.

- A viscous mixture may clog the tube, especially

when the flow has been disrupted by a position change.

A food pump, on the other hand, ensures a constant flow rate regardless of the patient's position or viscosity of the feeding. An example of a commercial food pump is the Flexiflo Enteral Nutrition Pump (Ross Laboratories, Columbus, Ohio).

Selection of the feeding method When the tube feedings are ordered, the physician may indicate a specific method for administering the feeding solution; it not, the nurse may decide which method is best for the patient. Factors to be considered in determining the best method to use include the patient's condition, the viscosity of the feeding mixture, and the type of equipment available.

The patient's condition is an important factor. For example, comatose patients must have a slow delivery of solution to prevent gastric distention and vomiting, with possible tracheal aspiration. A food pump set at a slow speed or a gravity drip method suitable for slow delivery of the mixture should be used. Patients receiving duodenal or upper jejunal feedings require a slow, constant feeding rate; best results are obtained with a mechanical food pump set at a low speed. All patients, however, do not require a constant, slow delivery of the feeding mixture. Alert patients, free of gastrointestinal problems, can tolerate fairly rapid feedings if the amount given each time is small; any of the feeding methods described earlier can be adapted for their needs.

FIG. 5-6. Tube feeding with Davol Gastric Feeding Unit. (Davol Rubber Co., Providence, Rhode Island)

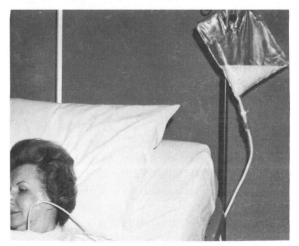

Failure to consider the viscosity of the feeding mixture can result in clogging the apparatus. A thick mixture is best given with a mechanical food pump or with a gravity drip apparatus having wide diameter tubing.

The type of equipment available greatly influences the method chosen; most hospitals have their own routine setups for tube feedings.

PREVENTION OF TUBE FEEDING COMPLICATIONS

Although tube feedings have proved to be of immense value, they can cause trouble if given incorrectly. Complications of tube feedings include the following:

Nausea

Vomiting

Aspiration pneumonia

Diarrhea

Hypernatremia and azotemia

In order to avoid untoward reactions from tube feedings, the nurse should keep the following in mind:

1. Check the position of the nasogastric tube prior to introducing the feeding mixture:
   a. The nurse can ask the patient to speak—if the tube is in the trachea, he will be unable to do so.
   b. The proximal end of the tube can be placed in a glass of water and if many bubbles appear as the patient exhales, the tube is in the respiratory tract. Remember that the absence of bubbles establishes that the tube is not in the breathing passage; it does not prove that the tube is in the stomach. Also remember that trapped air in the stomach can produce occasional bubbles.
   c. Another method involves listening with a stethoscope placed over the stomach while introducing 5 ml to 10 ml of air through the tube. If the tube is properly positioned, air will be heard entering the stomach. Belching will occur immediately if the tip of the tube is in the esophagus.
   d. A common check used to determine if the feeding tube is in the stomach is the withdrawal of a small quantity of gastric juice prior to each feeding. (It may be necessary to turn the patient or slightly withdraw or advance the tube to situate the tip below the gastric fluid level.)
   e. The tube should be taped securely to the nose to prevent it from partially sliding out unnoticed. (Should this happen, feedings may be inadvertently delivered into the esophagus and predispose to aspiration.)

2. Check the position of gastrostomy or jejunostomy tubes prior to introducing the feeding mixture by aspirating with a syringe to obtain gastric or intestinal fluids. (As stated earlier, gastrostomy and jejunostomy tubes need time to seal and are generally not safe for feedings until 3 days after insertion; the danger is leakage of solution into the peritoneal cavity.)

3. It is wise to check for gastric retention by aspirating the nasogastric or gastrostomy tube with an asepto or piston syringe prior to each intermittent feeding and every 6 hours during continuous feedings. The presence of a sizable volume of the previous feeding (such as 150 ml in an adult) is an indication to withold further feedings until the situation is reported to the physician. Some physicians recommend stopping the tube feeding for 2 hours if the aspirant is greater than 100 ml and then restarting it; if the residual remains high, the feeding should be reduced to a lower rate (as it should if there is abdominal cramping or nausea). Recall that gastric retention predisposes to regurgitation and aspiration of stomach contents.

4. Avoid bacterial contamination of the tube feeding mixture and administration apparatus. Do not prepare more than a 24-hour volume of the mixture at one time and keep it refrigerated until ready for use (unless, of course, the mixture is in a ready-to-use unopened can). Tube feedings should not be allowed to stand at or above room temperature for more than 2 or 3 hours in order to prevent bacterial overgrowth.[7] Administration equipment should be rinsed thoroughly after each intermittent use; equipment used for continuous feedings should be changed at regular intervals.

5. The tube feeding mixture should be neither too hot nor too cold, since such temperature extremes cause nausea and discomfort, especially if the feeding is given quickly. Refrigerated feedings should be warmed to room or body temperature by placing the container in a basin of hot water. The mixture should not be overheated, since this can curdle the protein in some feedings and clog the tubing.

6. It is wise to administer a small amount of water prior to giving the feeding mixture to assure patency of the tube and to moisten it (prevents feeding mixture from adhering to the wall of the tubing).

7. Ambulatory patients should be in the sitting position when receiving tube feedings. For bedfast patients the head of the bed should be elevated at least 30° at all times. This precaution is taken to prevent the aspiration of gastric contents if vomiting or regurgitation occurs; aspiration of the feeding mixture can cause pneumonia or strangulation. *Unconscious, extremely weak, or restrained patients require close attention, since the risk of aspiration is great in such persons; jejunostomy feedings are generally indicated for comatose patients (rather than nasogastric or gastrostomy feedings), since the likelihood of aspiration is less with this method.* If the patient has a cuffed tracheostomy or endotracheal tube, the cuff should not be deflated during the intermittent feeding, or within an hour afterwards, to prevent possible aspiration.

8. Continuous feedings should be started out slowly (such as 50 ml/hr) and at approximately half-strength.[11] If diarrhea or other untoward effects do not occur within 12 to 24 hours, the rate and concentration can be gradually increased until the desired levels are attained. (As stated earlier, isotonic solutions may be initiated at full strength.)

9. When the intermittent method is used, the amount given at each feeding is determined by dividing the total 24-hour dose by the designated number of feedings. Frequently 2000 ml is given over a 24-hour period, at intervals of 2, 3, or 4 hours. Individual feedings frequently vary between 150 ml and 250 ml. A single feeding should not exceed 300 ml unless it is to be given very slowly. As stated earlier, it may be necessary to start out with less than a full-strength solution until the patient develops tolerance for the feeding mixture.

10. The nurse should frequently assess the patient's ability to tolerate tube feedings; signs of intolerance include diarrhea, cramping, abdominal distention, vomiting, and glycosuria.

11. If nausea should occur, the feeding should be stopped and the patient given a rest period until further medical orders are received. A p.r.n. medication for nausea may be necessary. It may be helpful to dilute the mixture and administer it more slowly; other times a change in the content of the feeding mixture may be required.

12. If diarrhea occurs, the feeding should be discontinued until further medical orders are received. Medication, such as Lomotil or paregoric, may be ordered and given through the tube. Sometimes a modification is made in the feeding, such as decreasing its carbohydrate content.

13. Tube feeding flow rates should be checked every 30 minutes (when the continuous method is used) and the container should be time-taped. No attempt should be made to "catch-up" when the feeding falls behind; instead, increase the flow rate to the correct rate and make a new time-tape for the container. It is particularly important to monitor flow rate carefully when the solution is very concentrated (hyperosmotic), since rapid infusion of such fluids can cause difficulties in fluid balance.

14. *Additional water should be given as necessary to maintain a satisfactory urinary output (remember, the greater the protein and solute intake, the greater the need for water).* Extra water is particularly necessary when the patient has fever, decreased renal concentrating ability, or extensive tissue breakdown. Malfunctioning kidneys require more water to excrete a given amount of solute than do normal kidneys, and because aged patients frequently have impaired renal function, they require more water than do younger adults. *Hypernatremia and an elevated BUN may develop after an excessive solute load and inadequate water intake and lead to osmotic dehydration.* Schulte advocates allowing at least 0.5 ml of water for every milliliter of tube feeding to prevent this problem.[7] Early in protein overloading, the urine volume is large even though the water intake is inadequate; this can easily lead the staff to think that the water intake is adequate. In such instances, however, output actually exceeds intake. By the accurate recording of careful intake–output measurements, the nurse can unmask the true situation. *Thirst is an early indicator of need for extra water;* confused or unconscious patients should be observed carfully for inadequate water intake, since they are not aware of thirst. Clinical signs of dehydration include sticky tongue and mucous membranes, confusion, elevated plasma sodium level, thirst, and elevated body temperature. (The amount of additional water required with tube feedings varies with the osmolality of the tube feeding mixture; hypersmolar feedings require more water than less concentrated feedings.)

15. Urine output should be recorded and, unless abnormal fluid losses are occurring, the amount of fluid taken by tube should roughly equal the urinary output.

16. Constipation may sometimes be a problem when liquefied natural foods are administered very slowly into the stomach. The nurse should observe the frequency and character of bowel movements so that constipation can be detected before a fecal impaction occurs. Sometimes laxatives are added through the feeding tube to correct constipation; other times, a change in the content of the feeding mixture may be needed.

17. From 20 ml to 50 ml of lukewarm water should be administered through the tube after each feeding when the intermittent method is used (sometimes more than this amount is required to adequately clear the tubing and prevent clogging). Afterwards, the tube should be clamped and secured to the patient's clothing.

18. Even though the patient is not taking nourishment orally, one should respect his need for a clean mouth. Regurgitation can occur in patients receiving tube feedings, and when it does, the patient should be given a mouthwash. Even semialert patients can taste tube feedings. The nares should be cleaned and lubricated often to minimize discomfort caused by the tube.

19. The skin around the insertion site of a gastrostomy or jejunostomy tube should be cleansed and sterilely dressed; the tube should be anchored so as to prevent accidental dislodgement.

20. Gastric rupture secondary to overzealous tube feeding administration can be lethal. It can be prevented by frequently assessing for abdominal distention (manifested by increased abdominal girth and epigastric and left upper quandrant pain). Also, checking gastric contents prior to intermittent feedings or every 2 to 4 hours during continuous feedings is indicated to detect gastric distention.

Note: Tube feeding complications can be prevented or at least minimized if the nurse keeps the preceding points in mind when caring for patients receiving tube feedings.

INTAKE BY THE PARENTERAL ROUTE

The nurse's role in administering parenteral nutrients is discussed in Chapter 6.

MEDICAL THERAPY THAT CAUSES ELECTROLYTE DISTURBANCES

DIURETICS

HYPOKALEMIA The primary purpose of diuretics is to promote the excretion of sodium and water from the body. In varying degrees, most diuretics tend also to promote the excretion of potassium. Diuretics associated with hypokalemia include the thiazides, the mercurials, furosemide (Lasix), and ethacrynic acid (Edecrin) (Table 5-3).

TABLE 5-3   *Diuretic agents*

Nonproprietary name	Trade name	Comments
Thiazide-type diuretics		
Bendroflumethiazide	Naturetin	Mild, general purpose oral diuretics
Benzthiazide	Exna; Aquatag	Thiazides act by inhibiting Na^+ reabsorption in the ascending loop of Henle
Chlorothiazide	Diuril	
Cyclothiazide	Anhydron	May cause hypokalemia, hyperglycemia, hyperuricemia, and hypochloremic alkalosis
Hydroflumethiazide	Saluron	
Hydrochlorothiazide	Esidrix; HydroDiuril; Oretic	K^+ supplements or extra dietary K^+ are often necessary when thiazides are used routinely
Methyclothiazide	Enduron	Observe for arrhythmias and other symptoms of digitalis toxicity in digitalized heart patients (recall that hypokalemia intensifies the action of digitalis)
Polythiazide	Renese	
Trichlormethiazide	Naqua; Metahydrin	Thiazides cause lengthy lithium retention and thus patients taking lithium may require lower dosages of lithium carbonate
		Thiazides decrease urinary calcium excretion and sometimes increase serum calcium levels (may be used in treatment of patients with hypercalciuria)

(continued)

TABLE 5-3 *Diuretic agents* (continued)

Nonproprietary name	Trade name	Comments
Loop diuretics		
Ethacrynic acid	Edecrin	These are *potent* diuretics—indicated for refractory edema of congestive heart failure, cirrhosis, and nephrosis
Sodium ethacrynate	Sodium Edecrin	
Furosemide	Lasix	Both drugs block reabsorption of Na^+ from the ascending loop of Henle
		May be given by IV push for rapid action in emergencies
		Prone to cause hypokalemia (action not inhibited by alkalosis)
		Potassium supplements often needed
		Observe for arrhythmias in digitalized patients; drug-induced hypokalemia can cause fatal arrhythmias
		Daily weights are indicated to detect excessive diuresis; excessive and rapid weight loss may cause an acute hypotensive episode
		Ethacrynic acid and furosemide may be ototoxic
		Furosemide increases urinary excretion of calcium and is useful in the emergency treatment of hypercalcemia
Potassium-conserving diuretics		
Spironolactone	Aldactone	Triamterene acts on the distal renal tubule to depress the exchange of Na^+
Triamterene	Dyrenium	Spironolactone inhibits action of aldosterone (recall that aldosterone causes Na^+ retention and K^+ excretion)
		Both drugs tend to reduce K^+ excretion and may lead to hyperkalemia, thus, *K^+ supplements are contraindicated.* Salt substitutes containing K^+ are also contraindicated.
		Often combined with thiazides for effective diuresis—in this case, the hypokalemic tendency of the thiazides may offset the hyperkalemic tendency of triamterene and spironolactone (examples of such combinations are Dyazide and Aldactazide)
Osmotic diuretics		
Mannitol	Osmitrol	Osmotic diuretics may be used for several purposes:
		• Prophylaxis against acute renal failure (mannitol may be indicated in such diverse conditions as hemolytic transfusion reactions or traumatic injuries; its osmotic effect maintains urine flow within normal limits and protects the kidneys from noxious agents)
		• Differential diagnosis of acute oliguria (mannitol infusion in a patient with fluid volume deficit will increase urine flow rate; however, if either glomerular or tubular function is too severely compromised, mannitol will not increase urine flow rate)
		• Reduction of cerebrospinal or intraocular fluid pressures (mannitol increases osmolality of plasma and enhances the diffusion of water from cerebrospinal or intraocular fluids back to the ECF space)
		Contraindications to mannitol administration include marked pulmonary congestion, renal disease of sufficient severity to produce anuria, marked dehydration, and intracranial hemorrhage (unless craniotomy is to be performed)
Urea-Sterile	Ureaphil	Used for reduction of intracranial pressure, cerebrospinal fluid pressure, and introcular pressure
	Ureavert	100 ml of a 30% solution is usually given over 1 or 2 hours
		Check infusion site often; extravasation can cause necrosis
		Contraindicated in severe renal dysfunction, liver failure, marked dehydration, and active intracranial bleeding
		Monitor urine output closely
		Mannitol is usually the preferred drug when an osmotic diuretic is desired

Although the usual diet provides potassium in amounts ranging from 75 mEq to 125 mEq per day, losses of potassium by patients receiving any of the above diuretics can amount to several times this amount. Potassium deficit is even more apt to occur when the dietary intake is inadequate or when there are other abnormal losses of potassium from the body. Unfortunately, the body apparently has no adequate mechanism for the conservation of potassium, and the patient dying of a potassium deficit may lose 30 mEq to 40 mEq/day in the urine in spite of his dire need.

Potassium replacement Determination of the daily urinary excretion of potassium through analysis of the potassium content of the 24-hour urine specimen helps guide potassium replacement. A useful rule of thumb suggests that the patient receive a daily quota of potassium equal to his daily urinary excretion plus 10%. Potassium can be readily administered by use of one of the available pharmaceutical potassium supplements. The concentrations of potassium are variable, so read the labels carefully.

Some physicians prefer to reserve potassium supplementation for asymptomatic patients with serum potassium levels below 3 mEq/liter; those with symptoms or with additional risk factors (such as digitalis therapy) may require supplementation when the serum level drops below 3.5 mEq/liter.[12]

When potassium replacement is indicated, most physicians prescribe liquid potassium chloride supplements. Liquid supplements should be diluted as indicated by the manufacturers to avoid gastrointestinal irritation and a saline laxative effect. (The solution should be sipped slowly over a 5- to 10-minute period.) Because liquid supplements are not very palatable, patients often do not like to take them. Usually various brands are tried to find the least objectionable one. If the patient simply refuses to take a liquid supplement, the physician may prescribe slow-release potassium chloride tablets. These tablets contain potassium chloride in a wax matrix. Although they are improvements over the older enteric-coated tablets, they are still associated with gastrointestinal bleeding and ulceration in some patients. (Slow-release potassium chloride tablets should be taken with a full glass of water to help them dissolve.)

It should be remembered that hyperkalemia may result from overdosage of potassium supplements or from maintenance or therapeutic doses in patients with severe renal impairment. Potassium supplements are generally not indicated for patients receiving potassium-sparing diuretics (spironolactone and triamterene). When possible, diuretics should be administered intermittently rather than daily, allowing the patient to partially restore his potassium supply by dietary means. Regardless of the methods used to reduce the possibility of potassium deficit, the patient should be observed closely for undue fatigue, muscle weakness, nausea and vomiting, arrhythmias, and gaseous abdominal distention. All patients receiving diuretics associated with potassium loss should have occasional plasma potassium determinations.

Many foods are rich in potassium; they include those listed below:

- bananas
- oranges
- grapefruits
- dried figs and dates
- apricots
- raisins
- meats (avoid high-sodium varieties such as bacon, sausage, ham, and luncheon meats)
- nuts
- dried lentils
- sweet potatoes
- fresh tomatoes
- fruit juices (such as prune, grapefruit, pinapple, and grape)
- milk (intake may be limited if more than mild sodium restriction is necessary)

Dietary potassium sources may be sufficient for the mildly hypokalemic patient; however, those with moderate to severe deficits often need pharmacologic potassium supplements. It would be difficult for most patients to consume adequate dietary potassium sources without exceeding their recommended sodium or caloric limits. Some authorities recommend potassium-containing salt-substitutes as a potassium supplement (see Table 5-6). Patients should be cautioned to use salt substitutes sparingly if they are also taking other supplementary forms of potassium or are taking potassium-sparing diuretics. (*Hyper*kalemia could result.)

HYPERKALEMIA Potassium-conserving diuretics include spironolactone (Aldactone) and triamterene (Dyrenium). Aldactone is an aldosterone-blocking agent and, thus, promotes sodium loss and potassium retention, whereas Dyrenium has a unique mode of action, interfering with the exchange of sodium ions for potassium and hydrogen ions. Hyperkalemia can occur when potassium-conserving diuretics are used in patients with renal insufficiency, since such patients already have a tendency for potassium retention. Symptoms to be alert for include arrhythmias, paresthesias of the extremities, nausea, weakness, and intestinal cramping with diarrhea. If hyperkalemia is severe, cardiac arrest can occur. Only patients with adequate renal reserve should receive potassium-conserving agents. Periodic plasma potassium determinations should be obtained. Diuretics that allow retention of potassium are often used in conjunction with diuretics that cause potassium loss, since their combined use reduces the possibility of hyperkalemia.

LITHIUM TOXICITY It should be noted that thiazides reduce the renal clearance of lithium and greatly increase the risk of its toxicity. Lithium imbalance is described in Chapter 2.

SODIUM-RESTRICTED DIETS

Frequently the sodium intake of patients receiving diuretics is restricted, reducing the need for large doses of potentially harmful diuretics. Usually sodium restriction will not involve less than 1000 mg of sodium/day; more severe sodium restriction in patients receiving diuretics can cause acute sodium deficit, especially when another abnormal route of sodium loss is present (as in profuse sweating, vomiting, or diarrhea).

Low-sodium diets and diuretics may be used to treat a variety of conditions, such as congestive heart failure, nephrosis, hepatic cirrhosis, toxemia of pregnancy, and hypertension. Because low-sodium diets are in common usage, the nurse should know what constitutes such diets and be able to instruct patients in their use.

An average daily diet not restricted in sodium contains 6 to 15 g of salt, whereas low-sodium diets can range from a mild restriction to as low as 250 mg of sodium/day, depending on the patient's needs. It should be noted that the sodium content of diets is often expressed in grams of Na^+ rather than in grams of NaCl. One gram of Na^+ is 43 mEq while 1 g of NaCl is 17 mEq.[13] (Thus, a 4-g Na^+ diet is approximately equivalent in sodium content to a 10-g NaCl diet.)

The American Heart Association has prepared booklets describing 500 mg, 1000 mg, and mildly restricted sodium diets. The 500-mg sodium diet can be changed to a 250-mg diet by substituting low-sodium milk for regular milk. Food exchange lists have been devised to simplify sodium-restricted diets; they include list 1, milk; list 2, vegetables; list 3, fruit; list 4, breads; list 5, meat; list 6, fat; and list 7, free choice.

Also listed are seasonings and miscellaneous items allowed and those not recommended. These booklets are available to patients on a physician's prescription.

A mild sodium-restricted diet requires only light salting of food (about half the amount as usual) in cooking and at the table, no addition of salt to foods that are already seasoned (such as canned foods and foods ready to cook or eat), and avoidance of foods that are very high in sodium. Examples of foods high in sodium content include the following:

Sauerkraut and other vegetables prepared in brine

Bacon, luncheon meats, frankfurters, ham, kosher meats, sausages, salt pork, and sardines

Relish, horseradish, catsup, mustard, and worcestershire sauce

Processed cheese

Olives and pickles

Potato chips, pretzels, and other salty snack foods

Peanut butter

A definite relationship exists between the protein content of a diet and the degree of sodium limitation that it permits; thus, foods such as eggs, cheese, ordinary milk, meat, fish, poultry, and seafood must be used in measured amounts. The patient should be aware that most canned foods and ready-to-eat foods already have salt added and thus should be used only as their specific diets allow. Foods that may be used freely include most fresh vegetables and fruits and unprocessed cereals. Low-sodium milk, milk products, and bakery goods are available in most large cities. Table 5-4 lists the sodium and potassium content of selected foods and Table 5-5 lists the sodium and potassium content of selected beverages.

Numerous compounds containing sodium, some of which are listed below, are used by manufacturers of processed foods or in home preparation of foods to improve texture or flavor or to maintain freshness.

Sodium chloride (table salt; 1 tsp contains about 1955 mg of sodium)[14]

Sodium bicarbonate (baking soda; 1 tsp contains about 821 mg of sodium)[14]

Baking powder (1 tsp of commercial pyrophosphate baking powder contains about 486 mg of sodium)[14]

Monosodium glutamate (MSG; 1 tsp contains about 765 mg of sodium)[15]

Meat tenderizer (1 tsp of Adolph's meat tenderizer contains about 1745 mg of sodium)[14]

Soy sauce (1 tsp contains approximately 365 mg of sodium)[15]

TABLE 5-4 *Sodium, potassium, and caloric content of various foods*

Food	Sodium	Potassium	Calories	Food	Sodium	Potassium	Calories
Fast Foods	*mg*	*mg*		**Meats**			
Fish sandwich (Arthur Treachers)	836	248	440	Bologna, 1 slice	364	64	88
Hamburger, Big Mac (McDonalds)	963	387	541	Salami, dry 1 oz	540	102	112
Hamburger, Quarter Pounder with Cheese (McDonalds)	1206	471	518	Bacon, Canadian (broiled/ fried) 1 slice	442	91	65
Egg McMuffin (McDonalds)	911	222	352	Frankfurter, raw 1 average	550	110	128
Fruits				Turkey, light meat, roasted, 3½ oz	82	411	176
Banana, raw 1 small, 6″	1	370	85	Ham, Swifts (cured) 3½ oz	1078	250	224
Cantaloupe, raw ¼ melon (5″ diameter	12	251	100	**Cereals**			
Dates, dried 1 cup, pitted	2	1150	488	Rice Crispies, Kelloggs 1 cup	251	30	106
Grapefruit, raw (½ medium, (4″ diameter)	1	135	40	Product 19, Kelloggs 1 cup	413	59	106
Orange, Florida raw, 1 medium	2	311	71	Life, Quaker 1 cup	237	134	159
Raisins, dried, seeded, ½ cup	19	542	205	**Candy**			
Vegetables				Milk chocolate (Hersheys) 1 bar	61	214	302
Beans, common white, raw, ½ cup	19	1196	340	English toffee (Heath) 1 bar	90	50	220
Carrots, raw 1 large; 2 small	47	341	42	Breakfast bar, chocolate chip (Carnation) 1 bar	219	112	210
Potato, baked (without skin) 1 (2½″ diameter)	4	503	95	**Condiments**			
Tomato, raw 1 small	3	244	22	Catsup, 1 tbsp	156	55	16
Spinach, cooked ½ cup	45	291	21	Pickle, cucumber, dill 1 large	1428	200	11
				Steak sauce, (Lea & Perrins) 1 tbsp	149	64	18
				Mustard, yellow 1 tsp	63	7	4

(Pennington J, Church H: Bowes and Church's Food Values of Portions Commonly Used, 13th ed. Philadelphia, JB Lippincott, 1980)

Sodium benzoate (used as a preservative in many condiments)

Sodium cyclamate (an artificial sweetener)

Sodium propionate (used in pasteurized cheeses and some breads and cakes to inhibit mold growth)

Sodium alginate (used in many chocolate milks and ice cream for smooth texture)

Disodium phosphate (present in some quick-cooking cereals and processed cheeses)

Patients should be instructed to look for the words *sodium, salt,* or *soda* on food labels. The degree to which one's consumption of commercially prepared foods is limited varies with the severity of sodium restriction indicated for each patient. Patients should be aware that some foods do not list sodium as a content even though it is present; examples are standardized foods such as mayonnaise and catsup.

Medicines not prescribed by the physician should

TABLE 5-5 *Sodium, potassium, and caloric content of various beverages*

Beverage	Sodium	Potassium	Calories
	mg	mg	
Whole milk (3.25% fat) 1 cup	120	370	150
Whole, low-sodium milk 1 cup	5	617	149
Orange juice, canned, ⅔ cup	1	199	52
Tomato juice, canned ⅔ cup	200	227	19
Pineapple juice, canned ⅔ cup	1	149	55
Freeze dried coffee, 1 tsp	Trace	85	3
Tea, instant powder, 1 tsp	1	91	6
Gatorade, citrus, 1 cup	123	23	39
Pepsi-Cola, 12 oz	1	10	156
Coca-Cola, 12 oz	1	4	144
Pepsi-Cola, sugar-free, 12 oz	63	12	1
Tab, 12 oz	27	—	1
Seven-Up, 12 oz	4	—	144
Root beer, Frostie, 12 oz	28	79	154
Beer, (4.5% alcohol by volume) 12 oz	25	90	151

(Pennington J, and Church H: Bowes and Church's Food Values of Portions Commonly Used, 13th ed. Philadelphia, JB Lippincott, 1980)

be avoided, since some contain enough sodium to interfere with the desired intake. Among these are alkalinating agents (such as Alka-Seltzer), certain laxatives, salicylates, bromides, barbituates, cough syrups, and antibiotics. Some manufacturers provide low-sodium medications for patients requiring strict sodium restriction.

Toothpastes, tooth powders, and mouthwashes may be high in sodium content; therefore, patients on sodium-restricted diets should be instructed not to swallow these products and to rinse their mouths well after their use.

In certain communities, the drinking water may contain too much sodium for a sodium-restricted diet. Depending on its source, water may contain as little as 1 mg or more than 1500 mg of sodium/quart.[16] It may be necessary for patients to use distilled water when the local water supply is very high in sodium content. Most home installed water softeners contribute additional sodium to the water in exchange for calcium and magnesium ions and thus should not be used by patients requiring strict sodium restriction.

Soft drinks may be made with water high in sodium content (depending on where they are bottled); the exact sodium content of a beverage cannot be known unless the amounts of sodium in the water, syrup, and other ingredients are known.[16]

Salt substitutes are sometimes used to make low-sodium diets more palatable. These preparations contain potassium and should be used cautiously in patients taking potassium-sparing diuretics (such as spironolactone and triamterene), potassium supplements, or particularly in those who have renal impairment (and thus have an altered ability to excrete excess potassium). Potassium-containing salt substitutes may be useful in preventing hypokalemia in patients taking potassium-losing diuretics (such as furosemide or the thiazides). Salt substitutes containing ammonium chloride can be harmful to patients with liver damage. Salt substitutes and their sodium and potassium contents are listed in Table 5-6.

Occasional monitoring of the 24-hour urinary sodium excretion can be a useful compliance check on dietary sodium restriction.[17]

TABLE 5-6 *Salt substitutes and their sodium and potassium contents*

Brand name of salt substitute	Sodium		Potassium	
	mg/g	mEq/g	mg/g	mEq/g
Morton's Salt Substitute	1	0.044	493	12.62
Co-Salt	1	0.044	476	12.18
Adolph's Salt Substitute	2	0.09	333	8.51
Neocurtasal	100	4.40	469	12.00
Morton's Lite Salt	240	10.54	195	6.15
NaCl (reference)	410	18.00	0	0.00

(Halpern S: Quick Reference to Clinical Nutrition, p 132. Philadelphia, JB Lippincott, 1979)

PROLONGED USE OF LAXATIVES

The abuse of laxatives in our society is appalling. One reason for their widespread use is misunderstanding of the term *constipation*, which really refers to a hard, dry stool, difficult to evacuate. Even though the patient may not have had a bowel movement for several days, *he is not constipated unless his stools are hard, dry and difficult to evacuate.* Some persons normally have bowel movements every day; others may have them every 3, 4, or 5 days. If the latter persons have soft stools, they are not regarded as constipated.

It is the compulsion to have a daily evacuation that has led to the frequent use of laxatives. Many patients feel they must be "cleaned out" daily or their health will be impaired, a misconception that should be corrected by the nurse. Reliance on laxatives decreases the natural reflex activity of the colon; hence, *stronger* laxatives are required.

Should the patient actually be constipated, he should be provided a diet with these characteristics:

Generous quantities of fresh fruits, vegetables, and bran cereals

Increased water intake

Regular meal hours

A high-fiber diet, when there are no contraindications, can prevent the development of constipation if generous amounts of water are drunk. Stool softeners, such as Colace (dioctyl sodium sulfosuccinate), are useful physiologic tools for increasing the softness of the stools, and a glass of warm water or hot coffee first thing in the morning can stimulate the evacuation reflex.

Hypokalemia is the most frequent imbalance in patients using excessive laxatives.

ADRENOCORTICOSTEROIDS

Administration of adrenocorticosteroids may cause sodium and water retention and excretion of potassium, particularly when prolonged high doses are employed. For patients on such doses, the potassium intake should be increased and the sodium intake restricted because, if potassium loss is not replaced, metabolic alkalosis and potassium deficit may result. Persons taking corticosteroids and potassium-losing diuretics concomitantly should receive potassium supplements.

Prolonged administration of cortisone encourages negative nitrogen balance, and dietary treatment includes a generous intake of protein, a potassium supplement, and an adequate caloric intake to prevent weight loss. The use of cortisone can also cause decreased tolerance to carbohydrates, with hyperglycemia and glycosuria. Corticosteroids sometimes induce reversible Cushingoid changes (moon facies, acne, and hirsutism). These symptoms may occur in patients taking 25 mg or more of prednisone (or its equivalent) daily for longer than 2 weeks.

The hydrochloric acid and pepsin contents of the stomach are increased, a fact which may be related to the occurrence of ulcers in patients on steroid therapy. Patients are sometimes advised to take an antacid just before, and between, doses of corticosteroids.

Wounds heal more slowly in patients taking systemic corticosteroids because these medications retard formation of granulation tissue. Postoperative patients on steroids usually have their sutures in place longer than others because they are more likely to suffer wound dishiscence. High doses of vitamins (particularly vitamin C) are needed in postoperative patients to promote wound healing. Supplemental vitamin C also helps counter the bruises frequently evident in patients taking systemic corticosteroids.

Normally, the adrenal cortex produces about 25 mg of hydrocortisone/day; during stressful periods it can secrete as much as 300 mg/day.[18] When corticosteroids are supplied medically and then withdrawn, the adrenal cortex needs time to assume its normal function (it may take from 3 to 9 months for it to be able to secrete normal daily needs). It may require up to 2 years before the adrenal cortex is able to produce amounts needed during stress.[18] Patients being withdrawn from corticosteriods should wear a special alert tag or bracelet indicating a need for extra hydrocortisone during periods of stress. Early symptoms of acute adrenal insufficiency may include weakness, nausea, fainting, loss of appetite, and fever.

CATABOLIC EFFECTS THAT CAUSE NUTRITIONAL DISTURBANCES

IMMOBILIZATION

DISUSE OSTEOPOROSIS An increased rate of calcium excretion (in the urine and in the feces) occurs in patients immobilized by fractures or paralysis and in others at bedrest. One study of healthy adults showed that urinary calcium excretion was elevated throughout 30 to 36 weeks of bedrest, reaching the maximum at 7 weeks.[19] Renal stones may follow immobilization. (See discussion of nephrolithiasis in Chapter 10).

Bone mineral is lost during immobilization, sometimes causing elevation of total (and especially *ionized*) calcium in the bloodstream. *Symptomatic hypercalcemia* from immobilization, however, is rare; when it does occur it is limited virtually to persons with high calcium turnover rates (as in adolescents during a growth spurt or in patients with Paget's disease) and is never produced by immobilization alone.[20] Most cases of immobilization hypercalcemia occur after severe or multiple fractures or after extensive traumatic paralysis (as in quadriplegia). Immobilization hypercalcemia may remain clinically silent but is often associated with symptoms of anorexia, nausea, vomiting, constipation, weakness, emotional lability, polyuria, polydipsia, and abdominal pain. Symptoms usually begin rather abruptly a few weeks or months after the injury.[20] Laboratory examination shows an increased calcium level in the bloodstream; radiographs may show delayed fracture healing and osteoporosis.[21]

Disuse osteopenia has been thought to be due to mechanical factors; that is, either to absence of pressure transmitted to bone, or to absence of tension applied to bone by muscle, or both. The changes in mineral metabolism induced by bedrest appear to be reversed by quiet standing for 2 or more hours per day; however, vigorous supine exercise for as long as 4 hours daily seems to be ineffective.[22] This kind of evidence supports the concept that it is the absence of pressure forces (weight bearing) on the skeleton that is primarily responsible for disuse osteopenia.[22] The bone dissolution accompanying bedrest may occur to a greater extent in weight-bearing bones than in the rest of the skeleton; the process appears to be reversible when remobilization occurs. If remobilization is ineffective or impractical, the hypercalcemia may be treated with corticosteroids (to reduce intestinal calcium absorption and depress the effects of endogenous vitamin D).[23] Calcitonin, by inhibiting bone resorption, has been used to lower serum calcium in hypercalcemic crisis.[24]

Some authorities recommend a reduced dietary intake of calcium when immobilization hypercalcemia is present. Immobilized patients should be encouraged to ingest copious amounts of fluids to prevent calcium salts from precipitating within the kidney. (Phosphate therapy may be effective in reducing urinary calcium levels; it acts by binding calcium within the intestine and by reducing excretion of calcium in the urine.)

NEGATIVE NITROGEN BALANCE Nitrogen balance is a measure of protein metabolism, indicating that the rate of protein synthesis in the body equals that of protein degradation. Negative nitrogen balance refers to a state in which nitrogen loss exceeds intake; it indicates that protein catabolism exceeds protein anabolism. (Recall that approximately 1 g of nitrogen is contained in about 6.25 g of protein.) The immobilized patient develops a state of negative nitrogen balance after approximately 4 days of bedrest. By the tenth day of immobilization, the state of negative nitrogen balance reaches its peak, and gradually returns toward normal. It is important to encourage the intake of foods high in protein (meat, eggs, fish, milk) to compensate for the increased loss of protein. Commercial high-protein products are also available for complete dietary intake or supplemental feedings. (See discussion of elemental diets and high-protein tube feedings in this chapter.)

Clinically, negative nitrogen balance is manifested by the following symptoms:

Anorexia

Muscle weakness and wasting

Weight loss

Decreased state of awareness

Hypoalbuminemia with associated edema

Anemia

In addition to producing negative nitrogen balance, bedrest reduces the secretion of renin, aldosterone, and, to some extent, epinephrine and norepinephrine; it also predisposes to thrombosis in the deep veins of the pelvis and lower extremities.[25]

DECUBITUS ULCERS

A draining decubitus ulcer can quadruple the body's need for protein because the ulcer exudate is high in protein content. The fact that most patients with decubiti are immobilized further contributes to the increased protein need. Healing may be enhanced by increasing the daily protein intake to around 100 g; ensuring an adequate vitamin intake will also promote healing. Some physicians advocate zinc supplements, since it is known that zinc deficiency is associated with impaired wound healing. (Zinc deficiency is discussed in Chapter 2.)

It is often difficult for the patient to consume sufficient protein, since his appetite is usually poor. The nurse must use all her ingenuity to find foods rich in protein but acceptable to the patient. Tube feedings may be necessary if oral intake remains inadequate. The increased protein need must be considered as important as turning the patient often and keeping the skin clean and dry. It should be remembered that a poor nutritional state (especially protein-calorie deficit) is a common etiologic factor in the development of decubitus ulcers.

TRAUMA

Many types of severe trauma are accompanied by negative nitrogen balance, which is augmented by starvation. Trauma, such as burns, fractures, wounds, and crushing injuries, causes loss of protein through direct destruction of tissues. It contributes further to the loss through the so-called toxic destruction of protein, an increased catabolism brought on mysteriously by the trauma. It also promotes the accumulation of protein-rich fluid at the site of the injury, which contributes further to possible protein depletion. Immobilization made necessary by the injury also causes losses of protein and, in addition, of electrolytes.

Because of these losses and because of the requirements for optimal healing, the protein, electrolyte, and vitamin C intake of the injured patient should be increased; as much as 150 g of protein daily is frequently administered. (Of course, high-potassium foods should be withheld until adequate renal function has been established, since cellular breakdown releases large amounts of potassium into the bloodstream.) The diet should be high in calories to prevent ingested protein from being consumed for energy purposes.

The need for extra calories may be 40% to 100% greater than usual in patients with severe burns; major fractures of long bones can increase caloric demands by 10% to 25%. The use of elemental diets or other high-protein feedings (either orally or by tube feeding) has been effective in producing positive nitrogen balance in traumatized patients, as has TPN either through a central or peripheral venous line.

REFERENCES

1. Halpern S: Quick Reference to Clinical Nutrition, p 4. Philadelphia, JB Lippincott, 1979
2. Wieman T: Nutritional requirements of the trauma patient. Heart Lung 7, No. 2:284, 1978
3. Blackburn G: Techniques of nutritional assessment. In Nutritional Assessment, p 13. Cutter Medical Laboratories, Inc, 1980
4. Halpern S: p 5
5. Halpern, p 6
6. Sokal J: Editorial: Measurement of delayed skin-test responses. N Engl J Med 293, No 10:502, 1975
7. Condon R, Nyhus L (eds): Manual of Surgical Therapeutics, 4th ed, p 246. Boston, Little, Brown, & Co, 1978
8. Page C, Clibon U: Modern clinical nutrition: Part 12. A method of enterally feeding defined formula diets. Am J IV Ther Clin Nutr 9, No. 1:9, 1982
9. Schwartz S (ed): Principles of Surgery, 3rd ed, p 91. New York, McGraw-Hill, 1979
10. Bush J: Cervical esophagostomy to provide nutrition. Am J Nurs, Jan 1979, p 107
11. Griggs B, Hoppe M: Update: Nasogastric tube feeding. Am J Nurs, March 1979, p 483
12. Freitag J, Miller L (eds): Manual of Medical Therapeutics, 23rd ed, p 136. Boston, Little, Brown & Co, 1980
13. *Ibid*, p 23
14. Pennington J, Church H: Bowes & Church's Food Values of Portions Commonly Used, 13th ed, p 149. Philadelphia, Lippincott, 1980
15. American Heart Association: Your Mild Sodium Restricted Diet (revised), p 10
16. *Ibid*, p 9
17. Halpern, p 26
18. Newton D, Nichols A, Newton M: You can minimize the hazards of corticosteroids. Nursing 77, June 1977, p 30
19. Donaldson C, Halley S, Vogel J et al: Effect of prolonged bedrest on bone mineral. Metabolism 19, No. 12:1071, 1970
20. Maxwell M, Kleeman C (eds): Clinical Disorders of Fluid and Electrolyte Metabolism, 3rd ed, p 1016. New York, McGraw-Hill, 1980
21. Wolf A, Chuinard R, Riggins R, Walter R, Depner T: Immobilization hypercalcemia. Clin Orthop, No. 118:126, 1976
22. Donaldson, p 1082
23. Claus–Walker J, Carter R, Campos R, Spencer W: Hypercalcemia in early traumatic quadriplegia. J Chronic Dis 28:88, 1975
24. Wolf, p 128
25. Schwartz, p 4

BIBLIOGRAPHY

Gever L: Pharmacist on call: Administering potassium chloride supplements. Nursing 81, Oct 1981, p 32

Hodges R: Nutrition in Medical Practice. Philadelphia, WB Saunders, 1980

Orr G: Wade J, Bothe A, Blackburn G: Alternatives to total parenteral nutrition in the critically ill patient. Crit Care Med 8, No. 1:29-33, 1980

Intravenous therapy is a major component of patient care; millions of patients in hospitals receive this treatment each year. The nurse plays a major role in the administration of parenteral fluids. Exact nursing responsibilities are not uniformly defined and vary with geographical areas and individual hospitals. Many hospitals have wisely organized intravenous therapy teams to start, and assist in maintaining, infusions. Nevertheless, regardless of who starts the fluids, the nurse shares in the responsibility of assuring their safe and therapeutic administration. To ably assume this responsibility, one must understand basic principles of safe fluid administration and become familiar with parenteral fluids. If intelligent observations are to be made during their infusion, the purposes, contraindications, and complications associated with their use must be known.

ROUTES AND TECHNIQUES OF PARENTERAL FLUID ADMINISTRATION

INTRAVENOUS

INDICATIONS FOR USE Veins provide an excellent route for the quick administration of water, electrolytes, and other nutrients. The intravenous route is essential when nutrients are needed in a hurry, such as glucose in severe hypoglycemia, 5% sodium chloride in severe sodium deficit, or calcium gluconate in acute calcium deficit. Relatively large volumes of fluids can be given by the intravenous route, provided due care is exercised.

GENERAL PSYCHOLOGICAL CONSIDERATIONS Few persons are without some fear or dread of a needle being introduced into their veins, and normal fears are exaggerated in illness. Since some patients associate intravenous fluids with serious illnesses, they are disturbed when such therapy is employed. It is the nurse's responsibility to explain away as much of the fear as possible and point out that IV fluids are commonly used until oral intake is again possible. The nurse must always remember that although IV therapy

"I injected one hundred and twenty ounces, when like the effects of magic, instead of the pallid aspect of one whom death had sealed as his own, the vital tide was restored, and life and vivacity returned!" —Thomas Latta, 1831

6

Parenteral Fluid Administration—Nursing Implications*

---

*Sections of this chapter concerned with starting and maintaining intravenous infusions were prepared with the assistance of Darnell Thompson Roth, R. N., Intravenous Therapy Consultant, Alexian Brothers Hospital, St. Louis, Missouri.

is commonplace to the health care team, it is far from routine to the patient. A detailed explanation of why and how the fluids are given is indicated for some; others want only a brief account. In any event, the patient should *never* be approached with no explanation at all; the fear of not knowing what is to happen can be worse than the most painful venipuncture.

Only persons capable of skillful venipuncture should start fluids on anxious patients; just one traumatic experience may make IV therapy totally unacceptable to them. The nurse who starts fluids must always appear confident—patients sense insecurity and are understandably upset by it. An IV therapy department with competent IV therapy nurses provides many advantages, one of which is the skill developed by such therapists in performing venipuncture.

SELECTION OF SITE

Suitable superficial veins A number of superficial veins are available for venipuncture. Those most commonly used include the following:

Veins in the hand (metacarpal and dorsal venous plexus)

Veins in the radial area of the wrist

Veins in the forearm (basilic and cephalic veins)

Veins in and around the cubital fossa (antecubital, basilic, and cephalic veins)

Scalp veins in infants

Criteria for selection Selection of a vein depends upon a number of factors:

Availability of sites (depends upon condition of veins)

Size of needle to be used

Type of fluids to be infused

Volume, rate, and length of infusion

Degree of mobility desired

Skill of operator

Disease process

Potential surgery (*e.g.,* avoid arm on side of mastectomy)

Dominant extremity

Hand veins The early use of hand veins (Fig. 6-1) is important it parenteral therapy is to be prolonged. This allows each successive venipuncture to be made above the previous site, eliminating pain and inflammation caused by irritating fluids passing through a vein in-

jured by previous venipuncture. Because of their small diameter, hand veins do not accommodate large needles—a small-gauge scalp vein needle is sometimes used for venipuncture in the hand.

Small veins cannot accept hypertonic or otherwise irritating fluids because less blood is present in small veins to dilute such solutions. These peripheral veins collapse sooner in the presence of shock than do more centrally located veins. And finally, extravasation of blood may occur in venipuncture in this area, particularly when there is thin skin and inadequate connective tissue.

Forearm veins The cephalic vein flows upward along the radial border of the forearm and is an excellent site for venipuncture (Fig. 6-2). Also, the size of the vein will accommodate a large needle. The accessory cephalic vein joins the cephalic vein below the elbow, and it too is a good site for venipuncture. Both veins are frequently used for blood administration. When prominent, the median antebrachial vein can be used for venipuncture. The location of superficial veins of the forearm is somewhat variable and not always well defined.

Venipuncture in a forearm vein allows the patient some arm movement without the risk of puncturing the posterior venous wall.

Elbow veins The median cephalic and median basilic veins are found in the antecubital fossa; both veins are readily accessible to venipuncture because they are large and superficially located. In addition, they are kept from rolling and sliding by surrounding tissues. They will accommodate large needles, large volumes of fluids, and all but the most irritating intravenous fluids.

Because arteries in the antecubital area, though usually more deeply located, lie in close proximity to veins, it is easy to mistake an artery for a vein in this area. Aberrant arteries are not uncommon in the cubital fossa. (These arteries, more superficially located than usual, are found in one of ten persons.) Injection of fluids into an artery usually causes the patient to complain of sudden severe pain in the arm and hand, caused by arteriospasm. This is an indication to stop the infusion immediately.

When frequent blood specimens are necessary, it is wise to save the veins in the antecubital area for this purpose; large quantities of blood can be obtained from them. These veins can be used many times with-

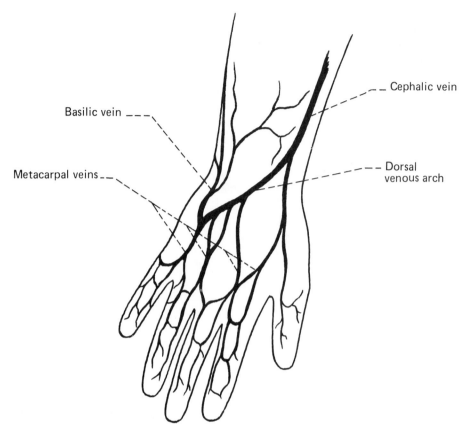

Basilic vein ---

Metacarpal veins ---

Cephalic vein

Dorsal venous arch

FIG. 6-1. **Superficial veins of hand.**

out damage if good technique is used. (It is extremely difficult to get sufficient blood from small veins.)

A disadvantage in using veins in the antecubital area is the restriction of elbow flexion during the infusion. Therefore, when long-term infusions are anticipated or the patient is uncooperative, it is best to use the veins in the forearm, because the patient can be moved and ambulated with less danger of dislodging the needle. The nurse should remember that damage to veins in the antecubital area can limit the use of sites below.

A right-handed person has more freedom if the infusion is given in the left arm; however, the need for multiple venipunctures is an indication to employ alternate sites in both arms.

Lower extremity veins Usually, lower extremity veins are not recommended for venipuncture because their use can result in dangerous complications. Thrombus formation at the venipuncture site occurs to some de-

gree in all venipunctures, but when ankle veins are used, the thrombus can extend to deep veins and may result in pulmonary embolism. Many institutions prohibit the use of lower extremity veins; individual hospital policy should be ascertained prior to usage.

GENERAL CONSIDERATIONS

1. When feasible, it is best to use veins in the upper part of the body.
2. When multiple punctures are anticipated, it is best to make the first venipuncture distally and work proximally with subsequent punctures.
3. Avoid venipuncture in the affected arm of patients with axillary dissection, as in radical mastectomy (embarrassed circulation affects the flow of the infusion, causing increased edema).
4. Avoid checking the blood pressure on the arm receiving an infusion because the cuff interferes with fluid flow, forces blood back into the needle, and may cause a clot to form.

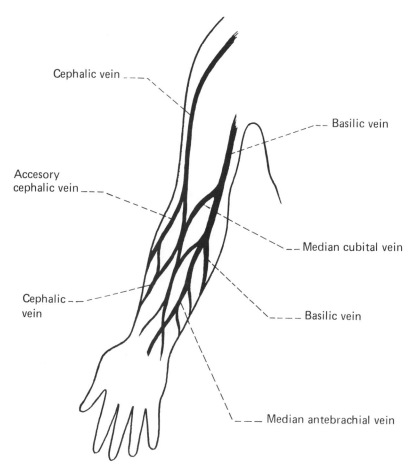

Cephalic vein

Basilic vein

Accesory cephalic vein

Median cubital vein

Cephalic vein

Basilic vein

Median antebrachial vein

FIG. 6-2. **Superficial veins of forearm.**

5. When the patient is on his side, the upper arm should be used for venipuncture. The lower arm has increased venous pressure which interferes with fluid flow and may cause clot formation in the needle.

6. Restraints should not be placed on the infusion site. If a restraint is applied above the IV site, extreme caution must be employed to avoid obstruction of venous flow. Restraints should never be used on extremities receiving infusions by means of volumetric infusion pumps. When the involved extremity must be immobilized, the restraint should be applied to the armboard only, never over the injection site, since it would act as a tourniquet and obstruct flow.

7. An armboard should be used when the venipuncture site is over an area of flexion (such as the wrist or elbow).

8. Hold the infusion bottle sufficiently high during ambulation of the patient and transport by wheelchair or stretcher to maintain a constant flow rate. When the patient is ambulating, the arm receiving the infusion should be placed across the abdomen to immobilize it and the IV pole should be rolled with the other hand.

9. Instruct the patient with a venipuncture in the hand or arm to avoid such movements as combing hair, shaving, brushing teeth, using the telephone, or cutting up food with the affected arm whenever possible.

METHODS FOR VENOUS ENTRY Fluids may be introduced into a vein by several means:

A metal needle

A plastic needle (catheter mounted on a needle)

A plastic catheter threaded through a metal needle

A plastic catheter introduced by means of a cut-down (minor surgical procedure)

SELECTION OF METHODS Short-term infusions are usually given through metal needles—either a regular straight needle or, most often, a scalp vein needle. The scalp vein needle is approximately three-fourths of an inch long and has attached plastic wings, used for holding the needle during venipuncture (Fig. 6-3). This needle has a thin wall and provides a larger lumen with a smaller needle diameter, and because the bevel is short, there is less danger of accidentally puncturing the posterior wall of the vein.

Scalp vein needles are available in two types: one has a variable length of plastic tubing attached to an adaptor that accommodates an administration set, and the other has a short length of tubing ending in a resealable injection site (sometimes referred to as a "heparin lock" or an intermittent infusion set; Fig. 6-4). An intravenous catheter can be converted to an intermittent infusion set by attaching an injection cap (designed for that purpose) to the catheter. Patency of the needle or converted catheter is maintained by periodically flushing the device with a dilute solution of heparin. (Heparin is commercially available in strengths of 10 units/ml and 100 units/ml.) Each time medication is injected into the resealable site, it is important to

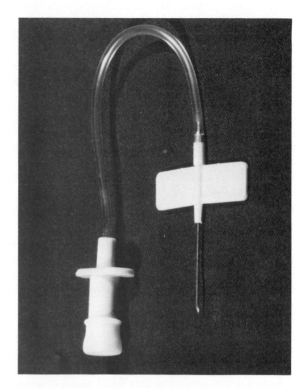

FIG. 6-4. Bard intermittent infusion set. Note the resealable injection site. (C. R. Bard, Inc., Murray Hill, New Jersey)

cleanse the area thoroughly with a sterile alcohol wipe. Prior to injecting the medication, it is imperative to confirm that the needle is still in the vein. (Never automatically inject medication without first verifying proper positioning of the needle.)

Another device available for intermittent infusions utilizes a stylet in the lumen of the catheter when IV fluids are not needed; it is removed when infusions are to be resumed. Such a device reduces the need for "keep-open" infusions and allows the patient more freedom of movement. A new sterile stylet is inserted into the catheter after each intermittent injection of medication, using scrupulous aseptic technique. Some authorities do not advocate use of this device because of the increased potential for contamination.

The size of the needle to be used depends on the vein as well as on the type of solution. Scalp vein needles range in size from a 25 gauge to a 16 gauge. (The smaller the gauge number, the larger the internal diameter of the needle.) The gauge of the needle should be appreciably smaller than the lumen of the vein to be entered. Thus, hemodilution of the infusing solution is ensured. When a large needle occludes the flow

FIG. 6-3. Pediatric scalp vein infusion set (Saftiwing Winged Infusion Set; Courtesy of Cutter Laboratories, Berkeley, California)

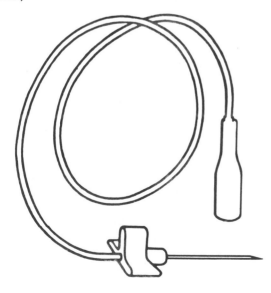

of blood, irritating solutions will probably produce chemical phlebitis. If a small needle is used, irritating fluids can mix readily with blood, decreasing the chance of phlebitis. A large-gauge needle is required for high viscosity fluids, such as blood. (Blood usually is administered through an 18- or 19-gauge needle.)

A plastic needle is a catheter mounted on a metal needle. When the venipuncture is made, the catheter is slipped off the needle into the vein and the metal needle is removed. Plastic needles do not infiltrate as easily as metal needles because of their pliability. It is recommended that plastic catheters be radiopaque so that they can be located by radiologic survey in the event of catheter emboli.

An intracatheter is a catheter inserted through a metal needle and used when a longer catheter is desired. Often it is used to administer drugs or irritating solutions that may cause tissue necrosis if infiltration occurs.

A cutdown may be performed when veins become exhausted from prolonged therapy and when peripheral veins have collapsed from shock. The necessity of signing a surgical permission form for the cutdown procedure is determined by individual hospital policy. The advent of plastic intracatheters introduced through a needle (either into a peripheral vein or a central vein) has reduced the need for surgical cutdowns.

TECHNIQUES OF INSERTION

Venipuncture with a metal needle After a suitable site has been located, the next step is to distend the vein, usually accomplished by a tourniquet; it also helps to steady the vein when the tourniquet is placed no higher than 2 inches above the site of injection. It should be tight enough to impede venous flow while arterial flow is maintained, but never too tight—a common error.

Occasionally, it may be necessary to distend the vein by placing it in a dependent position for several minutes. Sometimes a light slap with the therapist's fingers over the proposed site of venipuncture helps; however, if the slap is severe the vein will constrict due to the painful stimuli. Exercising the muscles distal to the site of puncture is sometimes helpful. However, exercising should not be done when blood is being drawn for determination of serum electrolytes, since a false reading may result.

Seventy percent alcohol is frequently used to pre-pare the injection site; it is most effective if the skin is scrubbed with friction for a full minute with a "clean to dirty" circular motion. Unfortunately, skin preparation most often consists of a quick light wipe with the sponge, failing to significantly reduce the bacteria count. Povidone-iodine solution is highly desirable for cleansing the injection site because it provides effective bactericidal, fungicidal, and sporicidal activity. (Because of occasional allergy to iodine, the patient should be questioned about this before its use.) Aqueous benzalkonium chloride and other quarternary ammonium compounds are not effective skin antiseptics and should not be used for this purpose. Scientific evidence does not substantiate the need to shave hair from the injection site to reduce bacterial flora; in fact, the nicks incurred from shaving may actually predispose to infection. Shaving does, however, facilitate removal of tape from a very hairy site.

Generally, the bevel of the needle should be facing upward during insertion. However, the introduction of a large needle into a small vein may require the bevel to face downward; otherwise, the needle would pierce the posterior wall of the vein when the tourniquet is removed.

With the tourniquet in place, the needle should pierce the skin to one side of, and approximately one half to one inch below, the point where the needle will enter the vein. As the needle enters the skin it should be at about a 45 degree angle; after the skin is entered, the needle angle is decreased. Although it seems more logical to enter the vein from above with one quick thrust, there seems to be less flattening of the vein and decreased risk of perforating the posterior aspect of the vein wall when a lateral approach is used. The free hand is used to palpate the vein while the needle is being introduced. An experienced operator may feel a snap as the needle enters the vein; after this, less resistance is offered to the needle. At this point, one should proceed very slowly with the insertion of the needle, threading it into the lumen approximately one half to three fourths of an inch. The tourniquet is then released. Frequently a thin stream of blood is seen in the tubing when the needle enters the vein. If in doubt, pinch the tubing just above the needle and release it; usually this will cause a flashback of blood into the tubing.

Fluid should be allowed to run in, and the area should be observed for swelling. The presence of swelling indicates that the needle is not in the vein and fluid is entering the subcutaneous area. The infusion should be discontinued immediately when swelling is noted or if the venipuncture attempt was unsuccessful; a venipuncture must then be made in another area. A tourniquet should not be reapplied immediately to the same extremity; if applied too soon, a hematoma will develop and the patient will experience unnecessary pain and discomfort.

After successful venipuncture, the next step involves anchoring the needle comfortably and safely. A piece of tape should be diagonally wrapped around the needle hub to prevent "to and fro" motion of the needle, which predisposes to phlebitis; then, a piece of tape should be applied over the hub to further stabilize it. A sterile 2″ × 2″ gauze pad should cover the entry site. The Communicable Disease Control Center recommends that an antimicrobial agent be applied to the venipuncture site. Some hospitals use povidone-iodine sponges at the site, while others use an antibiotic ointment. Tape should be applied in such a way that the tubing may be changed easily. (Note that tape should *never* completely encircle the extremity since it could have a tourniquet effect.) A loop of tubing should be secured with tape independently of the needle so that an accidental tug on the tubing will not dislodge the needle or catheter. The date and time of insertion should be written on the tape to alert personnel to the need to remove the device within a safe period of time. Also, the gauge and length of the device used for venipuncture should be noted, as well as the name or initials of the person performing the venipuncture (Fig. 6-5). It is important that the gauge of the venipuncture device be noted on the tape. Should an emergency situation arise, it is readily apparent to those responding whether or not the device in place will accommodate viscous fluids.

An armboard may be necessary at times to immobilize the involved extremity and decrease the likelihood of infiltration. (One is always required when the site is in an area of flexion.) It should be applied in such a way as to allow normal anatomic flexion of the fingers. At no time should the fingers be flattened out on the armboard.

Plastic catheter mounted over a needle A venipuncture is performed in the usual manner, with the needle inserted far enough to ensure entry of the catheter into the vein. After the catheter is slid into the vein to the desired length, the needle is carefully removed. If the venipuncture is not successful, withdraw the catheter completely from the puncture site before reinserting

FIG. 6-5. Close-up view of venipuncture dressing labeling.

the needle; reinsertion of the needle with the catheter in place could sever the catheter, resulting in catheter embolism. (The manufacturer's directions should be followed closely.) A small sterile dressing should be applied to the puncture site after the catheter is anchored securely with tape.

Plastic catheter inserted through a needle After performance of a normal venipuncture, the catheter is inserted into the vein through the needle. If venipuncture is not successful, pull out the needle and catheter together. The catheter should not be pulled out first because the sharp edges of the needle could sever it as it is being withdrawn, resulting in catheter embolism. (The manufacturer's directions should be followed closely.)

Only an experienced operator should attempt venipuncture by this method because of the higher incidence of complications associated with its use (e.g., tissue trauma, infection, and severed catheter emboli).

When removing any plastic device from a vein, do not use scissors to remove tape, since it is possible to cut the device accidentally. Always check the length of the plastic device when it is removed.

Some feel that plastic catheters are about twice as prone to produce venous complications as metal needles. It has been suggested that because metal needles infiltrate more quickly than plastic devices, they are changed more frequently, thereby decreasing the occurrence of phlebitis. It is known that the longer the intravenous device remains in place, the higher the rate of positive cultures, thus emphasizing the importance of regular site rotation.

An antimicrobial agent should be applied to minimize the occurrence of catheter sepsis. The catheter should be anchored securely with tape, and a sterile dressing should cover the area; the dressing should be changed every 24 to 48 hours to inspect the site and apply an antimicrobial agent. The date of insertion and the gauge, length, and type of catheter used should be noted.

SITE ROTATION According to a recommendation made by the Center for Disease Control, a peripheral venipuncture device should not be left in place longer than 72 hours and preferably should be changed at 48-hour intervals. This agency further notes, however, that this guideline may need to be modified when the patient has few available veins. In the latter case, explicit documentation concerning the reason for not changing the device and the appearance of the site must be made on the chart.

HYPODERMOCLYSIS

Hypodermoclysis, the administration of a solution subcutaneously, compares unfavorably with the intravenous route and is used less and less because of its many problems and associated discomfort to the patient.

The types of fluids that can be given subcutaneously are few, since they must closely resemble the electrolyte content and tonicity of extracellular fluid (ECF) if they are to be absorbed. Some of the fluids generally considered reasonably safe for subcutaneous administration include:

Isotonic saline (0.9%)

Half-isotonic saline (0.45%) with 2½% dextrose

Lactated Ringer's solution

Half-strength lactated Ringer's with 2½% dextrose

Fluids that are contraindicated for subcutaneous administration include electrolyte-free solutions, hypertonic solutions, alcohol, amino acids, solutions of high molecular weight such as albumin, and those with a pH differing significantly from body pH (such as gastric replacement solution). Orders for the subcutaneous administration of any of these solutions should be questioned.

The subcutaneous administration of a 5% dextrose in water solution attracts electrolytes from the surrounding tissues and from the plasma, which enter the pool created at the infusion site. (This electrolyte movement occurs to make the solution absorbable.) Often the pool remains at the infusion site for some time before it is absorbed, and subsequently the plasma sodium level is decreased. If the patient is already suffering from sodium deficit, the ultimate result can be death. Aged hospitalized patients may already have low plasma sodium levels secondary to ECF losses and, since such patients develop sodium deficit more readily than younger adults, they should certainly not receive electrolyte-free solutions subcutaneously.

Probably the only advantage of hypodermoclysis over the intravenous route is that it is easier to start. It is most often used for obese patients, infants, and the aged, for all of whom intravenous fluids are difficult to start. Suitable sites include the subcutaneous tissues on the lateral aspect of the thigh or abdomen, fatty tissue of the subscapular region, and loose fatty tissue at the base of the breasts. Infections are not uncommon with hypodermoclysis and are usually proportionate to the distention that occurs (because of tissue ischemia) and to the length of time required for the infusion.

Most often, two needles are used and the fluid is infused into two sites at once; the inverted Y-tubing from the solution bottle separates and goes to two sites (Fig. 6-6). The rate of infusion depends upon how well the fluid is absorbed from the injection site. When the fluid is absorbed well, 250 ml to 500 ml can be given at one site in one hour to an adult, but after a short while, the fluid is not this readily absorbed and the tissues become hard and swollen. To hasten absorption, hyaluronidase (Wydase) may be injected into the rubber portion of the tubing near each injection site. One hundred fifty units (75 units to each site) will hasten the absorption of one liter of solution. If an order

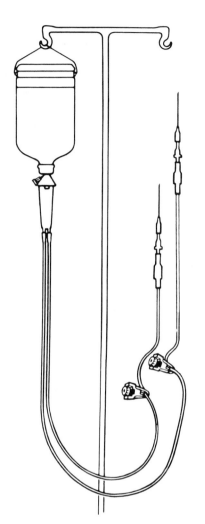

FIG. 6-6. Hypodermoclysis setup (Safticlysis Long-leg Setup; Courtesy of Cutter Laboratories, Berkeley, California)

for Wydase is not included with the fluid order, a call to the physician could be advantageous. The patient must be checked often when this route is used; a large amount of swelling can develop unless the flow rate is adjusted carefully. A small sterile gauze pad should be placed under the needle hub and another should cover the injection site. Following the removal of the needle, a light sterile dressing should be applied for at least 24 hours, since edematous injection sites are fertile fields for infections.

COMMENT Much discomfort is usually associated with hypodermoclysis. This route is undependable, especially when large amounts of fluids are needed in a hurry. Fluids are not absorbed well from the subcutaneous tissue when blood volume is severely reduced because of accompanying peripheral collapse. The fluids that can be given safely are quite limited and, ironically, the patients for whom hypodermoclysis is most often used (infants, the aged, and the obese) are the ones most prone to be harmed by them. Most authorities feel that a cutdown carries less risk than does hypodermoclysis.

DETERMINING FLOW RATE

PHYSICIAN'S ORDERS The physician should order the type of solution and, ideally, its rate of flow. In turn, the nurse is responsible for initiating most infusions and maintaining the proper rate of flow. Nurses must be aware of the composition of the prescribed solution, its desired effect, its usual rate of administration, and complications that may be associated with its use.

FACTORS INFLUENCING THE DESIRED RATE OF ADMINISTRATION The following are some of the factors considered in determining the best flow rate for the infusion:

Type of fluid
Need for fluids
Cardiac and renal status
Body size
Age
Patient's reaction during the infusion
Size of the vein

The desired rate for the infusion varies with the type of fluids used. For example, isotonic solutions can be given more rapidly than very hypertonic solutions. Although the other variables must also be considered, it is helpful for the nurse to know the usual infusion rates for various solutions. They are given in the descriptions of parenteral fluids in the next section of this chapter.

The patient's need for fluids influences the desired rate of administration; for example, a patient in hypovolemic shock needs fluids in a hurry. The infusion in this instance is much faster than usual. Because the heart and kidneys both play a major role in the utilization of fluids introduced intravenously, the presence of cardiac or renal damage can greatly alter the desired infusion rate. If the pumping action of the heart is inadequate, a rapid infusion of fluids could cause a dangerous fluid excess. The failure of the kidneys to excrete unneeded water and electrolytes can also result in excessive amounts of these substances in the body.

Aged patients almost always have some degree of cardiac and renal impairment; therefore, fluids are administered more slowly to them than to younger adults.

One of the best guides to safe flow rate is the patient's reaction to the infusion; the fact that persons respond differently to parenteral fluid infusions, just as they do to other medications, must never be forgotten. For this reason, the patient should be checked at least every 30 minutes during an infusion. The nurse should be aware of symptoms associated with the improper administration of various solutions so that she can know what to look for. Reactions associated with different parenteral fluids are described later in the chapter.

VARIATIONS OF DROP SIZE WITH DIFFERENT COMMERCIAL SETS Most nurses think in terms of "drops per minute" when considering rate of fluid flow. It must be remembered that commercial parenteral administration sets vary in the number of drops delivering 1 ml (Table 6-1). Unless one knows which administration set is to be used, it is more practical to consider the number of *milliliters* to be infused in 1 minute. From this figure, the number of drops per minute can be computed when the drop size of the administration set is learned.

Formula:
$$\text{drop factor} \times \text{ml/min} = \text{drops/min}$$

For example, to deliver 3 ml/min using a set with 10 drops to 1 ml, a flow rate of 30 drops/min would be necessary. To administer the same amount using a set

with 15 drops to 1 ml, a flow rate of 45 drops/min would be necessary. Drop size can vary somewhat according to the viscosity of the fluid being infused. Other factors affecting drop size include room temperature and height of the bottle; however, for practical purposes, the calibration of the IV set (gtts/min) should be accepted as valid. (Manufacturers list the drop factors on the administration set packages.)

CALCULATION OF FLOW RATE If the nurse knows the amount of fluid to be given in a prescribed time interval, plus the drop factor of the administration set to be used, the desired drops per minute can be easily calculated. The following formula is used:

$$\text{drops/min} = \frac{\text{total volume infused (ml)} \times \text{drop factor (drops/ml)}}{\text{total time of infusion in minutes}}$$

Sample Problem:

Infuse 1000 ml of 5% D/W in 2 hours (assume an administration set with 10 drops to 1 ml is to be used)

Total volume	=	1000 ml
Drops/ml	=	10
Total time of infusion in minutes	=	120

$$\frac{1000 \times 10}{120} = \text{approximately 80 drops/min}$$

To save the nurse's time, some handy calculators have been devised by manufacturers of parenteral fluids to determine the desired flow rate when the above factors are known, directions for which are included with the calculators.

A tape that indicates the hourly fluid rate should be placed on the bottle for convenient checking (Fig. 6-7).

An intravenous fluid flowsheet should be used for all patients receiving intravenous fluids (Fig. 6-8).

MECHANICAL FACTORS INFLUENCING GRAVITY FLOW RATE After the desired flow rate has been regulated, there are several mechanical factors that may alter it:

Change in needle position. A change in the needle position may push the bevel against or away from the venous wall. An adequate flow rate becomes diminished when the needle is pushed against the

TABLE 6-1 *Variation in size of drop in commercial administration sets—approximate no. of drops to deliver 1 ml*

Company	"Regular" set
Abbott Lab	15
Travenol Lab	10
McGaw Lab	15

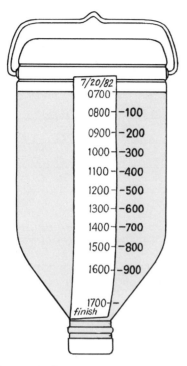

FIG. 6-7. Tape on the IV container indicates where the fluid level should be each hour.

vein, whereas it is increased when the needle moves away from the venous wall. Care must be taken to prevent speed shock by making sure the solution is flowing freely before adjusting the rate.

Height of the solution bottle. Because infusions flow in by gravity, a change in the height of the infusion bottle or bed can increase or decrease the rate—the greater the distance between the patient and the bottle, the faster the rate.

Patency of the needle. A small clot may occlude the needle lumen and decrease or stop the flow rate—when released, the rate increases. Clot formation may occur when an increase in venous pressure in the infusion arm forces blood back into the needle; causes of this include lying on the infusion arm, constriction with a blood pressure cuff, and the improper placement of parenteral solution containers when a continue-flow administration set is in use. A plugged needle should never be irrigated, since the dislodged clot could cause an infarction; if infected, it could cause spread of the infection to another part of the body. (To check for a plugged needle, kink the tubing a few inches above the injection site; then, pinch the tubing immediately above the needle. If resistance is met, and if there is no flashback of blood, the needle is probably plugged and should be removed.) Some

intravenous therapists favor the aspiration technique for removing fresh clots from plugged catheters.[1]

Venous spasm: A cold or irritating solution may retard flow rate by producing venous spasm. A warm pack placed proximal to the infusion site will relieve this condition.

Plugged air vent: A plugged air vent can cause an infusion to stop. Thus, patency of the air vent should be checked when no other cause is apparent for the stopped infusion.

Condition of final filter: Final filters can cause decreased flow rates should particulate matter block the filtering surface or if an air lock develops.

Crying in infants: This problem raises venous pressure and, thus, slows the rate of flow.

PARENTERAL NUTRIENTS

Nutrients that can be given intravenously include those listed below:

Carbohydrates

Proteins

Fat emulsions

Alcohol

Vitamins

Water

Electrolytes

Provision of nutrients intravenously is indicated when it is desired to put the gastrointestinal tract at relative rest, as in the presence of nausea, vomiting, diarrhea, peritonitis, ileus, or fistula. Intravenous nutrition is also indicated when the patient cannot take nutrients by an enteral route. In considering the nutritional needs of the patient exclusively in parenteral feedings, it should be borne in mind that the recommended energy need of an adult at bedrest is 1600 calories; this is a basal figure and does not allow for fever, high environmental temperatures, or other causes of increased metabolism.

It is difficult to restore the nutritionally depleted patient by the intravenous route using routine fluids; however, it is possible to maintain the state of nutrition fairly well with their use for a short period of time. The advent of hyperalimentation has made it possible to support life and maintain growth and development for prolonged periods. (See the discussion of parenteral hyperalimentation.)

Date	Type of Fluid	Additives	Time Started	Started By	Amount Received	Time Disc.	Disc By	Remarks
7/20	1000 ml D5W	KCl 10 mEg	0100	D.R.	1000	0900	N. M.	
7/20	100 ml D5W	CIMETIdiNE 300 mg	0600	D.R.	100	0620	D.R.	I.V.P.B.
7/20	1000 ml LR	—	0900	N.M.	1000	1700	M.H.	
7/20	100 ml D5W	cimetidine 300 mg	1200	N.M.	100	1220	N.M	I.V.P.B.
7/20	1000 ml LR	—	1700	M.H.				

FIG. 6-8. Intravenous fluid flowsheet.

CARBOHYDRATES

Included in the carbohydrates that can be administered and absorbed intravenously are glucose, fructose, and invert sugar. These are all monosaccharides and, therefore, ready for utilization by the body cells. Dex-

trose (glucose) is thought to be the closest to the ideal carbohydrate available because it is well metabolized by all tissues. Table 6-2 lists the caloric content and tonicity of various carbohydrate solutions.

It is impossible to meet daily caloric needs with isotonic solutions of carbohydrate solutions alone. For

TABLE 6-2 *Parenteral carbohydrate solutions*

Types of solutions	Calories/liter	Tonicity
Dextrose		
2½% Dextrose	85	Hypotonic
5% Dextrose	170	Isotonic
10% Dextrose	340	Hypertonic
20% Dextrose	680	Hypertonic
50% Dextrose	1700	Hypertonic
Fructose		
10% Fructose	375	Hypertonic
Invert sugar		
5% Invert sugar	190	Isotonic
10% Invert sugar	375	Hypertonic

example, to supply 1600 calories with a 5% dextrose solution would require 9 liters, a volume exceeding the tolerance of most patients. Mixtures of carbohydrate and alcohol are useful because of their rich caloric contribution and limited bulk.

Concentrated solutions, such as 20% or 50% dextrose, are useful for supplying calories for persons with renal insufficiency or who for other reasons are unable to tolerate large volumes. In order that the glucose be utilized, concentrated solutions must be administered slowly. When administered rapidly, such solutions act as a diuretic and pull interstitial fluid into the plasma for subsequent renal excretion. Such hypertonic solutions damage veins in direct proportion to their tonicity, and when used, they should be injected into large functioning veins so that they will be diluted by the relatively large blood volume.

If a hypertonic carbohydrate solution is infused too rapidly, hyperinsulinism may occur because the pancreas secretes extra insulin to metabolize the infused carbohydrate; discontinuance of the administration may leave an excess of insulin in the body. Symptoms of excess insulin include nervousness, sweating, and weakness. It is not uncommon for small amounts of isotonic carbohydrate solution to be given after hypertonic solutions to "cover" for the extra insulin and allow the return to normal secretion.

Potential substitutes for glucose include fructose (levulose) and invert sugar. Fructose has essentially the same caloric equivalent as glucose and may be less

irritating to the vein; it can be metabolized by adipose tissue independent of insulin. Fructose is contraindicated when lactic acidosis, or other forms of metabolic acidosis, are present or likely. (With fructose administration there is decreased conversion of lactate to pyruvate; a tendency to lactic acidosis occurs.) Fructose should not be administered rapidly. Invert sugar contains equimolar quantities of glucose and fructose; its caloric equivalent is the same as that of fructose. At similar rates of administration, considerably less invert sugar than dextrose is lost in the urine.

PROTEINS

When parenteral fluid therapy is necessary for more than 3 to 4 days, the patient should receive protein parenterally.[2] Protein is necessary for cellular repair, wound healing, growth, and for the synthesis of certain enzymes, hormones, and vitamins; proteins are available commercially as either protein hydrolysates* or crystalline amino acids.† Protein hydrolysates are derived from either casein or fibrin and are a mixture of essential and nonessential amino acids bonded together as polypeptides. Crystalline amino acids also contain essential and nonessential amino acids; however, they lack the polypeptide bonding and are thought of as a purer source of protein. (Peptides have been shown to cause nausea and vomiting.) Free amino acid preparations are reported to maintain slightly more positive nitrogen balance than are protein hydrolysate preparations.

Recall that nitrogen balance is the accepted measure for protein balance. A patient is in nitrogen balance if the rate of protein synthesis equals that of protein degradation; he is in negative nitrogen balance if protein catabolism exceeds protein anabolism. Patients may be kept in positive nitrogen balance when protein preparations are administered intravenously with sufficient calories to prevent breakdown of the protein for energy purposes. A healthy adult requires approximately 1 g of protein/kg of body weight daily to replace normal protein losses; healthy growing infants and

---

*Aminosol (Abbott Laboratories); Hyprotigen (McGaw Laboratories); Amigen, Travamin (Baxter-Trenavol Laboratories); C. P. H. (Cutter Laboratories).
†Aminosyn (Abbot Laboratories); FreAmine II (McGaw Laboratories); Travasol (Travenol Laboratories); Nephramine (McGaw Laboratories); Veinamine (Cutter Laboratories).

children require 1.4 g to 2.2 g of protein/kg of body weight. A liter of 5% amino acid solution contains 6.2 g of nitrogen; this is equivalent to 39 g of protein, the approximate daily requirement for an adult.[2] A 5% amino acid solution contains approximately 175 calories/liter.

The initial rate of infusion of protein hydrolysates should be slow (not more than 2 ml/min) to permit careful observation for adverse effects. A rate of 4 ml/kg body weight/hour should never be exceeded.[3] Excessively rapid infusion may cause nausea, vomiting, fever, chills, vasodilatation, urticaria, and abdominal pain. The infusion should be stopped if any of the above symptoms occur.

Protein solutions have a high NH_4^+ level and should be administered with special care to patients with hepatic insufficiency or emaciation; a rate slower than that used for other IV solutions should be employed. Seriously impaired renal function constitutes a contraindication to the administration of either protein hydrolysates or crystalline amino acid solutions, since the seriously impaired kidney cannot normally excrete nitrogenous wastes.

Supplemental medications should not be added to the solution or injected into the administration tubing without first checking their compatibility with the pharmacist. All protein solutions should be examined carefully before they are infused; either particulate matter or cloudiness should call for discarding the solution. Solutions should be used immediately after being opened since they are subject to spoilage.

FAT EMULSIONS

Fats supply more than twice the calories of proteins or carbohydrates; 1 g of fat yields 9 calories, while 1 g of carbohydrate or 1 g of protein yields only 4 calories. Fat emulsions for intravenous use have been in and out of vogue in the United States for a number of years. The presently available fat emulsion preparations in this country are Intralipid, 10% and 20% (Cutter Laboratories, Berkeley, CA) and Liposyn 10% (Abbott Laboratories, North Chicago, IL). The osmolarity of the 10% solutions is between 280 mOsm and 300 mOsm/liter, with a pH between 5.5 and 9.0; they provide 1.1 calories/ml.[4]

Emulsions of fat are used in conjunction with carbohydrates and amino acids to provide total parenteral nutrition (TPN) for patients requiring parenteral feedings for extended periods. No more than 60% of the total caloric intake of the patient should be made up of fats; carbohydrates and amino acids should comprise the remaining 40% or more of the caloric intake. When fat emulsions are used to correct a fatty acid deficiency, 8% to 10% of the caloric intake should be supplied by fats.

Fat emulsions are contraindicated in patients with a disturbance of fat metabolism such as pathological hyperlipemia, lipoid nephrosis, and acute pancreatitis if accompanied by hyperlipemia. Fat emulsions should be administered with caution in patients with severe liver damage, pulmonary disease, anemia, or blood coagulation disorders, or when there is danger of fat embolism.

General precautions to keep in mind when administering intravenous fat emulsions include the following:

Do not mix fat emulsions with electrolyte or other nutrient solutions.

Do not place additives in the fat emulsion container.

Do not use any bottle in which there appears to be an "oiling out" of the emulsion.

Do not use filters with fat emulsions.

Give slowly and increase the rate as the patient's reponse allows. The adult infusion rate for a 10% solution should be 1 ml/min for the first 15 to 30 minutes; if no untoward reactions occur, the rate can be increased so that 500 ml will be infused no faster than within 4 hours. The adult infusion rate for a 20% solution should be 0.5 ml/min for the first 15 to 30 minutes; if no untoward reactions occur, the rate can be increased so that 500 ml will be infused no faster than within 8 hours. (Follow manufacturer's instructions.)

Monitor liver function frequently; if tests indicate that liver function is impaired, the infusion of the fat emulsion should be discontinued.

Monitor the patient's ability to eliminate the infused fat from the circulation (the lipemia should clear between daily infusions).

One should observe carefully for the following adverse reactions that may occur with intravenous fat emulsion administration:

Dyspnea and cyanosis

Allergic reactions

Hyperlipemia

Hypercoagulability

Nausea and vomiting

Headache

Flushing

Increase in temperature

Sweating

Back or chest pain

Dizziness

Hepatomegaly and splenomegaly

Leukopenia

Deposition of a brown pigment in the reticuloendothelial system

Overloading syndrome (focal seizures, fever, leukocytosis, and shock)

Fat emulsions may be administered by a peripheral vein or central venous infusion. As stated earlier, fat emulsions should be infused separately from any other intravenous solution or medication. When it is necessary to administer a fat emulsion by conjunction with amino acids and hypertonic dextrose, it should be administered by means of a Y-tube or three-way stopcock near the infusion site. Because the lipid emulsion is less dense than the solution of dextrose and amino acids, it should be elevated more than the dextrose/amino acid preparation to prevent backflow. It is generally recommended that the rate of both solutions (fat emulsion and amino acid/dextrose solution) be controlled by infusion pumps for greater accuracy.

ALCOHOL SOLUTIONS

One gram of absolute ethyl alcohol yields 6 to 8 calories. Obviously an excellent source of calories, alcohol has been combined with carbohydrates to provide a high-calorie repair solution. When alcohol is infused with carbohydrate, it is apparently burned preferentially, thus permitting the glucose to be stored as glycogen. Alcohol spares body protein by providing readily accessible calories, and its sedating effect may be desirable for patients with pain. *Sedation can be achieved in the average adult without symptoms of intoxication by giving 200 ml to 300 ml of a 5% solution per hour.*

The nurse should be aware of the physiological and psychological effects of alcohol parenterally administered; these include dulling of memory, loss of ability to concentrate, an improved sense of well-being, increased respiration and pulse, and vasodilation. Alcohol solutions should *not* be employed in shock, impending shock, epilepsy, severe liver disease, or in patients with coronary thrombosis. Nausea and vomiting do not occur as frequently when alcohol is given parenterally as when a comparable amount is taken orally. However, whenever the rate of administration of alcohol given parenterally exceeds its metabolic destruction by the body, restlessness, inebriation, and coma can occur. Its major uses are probably in temporary situations in which a euphoriant effect is desirable or a high-energy solution is needed.

The parenteral administration of alcohol, particularly of hypertonic solutions, can cause phlebitis. Tissue necrosis can occur if the needle accidentally leaves the vein and permits solution to perfuse the surrounding tissue spaces. Care should be taken to see that the needle is carefully anchored in the vein and that the site is inspected frequently to detect infiltration.

Table 6-3 shows the caloric values and the tonicity of various solutions of alcohol.

PARENTERAL VITAMINS

Vitamins should be administered parenterally when there is inadequate oral intake or when parenteral fluid therapy is necessary for longer than 2 or 3 days. Although not food in themselves, vitamins are essential for the utilization of other nutrients. The need for vitamins is increased during periods of stress such as acute illness, infection, surgery, burns, injury, and convalescence.

The vitamins most frequently needed in parenteral alimentation are vitamin C and members of the B complex, all of which are water-soluble and stored by the body only in small amounts; they serve as coenzymes in the essential metabolic processes of the cells. Vitamin deficiency has been observed after only one week of parenteral administration of glucose and water alone. However, since most patients are on parenteral therapy for only limited periods, they do not require the fat-soluble vitamins A and D, which are better retained by the body. If IV therapy is prolonged, fat-soluble vitamins should be supplied weekly.

Because there is some waste of parenterally infused

TABLE 6-3 *Parenteral alcohol solutions*

Type of solution	Calories/liter	Tonicity
Alcohol 5% Dextrose 5%	450	Hypertonic
Alcohol 10% Dextrose 5%	730	Hypertonic

vitamins through urinary excretion, it is necessary to administer generous amounts to ensure an adquate intake. Hence, the patient who is on exclusive parenteral alimentation is usually given more than the daily vitamin requirement—sometimes as much as ten times the minimal requirement.

Vitamin C is particularly important in surgical patients to promote wound healing, whereas the B complex vitamins provide factors to aid carbohydrate metabolism and the maintenance of normal gastrointestinal function. (An occasional patient may show extreme sensitivity to the B complex vitamins.)

WATER

The patient on parenteral fluids exclusively can be provided with water by means of solutions of carbohydrates, the most common of which is 5% dextrose in water. Also useful are the various hypotonic electrolyte solutions, such as 0.45% sodium chloride or half-strength Ringer's. Isotonic electrolyte solutions provide no free water at all—they merely expand the ECF volume, ignoring the patient's need for water for renal excretion, insensible loss, and metabolic purposes; consequently, they should never be used to supply free water.

It should always be remembered that one should *never give water alone by a parenteral route*. If pure water were injected directly into the bloodstream, the red blood cells would absorb water, swell, and burst, with resultant damage to the kidneys and hemoglobinuria.

PERIPHERAL PARENTERAL NUTRITION

AMINO ACIDS USED ALONE IN PERIPHERAL VEINS

One peripheral method for parenteral nutrition involves the infusion of a 3% or 5% solution of amino acids with no nonprotein calorie source added. The basis for this method is adaptive starvation; ketones become the primary source of energy, causing protein to be spared.[5] This treatment is less hazardous than TPN through a central vein and is much less expensive and time-consuming for the medical staff.

This method provides few calories, reduces insulin formation, and mobilizes the body's own fat to meet energy needs. Fat mobilization and ketogenesis are efficient and desirable mechanisms in the reduction of bodily protein wasting. The goal of peripheral amino acid administration is not growth and synthesis but, rather, preservation of protein.[6] The preservation of body cell mass can be attained by peripheral amino acid infusion; restoration of depleted body cell mass can be attained only by hyperalimentation. Patients requiring long-term parenteral nutrition or those with severe undernutrition should receive hyperalimentation fluids through a central vein (even though this method is associated with greater technical, septic, and metabolic complications).

During the first few days of therapy, the patient should receive 1 or 2 liters of isotonic amino acids per day; this amount is gradually increased to 3 liters to 4 liters per day if the blood urea nitrogen (BUN) does not exceed 30 mg to 40 mg/dl, if the blood sugar falls toward 100 mg/dl and if ketones appear in the urine. Patients receiving solely isotonic amino acid solutions develop ketosis of starvation within 2 or 3 days; ketonuria should remain positive as a sign of adequate adaptation to a state of starvation. A significant rise in the BUN level for 3 consecutive days is an indication to discontinue the amino acid infusions; protein solutions must be used cautiously in patients with hepatic or renal failure. If adequate oral feedings cannot be resumed in 2 or 3 weeks, and particularly if the patient continues to display negative nitrogen balance, he may need to progress to hyperalimentation.

PERIPHERAL PARENTERAL NUTRITION WITH AMINO ACIDS, DEXTROSE, AND FAT EMULSIONS

Sometimes a fat emulsion is administered peripherally in conjunction with a 10% dextrose solution and a 5% amino acid solution for patients who are only moderately debilitated or cachectic. This regimen is safer and more convenient than parenteral nutrition through a central vein. Because the peripheral solution cannot be as concentrated as a solution administered through a central vein, it is necessary to use a larger fluid volume; thus, the peripheral method is not suited for patients with a fluid volume intolerance. (A highly concentrated fluid would cause phlebitis in a peripheral vein.)

It is generally recommended that the fat emulsion

be infused separately from the other solutions to avoid disturbing the emulsion's stability; it can, however, be infused simultaneously into the same vein by means of a Y-adaptor near the infusion site. Recall that the fat emulsion is less concentrated than the solution of dextrose and amino acids and thus helps to dilute the final infusate entering the vein (decreasing the incidence of thrombophlebitis). Because the lipid emulsion is less dense, it should be elevated more than the dextrose and amino acid preparation to prevent backflow. One should also remember that fat emulsions *cannot* be administered through a filter; if a filter is used for the dextrose and amino acid solution, the fat emulsion must be connected into the tubing *below* the filter. Once the seal on a fat emulsion flask is broken, the contents should be administered within 24 hours. Because a fat emulsion is an excellent bacterial cultural medium, special care should be taken to handle the emulsion flask and the IV set aseptically.

TPN THROUGH A CENTRAL VENOUS CATHETER

DEFINITION AND INDICATIONS

Total parenteral nutrition (also called hyperalimentation) refers to the infusion of large amounts of basic nutrients sufficient to achieve tissue synthesis and growth. Highly concentrated TPN solutions must be infused through an indwelling catheter into a large central vein, such as the superior vena cava. Patients requiring long-term parenteral nutrition or those with severe undernutrition should receive TPN fluids through a central vein, even though this method is associated with greater technical, septic, and metabolic consequences. Table 6-4 lists some conditions that may require the use of central TPN.

Because the incidence of serious complications with the use of central TPN is not low, it is important to establish the need for it. Many times tube feedings or peripheral vein nutrition will suffice.

CENTRAL TPN SOLUTIONS

Central TPN solutions are highly concentrated; a typical adult solution consists of approximately 25% dextrose and 3% to 4% protein, supplying close to 1000 calories/liter. Sometimes fat emulsions are infused

TABLE 6-4 *Conditions that may require total parenteral nutrition*

Gastrointestinal problems:
- Short-bowel syndrome
- Bowel obstruction
- Fistulas
- Pancreatitis
- Inflammatory bowel disease

Severe burns or trauma

Malignant disease (although the use for this purpose may be controversial, most people feel it is indicated if the patient is responding to chemotherapy)

Anorexia nervosa

Preparation of malnourished patients for surgical procedures

through a Y-tube near the infusion site to provide more calories.

Because potassium is needed for the transport of glucose and amino acids across the cell membrane, as much as 180 mEq of potassium may be given with 4000 calories. Magnesium, calcium, sodium, and phosphate are included as indicated. Some commercial preparations are available with electrolytes already added. In addition to electrolytes, water-soluble vitamins should be supplied daily; fat-soluble vitamins should be supplied weekly. (Trace elements are discussed later in this chapter.)

PREPARATION The TPN solution should be prepared by a pharmacist, preferably under a laminar flow hood (see Fig. 6-17). The solution should be infused as soon as possible after mixing because amino acids may undergo oxidation when the presence of hypertonic glucose, and glucosamines may form when amino acids remain in contact with hypertonic glucose for a long period of time.[7] Glucosamines can bind zinc and copper, causing them to be excreted in the urine; deficiencies of these trace elements can then occur.

STORAGE Storage of the TPN solution at 4° C, in the dark, for up to 48 hours is considered safe.[8] Admixtures that have been delivered to the nursing divisions for later administration should be stored in the medicine room refrigerator. Refrigerated admixtures should be removed from the refrigerator and allowed to reach room temperature prior to administration. Cold admixtures *should not* be warmed up in other manner (e.g., submerging in warm water) because such action could alter the stability of the solution. Each bottle should hang at room temperature no longer than 12 to 18 hours.[9]

ADMINISTRATION Initially in adults, 1 or 2 liters is administered daily; this is gradually increased according to the patient's tolerance to a maximum of 4 to 5 liters daily. Because the solution is hypertonic, it must be administered slowly and constantly over the 24-hour period. Too rapid rehydration may cause osmotic diuresis, which, untreated, can lead to dehydration, convulsions, and death. Patients particularly sensitive to hyperosmolarity, such as overt or latent diabetics, should be watched especially closely during hyperalimentation. All patients should be observed at least every 30 minutes. A mechanical infusion pump or controller should be used to administer the TPN solution. (Infusion pumps and controllers are described later in the chapter.) If the administration of the TPN solution should fall behind schedule, no attempt should be made to "catch up" by speeding up the infusion; a severe hyperosmotic reaction could result.

Most authorities recommend a bacterial filter to trap particulate matter, air, or bacteria inadvertently present in the TPN solution. A 0.45-micron filter will trap most bacteria except *Pseudomonas* and still allow flow by gravity; a 0.22-micron filter will block *Pseudomonas* but requires the use of an infusion pump.[10] A 0.22-micron filter is considered by most authorities to be the filter of choice for the administration of the TPN solution. (Recall that fat emulsions should never be infused through bacterial filters.) Before administering the TPN solution, the bottle should be held to the light and inspected for particulate matter; only clear solutions are suitable for administration.

COMPLICATIONS OF TPN

METABOLIC A number of potentially life-threatening complications may occur with TPN (Table 6-5).

Hyperglycemia is a common metabolic occurrence and is due to the high concentration of dextrose in the TPN solution. Factors predisposing to hyperglycemia include increased secretion of adrenal hormones in the stressed patient, increased age, underlying renal disease, peritoneal dialysis, and hypokalemia. (The mechanism by which hypokalemia causes glucose intolerance is not known.) Hyperglycemia, when severe, can cause osmotic diuresis and the hyperosmolar syn-

TABLE 6-5 *Metabolic complications that may occur with TPN*

Hyperglycemia

Hyperosmolar syndrome

Ketoacidosis

Postinfusion hypoglycemia

Essential fatty acid deficiency (if fat emulsions are not given)

Azotemia

Hyperammonemia (due to hepatic dysfunction)

Trace element deficiency (particularly zinc and copper)

Hypomagnesemia (due to insufficient magnesium in the TPN solution)

Hypermagnesemia (particularly prone to occur in patients with impaired renal function)

Hypophosphatemia (usually due to inadequate amount of phosphate in the TPN solution)

Hyperphosphatemia (particularly prone to occur in patients with impaired renal function)

Hypokalemia (due to excessive loss of potassium with inadequate replacement)

Hyperkalemia (particularly prone to occur in patients with impaired renal function)

Metabolic acidosis (not as common as it once was since manufacturers have substituted acetate salts of amino acids for the previously used hydrochloric salts of amino acids)[11]

Vitamin deficiencies or excesses

drome (dehydration, seizures, coma, and death). Postinfusion hypoglycemia (due to hyperinsulinism) can occur if the TPN solution is abruptly discontinued. (Recall that a large glucose load causes high insulin secretion.) It is important, then, that TPN solutions be discontinued *gradually* to allow the pancreas time to adapt to the decreased glucose load. If the TPN solution must be discontinued abruptly, a solution of 5% dextrose in one-quarter strength saline (with appropriate potassium) should be administered at the previous parenteral nutrition rate until insulin secretion decreases (approximately 12 to 24 hours).[12] The nurse should recall that symptoms of hypoglycemia include occipital headaches; cold, clammy skin; dizziness; rapid pulse; and tingling sensations in the extremities and around the mouth.

Amino acids can contribute to a high BUN level in

patients with severe renal damage. Hyperammonemia can be a problem in patients with severe hepatic damage and in infants.

Electrolyte disturbances that may occur with TPN are listed in Table 6-5. It should be noted that marked deficiencies of potassium, phosphate, and magnesium may occur with TPN if inadequate amounts of these substances are added to the solution; remember that anabolism causes these electrolytes to move into the cells. As mentioned earlier, hypokalemia can cause increased glucose intolerance. Hypophosphatemia may decrease the concentration of 2,3 diphosphoglycerate (DPG), causing the red cells to have an increased affinity for oxygen (decreasing tissue oxygenation).

There is a growing concern regarding trace element deficiencies in patients receiving TPN, especially copper and zinc. These two elements are of particular importance because they are cofactors for many enzymes. Symptoms of zinc deficiency associated with TPN administration may include diarrhea, abdominal pain, mental depression, alopecia, and skin eruptions. Copper deficiency syndrome can also occur with TPN administration, characterized by anemia, neutropenia, and hypoproteinemia. Oral supplements of zinc and copper may be used when the patient receiving TPN has sufficient gastrointestinal function for their absorption. Solutions of zinc and copper can also be prepared by the pharmacist for intravenous administration. Chromium deficiency may develop with long-term TPN administration; symptoms can include glucose intolerance, neuropathy, and encephalopathy. Although rare, manganese deficiency can occur in patients receiving TPN; symptoms can include weight loss, transient dermatitis, nausea, vomiting, and hypocholesterolemia. There is not general agreement on the required intravenous dosage of trace elements. It is generally recognized, however, that certain trace elements (particularly copper, zinc, manganese, and chromium) need to be considered for the maintenance of proper nutritional status of patients receiving TPN. There is a danger of excessive intake of trace elements when large supplements are added to infusates, since it has been demonstrated that commercially prepared solutions already contain variable amounts of trace elements as contaminants. It should be remembered that many trace elements may become toxic at high levels; the difference between therapeutic and toxic doses may be small.

INFECTIOUS Infections are a constant threat to patients receiving TPN through a central vein. Patients requiring TPN are often predisposed to infection as a result of malnutrition, frequent use of broad-spectrum antibiotics, and the presence of concomitant infections in wounds, the lungs, and the urinary tract. The indwelling subclavian catheter can serve as an entry port for organisms, and the TPN solution provides a rich culture medium for bacterial growth. Infectious complications of TPN are inversely related to the emphasis placed on aseptic technique in catheter placement and maintenance and on preparation and administration of the solution. Septicemia rates have been reported to be as high as 33% in institutions in which no uniform protocol for infection control is followed. The incidence of septicemia is much less (as low as 3%) in institutions using sound infection control practices.

An unexplained temperature elevation is an indication to halt the administration of the TPN solution until the source of the fever is found, or, at least, to start over with a new catheter, tubing, and solution. The possibility of catheter-induced septicemia must be considered. Many patients receiving TPN are concurrently receiving corticosteroids, cancer chemotherapy, or irradiation; these treatments may mask the signs of infection.

Organisms important in TPN-related septicemia include the *Candida* species and both gram-negative and gram-positive bacteria.

NURSING ASSESSMENT

Because of the numerous complications that can occur with TPN, it is necessary that the nurse assess the following:

> Daily body weights
>
> Fluid intake and output
>
> Vital signs (every 4 to 6 hours)
>
> Fractional urines for glucose and acetone (at least every 6 hours)

Accurate intake–output records and a weight chart are extremely valuable in assessing fluid balance. For example, a much larger urinary output than fluid intake can indicate osmotic diuresis. In regard to body weights, it should be remembered that no more than 0.2 kg to 0.5 kg/day weight gain can be attributed to lean body mass accumulation; any gain greater than this amount usually indicates fluid volume excess.

Vital signs should be evaluated at regular intervals. An elevated body temperature can indicate infection and must be carefully evaluated. Remember that elderly and chronically ill persons tend to be hypothermic; thus, a rise in body temperature to normal in these patients may be a sign of infection.

Urine must be tested for sugar and acetone every 6 hours; an attempt is made to keep the blood sugar below 200 mg/dl (ideally below 140 ml/dl) and the urine sugar below 2+. In patients with renal damage the urine glucose may not reflect the true blood sugar; the patient may have glucosuria with a normal blood sugar, or no glucosuria with an elevated blood sugar. In such patients, *blood* sugar sould be monitored frequently. Regular insulin is needed to control glucosuria of 3+ or 4+. Regular insulin can be added directly to the TPN solution or administered intermittently by the subcutaneous route. Hyperglycemia is more of a problem during the first week of therapy than in subsequent weeks. The most common cause of glucosuria is a sudden increase in the infusion rate.

Hyperglycemia may be the initial sign of sepsis.[13] Sepsis is a form of stress and can initiate gluconeogenesis and glycogenolysis by stimulating release of adrenocorticotropic hormone (ACTH) and catecholamines.[14] Glucosuria may occur up to 12 hours before a temperature elevation or other signs of developing sepsis![12]

The nurse should also be alert for signs of electrolyte disturbances, including those listed below:

Hypokalemia typically causes muscular weakness and cardiac arrhythmias.

Hyperammonemia can cause lethargy, seizures, and coma.

Hypomagnesemia can cause weakness, vertigo, positive Chvostek's sign, and convulsive seizures.

Hypermagnesemia (3 mEq to 5 mEq/liter) can cause hypotension, nausea, and vomiting; at a level of 5 mEq to 10 mEq/liter, drowsiness, muscular weakness, and hyporeflexia can occur, along with abnormal cardiac conduction.

Hypophosphatemia can be manifested by confusion, paresthesias, and hyperventilation.

Signs and symptoms of electrolyte disturbances are described more thoroughly in Chapter 2.

Since the patient is not eating, salivary secretion is diminished. Oral hygiene should be maintained; sour balls or lemon slices may periodically be sucked on to maintain salivary flow.

LABORATORY DATA Parenteral hyperalimentation requires frequent clinical evaluation of hepatic and renal function, as well as daily laboratory determinations of blood sugar, serum osmolality, and electrolyte levels (Na, P, Mg, Ca, K, Cl, and HCO_3) for the first week, and then as indicated. The nurse should be familiar with these tests and recognize abnormalities when they occur.

CATHETER AND TUBING CARE

Sterile dressing changes should be done every 48 hours, more frequently if necessary, but only by nurses skilled in this procedure. Danger of infection and air embolism is greater when a large central vein is cannulated. The dressing should be examined at least daily to ensure that it has not become wet or soiled. Surgical masks should be worn by all personnel and the patient during the dressing change. The insertion site should be inspected for signs of inflammation. An antimicrobial agent should be applied to the site, followed by an occlusion dressing. The catheter should be visually inspected for signs of cracking, splitting, kinking, and for the presence of clots or leakage.

It is recommended that the tubing be changed with each new bottle; during the tubing change, when the catheter is open to air, the patient should perform the Valsalva maneuver to increase positive pressure and, thus, decrease the chance of air being drawn into the vena cava (recall that venous pressure in the central vessels is low). The Valsalva maneuver is accomplished by having the patient bear down with his mouth closed (forced expiration against a closed glottis). If the patient has an endotracheal or tracheal tube, or cannot do the Valsalva maneuver, he may be ventilated with an Ambu bag, (and the tubing can be changed during a prolonged inspiratory phase).[15] Two nurses are required to perform this procedure.

It is generally recommended that the catheter not be used to infuse blood or obtain blood samples; medications should not be "piggybacked" or "pushed" into the central catheter.

Insertion of a central line catheter, under strict aseptic technique, remains the responsibility of the physician. Proper placement of the catheter should always be validated by radiography.

CENTRAL VENOUS CATHETERS FOR LONG-TERM USE Indwelling central venous catheters for long-term use (Figure 6-9) are being used with increasing frequency in many medical centers to administer TPN so-

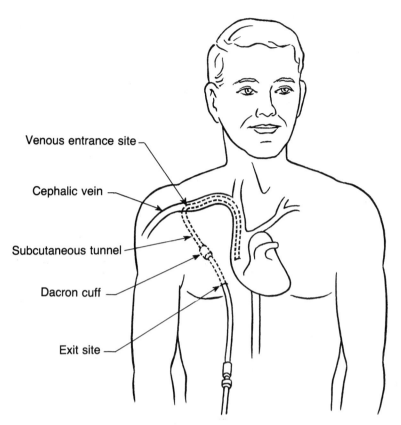

Venous entrance site

Cephalic vein

Subcutaneous tunnel

Dacron cuff

Exit site

FIG. 6-9
Indwelling central venous catheter for prolonged TPN or chemotherapy.

lutions, chemotherapeutic agents and other drugs, and blood products; they are also used as vehicles from which to draw blood samples.

The two types of long-term use catheters presently available are the Broviac catheter and the Hickman catheter. Both are made of silicone. The Broviac catheter has one dacron cuff; the Hickman has one or two. Fibrous tissue adheres to the dacron cuff in the subcutaneous tunnel and serves as a barrier to infection. The Hickman catheter is a modification of the Broviac catheter; its internal diameter is 1.6 mm as compared with 1.0 mm in the Broviac catheter.

The catheter is implanted in the operating room with the aid of fluoroscopy. A small incision is made in the deltopectoral groove, and the cephalic vein is isolated. A subcutaneous tunnel is then gently formed with a long forceps. When the tip of this instrument has reached a point between the nipple and sternum, its tip is exteriorized through a small incision (no larger than 1 cm). Forceps are then used to draw the catheter up the tunnel. (The dacron cuff is positioned between the two incisions and over an intercostal

space.) The catheter is then inserted into the vein and threaded into the lower part of the superior vena cava (at the entrance of the right atrium). Dressings are initially applied to both the insertion site and the exit site. Catheters may remain in place for months.

Central venous catheters may be used for continuous or intermittent infusions. When used for intermittent infusions, a catheter cap is used to plug the catheter in between administration of fluids or blood drawing. Some authorities prefer to use a cap with an injection port to plug the catheter; injections can be administered directly through the port, decreasing the chance of contamination and air embolism. A heparinized saline solution is used to flush the line between treatments. Clotting of blood in the catheter is a common problem that must be avoided. When not in use, the catheter is clamped with a smooth clamp and taped to the patient's clothing. Clamps with teeth must be avoided, since they may sever the catheter.

Catheter exit site care consists of cleansing with hydrogen peroxide, application of an antimicrobial agent, and covering with 2″ × 2″ gauze dressing. The catheter is taped to the patient's chest or clothing to prevent accidental dislodgement.

Removal of the catheter is usually relatively simple when performed correctly. The catheter is wrapped around the hand and firmly and steadily pulled with constant pressure; jerking may cause the catheter to break. The venous entry site closes after the catheter is removed, thus no bleeding occurs from this area. Slight bleeding occurs at the exit site due to passage of the dacron cuff. Occasionally the cuff remains adhered to the tissues; this is not a problem if the cuff is not infected. The exit site is cleansed and covered with a dressing and heals within a few days. Surgical removal may be necessary if the catheter is difficult to remove by pulling.

When indicated, patients and their families can be taught to care for the catheter at home.

TPN FOR OUTPATIENTS

Some medical centers are set up to provide home TPN programs. Selected patients, with the help of their families, can be taught to manage their own therapy as outpatients. They can receive TPN either during the day, overnight, or, with the aid of a special vest apparatus, while they perform their daily activities. Candidates for home TPN programs receive a psychosocial evaluation before being admitted to the program. A major factor to consider is the level of education the patient must have to manage his home care; substantial investigation of the home situation is required. Advocates of home TPN therapy state that it provides for increased patient comfort, fewer therapeutic and septic complications, psychological advantages, and financial savings. Substantial cost savings have been documented when parenteral nutrition patients are treated at home. Home TPN is third-party reimbursible in some instances and appears to be gaining in popularity with today's increasing emphasis on cost-containment.

Frequently used for home TPN is a cuffed silicone catheter that is implanted in the subclavian vein and threaded into the right atrium. The catheter enters through a subcutaneous tunnel in the anterior chest wall; a dacron cuff at the skin entrance site fixes the catheter in position and acts as a barrier to infection as fibrous tissue grows into it. This type of catheter is frequently used for hospitalized TPN patients also. (See the discussion on long-term catheters directly preceding this section.)

ELECTROLYTE SOLUTIONS

TYPES

A wide variety of electrolyte solutions are available for parenteral administration. Some of the more common solutions are listed in Table 6-6 for quick reference. Included in the table are the following:

  Electrolyte content
  Trademarks (brand names)
  Precautions for administration
  Usual rate of administration

Electrolyte solutions are considered isotonic if the total electrolyte content (anions + cations) approximates 310 mEq/liter; they are considered hypotonic if the total electrolyte content (anions + cations) is below 250 mEq/liter; and they are considered hypertonic if the total electrolyte content (anions + cations) exceeds 375 mEq/liter.

ADMINISTRATION OF POTASSIUM SOLUTIONS

Potassium may be given in the form of commercially prepared electrolyte solutions, or a potassium salt may be added as a supplement to an intravenous fluid, such as 5% or 10% dextrose in water. (Extreme care is required in measuring.)

The nurse should keep the following facts in mind when potassium-containing solutions are administered:

1. Small ampules containing concentrated solutions of potassium salts for addition to IV fluids are meant to be mixed with at least 500 ml to 1000 ml of solution. They should *never* be directly administered in concentrated form by IV push because of the danger of cardiac arrest.

2. When potassium chloride (KCl) is added to an IV solution, it should be *thoroughly mixed* by shaking the bag or bottle to prevent "crowning" or "layering" of the KCl. Crowning of the KCl allows the administration of a large bolus of the drug; phlebitis, tissue necrosis and, worst of all, cardiac arrest may occur. It is important to squeeze the medicine ports of plastic bags while they are in the upright position and then to mix the solution thoroughly.

(Text continues on p. 161)

TABLE 6-6 *Electrolyte solutions*

Solution	Tonicity	mEq/liter								Brand names	Reasons for use	Comments
		Na⁺ Na^+	K^+	Ca^{++}	Mg^{++}	NH_4^+	Cl^-	HCO_3^- Precursor*	HPO_4^{--}			
Sodium Chloride 0.45% (One-half isotonic saline)	Hypotonic	77					77				Supply daily salt requirements Supply water for excretory purposes	Available commercially with varying concentrations of carbohydrates
Sodium chloride 0.9%	Isotonic	154					154				Replace Na^+ and Cl^- Expands extracellular fluid volume Correct mild metabolic alkalosis	Sometimes used as a routine electrolyte replacement solution even though it supplies only Na^+ and Cl^- Does not supply free water for excretory purposes Cl^- is supplied in excess of normal plasma Cl^- level—excessive use of isotonic saline can cause Cl^- to replace part of the body HCO_3^- (metabolic acidosis)
Sodium chloride 3%	Hypertonic	513					513				Rapid correction of severe low-salt syndrome	Contraindicated unless severe salt depletion is present
Sodium chloride 5%	Hypertonic	855					855					Use with caution in edematous patients Administer a small volume at a slow rate (such as 200 ml over a minimum of 4h)
Ringer's	Isotonic	147	4	5			156				Replaces K^+ and Ca^{++} in addition to Na^+ and Cl^-	Contains an excess of Cl^- in relation to normal plasma Cl^- level
Lactated Ringer's (Hartmann's)	Isotonic	130	4	3			109	28			Routine electrolyte maintenance solution Correct metabolic acidosis Replace fluid lost as bile, diarrhea, and in burns	Electrolyte concentration closely resembles extracellular fluid

continued

*HCO_3^- equivalent may be bicarbonate, lactate, acetate, gluconate, citrate, or a combination of these.

TABLE 6-6 *Electrolyte solutions (continued)*

Solution	Tonicity	Na+	K+	Ca++	Mg++	NH4+	Cl-	HCO3- Precursor*	HPO4--	Brand names	Reasons for use	Comments
Gastric (Cooke and Crowley)	Isotonic	63	17			70	150			Electrolyte No. 3 Travenol Lab; Isolyte G McGaw Lab; Ionosol G Abbott Lab	Replace gastric fluid lost in vomiting and gastric suction	Should not be used as a routine maintenance solution; pH of solution is acid (3.3–3.7), similar to that of gastric juice; Contraindicated in hepatic insufficiency
Duodenal	Isotonic	138	12	5	3		108	50		Ionosol C-CM Abbott Lab	Replace duodenal fluid loss; Correct mild acidosis	Same precautions as for any K-containing solution
Duodenal (modified)	Hypotonic	79.5	36	4.5	3		63	60		Electrolyte No. 1 McGaw Lab Travenol Lab; Isonosol D Abbott Lab	Replace pancreatic and duodenal fluid losses; Supply water for excretory needs	Note that the K+ content is high
Sodium lactate 1/6 M (alkalinizing solution)	Isotonic	167						167			Correct severe metabolic acidosis	Contraindicated in liver disease, shock, and rightsided heart failure (lactate ions are improperly metabolized in these conditions); Contraindicated in alkalosis
Sodium bicarbonate 1.4%	Isotonic	167						167			Correct severe metabolic acidosis; Alkalinize urine (as in hemolytic reactions)	Contraindicated in metabolic and respiratory alkalosis
Sodium bicarbonate 5% (alkalinizing solution)	Hypertonic	595						595				Contraindicated when hypocalcemia is present; alkalinization of plasma may produce signs of tetany; Administer a small volume at a slow rate (such as 100 ml of a 5% solution over 2 hr); Administer with extreme caution to salt-retaining patients with cardiac, renal or liver damage

Solution	Tonicity	mEq/liter								Brand names	Reasons for use	Comments
		Na^+	K^+	Ca^{++}	Mg^{++}	NH_4^+	Cl^-	HCO_3^- Precursor*	HPO_4^{--}			
Ammonium chloride 0.9% (acidifying solution)	Isotonic					168	168				Correct severe metabolic alkalosis in children	Contraindicated in disturbed hepatic function, renal failure or any condition with a high NH_4^+ level
Ammonium chloride 2.14% (acidifying solution)	Hypertonic					400	400				Correct severe metabolic alkalosis in adults	Contraindicated in disturbed hepatic function, renal failure or any condition with a high NH_4^+ level
Balanced electrolyte (Fox)	Isotonic	140	10	5	3		103	55		Isolyte E McGaw Lab Plasma-Lyte Travenol Lab	Replaces gastrointestinal losses Burn treatment Postoperative fluid replacement	Electrolyte content similar to plasma except that it has twice as much K^+ (K^+ content is similar to that of intestinal juice)
Butler (Electrolyte No. 88)	Hypotonic	57	25		5–6		45	25–31	13	Electrolyte No. 2 McGaw Lab Travenol Lab Ionosol B Abbott Lab	Supply water Supply maintenance needs of Na^+, K^+, Mg^{++}, and Cl^- Replace fluid loss from the large intestine	Note relatively high K^+ content
Modified Butler (Electrolyte No. 48)	Hypotonic	25	20		3		22	26		Ionosol MB Abbot Lab	Supply water Used in pediatrics to treat dehydration of acidosis, diarrhea, and burns	
Balanced hypotonic solution	Hypotonic	40	13		3		40	16		Normosol-M Abbott Lab Plasma-Lyte 56 Travenol Lab Isolyte H McGaw Lab	Supply water Supply electrolytes for older children and adults	Note relatively high K^+ content

*HCO_3^- equivalent may be bicarbonate, lactate, acetate, gluconate, citrate, or a combination of these.

continued

TABLE 6-6 *Electrolyte solutions (continued)*

Solution	Tonicity	mEq/liter								Brand names	Reasons for use	Comments
		Na^+	K^+	Ca^{++}	Mg^{++}	NH_4^+	Cl^-	HCO_3^- Precursor*	HPO_4^{--}			
Balanced isotonic solution	Isotonic	140	5		3		98	50		Isolyte S McGaw Lab Normosol-R Abbott Lab Plasma-Lyte 148 Travenol Lab	Extracellular fluid replacement solution Correct mild acidosis	Contains no calcium; therefore can be administered simultaneously with blood
Electrolyte No. 75 (Talbot)	Hypotonic	40	35				40	20	15	Electrolyte No. 75 Travenol Lab Ionosol T Abbott Lab Isolyte M McGaw Lab	Supply water Supply maintenance electrolyte needs Correct K^+ deficit	Contains more K^+ than most maintenance solutions
Maintenance electrolyte solution	Hypotonic	40	16	5	3		40	24		Plasma-Lyte M Travenol Lab Isolyte R McGaw Lab	Supply water Supply maintenance electrolyte needs—contains Ca^{++} and Mg^{++} in addition to Na^+ and K^+	

*HCO_3^- equivalent may be bicarbonate, lactate, acetate, gluconate, citrate, or a combination of these.

Remember that KCl should never be added to an IV container in the hanging position![16]

3. It is wise to limit the potassium concentration in 1 liter of fluid to 20 mEq to 40 mEq, never more than 80 mEq, since an accidental rapid infusion rate is less dangerous when potassium content is moderate. (Note that some dextrose solutions are available commercially with varying concentrations of KCL, usually 10 mEq to 40 mEq/liter, already added.)

4. Rate of administration:
   - For usual IV replacement therapy, 5 mEq to 10 mEq/hr (suitably diluted) may be given as a constant infusion.
   - The maximal infusion rate should not exceed 20 mEq/hr (suitably diluted).
   - In severe cases of hypokalemia (plasma K less than 2.5 mEq/liter) it may be necessary to administer a higher dose of potassium (no more than 40 mEq/hr) while continuously monitoring the electrocardiogram (EKG).[16] Frequent plasma potassium determinations should be made during this treatment. Note that in the treatment of severe hypokalemia, it may be necessary to administer potassium in nondextrose solutions (such as isotonic saline), since the dextrose may cause a further fall in plasma potassium with ensuing cardiac arrhythmias. (Recall that dextrose administration favors a shift of potassium into the cells.)

5. Solutions containing potassium should be conspicuously labeled so that other personnel can readily note its presence (Figs. 6-10 and 6-11).

6. Potassium should be administered only after adequate urine flow has been established. A decrease in urine volume to less than 20 ml/hr for 2 consecutive hours is an indication to stop potassium infusion until the situation is evaluated. Urinary suppression may be due either to inadequate fluid intake or renal impairment; the rapid infusion of a hydrating solution (such as 5% D/W or a hypotonic electrolyte solution) should cause an increase in urine output if the problem is fluid volume deficit. Once urinary output is adequate, potassium infusion may be resumed. However, failure of the hydrating solution to increase urinary output indicates renal impairment and is an indication to withhold potassium. Recall that potassium is mainly excreted by way of the kidneys; when the kidneys are nonfunctional, a high potassium level builds up in the bloodstream (hyperkalemia).

7. A solution containing sizable amounts of potassium (30 mEq to 40 mEq/liter) is sometimes associated with pain in the vein it is entering, especially if infused into a vein where a previous venipuncture has been performed. Slowing the rate usually relieves this sensation. Because administration of a potassium solution into the subcutaneous tissues is painful, it is rarely administered by means of hypodermoclysis; when it is, the concentration should be no higher than 10 mEq/liter to avoid local pain and tissue damage. Care should be taken to avoid accidental subcutaneous infiltration of more concentrated potassium solutions, since *severe* tissue damage may result.

ADMINISTRATION OF CALCIUM SOLUTIONS

1. Calcium salts should not be added to IV solutions containing sodium bicarbonate since the precipitate calcium carbonate will form.

2. Calcium should not be added to IV solutions containing phosphate since calcium phosphate may be precipitated.

3. Solutions containing calcium may cause tissue sloughing if they are allowed to infiltrate into the subcutaneous tissues; thus, patients receiving calcium solutions should be observed frequently to prevent this complication.

4. Intravenous calcium administration usually is contraindicated in digitalized patients, since calcium ions exert an effect similar to that of digitalis and can cause digitalis toxicity with adverse cardiac effects.

5. Excessive or too rapid administration of calcium intravenously can cause cardiac arrest, preceded by bradycardia. Recall that hypercalcemia may cause the heart to go into spastic contraction.

6. Intravenous preparations of calcium salts suitable for prevention and treatment of hypocalcemia include calcium gluconate, calcium chloride, and calcium gluceptate. Calcium chloride is three times more potent than calcium gluconate; generally calcium preparations other than calcium chloride are preferred.

(Text continues on p. 163)

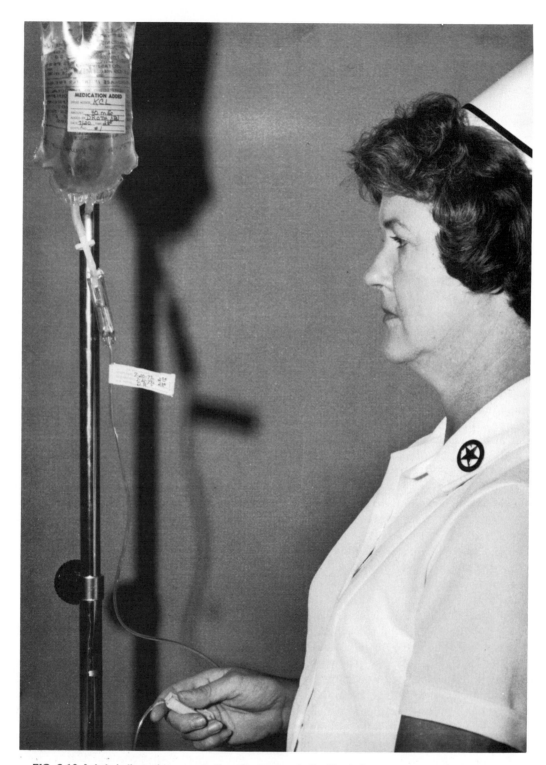

FIG. 6-10. Labels indicate the concentration of potassium in the IV solution and the time of tubing change.

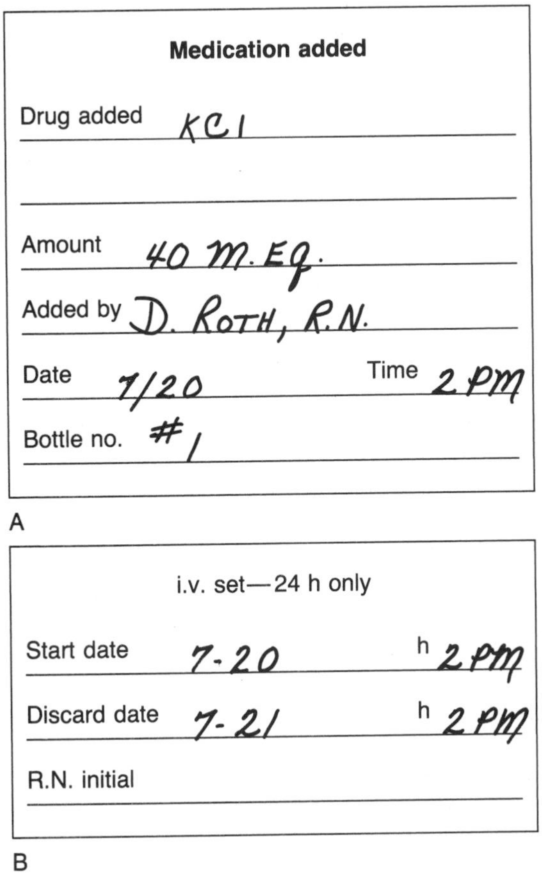

| **Medication added** |
| Drug added   *KCl* |
| Amount   *40 M.Eq.* |
| Added by *D. Roth, R.N.* |
| Date *7/20*   Time *2 PM* |
| Bottle no. *#1* |

A

| i.v. set—24 h only |
| Start date   *7-20*   h *2 PM* |
| Discard date   *7-21*   h *2 PM* |
| R.N. initial |

B

FIG. 6-11. (A) Close-up view of additive label. *(B)* Close-up view of tubing change label.

ADMINISTRATIONS OF MAGNESIUM SOLUTIONS

1. Question the use of magnesium solutions in patients with oliguria (99% of parenterally administered magnesium is excreted by the kidneys). A minimum of 100 ml urine output every 4 hours should be maintained.[17]

2. Avoid rapid administration of magnesium solutions since they may cause uncomfortable sensations of heat; more importantly, they cause coma, respiratory depression, and cardiac arrest. Equipment to maintain artificial ventilation must be available at all times.

3. Observe the patient's respirations and deep tendon reflexes. A respiratory rate below 12 to 14 per minute should be reported to the physician, as should poor to absent deep tendon reflexes (such as the knee jerk).

4. Have calcium gluconate immediately available to counteract serious symptoms of hyper-magnesemia, should they occur.

5. It may be necessary to reduce the dose of other central nervous system depressants (such as narcotics and barbiturates) when given in conjunction with magnesium sulfate.[17]

INCOMPATIBILITIES

The nurse is frequently required to add medications to parenteral fluids. There is a fast-growing number of medications and types of parenteral fluids on the market; thus, the number of possible combinations is astronomical. It is impossible, without help, for the nurse to know which medications can be safely mixed with certain intravenous fluids; such assistance must come from several sources:

Pharmacist

Manufacturer's directions

Publication of new admixture information

The nurse should keep the following points in mind when adding medications to parenteral fluids:

1. Thoroughly review the literature provided by the manufacturers of the IV fluid; most companies have prepared charts noting the compatibility of various medications with their solutions. Also, review the literature provided by the manufacturers of the additive.

2. When possible, it is best to add only one medication to each solution bottle, as the complex interaction between two additives may render the solution incompatible.

3. When in doubt, the nurse should check with the pharmacist; pharmacists often have more knowledge concerning drug incompatibilities and have ready access to publications concerning this problem. (Many hospitals have wisely added pharmacy-centralized intravenous additive programs in an attempt to provide safer admixture preparation.)

4. Use freshly prepared solutions whenever possible and monitor infusing fluids periodically for physical changes. Some incompatibilities do not become apparent until the solution has been mixed for awhile. A physical incompatibility may be mani-

fested by a haze, color change, or effervescence. Chemical incompatibilities are more difficult to detect because they do not always produce a visible change.

5. The formation of a precipitate when a medication is added to an intravenous fluid is an indication to discard the solution unless the directions accompanying the medication state otherwise. The intravenous administration of a solution containing insoluble matter may result in embolism or other damage to the heart, liver, and kidneys.

6. An administration set with a final filter should be used whenever possible to trap undetected precipitates.

7. The degree of solubility of an additive varies with the pH, as most incompatibilities are related to changes in pH. Solutions of a high pH seem to be incompatible with solutions of a low pH. (A chart listing the pH of certain drugs and parenteral solutions can help the nurse predict potential incompatibilities.)

8. Sodium bicarbonate and calcium salts should never be mixed since they form an insoluble precipitate (calcium carbonate).

9. Dilantin and Valium should not be mixed with any intravenous solution, since they precipitate out. They are both incompatible with any other drug in syringe or solution.

10. It is best not to mix multiple vitamin complexes with potassium, penicillin G, or ampicillin because the acidic ascorbic acid lowers the pH and causes degradation of the antibiotic.[18]

11. It is best not to mix heparin with antibiotics because heparin may affect the stability of certain antibiotics.[18]

12. Hydrocortisone should not be mixed with tetracycline or cephalothin, since a precipitate will form.[18]

13. It is best not to administer sodium bicarbonate and epinephrine jointly because sodium bicarbonate tends to neutralize epinephrine; they may be administered intermittently through the same tubing if it is flushed first.[18]

14. Although the current literature is conflicting, it has been suggested that the anticoagulant activity of heparin is decreased in the presence of dextrose.[18] It may be desirable, then, to use an electrolyte solution (such as isotonic saline) as a diluent for heparin infusions. If dextrose is desired, an increased dose of heparin may be needed to achieve the desired anticoagulant effect.[18]

If, after researching the literature, the nurse finds that an order for an additive is incorrect, the physician should be contacted for a change in orders.

SPECIAL EQUIPMENT FOR INFUSIONS

DROP SIZE REDUCTION

When potent medications are added to intravenous solutions, extreme precautions must be taken to prevent too rapid administration.

Manufacturers of intravenous equipment provide administration sets delivering from 50 drops to 60 drops/ml. These microdrip sets allow for greater control of fluid flow rate.

VOLUME CONTROL SETS

There are a number of commercial sets designed to administer limited amounts of solution in precise volume, all of which are extremely useful in the infusion of potent medications. They are also valuable in pediatric intravenous therapy. One type of volume control set is pictured in Figure 6-12.

BOTTLE ARRANGEMENTS

A series hookup may be used to add more fluid or to change fluids while an infusion continues (Fig. 6-13, A). The two bottles are connected by plastic tubing, with the air vent closed in the primary bottle and the secondary bottle vented—the secondary bottle will empty first. If the specific gravity is higher in the secondary bottle, there will be no mixing of solutions; there will be mixing if the specific gravity is higher in the primary bottle.

A parallel hookup may be used to administer two solutions alternately or simultaneously (Fig. 6-13, B). This type of hookup is more dangerous than a series hookup, because if either of the bottles empties completely and its clamp closure is not complete, large quantities of air will leak in by way of the empty bottle

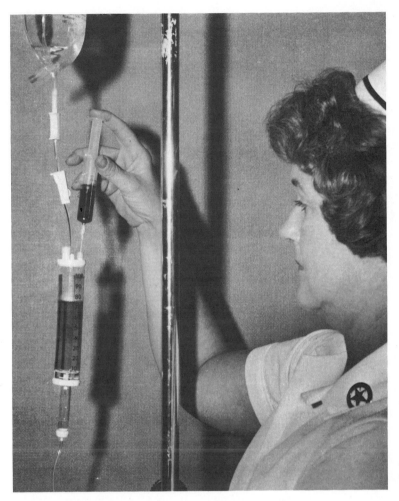

FIG. 6-12.
Metriset. (McGaw Laboratories, Santa Ana, California)

and form air bubbles in the infusion tubing. The nurse must exercise great care when using a Y-type set; complete clamp closure should be made when there is still a little fluid in the bottle. Well-meant intentions to administer every drop in the bottle are not warranted.

PLASTIC CONTAINERS

Plastic bags reduce the potential for air emboli because they are not vented; also, the possibility of contamination by way of an air filter is eliminated. They are lighter, easier to handle, and more easily stored and discarded than glass bottles. In addition, they are less likely to injure patients if accidentally dropped during administration. The use of plastic containers has reduced the incidence of particulate contamination be-

cause there are no rubber closures or glass particles. Plastic containers should never be written on with pen, pencil, or magic marker.

INTRAVENOUS PIGGYBACK SETUPS

Intermittent infusion of an intravenous medication into an established IV line is referred to as a "piggyback" and is a widely used procedure. This method has several advantages over the direct introduction of a concentrated medication into the vein by means of a syringe (IV push):

The medication can be diluted in a larger volume of diluent, thus reducing the possibility of chemical phlebitis.

The nurse does not have to be constantly present

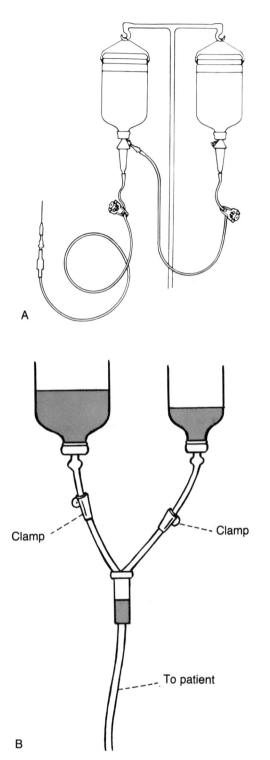

during administration of piggyback infusions (this is a definite advantage over the IV push method which frequently requires the nurse to spend 10 minutes or more at the bedside for safe injection of the medication).

The diluted medication can be delivered more slowly, allowing time for observation for adverse reactions before the entire dose is given.

To perform a piggyback infusion, the primary bottle is lowered by means of a hook below the secondary (additive) bottle (Fig. 6-14). The secondary setup is attached by means of a 1-inch 20-gauge needle into the injection site of the primary set after first thoroughly scrubbing the site with an alcohol sponge. Remember that the injection site at the distal end of the tubing may be exposed to drainage and excreta. The needle should be securely taped in place to avoid accidental dislodgement. A smaller gauge needle may break and enter the administration set; a longer needle can more easily puncture the tubing and allow the entrance of air and contaminants. When the clamp on the additive bottle is fully opened, its fluid will flow in owing to its greater hydrostatic pressure. (The rate of flow is determined by the clamp on the primary set.)

A convenient and safe method for piggybacking makes use of sets with check-valves to prevent air from entering the vein when the additive bottle empties and to prevent or minimize mixing of the primary and additive fluids. The valve automatically occludes flow from the primary bottle while the additive bottle is infusing and then shuts off the additive bottle when it empties, allowing the primary bottle to automatically resume flow (Fig. 6-15). The latter is of great value because the vein is kept open by the uninterrupted fluid infusion. (Some piggyback setups require the nurse to be present to switch the flow to the primary bottle; if not performed quickly, the needle clogs.) Rate of flow for both bottles is the same (unless readjusted), since both are controlled by the clamp on the primary set.

When this system is employed, it is necessary to calculate the volume of the piggyback diluent as part of the *total* volume to be infused over a 24-hour period in order to calculate the flow rate correctly. Otherwise, the primary solution will always be behind schedule. Example:

Primary solution, 1000 ml q 12 hr = 2000 ml
Piggyback solution, 50 ml q 6 hr = 200 ml
= 2200 ml total volume/24 hours

FIG. 6-13. *(A)* IV tandem setup (Saftiset Tandem Set). (Courtesy of Cutter Laboratories, Berkeley, California) *(B)* Bottles connected in parallel (Y-type) setup.

Clamp

Clamp

To patient

A

B

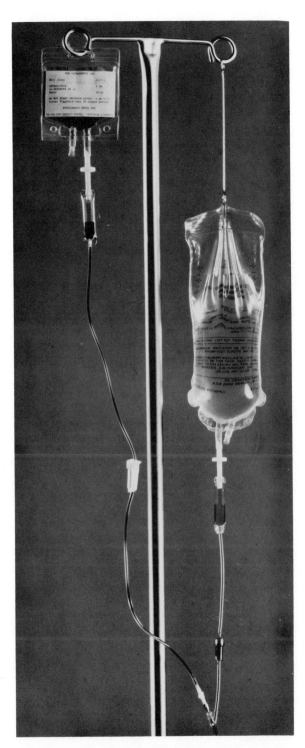

FIG. 6-14. Viaflex primary container with Minibag (piggyback setup). (Courtesy of Travenol Laboratories, Deerfield, Illinois)

(If the resulting flow rate would be contraindicated due to the recommended flow rate of the drug being infused, adjustment of the rate of the piggyback solution would have to be made.)

Remember that there is danger of air embolism when two vented containers are allowed to infuse simultaneously. This danger can be avoided by allowing only one bottle to flow at a time. If two vented solutions must run simultaneously through the same needle, care must be taken to ensure that the first container to drain is tightly clamped off before it completely empties. An unclamped empty container in a Y-type setup allows air bubbles to enter the bloodstream.

FINAL FILTERS

It has been recommended that intravenous fluids be administered through inline final filters to remove particulate matter (undissolved substances unintentionally present in parenteral fluids). This foreign matter may consist of particles of glass, rubber, metal, molds, bacteria, or drug precipitates. A number of filter devices are available commercially for this purpose.

Final filters are available as membrane filters or depth filters. The former are screen-type devices that are calibrated in pore size (such as 0.22 micron, 0.45 micron, or 5 microns); the latter are made of pressed fragmented material of nonuniform pore size. Depth filters are assigned a nominal rating according to the particle size that will be trapped 98% of the time. Membrane filters will block the passage of air (under normal pressure) when wet; depth filters will not. Both the 0.45- and 0.22-micron filters will block bacteria. A 0.45-micron filter will retain all particles larger than 0.45 micron; the 0.22-micron filter blocks all particles larger than 0.22 micron (absolute sterilization). Air-eliminating filters are commercially available and are gaining in popularity.

It is imperative that the nurse read and follow the manufacturer's directions carefully to avoid plugging or rupturing the filter. The latter is particularly dangerous since it may allow fragments of the filter to enter the administration set; also, the air blocking ability of the filter is lost when it is ruptured. Filters should be changed within 24 hours, or more often as indicated. *Caution: never attempt to infuse blood, blood components, or fat emulsions through such filters.*

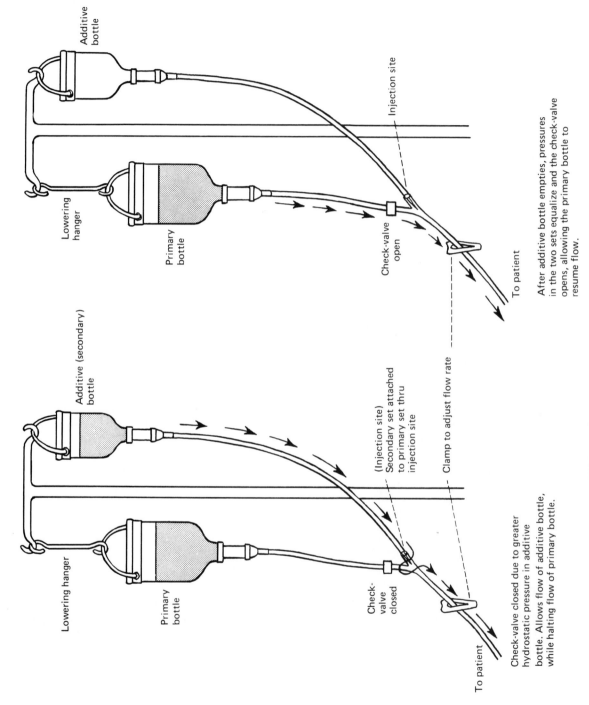

Additive bottle

Injection site

Lowering hanger

Primary bottle

Check-valve open

To patient

After additive bottle empties, pressures in the two sets equalize and the check-valve opens, allowing the primary bottle to resume flow.

Additive (secondary) bottle

(Injection site) Secondary set attached to primary set thru injection site

Clamp to adjust flow rate

Lowering hanger

Primary bottle

Check-valve closed

To patient

Check-valve closed due to greater hydrostatic pressure in additive bottle. Allows flow of additive bottle, while halting flow of primary bottle.

FIG. 6-15. **IV piggyback setup with back-check-valve.**

FILTER ASPIRATION NEEDLE

A filter aspiration needle (Fig. 6-16) is a device attached to the syringe used to draw the intravenous medication from its container (either a rubber-topped vial or a glass ampule). The filter needle traps particles 5 microns in size and larger (such as glass, rubber, or metal) and, thus, allows the use of a more particulate-free fluid. A new needle is applied before the medication is injected into the IV container (use of the filter needle for this purpose would allow the trapped particles to be injected into the intravenous fluid).

INFUSION PUMPS

A number of infusion pumps are available commercially and are being used in many patient care areas (neonatal, pediatric, and adult intensive care units; labor and delivery; radiology, surgery, and increasingly on general medical-surgical units). They are used to allow more accurate administration of fluids and critical drugs than is possible with routine gravity flow setups. Their precise rate of flow is particularly desirable in the administration of hyperalimentation fluids, chemotherapy, and other potent medications (such as dopamine, heparin, lidocaine, nitroprusside, oxytocin, pitressin, and aramine). Pumping mechanisms may be classified as linear peristaltic, rotary peristaltic, and piston and cylinder. Peristaltic pumps move fluid by conpressing the IV tubing; piston and cylinder pumps apply pressure to the fluid rather than to the tubing, much like a plunger pushes fluid through the barrel of a syringe.

Pumps that have flow rates calibrated in terms of ml/hr rather than in drops/min are called *volumetric infusion pumps.* These pumps provide greater accuracy than those measuring drops, since many factors can affect drop size (such as fluid viscosity and drop rate formation). Most pumps require indiviualized special administration sets, although a few use standard IV sets.

A *controller* is an electronic device used to regulate intravenous flow rates. It relies on gravity rather than exertion of pressure. Intravenous controllers are useful whenever the force of gravity is sufficient to provide the desired flow rate. They are appropriate for a large percentage of infusions that do not require the accuracy of a volumetric pump. (Controllers are limited by

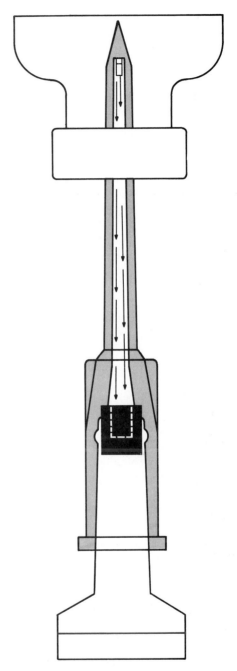

FIG. 6-16. Monoject Filter Aspiration Needle (Sherwood Medical Industries, St. Louis, Missouri; drawn from a brochure.)

the fact that drop rate is not a completely accurate reflection of volume infused due to variations in drop size.)

Circulatory overload is less apt to occur when infusion pumps are used to control the flow rate; also, needles are less apt to clog with the use of infusion pumps since the pressure generated by the pumps exceeds maximum venous pressure. Many infusion pumps may be used to administer fluids intra-arterially; some infusion pumps are occasionally used to administer slow gastrointestinal tube feedings, although special enteral infusion pumps are now available for this purpose.

Nursing considerations in the use of infusion pumps include the following:

1. Read the manufacturer's directions carefully prior to using any infusion pump or controller, since there are many variations in available models. (Instruction manuals should be readily available on all units using these devices.) It is best to learn how to safely operate each type of pump or controller under the direction of someone knowledgeable in its use, such as an inservice program presented by a representative of the manufacturer. It is particularly important to read the manufacturer's recommendations before administering blood with an infusion pump, since some models damage red cells and cause hemolysis.

2. Remember that some types of final filters may be damaged by the high pressures exerted by infusion pumps; thus, it is important to read directions furnished by manufacturers of specific filters prior to using them with infusion pumps. Some filters may cause rate inaccuracies.

3. Check the venipuncture site for infiltration, since most pressure pumps keep infusing fluid even though the tissue may be greatly distended, unlike the markedly diminished flow rate accompanying infiltration of a gravity-flow solution.

4. Check infusion rates closely when batteries are the power source, since low batteries may greatly alter the flow rate.

5. Closely monitor the fluid level in the container and adjust the pump as indicated to keep the infusion at the desired rate. Remember that all infusion pumps have some degree of inaccuracy. Also, remember that meters indicating the total volume infused may be misleading if there is a leak or if the tube is blocked, since meters count pump cycles rather than measuring the amount of fluid.

6. Check the function of occlusion detectors (present on a number of pumps and controllers) by periodically pinching the tubing shut several times each shift. Recall that an occlusion detector senses in-

creased back pressure due to pinched tubing, clogged final filter, clogged bottle airway, or an unopened tubing clamp.

7. Remember that any machine is subject to malfunction and, thus, it is necessary to continue to check the venipuncture site, flow rate, and amount of fluid in the container. Patients sometimes change the settings on the pump or controller, making it necessary to check these at regular intervals also. Although most pumps and controllers are equipped with alarms to signal trouble, it is unwise to rely too heavily on any mechanical device. Pumps and controllers were never devised to replace the nurse's responsibility in monitoring fluid administration. While they are helpful tools, they are no more than that. It is imperative that the nurse monitor patients on pumps and controllers as frequently as those on gravity flow infusions.

8. When using drop-sensitive pumps, always check the sensor eye to be sure it is clean and dry. The sensor eye should be cleansed only with water (detergents can cause a film that leads to malfunctioning of the sensor device).

9. Never purge an IV pump when it is connected to the patient.

COMPLICATIONS OF INTRAVENOUS FLUID ADMINISTRATION

Patients receiving parenteral fluids should be observed often to detect the early appearance of complications. The nurse should periodically check the rate of flow, the amount of solution in the bottle, the appearance of the injection site, and the patient's general response to the infusion.

Complications sometimes occurring with intravenous infusions include those listed below:

Pyrogenic reactions
Local infiltration
Circulatory overload
Thrombophlebitis
Air embolism
Speed shock

PYROGENIC REACTIONS The presence of pyrogenic substances (foreign proteins capable of producing fever) in either the infusion solution or the administration setup can induce a febrile reaction. Such a reaction is characterized by the following:

An abrupt temperature elevation (from 100° to 106° F [37.7° C to 41° C]) accompanied by severe chills (the reaction usually begins about 30 minutes after the start of the infusion)

Backache

Headache

General malaise

Nausea and vomiting

Vascular collapse with hypotension and cyanosis, which may occur when the reaction is severe

The severity of the reaction depends upon the amount of pyrogens infused, the rate of flow, and the patient's susceptibility. Patients having fever or liver disease are more susceptible than others.

If these symptoms occur, the nurse should stop the infusion at once, check the vital signs, and notify the physician. The solution, administration set, and the venipuncture device (aseptically capped), should be saved so that they can be cultured if necessary.

The wide use of commercially prepared solutions and administration sets has dramatically decreased the number of pyrogenic reactions. It must be remembered, however, that contaminants can enter the solution flask after the seal is broken. It is, therefore, a wise practice to indicate on the bottle the date and time that the seal was broken; according to the Centers for Disease Control, parenteral solutions should be in use no longer than 24 hours. This is important to remember when patients are on slow "keep-on" infusions.

For very slow infusions, it is wise to use 500 ml bottles to ensure change of the containers within a safe period of time. Administration sets should be changed at least every 24 to 48 hours, and any evidence of cloudiness or other particulate matter in a normally clear solution is an indication to discard it. The administration set should be changed after the administration of blood or other protein-containing solutions, since these substances become a growth medium for bacteria.

Prior to initiating an infusion, the nurse should squeeze plastic containers to detect leaks and inspect glass containers against light for cracks or bright reflections that penetrate into the wall of the bottle; if either are present, the solution is no longer sterile and must be discarded. Administration sets should also be routinely inspected for cracks or discolorations prior to use.

Unless the nurse uses the most careful technique, contaminants can be introduced when medications are added to the infusion fluid.

Because traffic generates air-borne contamination, medications should be added in an isolated area. Some hospitals are equipped with laminar flow clean air work stations; these stations provide a filtered air screen to reduce contamination of intravenous fluids during handling (Fig. 6-17).

A final filter is another safety device to prevent the infusion of bacteria and particulate matter. The filter is situated between the infusion tubing and the needle, allowing for final sterilization filtration immediately before the solution enters the bloodstream.

LOCAL INFILTRATION The dislodging of a needle and the local infiltration of solution into the subcutaneous tissues is not uncommon, especially when a small, thin-walled vein is used and the patient is active. Infiltration is characterized by the following:

Edema at the site of injection: compare the infusion area with the identical area in the opposite extremity; otherwise swelling may not be readily noticeable. Temperature of the skin around the insertion site is cooler than the rest of the skin because the IV fluid is cooler than the body.

Discomfort in the area of the injection (the degree of discomfort depends on the type of solution)

Significant decrease in the rate of infusion, or a complete stop in the flow of the fluid

Failure to get blood return into tubing when the bottle is lowered below the needle (this method is not always foolproof—sometimes the needle lumen is partially in the vein with its tip in the subcutaneous tissue)

Hypertonic carbohydrate solutions, potassium solutions, and solutions with a pH varying greatly from that of the body (such as protein hydrolysates, sixth molar sodium lactate, or ammonium chloride) often cause great pain if they infiltrate into the subcutaneous tissues. Tissue slough may result from the local irritation.

The infusion should be discontinued immediately when infiltration is apparent. When in doubt, an infiltration can be confirmed by applying a tourniquet (or applying pressure with the fingers) to restrict venous flow proximal to the injection site; if the flow continues, regardless of the venous obstruction, the needle is obviously not in the vein.

Close observation with early detection of infiltration will greatly reduce the severity of this all too com-

FIG. 6-17. Clean air center (laminar-flow hood) for mixing IV solutions. (Courtesy of Abbott Laboratories, North Chicago, Illinois)

mon complication. Below is a summary of factors that frequently contribute to infiltration:

Lack of proper patient education concerning care of extremity receiving infusion

Hyperactive patient

Improperly taped venipuncture device

Improper technique of person initiating therapy (such as pushing bevel of needle through the posterior wall of the vein)

Poor selection of venipuncture site (such as over an area of flexion)

Improper handling of extremity or equipment during transportation

CIRCULATORY OVERLOAD Overloading the circulatory system with excessive intravenous fluids may cause the following symptoms:

Rise in blood pressure and central venous pressure (CVP)

Venous distention (engorged neck veins)

Wide variance between fluid intake and output (higher intake than output)

Coughing

Shortness of breath, increased respiratory rate

Pulmonary edema with severe dyspnea and cyanosis

The nurse should be particularly alert for circulatory overload in patients with cardiac decompensation. If the above symptoms occur, the infusion should be slowed to a "keep-open" rate and the physician notified immediately. If necessary, the patient can be raised to a sitting position to facilitate breathing. Patients prone to circulatory overload may require CVP monitoring during infusions.

THROMBOPHLEBITIS The condition associated with clot formation in an inflamed vein is known as thrombophlebitis. Although some degree of venous irritation accompanies all intravenous infusions, it is usually of significance only in infusions kept going in the same site for more than 12 hours and when vesicant agents are infused. Thrombophlebitis at an infusion site may be manifested by:

Pain along the course of the vein

Redness and edema at the injection site (a red streak may form above the site)

Vein may feel hard, warm, and cordlike to touch

Flow rate may be sluggish due to venospasm

If severe, systemic reactions to the infection (tachycardia, fever, and general malaise)

Mechanical factors can produce thrombophlebitis. Needle movement can cause venous irritation when the infusion site is near a joint; careless technique venipuncture or in removing an infusion needle can seriously traumatize the vein.

Irritating solutions, such as alcohol, can be instrumental in causing thrombophlebitis. Hypertonic solutions are often associated with venous irritation; carbohydrate solutions in excess of 10% almost always produce this reaction.

Most water, electrolyte, and dextrose solutions are acidic, with the pH usually ranging from 3.5 to 6.7.[19] Solutions with an alkaline or acid pH are more frequently associated with thrombophlebitis than are the solutions that approximate body pH; reportedly, this problem can be reduced by adding buffers to raise the pH. Commercial products are available for this purpose; the small amount of basic solution will not alter the patient's plasma pH. However, increasing the pH of solutions containing additives can produce incompatibility problems, since some additives are stable only at a low pH.

Once thrombophlebitis is detected, the infusion is stopped and restarted in another site to allow the traumatized vein to heal. Usually, cold compresses are applied to the thrombophlebitic site, after which warm moist compresses can be employed to relieve discomfort and promote healing.

Nursing interventions Nursing interventions to minimize the occurrence of thrombophlebitis include the procedures below:

Changing the intravenous device as indicated (it is recommended that a peripheral venipuncture device not be left in place longer than 72 hours, and preferably no longer than 48 hours)

Infusing irritating fluids in large veins (large veins have a higher blood flow and thus can quickly dilute the irritant)

Generally speaking, irritating additives should be diluted with more fluid than nonirritating substances (keeping manufacturer's guidelines in mind)

Stabilizing the venipuncture device with proper taping and an armboard when the venipuncture is in an area of flexion

AIR EMBOLISM The danger of air embolism is present in all intravenous infusions, even though it does not occur frequently. Cannulation of central veins (for hyperalimentation or CVP measurement) is *far* more likely to be associated with air embolism than is cannulation of peripheral veins. The exact quantity of intravenous air than can lead to death in humans is unknown but appears to be related to the rate of entry. It is known that a much larger amount of air can be tolerated in the venous system than in the arterial system. Ordway extrapolated data from animal studies and concluded that the average lethal dose of air IV in humans would be between 70 ml and 150 ml/sec.[20] Yeakel reported a fatal episode following the sudden administration of 100 ml of air intravenously.[21] According to another source, normal adults can tolerate as much as 200 ml of air IV; but in seriously ill patients, as little as 10 ml may be fatal.[22]

There are measures the nurse should take to prevent the occurrence of air embolism.

Central veins

Have the patient perform the Valsalva maneuver (forced expiration with the mouth closed) during the seconds when the tubing is disconnected from the catheter for tubing change. Recall that there is a low pressure in the central veins that can pull air in during tubing changes or when connections in the administration apparatus are not air-tight. This low pressure is particularly pronounced in the hypovolemic patient; hypovolemia generates an increased "sucking force" in the central veins. Also, during the inspiratory phase of respiration, the intrathoracic pressure is below that of the atmosphere, making it possible for air to be drawn into the cannulated vein. The Valsalva maneuver increases the intrathoracic pressure as well as the mean pressure in the central veins, decreasing the danger of air being sucked in.

Be sure the connection between the catheter hub and the tubing is secure. Adequate taping of the tubing and catheter lessens the possibility of accidental disconnection. This is of particular importance when the patient is elevated from a flat to a sitting or standing position, since the danger of air embolism is greater when the patient is upright.

Peripheral veins

Do not elevate the extremity receiving the infusion above the level of the heart since this results in venous collapse and negative venous pressure. Negative pressure in the vein receiving the infusion can draw in air if there are any defects in the apparatus or if the solution flask empties.

Allow the infusion tubing to drop below the level of the extremity. This may help prevent air from entering the vein if the infusion flask empties unobserved.

Keep the clamp used to regulate the flow rate below the level of the heart. If the clamp regulating flow is placed above the level of the heart and is adjusted so that it only partially occludes the tubing, any existing defect in the tubing between the clamp and the level of the heart will allow air to enter the system.[23] On the other hand, if the clamp is placed below the level of the heart, a defect between the drip chamber and the clamp will cause leakage of fluid rather than entry of air (because of the greater pressure in the vein than in the tubing; Fig. 6-18).[23]

All infusions

Tightly secure all connections in the administration setup to prevent air from being drawn in. If a stopcock is part of the IV setup, the outlets not in use should be completely shut off. Avoid tightening connections so much that forceps are needed to loosen them (can cause cracking of the tubing or plastic IV device and create an entry site for air).

Inspect plastic bottles and tubing for cracks or other defects.

Discontinue an infusion before the bottle and tubing are completely empty.

Completely clamp off the first bottle to empty in a Y-type setup (parallel hookup); otherwise, air will be drawn into the vein from the empty bottle. The potential of air emboli exists when fluids from two vented containers run simultaneously through the same needle. The use of administration sets with check-valves for piggy-backing fluids prevents this problem (see discussion of piggybacking earlier in this chapter).

Read and follow the manufacturer's directions for safe use of infusion pumps since some types can pump air into the vein if the infusion bottle is allowed to empty.

The presence of an air embolism is manifested by the following:

Dyspnea and cyanosis

Hypotension

Weak rapid pulse

Loud continuous churning sound over the precordium (not always present)

Loss of consciousness

The occurrence of these symptoms in a patient receiving an infusion should lead one to suspect air embolism.

If an air embolism occurs, administration tubing should be promptly clamped. The patient should be immediately turned on his left side with his head down and the lower extremities elevated (left lateral and modified Trendelenburg position). This allows the air to rise into the right atrium and away from the pulmonary outflow tract. Oxygen is administered by mask to achieve a high oxygen concentration.

Note: The nurse should be aware that the presence of air in the administration set, no matter now small a bubble, is frequently a cause of apprehension in patients receiving intravenous fluids.

BLOOD EMBOLISM A blood embolism may result from the unwise irrigation of a plugged needle or catheter. Irrigation may dislodge the clot into the circulation, possibly resulting in an infarction. Also, the irrigating fluid may embolize small infected needle thrombi, resulting in septicemia. Plugged needles should be removed, not irrigated.

ADMINISTRATION OF BLOOD AND DEXTRAN

BLOOD

An anticoagulant must be added to blood to prevent clotting; most commonly used is citrate-phosphate-dextrose (CPD). Blood preserved in CPD can be used for a maximum of 21 days. A newer anticoagulant, citrate-phosphate-dextrose-adenine (CPD-A) allows storage for 35 days.[24]

In certain instances, blood cannot be used for its full storage period. For example, blood used for exchange transfusion of newborns may need to be less than 3 to 5 days old to avoid hyperkalemia; blood used

(Text continues on p. 176)

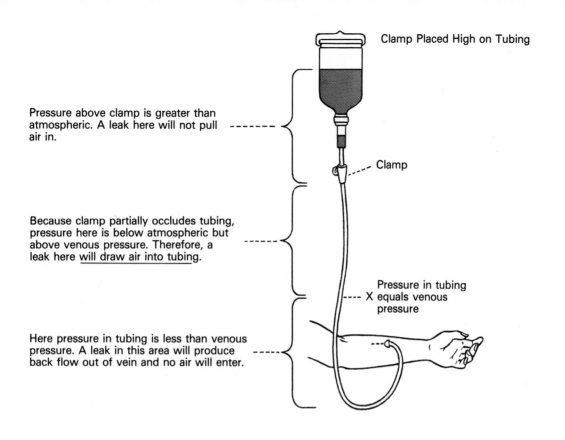

Clamp Placed High on Tubing

Pressure above clamp is greater than atmospheric. A leak here will not pull air in.

Clamp

Because clamp partially occludes tubing, pressure here is below atmospheric but above venous pressure. Therefore, a leak here will draw air into tubing.

Pressure in tubing X equals venous pressure

Here pressure in tubing is less than venous pressure. A leak in this area will produce back flow out of vein and no air will enter.

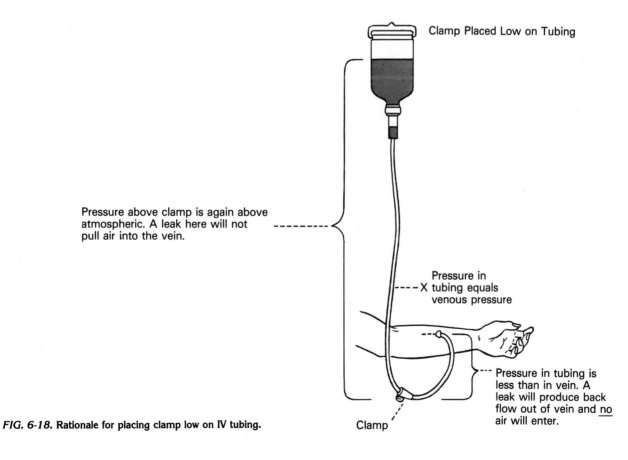

Clamp Placed Low on Tubing

Pressure above clamp is again above atmospheric. A leak here will not pull air into the vein.

Pressure in X tubing equals venous pressure

Pressure in tubing is less than in vein. A leak will produce back flow out of vein and no air will enter.

Clamp

FIG. 6-18. Rationale for placing clamp low on IV tubing.

to supply all of the plasma clotting factors and platelets must be less than 6 hours old.[25]

PACKED RED BLOOD CELLS Sometimes it is necessary to administer packed red blood cells, obtained by centrifuging whole blood and drawing off approximately 200 ml to 225 ml of plasma. This is the case when a patient with congestive heart failure or fluid volume excess requires blood. Advantages of packed cells, besides reduced volume, include reduced chemical content and reduced agglutinins. Potassium, sodium, and citrate content are substantially reduced, decreasing the danger of hyperkalemia, sodium overload, and citrate toxicity. Because most of the plasma is removed, the amount of anti-A and anti-B agglutinins is also reduced. Packed cells with a hematocrit of 70% can be infused with little difficulty. Packed red blood cells have the same storage life as whole blood.

COMPLICATIONS OF BLOOD TRANSFUSION Whole blood transfusions often are indicated in the presence of a significantly decreased blood volume. Blood replacement therapy is often lifesaving, but, without careful attention to its hazards, it may also be lethal. Because the nurse shares a large reponsibility in assuring safe blood administration, she should be familiar with complications associated with its use.

Stored blood and potassium excess Continual destruction of red blood cells occurs when blood is stored; at the end of 21 days only 70% to 80% of the original number of cells remain. The longer the blood ages, the higher its plasma potassium level becomes. One-day-old blood has a plasma potassium content of approximately 7 mEql/liter; blood stored for 21 days has a plasma potassium content of approximately 23 mEq/liter. Not only is potassium released from the destroyed red blood cells but there is also a transfer of potassium from the intact cells into the surrounding plasma. After the blood cells are infused into the recipient, they are reoxygenated, causing them to "take up" the leaked potassium, resulting in a reduced plasma potassium concentration.

Aged blood poses no problem for the normal patient but should not be given rapidly to patients with oliguria or anuria, since there is danger of causing potassium excess. Potassium excess may be recognized by the following symptoms:

Gastrointestinal hyperactivity (nausea, intestinal colic, and diarrhea)

Vague muscular weakness, first in the extremities, later extending to the trunk

Paresthesia of hands, feet, tongue, and face

Flaccid paralysis (involving respiratory muscles last)

Apprehension

Slowed pulse rate (may also be irregular)

Cardiac arrest and death when plasma potassium level reaches 10 mEq to 15 mEq/liter (due to marked dilation and flaccidity of the heart)

The incidence of cardiac arrest during surgery is correlated with the rapid infusion of large quantities of aged blood.

Citrated blood and hypocalcemia Rarely, symptoms of calcium deficit can be caused by the rapid administration of large volumes of citrated blood. It should be recalled that acid-citrate-dextrose (ACD)- and CPD-preserved bank blood contain citrate in excess of that needed to combine with the calcium in the blood collected. Under certain circumstances, the excess citrate, when infused into the bloodstream, can combine with ionized plasma calcium and cause tetany. The excess citrate normally causes no difficulty since the liver can remove it from the blood within a few minutes; also, binding of ionized calcium by citrate causes rapid mobilization of calcium from the bones.

Patients with severe liver disease cannot tolerate usual administration rates because the liver is unable to remove citrate from the blood quickly enough to prevent a reaction. Also, patients with inadequate bone stores of calcium (such as young children and osteoporotic adults) are more susceptible to citrate toxicity during massive transfusions. (It may occur in infants receiving exchange transfusions with citrated blood.)

Citrate-induced ionized calcium deficit may cause the following symptoms:

Tingling of the fingers and circumoral region

Hyperactive muscular reflexes

Muscle spasms

Bradycardia, arrhythmias, and hypotension

If citrate toxicity is anticipated or suspected, the patient is given calcium gluconate or calcium chloride to correct the deficit of ionized calcium. (The physician injects the calcium preparation into a vein remote from the transfusion site.) One source recommends that calcium chloride be given intravenously prior to or concomitant with whole blood transfusions exceeding two in 4 hours.[26]

Another potential complication of the citrate anticoagulant (as it is metabolized to bicarbonate) is alkalosis and concomitant hypokalemia.[27] Rarely does the alkalosis require treatment.

Circulatory overload Discussed earlier in the chapter as a possible complication of all intravenous infusions, circulatory overload is particularly likely to occur in massive blood replacement for hypovolemia or when blood is given to a patient with normal blood volume. In addition to inducing pulmonary edema, the increased blood volume may cause hemorrhage into the lungs and the gastrointestinal tract.

To guard against the occurrence of circulatory overload, many physicians order the monitoring of venous pressure during the rapid replacement of large volumes of blood. (The procedure is discussed in the last section of this chapter.) Aged patients and patients with cardiac damage should be closely observed for circulatory overload even when small volumes of blood are given; venous pressure monitoring is frequently used to protect such patients from overload. Administration of packed cells decreases the likelihood of fluid overload.

Allergic reactions Allergic reactions to blood are reported to occur in 1% to 4% of all transfusions. Symptoms may include flushing, chills, pruritus, and urticaria; rarely do bronchospasm, angioneurotic edema, and vascular collapse occur. (Urticaria occurs only in allergic blood reactions, not in the other types of transfusion reactions.) The allergic reaction results from transfer of donor antigen to a sensitive recipient; frequently the donor has just ingested the antigen prior to donating blood. Reactions are much more apt to occur in patients with a history of asthma, hay fever, or atopy; it has been suggested that such patients be transfused with washed packed plasma-free red blood cells to decrease the likelihood of an allergic reaction. Symptoms usually do not occur until 250 ml of whole blood or 125 ml of packed red blood cells have been transfused; they may not occur until the transfusion has been completed.[28] Blood administration is often slowed if a minor reaction occurs; it is stopped immediately if severe symptoms are present. Antihista-

mines usually relieve mild symptoms; but the rarer, more severe reactions require epinephrine or steroids. Sometimes antihistamines are used prophylactically when an allergic reaction is considered likely.

2,3 DPG depletion 2,3 DPG is a substance in red cells that is necessary for the efficient delivery of oxygen to the tissues. Fortunately, 2,3 DPG deteriorates slowly and is still near normal by the second week of blood storage.[27] The decreased level of 2,3 DPG in stored blood can be a problem when massive transfusions are given, since a decreased level of 2,3 DPG causes decreased tissue oxygenation. Arterial blood gases should be monitored as necessary.

Hypothermia Cold solutions are quickly warmed as they mix with circulating blood during usual administration rates. However, infusion of 3 to 5 units of cold blood (at refrigerator temperature) within 1 to 2 hours may lower the recipient's body temperature by 5° to 7° F (3° to 4° C).[29] Hypothermia can result in diminished cardiac output and metabolic acidosis and can trigger cardiac arrest. One study in a New York hospital showed a significant decrease in cardiac arrest (from 58.3% to 6.8%) when cold blood was warmed to body temperature during rapid massive infusion (6 pints or more/hr). During rapid blood replacement, it is necessary to warm the blood to 37° C either by a mechanical warmer or by passing the blood through a coil of tubing submerged in a basin of 37° C water. Warming should never be attempted by placing the container in hot water.

Acidosis Although CPD-preserved blood is less acidic than blood preserved with ACD, there is still a risk of acidosis during massive transfusions. This is not a problem in patients who are not acidotic prior to transfusion. It is usually a problem only in patients who are unable to maintain proper acid–base balance (such as in exchange transfusions of premature infants with respiratory problems.)[30] Patients with preexisting acidosis who are to receive rapid transfusions (1 unit of blood every 20 minutes or faster) may require the administration of IV sodium bicarbonate.

The pH of 2-week old bank blood is 6.5 owing to leakage of lactate and pyruvate into the plasma (as a result of red cell hypoxia during storage).[26]

Hyperammonemia Ammonia concentration in stored blood begins to rise after 5 to 7 days and reaches high levels after 2 to 3 weeks storage. Normal patients can tolerate the extra ammonia with no difficulty; however, patients with severe liver disease should be given blood not more than 5 to 7 days old.[31]

Febrile reactions Minor febrile reactions are characterized by fever (rarely exceeding 103° F [39.5° C]), flushing, chills, and headache. Usually they are due to sensitivity to donor white cells; less frequently they are caused by sensitivity to platelets and plasma proteins in patients who have received numerous transfusions in the past. Symptoms do not usually occur until more than 250 ml of blood has infused. It may be necessary to administer blood that has had the white cells and platelets removed for patients with a history of severe reactions. Febrile reactions may also result from contamination of the administration setup. The patient does not appear toxic and is usually treated with aspirin to relieve the fever; therapy is essentially symptomatic.

Such a reaction is sometimes confused with an incompatible blood reaction; hence, it should be noted that febrile reactions tend to occur toward the end of the transfusion or even after it is completed (an acute hemolytic reaction usually occurs during the infusion of the first 100 ml of incompatible blood). Also the back pain so characteristic of a hemolytic reaction is often absent in a febrile reaction.

Bacterial contamination Before blood is used it should be inspected for bacterial growth signs, such as discoloration or gas bubbles. If these are present the blood should be discarded. If blood used for transfusion is grossly contaminated with gram-negative organisms, the recipient usually develops a high fever, intense flushing, vomiting, diarrhea, headache, and symptoms of shock which may prove fatal. As little as 50 ml of contaminated blood can precipitate this reaction. Blood samples should be drawn and cultured; the patient is usually treated with appropriate antibiotics, steroids, and fresh uncontaminated blood. (It has been reported that approximately 0.1% of all units of whole blood are contaminated with cold-growing organisms.[32]

Serum homologous hepatitis There are several viruses that cause hepatitis. At present, blood banks are able to test for only one type in donor blood—hepatitis B.

Because of this limitation, the majority of units that can transmit hepatitis will not be detected. Prevention of hepatitis transmission is largely reliant on careful screening of potential donors. Persons known to have had hepatitis are, of course, not allowed to give blood. (Nonetheless, not all persons with hepatitis have had the condition diagnosed; thus, the danger always exists.) Statistics indicate that paid donors cause many more cases of serum hepatitis than unpaid donors. (Drug addicts often sell blood to raise money for more drugs; serum hepatitis is fast spreading among them because most use unsterile equipment for intravenous injections.) When serum hepatitis occurs in a blood transfusion recipient, it is necessary to trace the donors who contributed the blood and to eliminate them from the donor pool.

The incubation period for serum hepatitis is from 35 days to 120 days; symptoms include malaise, anorexia, vomiting, abdominal discomfort, enlarged liver with tenderness, diarrhea, headache, fever, and jaundice. Serum hepatitis is particularly serious in the very young and elderly and in patients already debilitated by serious illness or injury.

Pooled plasma products (such as Factor VIII concentrate, and Factor II-VII-IX-X complex) are associated with a greater risk of hepatitis transmission than is whole blood.[33] Blood components virtually free of hepatitis transmission include albumin, plasma protein fraction, and immunoglobulin preparations.[33]

Acute hemolytic reaction The most dreaded reaction to blood transfusion is that caused by the administration of grossly incompatible whole blood or packed cells. It is reported to occur in 1 out of every 3000 transfusions.[32] A transfusion of incompatible blood causes rapid cell agglutination in the recipient and eventual intravascular hemolysis. The incompatible transfusion is most often caused by the careless administration of the wrong blood; it can also be the result of inadequate crossmatching. Some hemolytic reactions are caused by the administration of blood accidentally hemolyzed from improper handling. Symptoms of an acute hemolytic reaction usually occur during the transfusion of the first 100 ml (sometimes as little as 5 ml to 10 ml) of blood and may include the following:

Lumbar and flank pain (flank pain caused by hemoglobin precipitation in the renal tubules)

Tachypnea and tachycardia

Feeling of constriction in the chest or precordial pain

Urge to defecate or urinate

Fever and severe shaking chills (a hemolytic reaction may cause considerable fever and general toxicity; these effects, in the absence of renal shutdown, are rarely fatal)

Jaundice (usually does not occur unless more than 3000 ml of blood is hemolyzed in less than a day)

Later, hemoglobinuria and acute renal tubular necrosis may occur

The only signs of serious transfusion reaction in an anesthetized or otherwise unconscious patient may be increasing tachycardia, shock, and oozing of blood at the site of operation (bleeding may follow disseminated intravascular coagulation, which presumably follows the release of erythrocytic thromboplastic substances)

Because an acute hemolytic reaction becomes evident during transfusion of the first 20 ml to 100 ml of blood, it is highly desirable that the patient be carefully observed during this part of the infusion. If the transfusion is stopped early, acute renal tubular necrosis and death rarely occur; if more than several hundred ml are infused, renal shutdown and death are common. Renal shutdown is due to blockage of the tubules with hemoglobin and to the powerful vasoconstriction caused by toxic substances released from the hemolyzed blood.

The blood transfusion should be discontinued immediately when a hemolytic reaction is evident. The blood bottle and set should be refrigerated so that further compatibility tests may be made. All urine should be saved and observed for discoloration; urine is sent to the laboratory for evaluation of hemoglobin content. The onset of a hemolytic reaction may be delayed when factors of less moment than those involving the ABO system are present. For example, a patient who has received multiple transfusions in the past may have become sensitized to Rh or to the minor factors, and his hemolytic reaction might have a delayed onset. The slow developing type of hemolytic reaction may be characterized by jaundice appearing hours or days after the transfusion.

Hemolytic reactions are more prone to occur in patients who have had past transfusions and tend to be proportinate to the number of such transfusions, regardless of when given.

Treatment of an incompatible blood reaction usually consists of rapid administration of dilute fluids to promote diuresis. Alkaline fluids, such as sixth molar sodium lactate, increase the solubility of hemoglobin and aid in its excretion. An osmotic diuretic (usually mannitol) may be given intravenously to promote fluid excretion by the tubules and to overcome renal vasoconstriction; further treatment is dependent upon the amount of renal damage present. It is impossible to overstress the need to check and recheck all labels for both donor and patient, as errors in labeling still constitute a frequent cause of reaction.

SAFE BLOOD ADMINISTRATION To help ensure that blood transfusions are as safe as possible, the nurse should keep the following points in mind:

1. Be aware of the complications associated with blood administration; keep their descriptions in mind while observing the patient.
2. *Give the right blood to the right patient.*
   - Read the labels identifying the blood and check them carefully with the patient's full name, obtained from the wrist identification band. The bed card should not be relied on solely to identify the patient; these cards are often not up to date. (An accurate numbering system such as the Hollister Blood Identification System is utilized by many institutions.)
   - The patient should be called by name and his response observed; this practice in itself is not foolproof because some patients respond to any name.
   - Location of the patient should not be relied upon as the sole means of identification; patients are moved frequently in most hospitals.
   - Particular precautions are necessary when the patient has a common name; many errors have been made by not checking further than the last name and first initial. This is not to imply that two patients with the same uncommon name cannot be confused; it *has* happened. Most importantly, the label on the container should be matched with the patient's wrist identification band; the name and number on both should be identical.
   - *Cross-checking all identifying data with another nurse prior to the initiation of a transfusion is definitely considered prudent.*
3. Blood sent to nursing divisions should be started within 30 minutes, because *rapid deterioration of red blood cells occurs after blood has been exposed to room temperature for more than 2 hours.* Ward refrigeration is inadequate for storing blood

because it is not controlled and has no alarm to signal fluctuation of temperature. (Accidental freezing of blood renders it unsuitable for use.)

4. The patient's temperature should be taken before the transfusion to serve as a baseline for later temperature comparisons. An elevated temperature during a transfusion may indicate either a pyrogenic or an incompatible blood reaction and, for this reason, the temperature should be checked hourly during the blood administration and for several hours afterwards.

5. The blood should be inspected before use for discoloration or gas bubbles; if these are present, the blood is probably contaminated and must be discarded.

6. Blood should always be given through a filter to remove the particulate matter formed during storage. Stored bank blood contains cellular degradation debris that can cause a condition known as posttransfusion lung syndrome when multiple units have been transfused through a standard (170 micron) mesh filter. A micropore filter (40 microns and under) should be used when more than 4 units of whole blood are to be transfused in one day.

7. Do not start blood with 5% dextrose in water or hook blood in series of parallel with it. Isotonic saline is compatible with blood and is often used to start blood.

8. Do not use a calcium-containing solution (such as lactated Ringers) to start citrated blood or hook it in series or parallel with citrated blood, as calcium ions may cause the blood to clot and clog the infusion apparatus.

9. If it is necessary to warm blood before administration, a blood warmer can be used, as described earlier. Hot water should never be used to heat blood because excess heat destroys red blood cells.

10. The first 50 ml of blood should be delivered slowly while the patient is observed closely. If no adverse reactions occur, the rate may be increased as ordered. Frequent observations should continue during the transfusion.

11. Unless the patient has severe hypovolemia, blood transfusions should be given no faster than 500 ml every 30 minutes. The presence of factors such as cardiac, renal, or liver damage may necessitate much slower rates. A patient with normal blood volume should receive blood at a slow rate to prevent circulatory overload. Usually a unit of blood can be given in less than a 2-hour period. Remem-

ber that blood cells deteriorate rapidly after exposure to room temperature for more than 2 hours.

12. If not contraindicated by the patient's condition, 50 ml to 100 ml of isotonic saline may be added to a unit of packed red blood cells to facilitate ease of delivery.

13. The volume of blood/blood components (and isotonic saline if used) should be tabulated when calculating the patient's daily fluid intake.

Blood can be given under pressure by applying a pressure sleeve to the plastic blood container (Fig. 6-19). Air, forced into the sleeve by means of a pressure bulb, exerts force on the flexible container and increases the flow rate.

DEXTRAN

Dextran is a polymer of glucose having a large molecular weight. It is available as low-molecular-weight (LMW) dextran and as medium-molecular-weight (MMW) dextran.

LOW MOLECULAR WEIGHT DEXTRAN LMW dextran (dextran 40)* has been advocated as a plasma volume expander in the treatment of shock and burns and for prevention of thromboembolic disease, frostbite, and other conditions associated with vascular insufficiency.[34] LMW dextran enhances blood flow, particularly in the small vessels, by increasing blood volume, decreasing blood viscosity, and reducing the aggregation of red blood cells in the capillaries.

This substance has a mean molecular weight of 40,000. Each gram of LMW dextran exerts the osmotic pressure of 25 ml of plasma.[35] Thus, 500 ml of 10% solution will pull into the plasma space approximately 750 ml of additional fluid. However, the effectiveness of LMW dextran as a plasma volume expander is short lived, since its low-molecular size allows it to leave the intravascular space rapidly.

LMW dextran is available in a 500 ml container as a 10% solution in either isotonic saline or 5% dextrose in water. While it is stable for long periods at a temperature under 25° C (77° F), elevated or fluctuating temperatures can cause crystallization.[36]

*Gentran 40 (Travenol Laboratories); LMD (Abbott Laboratories); Rheomacrodex (Pharmacia Laboratories); Dextran 40 (Cutter Laboratories).

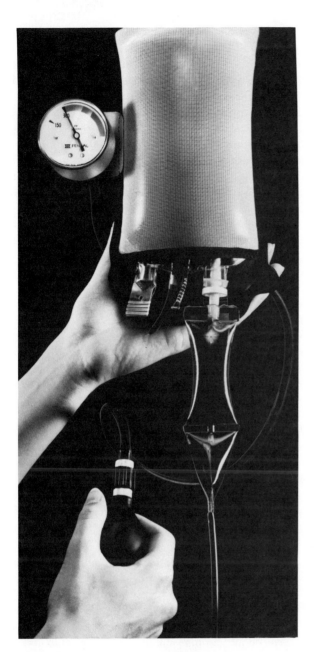

FIG. 6-19. Plastic blood bag with pressure infusor. (Courtesy of Fenwal Laboratories, Morton Grove, Illinois)

Nursing considerations A number of adverse reactions can occur with the adminsitration of LMW dextran:

Generalized urticaria

Chills and fever

Nausea and vomiting

Hypotension

Anaphylaxis

The patient should be observed closely for anaphylaxis during the first 30 minutes of the infusion; allergic reactions have been known to appear after the administration of as little as 10 ml of dextran. If any of the above symptoms appear, the infusion should be stopped immediately; adrenalin may be necessary to control a severe reaction. In mild reactions, withdrawal of the dextran will usually suffice.

Particular caution is necessary when dextran is administered to patients with heart disease or renal disease because of the danger of congestive heart failure and pulmonary edema. Careful monitoring of venous pressure during rapid dextran administration is sometimes indicated to guard against circulatory overload; a precipitous rise is an indication to stop the infusion immediately. If venous pressure is not monitored, the infusion must be given more slowly and the patient observed carefully for signs of circulatory overload. The physician should indicate the desired flow rate.

Recommended dosages vary. Manufacturers state that 20 ml/kg of body weight can be given daily and that the first 500 ml can be given rapidly. Some authorities recommended that the rate of infusion of LMW dextran should be adjusted to the hourly urinary output and the specific gravity of the urine. *It is imperative that adequate urine volume be maintained while LMW dextran is being administered since, during periods of oliguria, it can precipitate in the renal tubules, causing obstruction and acute renal failure.*

A transient increase in bleeding time is sometimes seen several hours following the administration of dextran (particularly if more than 1000 ml has been given). For this reason, patients should be observed for bleeding tendencies. Dextran should be used with caution in patients with thrombocytopenia and severe bleeding disorders.

Blood samples for typing, crossmatching, and other tests should be obtained before the dextran infusion is started (dextran can interfere with the accuracy of some laboratory tests).

MIDDLE MOLECULAR WEIGHT DEXTRAN MMW dextran has a molecular weight close to that of serum albumin—about 70,000; it has a colloidal osmotic pressure equivalent to that of plasma. Because the average molecular size of MMW dextran (dextran 70*) is

*Gentran 75 (Travenol Laboratories); Macrodex (Pharmacia Laboratories); Dextran 70 (Cutter Laboratories); Dextran (Abbott Laboratories).

larger than that of LMW dextran, movement out of the bloodstream is slower, and plasma volume expansion lasts longer (approximately 4 hours after the end of the infusion).[36] MMW dextran has been used primarily for volume expansion in the emergency treatment of hypovolemic shock and burns; it has also been used to prevent thrombophlebitis and reclotting of arteriovenous shunts for hemodialysis.

MMW dextran is available in 500 ml containers as a 6% solution in isotonic saline. The recommended dosage of MMW dextran for volume expansion is up to 20 ml/kg body weight in 6 hours; the usual adult dose is approximately 500 ml/day.[36] When used for antithrombotic effects, 10 ml to 15 ml/kg body weight/day may be infused (for a period of 3 to 5 days).[36]

Adverse effects and nursing considerations are generally the same as those described above for LMW dextran; however, sludging in the renal tubules and development of acute renal failure is not reported with MMW dextran.

CENTRAL VENOUS PRESSURE MEASUREMENT

Central venous pressure (CVP) refers to the pressure in the right atrium or vena cava and provides information about the following parameters:

Blood volume

Effectiveness of the heart's pumping action

Vascular tone

Pressure in the right atrium is usually 0 cm to 4 cm of water; pressure in the vena cava is approximately 4 cm to 11 cm of water.

A low CVP may indicate decreased blood volume or drug-induced vasodilatation (causing pooling of blood in peripheral areas). A high CVP may indicate increased blood volume, heart failure, or vasoconstriction produced by vasopressors (causing the vascular bed to become smaller). More important than absolute values are the upward or downward trends in CVP; these trends are determined by taking frequent readings (often every 30 to 60 minutes). It is always important to evaluate CVP in reference to other available clinical data such as blood pressure, pulse, respirations, breath and heart sounds, fluid intake, and urinary output. For example; a rise in CVP paralleling that of systolic blood pressure is an indication of adequate fluid volume replacement; a low CVP persisting

after fluid volume replacement may be a sign of continued occult bleeding. Sometimes the rate of an intravenous infusion is titrated according to the patient's CVP; when this is necessary, the physician should designate the desired limits so that the nurse can adjust the flow rate accordingly.

Equipment used to measure CVP is relatively simple, consisting of a CVP catheter, water manometer, three-way stopcock, and a routine IV setup. After the physician has threaded the CVP catheter into the vena cava, the catheter is connected by way of the stopcock to the manometer and the infusion apparatus. The manometer is attached to an IV pole with the zero point at the level of the patient's right atrium (mid-axillary line; Fig. 6-20).

Readings should be made, if possible, with the patient flat in bed; if not, they should be made with the patient in the same position each time (the position should be indicated when charting the pressure). When the catheter is properly positioned in the vena cava, the fluid should fluctuate 3 cm to 5 cm in the manometer with respirations. If the patient is being ventilated on a respirator, it should be disconnected temporarily for the brief period it takes to read the CVP, since a respirator will cause a false high reading. If the respirator cannot be discontinued, it should be noted that readings were made with it in operation (a trend can still be noted although the absolute readings are not accurate). Sometimes methylene blue or vitamin B complex is added to color the IV solution and facilitate reading the fluid level in the manometer. If order to read venous pressure, it is necessary to follow the procedure below:

1. Turn the stopcock so that the solution will flow from the container to the manometer, allowing it to reach a level of 30 cm (see system 2 in Fig. 6-21).
2. Then, turn the stopcock to direct manometer flow to the patient (see system 3 in Fig. 6-21). The fluid level should drop, reaching a reading level in about 15 seconds.
3. The reading should be made at the upper level of the respiratory fluctuation of fluid in the manometer (fluid falls slightly on inspiration and rises slightly on expiration).
4. Turn the stopcock to allow resumption of the infusion to keep the catheter patent and to supply needed fluids (see system 1 in Fig. 6-21).

Meticulous aseptic technique should be used for dressing changes, catheter care, and tubing changes.

(Text continues on p. 184)

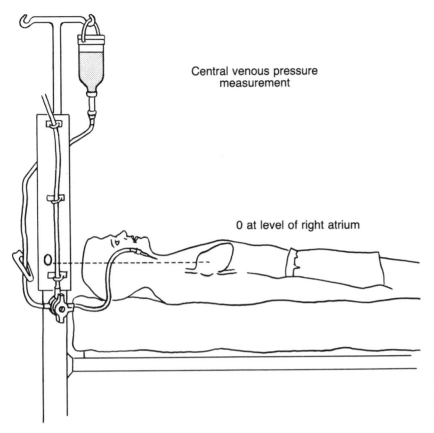

Central venous pressure
measurement

0 at level of right atrium

FIG. 6-20.
**Central venous pressure measurement
with zero-point of manometer at level of
right atrium.**

FIG. 6-21. Fluid flow systems in central venous pressure measurement. *System 1* allows flow
from the container to the patient (routine infusion); *System 2* allows flow from the container
to the manometer (allows manometer fo.fill); and *System 3* allows flow from the manometer
to the patient (allows reading of CVP). (Hudak C et al: Critical Care Nursing, 2nd ed, p. 187
Philadelphia, JB Lippincott, 1977)

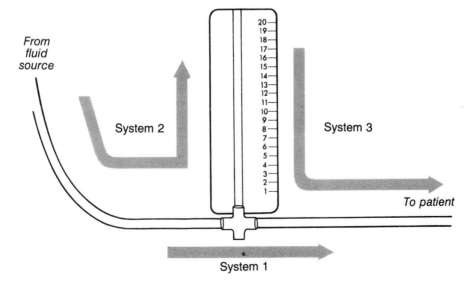

*From
fluid
source*

System 2

System 3

To patient

System 1

Check frequently for signs of redness, swelling, and infection at the insertion site. The connections should be checked frequently to be sure they are secure (to prevent the occurrence of air embolism). Remember that air embolism is more likely to occur when a catheter is placed in the central veins where pressure is low.

REFERENCES

1. Clifton M: Irrigate a plugged catheter? Never! Natl IV Ther Assoc J 3, No. 1:26, 1980
2. Goldberger E: A Primer of Water, Electrolyte and Acid–Base Syndromes, 6th ed, p 405. Philadelphia, Lea & Febiger, 1980
3. Gahart B: Intravenous Medications, 3rd ed, p 203. St Louis, CV Mosby, 1981
4. Grant J: Handbook of Total Parenteral Nutrition, p 96, Philadelphia, WB Saunders, 1980
5. Bivens B, Rapp R: Modern clinical nutrition. Part 4: Central versus peripheral nutrition—the controversy. Am J IV Ther Clin Nutr 7, No. 3:30, 1980
6. *Ibid*, p 33
7. Maxwell M, Kleeman C (eds): Clinical Disorders of Fluid & Electrolyte Metabolism, 3rd ed, p 429. New York, McGraw-Hill, 1980
8. Laegeler W, Tio J, Blake M: Stability of certain amino acids in a parenteral nutrition solution. Am J Hosp Pharm. 31:776, 1974
9. Grant, p 70
10. Maxwell & Kleeman, p 430
11. *Ibid*, p 411
12. Grant, p 131
13. Maxwell & Kleeman, p 443
14. Grant, p 130
15. Colley R, Wilson J: Meeting patient's nutritional needs with hyperalimentation. Nursing 79, May 1979, p 83
16. Simone P, Linkewich J: Guidelines for administration of parenteral drugs. Am J IV Ther Clin Nutr 8, No. 4:24 1981
17. Gahart, p 129
18. Frenier E: Problems of I.V. incompatabilities. Am J IV Ther Clin Nutr 3:22, 23, 1976
19. Maxwell & Kleeman, p 488
20. Ordway C: Air embolus via CVP catheter without positive pressure. Ann Surg 179:479, 1974
21. Yeakel A: Lethal air embolism from plastic blood storage containers. JAMA 204:267, 1969
22. Freitag J, Miller L (eds): Manual of Medical Therapeutics, 23rd ed, p 293. Boston, Little, Brown & Company, 1980
23. Gottlieb J, Ericsson J, Sweet R: Venous air embolism: A review. Anesth Analg 44, No. 6:776, 1965
24. Kasprisin D, Kasprisin C: Introduction to Transfusion Therapy, p 31. Garden City, Medical Examination Publishing Company, 1980
25. *Ibid*, p 32
26. Condon R, Nyhus L: Manual of Surgical Therapeutics, 4th ed, p 338. Boston, Little, Brown & Company, 1978
27. Kasprisin & Kasprisin, p 70
28. Condon & Nyhus, p 332
29. Condon & Nyhus, p 337
30. Kasprisin & Kasprisin, p 69
31. Condon & Nyhus, p 339
32. *Ibid*, p 333
33. Kasprisin & Kasprisin, p 71
34. Maxwell & Kleeman, p 484
35. *Ibid*, p 483
36. *Ibid*, p 485

BIBLIOGRAPHY

Bjeletich J, Hickman R: The Hickman indwelling catheter. Am J Nurs, Jan 1980, p 62

Grundfest S, Steiger E: Experience with the Broviac catheter for prolonged parenteral alimentation. J Parent Enter Nutr 3, No. 2:45, 1979

Ostrow L: Air embolism and central venous lines. Am J Nurs Nov 1981, p 2036

Riella M, Scribner B: Five year's experience with a right atrial catheter for prolonged parenteral nutrition at home. Surg Gynecol Obstet 143:205, 1976

Schaefer N: Hickman catheter. Nat IV Ther Assoc J. 4, May/June 1981

Turco K: Hazards associated with parenteral therapy. Am J IV Ther Clin Nutr 8, No. 10:9, 1981

West K: Modern clinical nutrition: Part 2. The importance of trace elements in total parenteral nutrition. Am J IV Ther Clin Nutr 8, No. 9, 1981

7

Fluid Balance in the Surgical Patient

THE BODY'S RESPONSE TO SURGICAL TRAUMA

Although a surgical procedure may be lifesaving, the body responds to it as trauma. Postoperative responses bear a direct relationship to nursing care.

ENDOCRINE RESPONSE (STRESS REACTION)

PERIOD OF FLUID RETENTION AND CATABOLISM

The stress reaction described by Selye, representing a response to surgical trauma, is present for the first 2 to 5 days postsurgery. The intensity of changes depends on the severity and duration of the trauma.

A scale of 10 has been proposed to grade the severity of trauma. An appendectomy or a simple herniorrhaphy is rated low on the scale, around 1 or 2, whereas severe trauma, such as deep burns or pelvic evisceration, rates high on the scale. In the middle of the scale are abdominal surgical procedures such as subtotal gastrectomy or colectomy. Postoperative apprehension and pain enhance the stress reaction, and extreme preoperative apprehension can initiate the stress reaction before surgery. Anesthesia also constitutes a form of stress.

The endocrine responses may be briefly outlined as follows (Fig. 7-1):

1. Increased adrenocorticotrophic hormone (ACTH) secretion from the anterior pituitary.
2. Increased mineralocorticoid and glucocorticoid secretion from the adrenal cortex (in response to stimulation by ACTH)
   - Mineralocorticoids (desoxycorticosterone [DOCA] and aldosterone) cause
     Na^+ retention
     Cl^- retention
     K^+ excretion
   - Glucocorticoids (mainly hydrocortisone) cause
     Na^+ retention
     Cl^- retention
     K^+ excretion
     Catabolism
       Protein breakdown
       Gluconeogenesis and elevated blood sugar
     Fat mobilization
     Drop in eosinophil count
3. Increased antidiuretic hormone (ADH) secretion

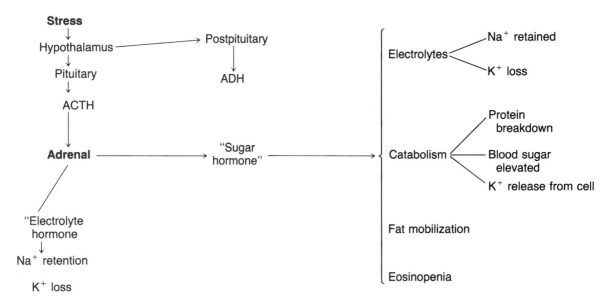

FIG. 7-1. Effect of stress on electrolyte excretion. (Adapted from Statland H: Fluids and Electrolytes in Practice, 3rd ed, p 45. Philadelphia, JB Lippincott, 1963)

from posterior pituitary, causing decreased urinary output

4. Vasopressor substances (epinephrine and norepinephrine) secreted from the adrenal medulla to help maintain blood pressure, a response stimulated by fear, pain, hypoxia, and hemorrhage

The body's response to stress appears purposeful. For example, sodium retention, chloride retention, potassium loss, and increased antidiuretic hormone (ADH) secretion help to maintain blood volume. Sodium and chloride retention cause water retention, cellular potassium loss releases cellular water into the extracellular space, and ADH secretion causes decreased fluid excretion by way of the kidneys. Glucocorticoids cause protein breakdown, make amino acids available for healing at the site of trauma, and cause conversion of protein and fat to glucose (gluconeogenesis), creating a ready supply of glucose for use during the stress period. (The elevated blood sugar may be mistaken for diabetes mellitus.)

As a result of stress, the early postoperative patient loses more urinary nitrogen than normal, even though his protein intake is low or absent. Because the body nitrogen losses exceed intake, the patient is said to be in negative nitrogen balance. Laboratory findings during the stress period include reduced eosinophil count and elevated level of serum 17-hydroxycorticosteroid hormones; both indicate increased adrenal activity.

The changes described here are *normal* responses

to trauma and do not usually require corrective measures.

PERIODS OF DIURESIS AND ANABOLISM After the second to fifth postoperative days, adrenal activity is decreased and a mild water and sodium diuresis occurs. The body also begins to retain potassium.

Anabolism, the building up of body protein, usually begins by the seventh to the tenth postoperative day. At this time the urinary nitrogen losses are decreased even though the patient is consuming protein foods. Thus, the renal excretion of nitrogen no longer exceeds the nitrogen intake from protein foods, and the patient begins to gain weight, provided oral intake is adequate.

TISSUE INJURY

Trauma results in the formation of edema in and about the operative site for the first few days after surgery. The fluid closely resembles plasma; its volume is roughly proportionate to the amount of tissue trauma. Traumatic edema is functionally sequestered as a "third-space" and cannot be readily mobilized to meet body needs. While the amount of fluid lost in edema is not in itself significant, it may enhance the extracellular fluid (ECF) volume deficit created by peritonitis, hemorrhage, or other complications. The edema fluid is reabsorbed and excreted during the diuretic phase of the stress reaction.

IMMOBILIZATION

Despite the emphasis on early ambulation, postoperative patients are much less physically active than normal. Immobilization favors increased renal nitrogen and calcium excretion and negative nitrogen balance, with muscle atrophy occurring from disuse. However, except for weight loss and temporary weakness, immobilization for a few days does not significantly affect metabolism. (The effects of long-term immobilization are described in Chapter 15.)

STARVATION EFFECT

Most patients eat inadequately, or not at all, during the first few postoperative days; thus, a starvation effect is induced. The majority of patients receive only routine dextrose and electrolyte fluids during this period, furnishing limited calories (e.g., a liter of 5% dextrose solution contains only approximately 170 calories). Accompanying starvation is a daily weight loss of about ½ lb, reflecting a decrease in lean and fatty tissue mass. Renal nitrogen excretion is increased as a result of lean tissue catabolism.

The weight loss following surgery is generally constant as revealed by accurate weighing procedures and recording the findings on a weight chart. A weight gain, in the face of starvation, indicates fluid retention. Usually the patient will regain his normal weight in 2 or 3 months if preoperative nutrition was good, operative trauma was of only moderate severity, and there was prompt return of gastrointestinal function.

PREVENTION OF POSTOPERATIVE COMPLICATIONS BY CAREFUL PREOPERATIVE ASSESSMENT AND PREPARATION

NUTRITION

A patient in good nutritional condition preoperatively withstands postoperative negative nitrogen balance and early starvation without serious effects. On the other hand, the nutritionally depleted patient goes to surgery under a serious handicap—a poor tolerance for operative stress. Increased susceptibility to infection results from a diminished ability to form antibodies and from the superficial atrophy in the mucous membrane linings of the respiratory and gastrointestinal tracts that often accompanies malnutrition. Hypoproteinemia follows prolonged negative nitrogen balance and increases susceptibility to shock from hemorrhage. Diminished supplies of protein and vitamin C retard wound healing.

When surgery is elective, the patient with a real or potential fluid balance problem is hospitalized early for preoperative evaluation and buildup. During this time the patient is given a well-balanced diet to provide the body with substances to make its own repairs and to help the patient weather the impending surgical trauma. Specific oral or parenteral medications may be deemed necessary after evaluation of clinical and laboratory findings. Baseline electrolyte studies are required for all patients scheduled for major surgery.

A primary nursing responsibility in the preoperative period is getting the patient to eat—sometimes difficult, especially when the patient's illness is such that his appetite is diminished. Fear and depression, common before surgery, may also deter the patient's desire to eat. (For nursing measures to promote eating, see Chapter 5). The benefits of activity are sometimes overlooked in the preoperative period; activity stimulates appetite and sleep, as well as general well-being.

Another important nursing responsiblity is reporting inadequate oral intake, as when all efforts fail to promote eating. Nasogastric tube or parenteral feedings may then have to be given. (For nursing responsibilities in nutritional assessment and in administering tube feedings, see Chapter 5). Anorexia accompanies malnutrition; correction of malnutrition with tube feedings often restores the patient's appetite. (For nursing responsibilities in parenteral nutrition, see Chapter 6.)

EMOTIONAL RESPONSE

Emotional stimuli may produce changes in electrolyte metabolism, which are mediated through the hypothalamus and anterior pituitary gland with the adrenal cortex as the target organ. Thus, the patient's attitudes toward surgery may significantly affect his postoperative course. Some fear of surgery is natural; undue fear and apprehension, however, may initiate the adrenocortical stress reaction. Discomfort resulting from venipuncture or insertion of a gastric tube, as part of the preoperative preparation, may contribute to the stress reaction. Most nurses can recall more than one postponement of surgery because the patient was not emotionally ready for the experience.

The nurse has an ideal opportunity to observe the patient's behavior and to detect signs of apprehension or severe depression that may be missed by the surgeon. Patients display fear in different ways: some refuse to discuss the oncoming surgical event; others can talk of nothing else. Significant behavior observations should be discussed with the physician.

No nursing function is more important than providing emotional support for the surgical patient. The most effective support comes from persons who have a sincere interest in the patient's welfare and a respect for his feelings. Thoughtful explanations before new procedures and experiences do much to relieve fear. Inspiring confidence by performing all nursing functions with skill and confidence is also a form of emotional support. Willingness to listen when the patient feels like talking helps; many patients find it easier to verbalize fears to an understanding nurse than to a relative or close friend. This is understandable when one considers the patient's desire to spare his loved ones additional worry. Also, some patients regard fear of surgery as a weakness and prefer to hide such fears from those close to them.

BODY WEIGHT

All surgical patients should have an admission weight recorded on the chart. The preoperative body weight serves as a baseline for comparison with subsequent body weight measurements. Obtaining the weight of an ambulatory patient presents no problem; the patient confined to bed can be weighed on a bed scale. (For procedures in weighing the patient, see Chapter 4.)

INTAKE–OUTPUT MEASUREMENTS

Patients requiring preoperative electrolyte studies are placed on the nursing intake–output measurement list. An accurate record of fluids gained and lost from the body is of great importance in detecting inadequate intake and abnormalities in renal function and fluid balance. (For nursing reponsibilities in measurement of fluid intake and output, see Chapter 4.)

MEDICATIONS

STEROIDS The adrenals normally produce about 25 mg of hydrocortisone a day; under stress, however, they may need to increase their output up to ten times above this amount. A patient's ability to withstand surgical stress depends upon his adrenals' ability to secrete extra hydrocortisone. It is wise to routinely ask all surgical patients if they have taken cortisone or any other steroid preparation for more than 1 or 2 weeks within the past 6 to 12 months. When a patient is on steroid therapy there is less need for adrenal secretion and the glands tend to atrophy from disuse; after withdrawal of steroid therapy, the adrenals gradually resume their function. However, if steroids are suddenly withdrawn and the patient is subjected to massive trauma, such as a major surgical procedure, the atrophied adrenal glands may be unable to respond to the stress signal; adrenocortical failure may follow cessation of adrenocortical substitution therapy. Symptoms include hypotension, nausea and vomiting, thready pulse, subnormal temperature early with later hyperpyrexia, hallucinations, confusion, stupor, or coma. The reaction usually occurs in the first 24 hours.

The actual degree of adrenal suppression caused by steroid administration is related to the dose and duration of use. Although there are variations in degrees of adrenal suppression, patients who are currently receiving steroids, or who have received steroids for more than 1 or 2 weeks within the preceding 6- to 12-month period, are usually considered as having adrenal suppression. Such patients require appropriate perioperative glucocorticoid supplementation. One should remember that short-term *excess* of glucocorticoids is relatively harmless, but short-term *deficiency* during stress can be fatal.[1]

Today, because of medical specialization, one patient may have two or three physicians prescribing medications at the same time. The fact that the patient may recently have received steroids is often overlooked, and, in fact, many patients do not know what medicines they have taken; if this is the case, the nurse can ask for a description of the medication and why it was given. Conditions for which steroid therapy is often used should be kept in mind (e.g., rheumatoid arthritis, asthma, dermatitis, and ulcerative colitis). When in doubt, one should check with those who prescribed medications for the patient.

DRUG ALLERGIES OR IDIOSYNCRASIES The physician or the nurse should always ask the prospective surgical patient if he knows of any drug allergies, sensitivities, or idiosyncrasies he may have. If the patient does not understand the question, typical symptoms of sensitivity, such as urticaria, asthma, and the like, can be mentioned. One can also ask the patient if a physician has ever cautioned him to avoid a specific medication because of an unusual reaction to it. In questioning the newly admitted patient, the nurse will do well to remember that he is often upset and may have difficulty remembering.

Some patients assume that their physician remembers any drug allergies from office interviews prior to hospital admission, and unless specifically asked, they may not volunteer information. Allergies must be discovered before the patient is sedated or anesthetized; it is too late to ask when he is unconscious or semireactive after surgery, since most medications are ordered in the immediate postoperative period. Failure to ascertain the presence of allergies may be disastrous. The most dreaded allergic reaction is anaphylactic shock; other less dangerous reactions include skin eruptions and asthma.

Idiosyncratic reactions to drugs deserve consideration. For example, a narcotic may cause more depression in one patient than in another of equal weight and age. A dose creating the desired effect in one patient may overwhelm the patient who is unusually reactive to the drug. The aged are particularly sensitive to narcotics and should be given much smaller doses than younger adults; this is particularly important in aged surgical patients (see Chapter 18).

PULMONARY VENTILATION

RISK FACTORS FOR PULMONARY COMPLICATIONS Pulmonary complications are the most frequent cause of death and morbidity in the patient following anesthesia and surgery.[2] Inadequate pulmonary ventilation is common in postoperative patients and can lead to hypoxemia, respiratory acidosis, atelectasis, and pulmonary embolism. Listed below are some of the risk factors for postoperative ventilation problems:

Heavy cigarette smoking

Obesity

Advanced age

Thoracic or upper abdominal operative site

Extended length of operation

Smokers should be encouraged to stop smoking at least several weeks prior to elective surgery to decrease the likelihood of postoperative pulmonary complications. Smoking increases the rate of mucus secretion, impairs mucociliary clearance, damages the ciliated epithelium, and reduces the amount of surfactant. In addition, it is thought that smoking predisposes to pulmonary embolism.

Elective operations, particularly those involving the upper abdominal area, may need to be delayed to allow obese patients time to attempt weight reduction. A marked increase in postoperative pulmonary complications has been reported in obese patients.

Decreased pulmonary function is common in the aged. It has been demonstrated that vital capacity is reduced with an increased residual volume, since rigidity of the chest wall interferes with normal pulmonary excursions. Since diminished respiratory function interferes with carbon dioxide elimination, many aged persons may be in a state of impending respiratory acidosis. The degree of postoperative hypoxemia has been related directly to age. The preponderance of impaired pulmonary function in the elderly apparently increases risk for postoperative pulmonary complications. Patients with thoracic or high abdominal incisions are particularly prone to develop ventilatory problems. Significant pulmonary complications have been reported to occur in 20% to 80% of all patients following abdominal or thoracic operations. Studies indicate that as the duration of anesthesia increases, pulmonary complications tend also to increase. Preexisting pulmonary conditions are major factors in predisposing to post-operative pulmonary complications such as atelectasis, pneumonia, and pulmonary embolism.

NURSING INTERVENTIONS As stated above, smoking should be discouraged in preoperative patients for at least several weeks before surgery to diminish the incidence of postoperative pulmonary complications. In addition, provided enough time exists, obese persons should be encouraged to lose weight sensibly before elective surgery. During the immediate preoperative period, the nurse should teach the patient how to leg exercise, deep breathe, and cough effectively, explaining that these activities are preventive measures against pulmonary complications. If blow bottles are to be used postoperatively, the patient should be allowed to practice their use preoperatively. Once the patient is properly motivated and knows what is expected of him, greater success may be encountered in carrying through the common postoperative regime of "coughing, leg exercising, and deep breathing" every hour. (Nursing interventions for prevention of postoperative pulmonary complications are discussed in greater depth later in this chapter.)

CHRONIC ILLNESSES

Certain chronic illnesses greatly increase the hazard of postoperative water and electrolyte imbalances. Such illnesses include the following:

Heart disease
Diabetes mellitus
Pulmonary disease
Adrenal insufficiency
Renal disease

It has been estimated that about one third of surgical patients over 35 years of age have evidence of heart disease, hypertension, or diabetes mellitus.[3]

HEART DISEASE Patients with cardiovascular disease are more susceptible to the cardiovascular stresses associated with general anesthesia and surgery (hypotension, hypoxemia, sepsis, and thromboembolism). Review of the electrocardiogram (EKG) and careful history and physical examination are required to identify factors that predict serious postoperative cardiac complications. General anesthesia and surgery involve at least a 1 in 3 chance of a perioperative myocardial infarction (MI) in the patient who has sustained an MI during the *previous 6 months*; surgery for patients in this group should be deferred if at all possible.[4]

DIABETES MELLITUS Obviously, patients with diabetes mellitus should be in the best possible balance at the time of operation. Therefore, unless the operation is urgent, it should be delayed until the diabetic state, nutritional status, hydration, and electrolyte status are controlled. Serum glucose levels should be closely monitored during the perioperative period. Close cooperation between the internist, surgeon, and anesthesiologist is required for optimal management of diabetic surgical patients. One should also remember that the diabetic patient is particularly susceptible to gram-negative septic shock.

PULMONARY DISEASE Patients with pulmonary disease are at special risk for such complications as atel-

ectasis, pneumonia, and pulmonary embolism. Some indicators of significant operative risk in patients with pulmonary disease include elevated preoperative arterial Pco_2 (greater than 45 mm Hg), severe hypoxemia before operation (arterial Po_2 less than 55 mm Hg), and maximal volume ventilation (MVV) less than 50% of predicted or less than 50 liters/min.[5]

ADRENAL INSUFFICIENCY Patients with known or suspected adrenal insufficiency should be supplied with adequate doses of hydrocortisone preoperatively to prevent precipitation of adrenal crisis during the stress of surgery.

RENAL DISEASE Patients with chronic renal disease may present a number of problems during surgery. Preoperative preparation should include adequate fluid administration to furnish the greater-than-normal urine volume required by the diseased kidney to excrete metabolic wastes, taking care not to overload the circulatory system and cause congestive failure. Anemia and abnormal coagulation factors should be corrected preoperatively, as should metabolic acidosis. Respiratory acidosis caused by anesthesia may be lethal when added to an already existing metabolic acidosis. Hemodialysis may be required the day prior to surgery in patients with severe renal damage. Hyperkalemia, due to tissue trauma, increased catabolism, and failure of the diseased kidney to adequately remove potassium, should be anticipated following surgery.

IMMEDIATE PREOPERATIVE PREPARATION

ENEMAS A cleansing enema may be ordered either the night before or on the morning of abdominal or rectal surgery. Occasionally one still sees an order for "tap water enemas until returned clear"; fortunately, such an order is seen less and less. As many as five to ten enemas may be needed before the solution is "returned clear"; sodium and potassium may be lost with the enema return, decreasing stores of these valuable electrolytes when the patient can ill afford to lose them. Many surgeons feel that one cleansing enema, properly given, suffices.

WITHHOLDING ORAL FLUIDS Some physicians allow the patient nothing by mouth for at least 8 hours prior to surgery to reduce the risk of vomiting and aspiration on induction of anesthesia. Usually the pa-

TABLE 7-1 *Intravenous fluids used for maintenance needs and correction of imbalances**

Type of fluid	Electrolyte content mEq/liter		Comments
5% Dextrose in water	None		Supplies 170 calories and free water for excretory purposes
5% Dextrose in lactated Ringer's solution	Na⁺	130	Sometimes used as a routine maintenance solution; electrolyte content approximates that of plasma (does not supply Mg). Dextrose supplies 170 calories
	K⁺	4	
	Ca⁺	3	
	Cl⁻	109	
	Lactate	28	
Isotonic saline (0.9% sodium chloride)	Na⁺	154	Used to supply sodium chloride—expands extracellular fluid volume (not a good routine maintenance solution). Sodium and chloride content exceeds that found in plasma; excessive Na can cause edema, and excessive Cl can cause metabolic acidosis
	Cl⁻	154	
Balanced isotonic solution	Na⁺	140	Used as an extracellular fluid replacement solution. Manufactured commercially under a variety of trade names. Note that it contains no Ca; can be given simultaneously with blood. Supplies Mg
	K⁺	5	
	Mg⁺	3	
	Cl⁻	98	
	Acetate	27	
	Gluconate	23	
Gastric replacement solution	Na⁺	63	Used to replace gastric fluid loss. Acid pH of 3.3 to 3.7 is irritating; should be administered in a large functional vein. Manufactured under a variety of trade names. Contraindicated in hepatic insufficiency
	K⁺	17	
	NH₄⁺	70	
	Cl⁻	150	
Duodenal replacement solution	Na⁺	138	Given to replace duodenal fluid loss and to correct mild acidosis. Manufactured under a variety of trade names. Supplies Mg
	K⁺	12	
	Ca⁺	5	
	Mg⁺	3	
	Cl⁻	108	
	Lactate	50	
Butler solution (electrolyte 88) (balanced hypotonic solution)	Na⁺	57	Given to replace fluid lost from the large intestine. Hypotonic fluid, supplying electrolytes and free water for excretory purposes. Manufactured under several trade names. Supplies Mg
	K⁺	25	
	Mg⁺	5–6	
	Cl⁻	49–50	
	Lactate	25	
	HPO₄⁻	13	

*The reader is encouraged to see Table 6-6 for a more extensive tabulation of intravenous fluids.
(Metheny, N: Water and electrolyte balance in the postoperative patient. Nurs Clin North Am 10:55, 1975)

After the fluid retention of stress has subsided, a larger amount of fluid is given. If oral intake is still prohibited, an attempt must be made to supply body needs solely with parenteral fluids. Magnesium replacement may be necessary when parenteral fluid administration is prolonged. Magnesium deficit is not as rare as was once thought; prolonged administration of magnesium-free fluids dilutes the plasma magnesium level and may produce symptoms of deficit, particularly when magnesium loss has resulted from gastric suction. Parenteral vitamin preparations of the B complex group and vitamin C should be given daily when parenteral therapy is necessary for more than 2 days. When parenteral therapy must be prolonged, other nutrients must be furnished. (Parenteral nutrition is discussed in Chapter 6.)

ORAL INTAKE Many physicians prefer that patients undergoing gastrointestinal suction receive nothing by mouth; others allow "ice chips sparingly" to relieve thirst. The term "sparingly" is open to interpretation by the staff, and more ice chips may be given than was intended by the physician because of a thirsty patient's constant plea for more ice.

Drinking plain water causes a movement of electrolytes into the stomach to make the solution isotonic; before the water and electrolytes can be absorbed they are removed by the suction apparatus. This process can deplete the body of valuable electrolytes, primarily sodium, chloride, and potassium. Profound states of metabolic alkalosis or of sodium deficit have been caused by the unwise practice of giving plain water to a patient undergoing gastric suction. *If ice chips are to be given, they should be limited carefully.* Some physicians allow approximately an ounce of plain ice chips per hour to relieve oral dryness.

Once oral feedings are allowed, the patient should be returned to a full diet as early as possible, because good nutrition decreases both the duration and the complications of convalescence.

POSTOPERATIVE PROBLEMS IN WATER AND ELECTROLYTE BALANCE

WATER INTOXICATION

Water excess (sodium deficit) is also referred to as water intoxication or hyponatremia, an imbalance most likely to occur in the first one or two postoperative days while the water retention effect of stress (transient excessive ADH secretion) is still present. Excessive administration during this period of water-yielding fluids, such as 5% glucose in water, predisposes to this condition. Symptoms of water excess (sodium deficit) include the following:

Behavior changes
- Inattentiveness
- Confusion
- Hallucinations
- Shouting and delirium
- Drowsiness

Acute weight gain

Neuromuscular changes
- Cramping of exercised muscles
- Isolated muscle twitching
- Weakness
- Headache
- Blurred vision
- Incoordination
- Elevated intracranial pressure may occur with hypertension, bradycardia, decreased respiration, projectile vomiting, and papilledema

- Convulsions
- Hemiplegia

The nurse should suspect water excess (sodium deficit) when several of these symptoms occur in the early postoperative period. Behavior changes are usually noticed first.

Prevention of body fluid disturbances demands the study of daily accurate body weight measurements; a sudden weight gain in the early postoperative period is an indication to decrease fluid intake. Fluid intake during the water retention of stress should not greatly exceed body fluid losses.

Mild water excess can be corrected by prohibiting further water intake; however, brain damage or death may supervene if the condition is allowed to go untreated.

DIMINISHED POSTOPERATIVE VENTILATION AND RESPIRATORY ACIDOSIS

Normally, carbon dioxide is given off by the lungs during exhalation. Respiratory acidosis (primary carbonic acid excess) occurs when the lungs retain carbon dioxide, owing to decreased respiration depth or blockage of oxygen–carbon dioxide exchange at the alveolar level. The surgical patient may develop respiratory acidosis for one or several reasons:

Depression of respiration by anesthesia

Blockage of oxygen–carbon dioxide exchange in the lungs due to atelectasis, pneumonia, or bronchial obstruction

Depression of respiration with too frequent or too large doses of narcotics

Shallow respiration because of abdominal distention and crowding of the diaphragm

Shallow respiration as a result of pain in the operative site or large cumbersome dressings

As stated earlier, patients having thoracic or high abdominal incisions are particularly prone to develop ventilatory problems. For example, vital capacity is decreased by 40% of normal on the day after subtotal gastrectomy. However, this loss of reserve still allows for adequate oxygen and carbon dioxide exchange if preoperative ventilation was normal. In addition to causing respiratory acidosis, decreased ventilation interferes with the correction of metabolic acidosis. (Recall that the lungs attempt to compensate for metabolic acidosis by eliminating more carbon dioxide.)

Indiscriminate use of oxygen in the postoperative period increases the chance of overlooking respiratory acidosis. Cyanosis is usually the chief criterion for detecting inadequate ventilation; oxygen therapy may prevent cyanosis and keep the skin color pink even though respiratory acidosis is progressing.

NURSING INTERVENTIONS The nurse can perform various procedures to help improve pulmonary ventilation in the postoperative patient and prevent respiratory acidosis:

Encouraging the patient to cough, breathe deeply, and leg exercise at regular intervals (unless contraindicated)

Administering sufficient narcotics to make coughing tolerable, yet not enough to produce shallow respirations. This requires good nursing judgment; the timely use of analgesics enables the patient to expand his thoracic cavity with less discomfort.

Positioning the patient in high Fowler's, when permitted, to allow for better gas distribution throughout the lungs. The sitting position allows for greater lung expansion because gravity pulls the abdominal organs away from the diaphragm.

Splinting the patient's incision with a pillow while he coughs, so that he is less fearful and less apt to splint himself by muscular contraction, which limits ventilation. The patient should be taught to splint his own incision when assistance is not available.

Turning the patient at regular intervals

Ambulating the patient as early as possible. More than just a few steps to a bedside chair are necessary to prevent deep vein thrombosis and pulmonary embolism.

Teaching the patient to do leg exercises to prevent deep vein thrombosis and pulmonary embolism. The nurse should perform passive leg exercises for the patient if he is unable to do active exercises.

Assuring that dressings or restraints are not so tight that they will interfere with normal respiratory excursions

Some physicians order intermittent positive pressure breathing (IPPB) treatments to improve pulmonary ventilation; others may prefer the use of blow bottles or similar devices.

ILEUS

Peristalsis is inhibited for 2 to 4 days after intra-abdominal manipulation; for this reason, patients with abdominal surgery usually receive nothing by mouth until bowel sounds indicate the return of peristalsis. Gastric intubation is often employed early in the postoperative period to prevent the bowel from becoming greatly distended with swallowed air and gastrointestinal fluids.

If paralytic ileus persists longer than normal, the patient must be maintained on intravenous fluids and gastric suction until peristalsis returns (as evidenced by passage of flatus). Factors that may prolong the expected period of ileus postoperatively include bacterial and chemical peritonitis, excessive handling of the intestines during surgery, advanced age, or actual mechanical obstruction. Oral intake should be withheld in any postoperative patient who is vomiting; the nature of the vomitus, as well as the degree of abdominal distention, should be noted and reported to the surgeon.

Large amounts of water and electrolytes may be sequestered into the bowel. The amount of fluid "lost" in this manner is not revealed by body weight change. Clinical signs of fluid volume deficit, decreased urinary volume, and increased urinary specific gravity help indicate the amount of fluid trapped in the bowel. (See the discussion of bowel obstruction in Chapter 9.)

IRRIGATING SUCTION TUBES

Many feel that the fluid used to irrigate gastrointestinal tubes should be isotonic saline rather than plain water, since the latter can promote electrolyte loss from the gastrointestinal mucosa. (Some controversy exists over whether sodium losses are significant with irrigation with tap water.) Sometimes there are specific orders for the frequency of the irrigation and the amount of solution to be used; other times the order is general and leaves the frequency of the irrigation and the amount of fluid to the nurse's discretion. In either event, the amount of fluid instilled should be removed; any difference between the amount instilled and amount removed should be recorded on the intake–output record.

Some physicians prefer hourly irrigation of the tube with 30 ml to 50 ml of air to prevent clogging; air does not interfere with electrolyte balance or with intake–output record keeping. However, when blood or thick secretions are occluding the tube, it may be necessary to use isotonic saline. In addition, the use of air may lead to uncomfortable abdominal distention and bloating.

IMBALANCES ASSOCIATED WITH SPECIFIC BODY FLUID LOSSES

Metabolic alkalosis is most commonly seen in surgical patients as a result of the loss of large amounts of gastric secretions, either through vomiting or gastric suction. It is closely associated with potassium deficit, produced by the excessive loss of potassium-rich intestinal secretions or by prolonged parenteral therapy without potassium replacement. Potassium deficit is enhanced by the stress reaction in the early postoperative period.

Metabolic acidosis (primary base bicarbonate deficit) follows excessive loss of alkaline intestinal secretions, bile, and pancreatic juice. (See Chapter 9, Table 9-2.)

SHOCK

PATHOPHYSIOLOGY

Shock is a clinical syndrome indicating inadequate tissue perfusion secondary to decreased effective circulating blood volume. To understand the different kinds of shock, it is necessary to briefly review the basics of hemodynamics. Circulatory integrity involves the following basic components:

  The heart (pump)

  The blood volume (pumped fluid)

  The vascular bed, made up of
  • Resistance vessels (arteries and arterioles)
  • Exchange vessels (capillary network)
  • Capacitance vessels (veins and venules)

Shock results from dysfunction of *one or more* of these separate but interrelated components. In the shock state, insufficient perfusion can result from a depressed cardiac output due to the following:

  Inadequate venous return due to decreased blood volume, as seen with internal or external blood loss; acute dehydration; or "third-space" loss of fluid as occurs in intestinal obstruction, burns, and massive soft-tissue damage

  Inadequate venous return due to pooling of blood in an abnormally dilated vascular bed, as seen in gram-negative septic shock, anaphylactic shock, high spinal anesthesia, and spinal cord transection

  Failure of the heart as a pump, which can be due to a variety of causes including tachyarrhythmias, pericardial tamponade, myocardial infarction, myocarditis, heart block, and severe mitral or aortic stenosis

  Impeded blood flow, such as is seen in pulmonary embolism or vena cava obstruction

As stated earlier, there may be certain elements of each of these in any given patient.

CLASSIFICATION

Shock in surgical patients is usually of the hypovolemic type; however, a number of associated problems may cause other forms of shock. A commonly accepted classification of shock is as follows:

  Hypovolemic shock

  Neurogenic shock

  Septic shock

  Cardiogenic shock

HYPOVOLEMIC SHOCK Hypovolemic shock is due to actual internal or external blood loss (hemorrhage), acute dehydration (as in severe diarrhea or vomiting), or "third-space" losses of plasma (as in intestinal obstruction or severe burns). If blood volume is reduced by more than one third and the deficiency is maintained, the shock that ensues will be fatal.[7]

Symptoms Symptoms of hypovolemic shock are listed below:

  Poor capillary refill time (may take several seconds) as noted by pressing over fingernails

  Anxiety due to hyperactivity of the sympathetic nervous system when blood loss is sustained

  Decreased systolic blood pressure of 10 mm Hg or more when position is changed from lying to standing or sitting (early sign)

  Decreased pulse pressure, often less than 20 mm Hg. When blood pressure changes are due to blood loss, the systolic pressure falls more rapidly than the diastolic pressure.

  Eventually, blood pressure remains low even in flat position as the body's defense mechanisms are unable to cope with hypovolemia. Systolic pressure generally does not fall significantly until a blood volume deficit of at least 15% to 25% has been sustained.

  Collapse of neck veins (see discussion of this subject in Chapter 4)

  Respiratory alkalosis in early shock, caused by hyperventilation

  Pulse rate usually increased by more than 20 per minute when position is changed from lying to standing or sitting. Later, pulse remains rapid and thready in all positions. With slow hemorrhage, as much as 1000 ml of blood may be lost without significant increase in pulse rate as long as the supine position if maintained.[8]

Blood loss is indicated by a decreased hematocrit level. Recall that it takes 4 to 6 hours for interstitial fluids to move into the bloodstream after a hemorrhage; therefore, a hematocrit value will not reflect a very recent hemorrhage.

Decreased urinary output. Increased ADH and aldosterone secretion causes water and sodium retention by kidneys.

Pale, cool, moist skin caused by peripheral vasoconstriction

Metabolic acidosis in late shock resulting from lactic acid buildup caused by hypoxia

Central venous pressure (CVP) usually below normal. Isolated CVP reading may mean relatively little, but the change caused by fluid replacement is quite significant.

Central venous pressure The CVP is usually low in hypovolemic patients owing to a diminished return of blood to the right side of the heart. Basically, a substantially lowered CVP is treated by fluid replacement until it returns to normal. (Recall that the normal pressure in the right atrium is 0 cm to 4 cm of water and that the normal pressure in the vena cava is approximately 4 cm to 11 cm of water.) Right-sided heart failure or obstruction of the pulmonary circuit may cause high CVP despite hypovolemia. Nursing responsibilities in the measurement of CVP are discussed in Chapter 6. The management of severe shock may necessitate monitoring the left side of the heart by means of a Swan-Ganz catheter. (Monitoring of the left side of the heart is discussed in Chapter 11.)

Positioning Elevating the legs to a 45 degree angle (Fig. 7-2) temporarily releases about 500 ml of blood and, thus, partially relieves the effect of moderate hypovolemia; in severe hypovolemia there may be insufficient blood in the extremities to be of value. Trendelenburg position is contraindicated in most instances since it allows the abdominal viscera to interfere with respirations by pressing against the diaphragm; also, the abdominal contents press against the vena cava, interfering with venous return.

Fluid replacement Many physicians feel that hypovolemia in hemorrhagic shock is best treated with a judicious mixture of electrolyte solutions and whole blood. As a rule, resuscitation is begun with a balanced isotonic electrolyte solution, and whole blood is

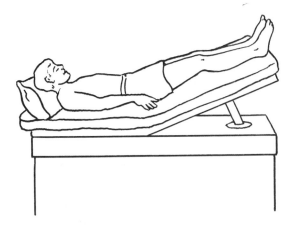

FIG. 7-2. Elevation of legs during shock.

added when serial measurements show that the hematocrit has fallen below 30%.[9] This approach provides an effective balance between electrolytes, red cell mass, and colloid to allow adequate volume distribution and oxygen-carrying capacity.[9] Parameters to be monitored during fluid resuscitation include blood pressure, pulse, urinary output, hematocrit, and, as indicated, CVP.

A typical regime for replacement therapy in the emergency room for the patient in hemorrhagic shock allows for the rapid infusion of lactated Ringer's solution (e.g., 1000 ml–2000 ml in 45 minutes); if the blood pressure, pulse, and urinary output return to normal, and if the hematocrit is greater than 30%, the solution is continued at a maintenance level. If, however, blood loss has been severe or hemorrhage is continuing, the rise in blood pressure and decrease in pulse rate will usually be transient; at this point, whole blood should be readied for administration. (Blood should have been drawn for typing and cross-matching at the same time the electrolyte solution was started.) The amount of blood to be given is dependent on the degree of blood loss.

Considerable controversy exists about the choice of fluids for treatment of hypovolemia. The physician has a choice of electrolyte solutions, whole blood or packed cells, dextran, and plasma or albumin. A brief review of each of these follows. (Nursing responsibilities in the administration of these substances are discussed in Chapter 6.)

Electrolyte solutions Many physicians advocate the use of a balanced electrolyte solution (such as lactated Ringer's) as a principal fluid in initial shock therapy. Its advantages are that it is readily available and inexpensive, it is virtually free of reactions, it appears to reduce the amount of blood needed for correction of blood loss, it is effective in controlling hypovolemia, and it does not aggravate other fluid and electrolyte problems that may be present. The literature reports significantly increased survival during shock resuscitation with the use of lactated Ringer's solution as compared with the use of colloids.[10] Since balanced electrolyte solutions diffuse out of the intravascular space rather rapidly, it may take three to four times as much of such solutions to replace the volume of lost blood. Care must be taken to observe the hematocrit during fluid administration to avoid diluting the red cell mass beyond the level needed for normal peripheral oxygenation. Electrolyte solutions are only substitutes for blood in the initial resuscitation of the patient in hemorrhagic shock; major blood loss always necessitates the administration of blood as soon as it is available. Only minor blood loss can be completely replaced by the sole use of electrolyte solutions.

Isotonic solution of sodium chloride (0.9% NaCl) is not a fluid of first choice because it is *not* a balanced or physiologic solution. It contains 154 mEq of sodium and 154 mEq of chloride/liter, approximately 10 mEq of sodium and 40 mEq of chloride in *excess* of the normal extracellular concentrations.

Electrolyte-free solutions (such as 5% dextrose in water) should *never* be used for emergency fluid replacement in hypovolemia; these solutions may rapidly dilute the contracted ECF, producing the threat of "water intoxication." Contributing to the threat of water overload is the increased ADH action associated with acute stress states.

Blood Whole blood is still a mainstay of fluid replacement in major hemorrhage. Its major advantage is its oxygen-carrying capacity. As stated earlier, blood administration is usually preceded by the administration of a balanced electrolyte solution, allowing time for typing and crossmatching and actually decreasing the volume of blood needed for resuscitation. Blood transfusions are limited by cost, limited availability, delay required for typing and crossmatching, and most importantly, a number of potentially serious complications. (The reader is referred to the section on blood and blood products administration in Chapter 6.)

Dextran Dextran is an effective plasma volume expander and is thought to decrease blood viscosity and improve microcirculatory flow. The longest effect of dextran in maintaining an expanded volume has been shown to be 24 to 48 hours.[11] Problems associated with dextran administration may include occasional allergic reactions, defects in clotting mechanism when more than one liter is given per day, and "tubular burn" if dextran is excreted in an inadequate volume of urine. No more than one liter of dextran should be given each day.

Plasma and albumin The risk of hepatitis transmission from plasma, especially when pooled but even when prepared individually, has severely limited its use as a volume expander. In addition, the effect of plasma or albumin as a blood volume substitute is transient; plasma dispersal from the intravascular space to the extravascular space may proceed at a rate of 500 ml/hr.[12] Plasma is seldom administered except when given in the fresh, frozen form to replenish clotting components.[13]

Human serum albumin is free of the risk of hepatitis transmission and is readily available. However, when given in large volumes during resuscitation, transudation into the pulmonary interstitium may occur, perhaps contributing to adult respiratory distress syndrome (ARDS). (This syndrome is described later in the chapter and in Chapter 14.) Those who advocate the use of colloids in resuscitation strive to give enough to restore hemodynamics without overloading the pulmonary interstitium.

Oxygen administration Oxygen is not used routinely in the treatment of shock, since the oxygen saturation in most patients with uncomplicated hypovolemic shock is generally normal. However, in hypovolemic patients in whom the oxygen saturation is *not* normal, the *initial* use of oxygen may be very important. This may occur in patients with obstructive lung disease but more frequently the oxygenation problems arise from the patient's injury (such as pneumothorax, pulmonary contusions, or aspirated gastric secretions). When necessary, oxygen is best administered through a loose-fitting face mask; if a controlled airway is indicated for other reasons, an endotracheal tube is an ideal route. When in doubt as to the adequacy of arterial oxygenation, it is best to administer oxygen until the patient can be assessed more thoroughly.

In considering the need for oxygen therapy, one must consider the hemoglobin level in the patient who has sustained major blood loss. A normal hemoglobin of 15 g/dl provides transport for 20 vol% of oxygen;

when the hemoglobin is decreased to 7 g/dl, only 10 vol% of oxygen can be transported, seriously compromising tissue oxygenation.[14] Oxygen administration will improve hemoglobin saturation and oxygen delivery to the tissues. One should remember that cyanosis, as a sign of poor oxygenation, may be absent in the patient with a hemoglobin concentration below 5 g/dl.[14]

Acid–base problems Acid–base changes in shock have been divided into four states:

1. Normal condition
2. Respiratory alkalosis
3. Metabolic acidosis
4. Combined acidosis (both metabolic acidosis and respiratory acidosis)[15]

Extreme degrees of hyperventilation may be caused not only by anxiety, pain, or hysteria, but also by trauma, sepsis, or shock; hyperventilation, in turn, causes respiratory alkalosis. As shock progresses with its diminished tissue perfusion, metabolic acidosis ensues. Recall that decreased tissue perfusion leads to hypoxia, causing the body to resort to anaerobic (non-oxygen-consuming) metabolism; the result is excessive formation of lactic acid and thus metabolic acidosis.

The most frequent acid–base abnormality seen initially in shock patients is metabolic acidosis with a partially compensated respiratory alkalosis. If treatment of shock is delayed or unsuccessful, the patient may eventually develop the lethal combination of metabolic acidosis and respiratory acidosis due to respiratory failure.

Adult respiratory distress syndrome Shock can act as an insult to the lungs and produce a condition called adult respiratory distress syndrome (ARDS); this condition is sometimes referred to as "shock lung" or "traumatic wet lung." The syndrome may range in clinical severity from mild dysfunction to progressive, eventually fatal, pulmonary failure.

It is known that during shock the body gives top priority to certain vital organs (such as the brain and heart) at the expense of other organs less essential for immediate survival; thus, it has been hypothesized that the pulmonary changes following shock are, at least in part, due to ischemic injury to the lung. The injury is apparently at the alveolar–capillary membrane; there is a leakage of proteinaceous fluid from the intravascular space into the interstitial and alveolar spaces. The prominent derangements in pulmonary function associated with ARDS include hypoxia, which is unresponsive to increased inspired oxygen concentrations; "stiff lung" (decreased pulmonary compliance); and a fall in the amount of air remaining in the lungs after a normal expiration.[16] The hypoxemia indicates ventilation/perfusion imbalance and shunting. The decreased pulmonary compliance is manifested by progressively increased airway pressure needed to achieve adequate ventilation. In addition, ARDS is accompanied by chest radiograph changes (diffuse infiltrates that may progress to widespread areas of consolidation).

Management of ARDS is discussed in Chapter 14. Early and complete correction of hypovolemia is a major factor in the *prevention* of this complication of shock.[17] However, *excessive* fluid administration contributes to the respiratory complications in the immediate postoperative state.

Prevention of renal insufficiency The surgical patient is particularly susceptible to suboptimal tissue perfusion due to preoperative fluid restrictions, surgical blood loss, third-space trapping of ECF at the site of surgical trauma, and the use of anesthetic agents.

Oliguria in the surgical and postoperative patient is due to a combination of events: hypovolemia with its associated diminished renal blood flow and glomerular filtration rate, and increased ADH and aldosterone secretion as a result of the hypovolemia and stress (recall that ADH causes water retention and that aldosterone causes retention of both sodium and water). Oliguria during surgical treatment is generally reversible; however, the concurrent occurrence of toxic shock or blood transfusion reaction may act as an additive event and cause acute tubular necrosis. (Renal failure is discussed in Chapter 10.)

It should be noted that diminished renal blood flow is *not* a necessary consequence of surgical procedures. In fact, if the patient receives adequate fluid administration of salt-containing fluids *before* surgery, decreases in renal blood flow can be prevented, or at least minimized. In contrast, administration of the same fluids *after* induction of anesthesia will have minimal effect in increasing renal blood flow.[18] Thus, it is important that parenteral fluids be started before surgery and an adequate urine volume established *prior* to induction of anesthesia. Urine flow should be at least 50 ml/hr (preferably 75 ml–100 ml/hr).[18]

NEUROGENIC SHOCK Neurogenic shock follows serious interference with the balance of vasodilator and vasoconstrictor influences to both arterioles and venules. In neurogenic shock there is loss of sympathetic control of resistance vessels, with resultant dilatation of arterioles and venules; shock is produced by the decrease in effective circulatory volume. This is the shock seen with high spinal anesthesia, syncope upon sudden exposure to unpleasant stimuli, and even acute gastric dilatation.

Clinically, it is manifested quite differently from hypovolemic shock; while the blood pressure may be extremely low in both hypovolemic shock and neurogenic shock, the pulse rate is often slower than normal in the patient with neurogenic shock (as opposed to rapid pulse in the hypovolemic patient). Also, the patient with neurogenic shock often has dry, warm, and even flushed skin. In neurogenic shock, the blood volume is apparently normal, but its distribution pattern is changed owing to an increased reservoir capacity in the arterioles and venules. As a result of peripheral pooling of blood, the venous return to the right side of the heart is diminished, causing a reduction in cardiac output.

Failure to treat neurogenic shock can cause all the ravages of hypovolemic shock, including damage to the kidneys and brain due to decreased blood flow. Fortunately, neurogenic shock is usually rather easy to treat. High spinal anesthesia shock can be treated with a vasopressor to produce peripheral vasoconstriction; acute gastric dilatation can be relieved by insertion of a nasogastric tube. In milder forms of neurogenic shock, as in fainting from emotional stimuli, merely removing the patient from the source of the unpleasantness may be sufficient. Similarly, relieving acute pain usually corrects the shock state due to this cause. The patient in neurogenic shock may respond favorably to the head-down position.

Of course, when neurogenic shock is due to injury, its treatment becomes more complicated. If due to spinal cord transsection with associated loss of blood and cerebrospinal fluid in the injured area, it becomes necessary to balance fluid replacement with vasopressor therapy.

SEPTIC SHOCK Septic shock is the result of peripheral pooling of blood in capacitance (primarily venous) vessels, causing decreased effective circulating blood volume without actual blood loss. This form of shock is most often due to gram-negative bacteria sepsis; other causative organisms may include gram-positive bacteria and—far less often—viruses, parasites, and fungi. The highest incidence of septic shock occurs during the seventh and eighth decades of life.[19] It is more likely to occur in patients debilitated from other conditions, such as malignancies, diabetes mellitus, uremia, and cirrhosis. The mortality rate remains in excess of 50%.[19] Despite improved antimicrobial therapy, there has been an increase in the incidence of gram-negative sepsis; this increase has been attributed to more extensive operations on elderly and poor-risk patients; the more frequent use of steroids, immunosuppressive, and anticancer drugs; and the widespread use of antibiotics, with the resultant development of more virulent and resistant organisms.[19] Gram-positive infections are less apt to cause shock because they can be more effectively controlled by antibiotics.

Gram-negative septic shock Gram-negative sepsis as a cause of shock is more serious than that seen with gram-positive sepsis. The most common causative gram-negative organisms are *Escherichia coli, Klebsiella aerobacter, Proteus, Pseudomonas,* and *Bacteroides.*[20] The most frequent source of gram-negative infections is the genitourinary system; many patients have had an associated operation or instrumentation of the urinary tract. The second most frequent site of origin is the respiratory system. This form of shock has also been seen following peritonitis, biliary tract infections, burns, septic abortions, and postpartum infections. Indwelling venous catheters for hyperalimentation and monitoring are used more frequently than in the past and are an increasing source of contamination. (Nursing responsibilities in caring for central catheters are described in Chapter 6.)

Clinical manifestations Symptoms include a temperature above 101° F (38.3° C) and chills. The hemodynamic abnormalities that follow gram-negative shock are not completely understood but are probably initiated by endotoxins released from the cell walls of the gram-negative bacteria.[20]

The development of mild hyperventilation (with respiratory alkalosis) and change in sensorium may be the earliest signs of gram-negative sepsis; these symptoms may precede the usual symptoms by several hours to several days.[21] Although this is thought to be a primary response to bacteremia, the exact mechanism is unknown; when it occurs, a source of infection should be sought.

Clinically, the shock state in gram-negative sepsis may be characterized by hypotension, peripheral vasoconstriction, and cold, clammy extremities; earlier in its course, however, there may have been warm, dry extremities. Patients may be of two types: those with normal blood volume prior to onset of shock (e.g., a young, previously healthy woman with septic abortion) and those with hypovolemia prior to onset of shock (e.g., a patient with intestinal obstruction and an associated severe fluid volume deficit). These two types manifest different hemodynamic patterns. If therapy to combat sepsis is delayed or is unsuccessful, the patient may develop circulatory failure and resistant metabolic acidosis.

Generally the white blood cell (WBC) count is elevated; however, it may be normal or low in debilitated patients, in patients on immunosuppressive drugs, or in patients with overwhelming sepsis. There is usually a left shift in the WBC differential.

Pulmonary insufficiency is often seen in patients with septic shock. Early in the course of shock, one may note mild hypoxia with compensatory hyperventilation and respiratory alkalosis. Later, there may be rapid deterioration of pulmonary function with severe hypoxemia and resultant metabolic acidosis.

Treatment The best treatment of septic shock is *prevention*; that is, treating the infection before the onset of shock. Use of appropriate antibiotics is mandatory, as is the debridement or drainage of the site of the infection when possible. Adjunctive forms of therapy include fluid replacement and steroid administration as indicated.

Fluid replacement varies with the patient's underlying disease process. If the patient has incurred "third-space" losses (trapped fluid not immediately accessible for body use) from peritonitis, burns, or bowel obstruction, the replacement fluid should be a balanced electrolyte solution. Enough fluid must be given to maintain an effective circulating volume; however, care must be taken to avoid fluid overload. This is best accomplished by monitoring at least the CVP; if function of the left side of the heart is in question, a Swan-Ganz catheter should be inserted for measurements of pulmonary artery (PA) and pulmonary capillary wedge (PW) pressure. Many patients show improvement after appropriate fluid replacement and antibiotic therapy.

The use of corticosteroids in the treatment of septic shock is controversial. Although there is no direct evidence that steroids are helpful in these cases, some improvement in cardiac, pulmonary, and renal function has been reported. Some physicians elect to use steroids only if the patient fails to respond adequately to antibiotics and fluid therapy.

Since many patients with sepsis may develop significant pulmonary insufficiency, the use of a controlled airway (such as an endotracheal tube) and assisted ventilation may be required. (ARDS is discussed in Chaper 14.)

Fever in the septic patient can contribute to the deleterious cellular metabolic effects of shock and thus must be treated; the patient should be kept at normal temperature by use of a hypothermic blanket and administration of aspirin or acetaminophen.

CARDIOGENIC SHOCK Cardiogenic shock is due to failure of the heart as a pump, with resultant inadequate output and inadequate tissue perfusion in spite of normal blood volume. Causes may include primary myocardial dysfunction (as in myocardial infarction or serious arrhythmias); mechanical restriction of cardiac function or venous obstruction, such as occurs in cardiac tamponade, tension pneumothorax, or vena cava obstruction; and myocardial depression (resulting from a variety of causes).

Management of postoperative patients with this type of hypotension must be accompanied by accurate hemodynamic monitoring. (Nursing responsibilities in hemodynamic monitoring are discussed in Chapter 11.) Typically, the classic findings in cardiogenic shock are a CVP that is elevated or rises briskly with fluid administration and a depressed cardiac output that fails to respond to fluid administration. Primary treatment of cardiogenic shock must be directed at the cause; for example, cardiac tamponade should be relieved by pericardiocentesis. (The reader is referred to Chapter 11 for a more thorough discussion of cardiogenic shock.)

REFERENCES

1. Condon R, Nyhus L: Manual of Surgical Therapeutics, 4th ed, p 283. Boston, Little, Brown & Co, 1978
2. Risser N: Preoperative and postoperative care to prevent pulmonary complications, Heart Lung, Jan–Feb 1980 p 57
3. Feinberg L: Perioperative care of patients with cardiac disease. Postgrad Med, Feb 1980, p 27
4. *Ibid*, p 228

5. Luce J: Preoperative evaluation and perioperative management of patients with pulmonary disease. Postgrad Med Jan 1980, p 203

6. Shires T, Canizaro P: Fluid, electrolyte, and nutritional management of the surgical patient. In Schwartz S, Shires G, Spencer F, Storer F (eds): Principles of Surgery, 3rd ed, p 83. New York, McGraw-Hill, 1979

7. Schumer W, Nyhus L: Treatment of Shock: Principles and Practice, p 23. Philadelphia, Lea & Febiger, 1974

8. Shires T, Shires G, Canizaro P, Carrico J: Shock. In Schwartz S, Shires G, Spencer F, Storer, F (eds): *Principles of Surgery, 3rd ed,* p 137. New York, McGraw-Hill, 1979

9. Schumer & Nyhus, p 35

10. Shoemaker W, Hauser C: Critique of crystalloid versus colloid therapy in shock and shock lung. Crit Care Med 7, No. 3:119, 1979

11. Shires et al: p 166

12. *Ibid,* p 165

13. Gann D: Endocrine and metabolic responses to injury. In Schwartz S, Shires G, Spencer F, Storer F (eds): Principles of Surgery, 3rd ed, p 33. New York, McGraw-Hill, 1979

14. Condon & Nyhus, p 9

15. Schumer & Nyhus, p 41

16. Shires et al, p 151

17. Shoemaker & Hauser, p 117

18. Maxwell M, Kleeman C: Clinical Disorders of Fluid and Electrolyte Metabolism, 3rd ed, p 765. New York, McGraw-Hill, 1980

19. Shires et al, p 174

20. *Ibid,* p 175

21. *Ibid,* p 176

BIBLIOGRAPHY

Byyny R: Preventing adrenal insufficiency during surgery. Postgrad Med, May 1980

Elwyn D: Nutritional requirements of adult surgical patients. Crit Care Med 8, No 1, January 1980

Metheny N: Preoperative fluid balance assessment. AORN J 33, No 1, January 1981

Burns cause a series of major water and electrolyte changes. The purpose of this chapter is to explore these changes and their implications for nursing care. A background discussion of physiologic changes accompanying burns precedes the discussion of treatment and nursing care.

EVALUATION OF BURN SEVERITY

The severity of water and electrolyte changes is largely dependent on the *burn depth* and the *percentage of body surface* involved.

BURN DEPTH

Burns may be classified as first, second, or third degree, according to the depth of skin damage. A more recent classification of burns is partial-thickness and full-thickness: deep partial-thickness burns extend into the dermis and full-thickness burns extend into the subcutaneous tissue or deeper. Factors considered in determining burn depth (Table 8-1) include the following:

Nature of burning agent plus length of exposure to it

Appearance of the burned surface

"Feel" of the burned surface with the gloved hand

Amount of sensation remaining

First-degree burns are characterized by simple erythema, with only microscopic destruction of superficial layers of the epidermis. This type of injury is of little clinical significance since the skin's water barrier is not disturbed. First-degree burns are not considered in the planning of fluid replacement.

Second-degree burns extend through the epidermis into the dermis; this type of injury is characterized by damaged capillaries and the appearance of blebs containing fluid. Some skin elements remain viable, from which epithelial regeneration can occur.

In third-degree (full-thickness) burns, there is destruction of all skin elements, spontaneous regeneration of epithelium is not possible. Clinically the burn appears dry, hard, inelastic, and translucent, with thrombosed veins.

8

Fluid Balance in the Badly Burned Patient

TABLE 8-1 *Diagnosis of burn depth*

	Degree	Nature of burn	Symptoms	Characteristics	Course
Epidermal burn	First	Sunburn	Tingling Hyperesthesia Painful Soothed by cooling	Reddened, blanches with pressure	Complete recovery within a week Peeling
Intradermal burn (partial-thickness)	Second	Scalds (spills) Flash flame	Painful Hypesthesia Sensitive to cold air	Blistered, mottled red base, broken epidermis, weeping surface Feels moist, soft and pliable to touch Edema Hair stubble visible	Recovery in 2–3 weeks Some scarring and depigmentation
Subdermal burn (full-thickness)	Third	Fire Electrical Scalds (immersion) Chemical contact	Painless Shock	Dry; pale white or charred Feels hard and dry to touch Edema Hair absent	Eschar (nonviable skin) sloughs Grafting necessary Scarring and loss of contour and function

PERCENTAGE OF SURFACE INVOLVED

The "rule of nines" is commonly used to estimate the severity of burns in adults. It divides the body surface into areas of 9% or its multiples:

Head	=	9%
Each arm	=	9%
Each leg	=	18%
Front of torso	=	18%
Back of torso	=	18%
Genitalia	=	1%

The proportion of each of these areas with either second- or third-degree burns is estimated; summation of these estimates represents the percentage of total body surface area burned. Unless used cautiously, this method can result in dangerously high estimates. A more detailed breakdown of body surface areas is beneficial, since it contributes to a more complete understanding of percentage of surface involved (Table 8-2).

After the percentage of partial-thickness and full-thickness burns is estimated, the therapeutic approach is planned. Factors other than the percentage of burned body surface area can influence the severity of the burn. For example, burns are made more serious by the presence of prior renal, cardiac, or metabolic disorders; concurrent injuries; burns of the face, hands, or genitalia; respiratory burns; and extreme age variation (very young children and the elderly). Mortality associated with burns has been markedly reduced in the past 25 years. In several series, 50% of patients between the ages of 8 and 45 have been reported to survive burns of 75% of the total body surface. However, persons over the age of 65 with burns of more than 25% total body surface have a guarded prognosis.[1] (Recall that chronic disease states in the aged interfere with appropriate physiologic responses to major burns).

WATER AND ELECTROLYTE CHANGES IN BURNS

Body fluids are lost in severe burns in several ways:
Plasma leaves the intravascular space and becomes trapped as edema fluid
Plasma and interstitial fluid are lost as exudate
Water vapor is lost from the denuded burn site
Blood leaks from the damaged capillaries

Sodium and water shifts into the cell ("sick-cell" syndrome)

PLASMA-TO-INTERSTITIAL FLUID SHIFT Intravascular water, electrolytes, and protein are lost through damaged capillaries at the burn site. Clinically, the shift results in edema; the magnitude of this shift depends upon the burn depth and the percentage of surface area involved. Consider a specific example: an adult with a body surface of 1.75 m² has about 10.5 liters of extracellular fluid (ECF). If he sustains a 50% burn, the volume of edema fluid formed during the first day or two would approximate 5.25 liters, a quan-

TABLE 8-2 *Proportional contribution of body parts to total surface area: variations with age*

Body part	Percentage of body surface					
	Age in years					
	Birth	1	5	10	15	Adult
Head	19	17	13	11	9	7
Neck	2	2	2	2	2	2
Anterior trunk	13	13	13	13	13	13
Posterior trunk	13	13	13	13	13	13
Right buttocks	2.5	2.5	2.5	2.5	2.5	2.5
Left buttocks	2.5	2.5	2.5	2.5	2.5	2.5
Genetalia	1	1	1	1	1	1
Right upper arm	4	4	4	4	4	4
Left upper arm	4	4	4	4	4	4
Right lower arm	3	3	3	3	3	3
Left lower arm	3	3	3	3	3	3
Right hand	2.5	2.5	2.5	2.5	2.5	2.5
Left hand	2.5	2.5	2.5	2.5	2.5	2.5
Right thigh	5.5	6.5	8	8.5	9	9.5
Left thigh	5.5	6.5	8	8.5	9	9.5
Right leg	5	5	5.5	6	6.5	7
Left leg	5	5	5.5	6	6.5	7
Right foot	3.5	3.5	3.5	3.5	3.5	3.5
Left foot	3.5	3.5	3.5	3.5	3.5	3.5

(Modified from Lund C, Browder N: The estimation of area of burns. Surg Gynec Obstet 79:352, 1944)

tity exceeding the total plasma volume of the patient. Obviously all of the edema fluid is not derived from plasma; some of it comes from the body cells and some from administered fluids.

Proportionately greater amounts of water and electrolytes than of protein are lost from the plasma. Protein molecules are larger and, thus, fewer escape through the damaged capillaries. As a result, the circulating plasma protein becomes more concentrated, the increased osmotic pressure draws fluid from undamaged tissues in all parts of the body, and generalized tissue dehydration results. Sometimes it is difficult to visualize the presence of severe dehydration in a patient so obviously edematous; one must remember that the edema represents trapped fluids unavailable for body use. The increased capillary permeability lasts for approximately 24 hours; at this time, the capillary walls seal and compensatory changes set in.[2]

BURN EXUDATE A protein-rich fluid is lost through the leakage of approximately equal parts of plasma and interstitial fluid from the burned surface. Visible fluid loss by way of the surface is mostly limited to second-degree burns, and the amount of fluid lost in this manner is proportional to the percentage of second-degree burns. Such losses do not increase appreciably in burns involving 50% or more of the body. Burn exudate has approximately two-thirds as much protein as plasma.

WATER VAPOR AND HEAT LOSS The intact skin serves as a barrier against the loss of water and heat. When skin is destroyed by a burn, increased water and heat loss result, and the larger the burned surface, the greater the loss of water vapor and heat. Burned infants are particularly vulnerable to heat loss.

The average water vapor loss from a major burn wound is thought to be from 2.5 liters to 4.0 liters per day. Maintenance of an environment saturated with water vapor decreases the transcutaneous water loss; for example, wounds covered with dressings of aqueous topical agents and dry covers lose only half the amount of water vapor lost through exposed wounds. Some physicians feel that early coverage of burn wounds with a temporary biologic dressing (such as pig skin) decreases evaporative water and heat loss. Patients treated by exposure often shiver and complain of feeling cold; external heat sources should be provided for comfort. Ideally, the room temperature should be 82° to 84° F (28° to 29° C). Shivering increases the rate of metabolism, as do fever and infection (causing increased loss of water from the body).

WATER AND ELECTROLYTE CHANGES IN MAJOR BURN PHASES The nurse should be aware of the water and electrolyte changes occurring in burns so that she can recognize significant changes in the patient. Observations are more meaningful when one has at least some idea of what to look for. An outline of expected water and electrolyte changes is presented in Table 8-3. (The reader is referred to earlier chapters for detailed descriptions of these imbalances.)

PHYSIOLOGIC BASIS FOR TREATMENT AND NURSING CARE DURING THE FLUID ACCUMULATION PHASE

The adequacy of early burn treatment largely depends upon the physician's and nurse's understanding of physiologic derangements caused by burns, the organization of equipment, and the ability to act quickly and skillfully.

TABLE 8-3 *Water and electrolyte changes in different burn phases*

Phase	Water and electrolyte changes	Comments
Fluid accumulation phase (shock phase) First 48–72 hours in most patients; may last longer in the elderly	Plasma-to-interstitial fluid shift (edema at burn site)	Plasma leaks out through the damaged capillaries at the burn site—edema forms
	Generalized dehydration	Undamaged tissues give up fluids to help increase plasma volume—part of it leaks through the damaged capillaries and helps form edema
	Contraction of blood volume	Loss of plasma causes a decreased circulatory volume; a rise in pulse rate ensues
	Decreased urinary output	Secondary to: Decreased renal blood flow Increased secretion of ADH Sodium and water retention caused by stress (increased adrenocortical activity) Severe burns may cause hemolysis of red blood cells; the ruptured cells release Hb and it is excreted by the kidneys (hemoglobinuria can cause severe renal damage)
	Cellular uptake of sodium and water	The so-called sick-cell syndrome (an alteration that burn injuries share with other major, nonthermal forms of trauma)[2]
	Potassium excess	Massive cellular trauma causes the release of K^+ into the extracellular fluid (recall that most of the body's K^+ is located *inside* the cells; only small amounts are tolerated in plasma)
	Sodium deficit	Large amounts of Na^+ are lost in the trapped edema fluid and in the exudate (recall that Na^+ is the chief extracellular ion and large amounts are lost when extracellular fluid is lost)
		Research work done under the direction of C. A. Moyer indicates sodium deficit in unburned tissues is closely linked with burn shock (partial correction of Na^+ deficit relieved burn shock symptoms in some patients even though a substantially decreased blood volume persisted)
	Metabolic acidosis (base bicarbonate deficit)	Develops within a few hours due to: Accumulation of fixed acids released from injured tissues Ineffective tissue perfusion
	Calcium deficit	May occur after 12–24 hours in extensively burned patients—may be particularly marked in infants and children (presumably due to saponification of fat in the burned subcutaneous adipose tissue)
	Hemoconcentration (elevated hematocrit)	Relatively greater loss of liquid blood components in relation to blood cell loss

(continued)

TABLE 8-3 *Water and electrolyte changes in different burn phases* (continued)

Phase	Water and electrolyte changes	Comments
Fluid remobilization phase (stage of diuresis) Starts 48–72 hours post burn	*Interstitial fluid-to-plasma shift*	Edema fluid shifts back into the intravascular compartments
	Hemodilution (hematocrit decreased)	The blood cell concentration is diluted as fluid enters the vascular compartment and, in addition, at this time a decrease in the number of cells becomes evident (destruction of red blood cells at the burn site causes anemia—as much as 10% of the total number of RBCs may be destroyed)
	Increased urinary output	Fluid shift into the intravascular compartment, increases renal blood flow and causes increased urine formation
	Sodium deficit	Sodium is lost with water when diuresis occurs
	Potassium deficit (may occasionally occur in this phase)	Beginning about the fourth or fifth postburn day, K^+ shifts from the extracellular fluid into the cells
	Metabolic acidosis	
Convalescent Phase	*Calcium deficit*	Since calcium may be immobilized at the burn site in the slough and early granulation phase of burns, symptoms of calcium deficit may occur (recall that for some unknown reason calcium rushes to damaged tissues)
	Potassium deficit	Extracellular K^+ moves into the cells, leaving a deficit of K^+ in the extracellular fluid
	Negative nitrogen balance (present for several weeks following burns)	Secondary to: Stress reaction Immobilization Inadequate protein intake Protein losses in exudate Direct destruction of protein at the burn site
	Sodium deficit	

NEED FOR EARLY TREATMENT

The shift of fluid from plasma to the interstitial space is rapid and well underway by the end of the first hour. The maximal speed of edema formation is reached by the end of the first 8 to 10 hours; the shift continues until the 36th to 48th hour. By this time, the capillaries have healed sufficiently to prevent further fluid loss. The decreased plasma volume can lead to hypovolemic shock and renal depression caused by decreased renal blood flow unless quickly corrected by fluid replacement therapy. Oliguria or anuria are particularly threatening during this phase because of the excessive amounts of potassium flooding the ECF. Potassium is mainly excreted in urine; decreased urinary output causes a dangerous excess to build up in the bloodstream. The sodium deficit requires prompt attention as does the acidosis so frequently present. Acute tubular necrosis may result from hypovolemia, hemoglobinuria, or myoglobinuria and is an entirely preventable complication. Prevention consists of adequate fluid replacement and, when necessary, the administration of mannitol to flush out the tubules.

INITIAL PATIENT EVALUATION

There are pertinent questions the nurse should consider when the patient is first seen:

1. When did the burn occur? The degree of fluid shift is related to the length of time the burn has been present.
2. What was the nature of the burning agent? (Notice in Table 8-1 that burns are classified in relation to burning agents frequently associated with them.)
3. What was the length of exposure to the burning agent? (Questions 2 and 3 are intended to help the physical establish the burn depth—appearance of the burns on admission is often misleading.)

4. Were any medications given prior to hospital admission? (Sometimes narcotics are given at the scene of the accident; it is important to avoid repeating drugs too soon, especially if respiratory tract burns or shock are present.)

5. Was the burn sustained in an enclosed area where heat and fumes were inhaled? This question is highly significant in establishing the likelihood of respiratory burns.

6. Are there any preexistent illnesses, such as cardiac or renal damage or diabetes, that will require therapy in addition to burn treatment? Failure to ascertain the presence of such illnesses is not uncommon in the initial rush.

7. What is the normal preburn weight? The preburn weight is a baseline for comparison for later weight changes; the weight is also instrumental in determining drug doses.

8. Is pain present? If so, how severe? It should be remembered that severe pain can cause a drop in blood pressure and further complicate the patient's condition.

9. Is the patient known to have any drug allergies?

10. What is the status of tetanus immunization? All patients with severe burns should receive prophylaxis against tetanus.

11. Are there any associated injuries requiring treatment (e.g., head injuries or fractures)?

While asking these questions, the nurse can be busy with other activities, such as readying fluid equipment, removing loose clothing not stuck to the burns, and removing constrictive jewelry before edema becomes severe. The unit that is to receive the patient after initial treatment should be notified so that every necessary preparation can be made.

The physician will usually request that blood be drawn for determination of electrolytes, hemoglobin, hematocrit, glucose, and urea nitrogen. Arterial blood may be drawn for determination of pH, P_{CO_2}, and P_{O_2}.

OBSERVING FOR BURN SHOCK

The nurse should be particularly alert for symptoms of burn shock:

Extreme thirst, caused by generalized cellular dehydration

Increasing restlessness; persistent slight to moderate restlessness may be due to apprehension and discomfort produced by the burn rather than to inadequate fluid resuscitation

Sudden high pulse rate; heart beats faster to compensate for decreased blood volume

Respiratory rate is often increased, but the character of breathing should be normal unless there are complicating factors

Blood pressure, when measurable, is either normal or low (the body may be too extensively burned to permit application of a blood pressure cuff). The supine burned patient can often tolerate a large fluid volume deficit with little or no change in blood pressure; however, upon sitting or standing, hypotension and even syncope may occur—thus, the badly burned patient should be left supine for at least the first 48 to 72 hours

Cool pale skin in unburned areas indicative of cutaneous vasoconstriction—a compensatory mechanism that helps preserve normal blood flow

Oliguria due to decreased renal blood flow and to increased levels of antidiuretic hormone (ADH) and aldosterone

Delirium or coma; presumably these symptoms are due to inadequate cerebral blood flow and are serious signs

Seizures sometimes occur; may be the result of cerebral ischemia

Fortunately, burn shock is slow in onset and can be prevented or corrected by the intravenous administration of fluids in sufficient volume to maintain an adequate urinary output.

INITIAL URINE OBSERVATIONS

An indwelling urinary catheter is usually inserted when burns involve 20% or more of the body surface; the catheter should be connected to a device for hourly measurement of the urine flow. If the patient voids before the catheter is inserted, the urine should be measured and saved. The urine should be observed for discoloration resulting from blood; if present, the patient probably has severe third-degree burns. Hemolysis of red blood cells at the burn site (caused by trauma) causes the release of hemoglobin and, consequently, hemoglobinuria. Thermal trauma (particularly electrical burns) may also destroy muscle tissue, releasing myoglobin into the bloodstream to be filtered out by the kidneys.

INITIAL OBSERVATIONS AND TREATMENT OF RESPIRATORY BURNS

The nurse should observe the patient closely for symptoms of respiratory burns, which include the following:

Increased respiratory rate

Singed nasal hair

Burns around mouth

Hoarseness

Red painful throat

Dry cough

Moist rales and dyspnea

Stridor

Bronchospasm with prolonged wheezing

Pulmonary complications may be noted immediately in the burned patient, or detection may be delayed until swelling of upper respiratory structures exposed to hot gases occurs. Burns about the face and neck may produce sufficient edema to embarrass respiration; soft tissue edema is maximal between 24 to 48 hours. Upper airway obstruction is usually heralded by tachypnea, progressive hoarseness, and increased difficulty in clearing bronchial secretions (due to vocal cord swelling). In these instances, an airway *must* be established. If possible, an endotracheal tube is inserted; if not, a tracheostomy is performed.

The head of the bed should be elevated (unless contraindicated) and the patient should be encouraged to deep breathe and cough every 20 minutes. It also is important that the patient be turned at least hourly. Fluid should be suctioned frequently from the respiratory tract to prevent its accumulation. In addition, humidifiers and bronchodilators are used to loosen secretions and improve ventilation; oxygen should be administered to decrease anoxia. Use of mechanical ventilation may be necessary.

Prophylactic antibiotics are given to prevent infection of the lungs and consequent increased edema. Some physicians prescribe intravenous corticosteroids to decrease the pulmonary inflammatory reaction. Lastly, parenteral fluids are administered cautiously to avoid overloading the circulatory system and causing pulmonary edema. The desired urinary volume is slightly less when respiratory tract burns are present.

If the patient has sustained excessive exposure to smoke, carbon monoxide poisoning may result. Carbon monoxide poisoning is manifested by hypoxia, which may range from pronounced tachypnea to respiratory arrest. Diagnosis is confirmed by analyzing the carboxyhemoglobin concentration in the blood. Treatment consists of the administration of 100% oxygen and ventilatory support if needed.

It should also be noted that a circumferential burn of the thorax may interfere with respiratory excusions.

INTRAVENOUS FLUID REPLACEMENT

Intravenous fluids are lifesaving in the treatment of moderate and severe burns. Burns involving 20% or more of the body surface (less in the very young or the aged) usually require intravenous fluid therapy. The aim of early fluid therapy is to give the least amount of fluids necessary to maintain the desired urinary output and keep the patient relatively free of burn shock symptoms.

KINDS OF INTRAVENOUS FLUIDS USED IN BURN TREATMENT Physicians are of widely varied opinions about the kinds of intravenous fluids to be used in burn treatment. It is generally agreed, however, that the keystones of therapy are sodium and water.[3] Many physicians favor beginning treatment with a balanced salt solution, such as lactated Ringer's (LR) solution. It not only supplies sodium, it helps to correct the metabolic acidosis associated with burns: 1 liter of LR contains 130 mEq of sodium and 28 mEq of lactate (a bicarbonate precursor). Isotonic saline (0.9% sodium chloride) can also be used to supply sodium: 1 liter of isotonic saline contains 154 mEq of sodium and 154 mEq of chloride. It has the disadvantage of supplying an excessive amount of chloride ions, which contribute to the acidosis.

The use of colloids (plasma or plasma substitutes) is advocated by some physicians who reason that plasma or other colloids must be given to replace the plasma leaked through injured capillaries into the burn site. Excessive administration of salt solutions (such as LR) further contributes to dilution of the colloidal content of plasma; the decrease in colloidal pressure contributes to edema formation. Although many physicians usually incorporate colloids into the fluid regimen, the exact timing of their administration cannot be set rigidly. They are generally prescribed after the first 24 hours of fluid therapy.[3] Burke states that colloids should be administered before the plasma albumin drops to, or below, 2 g/dl.[4] Plasma contains sodium (sometimes as much as 200 mEq/liter) and thus helps correct sodium deficit. However, it should be remembered that the use of plasma is associated with the risk of serum hepatitis (see Chapter 6). A frequently used colloid, Plasmanate, is a plasma protein fraction that has good expansion properties without the disadvantage of possible viral hepatitis transmission (heat-treated to destroy hepatitis virus). Albumin is also an often-used colloid in burn treatment.

Blood is usually not included in burn fluid therapy unless the patient has been bleeding from an associated injury; it should be given if the hematocrit is less than 30%.[3]

Hypertonic albuminated fluid-demand resuscitation (HALFD) is advocated by Jelenko.[5] This fluid regimen consists of the administration of a hypertonic albuminated salt solution. The fluid contains 240 mOsm of sodium and 120 mOsm each of chloride and lactate; to that is added 1.5 g of fresh albumin/liter.

The use of a hypertonic lactated solution (HLS) is advocated by some physicians for the treatment of burned patients. One liter of HLS contains 250 mEq of sodium, 100 mEq of lactate, and 150 mEq of chloride; it supplies sodium in a minimal fluid load, causing a sustained hypernatremia and increase in serum osmolality. Proponents of this therapy feel that it is associated with less tissue edema and fewer pulmonary complications than seen with other forms of treatment.[6]

Some physicians favor the use of dextran solutions as plasma volume expanders in the treatment of burns. (Dextran is discussed in Chapter 6.)

A nonelectrolyte solution, usually 5% dextrose in water, may be used to replace the normal insensible water loss and the increased transcutaneous water vapor loss. Although free water is lost from the burn site, it is generally not replaced in the first 24 hours of treatment, since during this early phase, water contributes little to the maintenance of normal cardiovascular function. (Note that the formulas listed in Table 8-4 allow for its replacement in the second 24 hours of treatment.)

BURN FORMULAS FOR PLANNING INTRAVENOUS FLUID THERAPY Whenever intravenous fluids are given there is a danger of giving too much or not enough; both hazards are always present in burn therapy. To serve as a guide for the amount to be given to burned patients, numerous formulas have been devised. Several of the more commonly used formulas are present in Table 8-4.

MONITORING FLUID REPLACEMENT THERAPY It is essential to remember that any formula serves only as a *guide* because there are many variations among individual patients; frequent clinical assessment is mandatory to adequately tailor fluid replacement to individual needs.

Successful fluid resuscitation of the burned patient is signaled by the following:
- Adequate urine volume
- Normal sensorium
- Near normal vital signs
- Normal central venous pressure (CVP)

(These criteria are discussed below.)

Urine volume Hourly urine volume by means of a Foley catheter is generally considered to be the best clinical sign of the adequacy of fluid resuscitation. The ideal range is thought to be 30 ml to 50 ml/hr in the adult and 1 ml/kg/hr for a child weighing less than 30 kg.[3] Commercial devices are available for measurement of small urine volumes. When dealing with small urine volumes, any error can be significant. Absent or decreased urinary output can be due to inadequate fluid replacement, gastric dilatation, or renal failure. Remember that a clogged catheter may falsely indicate oliguria. Patency of the catheter should be checked before assuming oliguria is present. Extreme care should be taken to record accurately the amount of fluid instilled as an irrigant and the amount of fluid removed. If the oliguria is due to inadequate fluid replacement, the urine volume will increase when the fluid flow rate is increased. If it is due to renal failure, the output will remain small. In addition to oliguria, acute renal failure will manifest itself as a low urine osmolality.

It should be noted that patients who have suffered a hypovolemic insult from burns (not sufficient to produce acute tubular necrosis) may develop a transient polyuric renal failure.

The physician should indicate the desired urinary volume plus the variations in either direction to be reported. The desired urinary volume should be realistic and approximate the minimum, not the maximum. Attempts to increase fluid input sufficiently to cause large urine volumes in the aged or the very young are dangerous.

TABLE 8-4 *Formulas for use in planning intravenous fluid replacement in burned patients*

Name	Type and volume of fluid First 24 hours	Second 24 hours	Special monitoring recommendations
Parkland	Ringer's lactate: 4 ml/kg/% burn one half in first 8 hr, one half in next 16 hr	Colloid: 0.5 ml/kg/% burn D₅W 2000 One half of first day's Ringer's lactate*	Urine output 30–50 ml (adults) 1 ml/kg/hr (children <3 yrs) 15–25 ml/hr (children >3 yrs)
Hypertonic resuscitation	Na⁺ 250 mEq/liter Cl⁻ 150 mEq/liter Lactate 100 mEq/liter	D₅W	Adjust formula to urine output 30–40 ml/hr (adults), mental state, peripheral capillary filling, vital signs
New Brooke	Ringer's lactate: 2–3/kg/% burn one half first 8 hr, one half next 16 hr	D₅W one half of first 24-hr total + colloid 0.3–0.5 ml/kg/%	Urine output 30–50 ml/hr
Hypertonic albuminated (Jelenko)	1000 ml 5% D₅W and 120 mEq NaCl 120 mEq Na lactate 12.5 g albumin	Most patients complete resuscitation 24 hr; no recommendations for second 24 hr	Infuse so that mean arterial BP is 60–110 mm Hg, urine output 30–50 ml/hr

*Dr. Baxter no longer recommends the use of Ringer's lactate on the second day.
(Munster A: *Burn Care for the House Officer,* p 19. Baltimore, Williams & Wilkins, 1980)

Lastly, *accurate* recording of intake–output is a necessity for assessing the patient's fluid balance status during burn treatment.

Sensorium It is important to assess the patient's sensorium frequently, since this is one measure of the adequacy of cardiovascular functioning. Is he alert and lucid? Does he respond in a normal fashion? With adequate fluid replacement, sensorium should remain normal unless other known factors (e.g., head injury, drug intoxication, carbon monoxide poisoning, arterial hypoxia) preclude this.

Vital signs Vital signs should be taken at least hourly in the newly burned patient. Blood pressure should be near normal; as a general guideline, a systolic pressure less than 100 mm Hg or greater than 200 mm Hg should be reported to the physician.[7] If the blood pressure cannot be easily measured by cuff owing to peripheral edema, an arterial line should be considered. One should, of course, remember that the patient's baseline "normal" blood pressure must be considered when evaluating his current status.

Temperature is usually slightly elevated in the burned patient; as a general guideline, a reading greater than 101.5° F (38.6° C) should be reported to the physician.[7]

Tachycardia is common in the burned patient but should diminish with fluid replacement. As a general guideline, a pulse greater than 160/min or less than 60/min should be reported to the physician.[7] Peripheral pulses in the burned extremities should be checked every few hours; it should be remembered, however, that overlying soft-tissue edema may prevent palpation of underlying arteries.

Capillary filling of unburned skin Capillary refill time of unburned skin should be carefully observed. Warm, pink skin that displays a normal capillary filling time after blanching is a sign of a physiologically intact circulation.[8] One must remember that vasoconstriction of unburned skin is a normal compensatory response to help preserve normal blood flow during the *early* hours of a severe burn, causing unburned skin to be cool and pale during this period. The nail beds are good sites for observing capillary refill time.

Central venous pressure The CVP should remain within normal limits in the adequately treated burned patient. CVP measurement is frequently used to monitor the effects of intravenous fluid replacement therapy, particularly in infants, the aged, and patients with cardiac and renal disease. It is generally indicated when the burned surface is greater than 50% or 60%. The margin for therapeutic error is small when the patient has deep extensive burns. Frequent checks of venous pressure allow more aggressive fluid replacement therapy without the risk of circulatory overload. Normal venous pressure is from 4 cm to 10 cm of water, and a level of 15 cm to 20 cm represents a significant elevation. When venous pressure becomes elevated above a point designated by the physician, the fluid infusion rate should be curtailed. If the patient is responding poorly to fluid replacement therapy, a Swan-Ganz catheter should be considered to measure pulmonary artery pressure, wedge pressure, and cardiac output.

In summary, the nurse must be alert for symptoms of inadequate or excessive fluid administration. Inadequate fluid therapy in burned patients is indicated by the following:

Decreased urinary output

Thirst

Restlessness and disorientation

Hypotension and increased pulse rate

(Burn shock is described earlier in this chapter.)

The following are indicative of circulatory overload:

Elevated CVP

Shortness of breath

Moist rales

Increased blood pressure

ORAL ELECTROLYTE SOLUTIONS

Oral electrolyte solutions may be the sole source of fluids for the patient with minor burns and may be used in conjunction with intravenous fluids in moderately burned patients. Oral fluids should not, of course, be administered if the gastrointestinal tract is not functional. (Recall that paralytic ileus, gastric dilatation, and nausea are frequent complications of severe burns.) Some physicians prefer not to give fluids orally for the first day or two or until bowel sounds are heard; fluids are offered slowly at first and increased as tolerated.

Thirst is an early symptom following burns. The patient permitted unlimited quantities of plain water is in danger of developing water intoxication (sodium deficit) because of simple dilution.

The nurse must explain to the seriously burned patient that plain water (usually in the form of ice chips) should be taken in limited amounts, if at all. Physicians favoring the use of oral electrolyte fluids vary as to the exact contents and proportions of the solutions they recommend. One solution consists of 1 tsp of sodium chloride and ½ tsp of sodium bicarbonate in 1 liter of chilled water. Occasionally isotonic saline or sixth molar sodium lactate is used.

The solution should be prepared carefully; *errors between teaspoons and tablespoons can be serious.* Oral electrolyte solutions have a definite taste and some patients find them difficult to accept. Measures that help to make the solution more palatable include chilling it, making ice chips from it, flavoring it with lemon, or disguising it in juices (when allowed). Orange juice and other potassium-containing fluids should be withheld until renal function is established and the physician approves their use.

Oral solutions are not given in the presence of the conditions below:

- Mental confusion (there is danger of aspirating fluid into the lungs)
- Acute gastric dilatation

Gastric dilatation resulting from stress is not uncommon in burned patients. When it is present, fluids taken orally become trapped in the distended stomach. Thus, even though oral fluids are swallowed, they are not available for body use and urine formation.

OBSERVATIONS FOR GASTROINTESTINAL COMPLICATIONS

Reflex paralytic ileus will develop some time during the first 24 postburn hours in most patients with more than 20% total body surface area burns. Usual symptoms include nausea, effortless vomiting, hiccoughing, and abdominal distention. Vomiting carries a high risk of tracheal aspiration and must be prevented. Some physicians favor placing a nasogastric tube in all patients with major burns in order to decompress the stomach until normal bowel sounds have returned. Another purpose of insertion of a nasogastric tube is to allow for the observation of gastric secretions. Patients with major burns are at risk of hemorrhagic gastritis due to stress. Bleeding from the stomach may manifest itself in gastric secretions or, if a slow ooze, by guaiac-positive stools and a gradual decrease in the hematocrit. Gastric aspirates should thus be observed frequently for signs of bleeding; antacids may be added through the nasogastric tube at hourly intervals to deter superficial erosions of the gastric mucosa. A common practice is to add an ounce of antacid through the tube every hour and clamp the tube for 30-minute intervals. The incidence of Curling's ulcers, a once common burn complication, has been markedly diminished over recent years due to the use of antacids, enteral feedings, and when indicated, cimetidine.

The presence of abdominal distention after the first few postburn days may indicate the presence of invasive wound sepsis, since ileus commonly occurs with this condition. Ileus due to wound sepsis may be associated with gastroduodenal perforation. In the past, 85% of the cases of upper gastrointestinal hemorrhage were associated with bacteremia.[9] The decreased incidence of gastrointestinal ulcers in burned patients has been attributed to reduced frequency of major sepsis, prophylactic use of antacids by means of nasogastric tubes, and improved provision of nutritional supplements (allowing for quicker healing of small eroded areas).

CONTROL OF PAIN

The amount of pain present varies with the depth of the burn, the extent of surface area involved, and the patient's pain threshold. Third-degree (full-thickness) burns are painless because the nerve endings are destroyed; however, pain is experienced around the periphery of third-degree burns where first-degree and second-degree burns are present. As a rule, the requirement for analgesics is inversely proportional to the depth of the initial injury.

After 48 hours, the requirements for analgesics are greatly diminished, except when wounds are being actively debrided during the waking state.[10] Burned patients complaining of severe pain may be given small doses of meperidine or morphine through the intravenous cannula put in place for fluid administration. Fluid resuscitation should be underway before analgesics are administered. The intravenous route assures rapid and dependable concentrations of the drug in the central nervous system. The subcutaneous or intramuscular route should *never* be used to administer

narcotics to a burned patient with circulatory collapse. Peripheral tissue perfusion is erratic when shock is present; thus, absorption of the drug may not occur or may be delayed. Failure to achieve the desired effect may prompt repeated dosess by the same route. When peripheral circulation is improved after fluid replacement, there may be a rapid absorption and cumulative overdosage of the narcotic, resulting in respiratory narcosis and depression. The administration of a large dose of an analgesic to an inadequately resuscitated patient in or near burn shock can be lethal.

An 18-year-old boy received superficial intradermal burns over 80% of his body in a flash fire and explosion at a motel. He was initially alert and oriented when seen by a local physician who prescribed no treatment. He was taken 25 miles by automobile to his family physician; by this time 4 hours had elapsed since his accident. He was given 100 mg of meperidine subcutaneously (but no intravenous fluids) and placed in an ambulance for transfer to St. John's Mercy Hospital—a distance of 75 miles. When he arrived, 8 hours after the accident, he was comatose and cyanotic, with infrequent, gasping shallow respirations. An endotracheal tube could not be inserted because of facial and cervical edema. Although tracheostomy and cannula phlebotomy were performed immediately, spontaneous breathing never resumed and the pupils remained fixed and dilated. There was persistent oliguria until his death in coma, 20 hours later. *(Comment: this patient had a potentially curable lesion; most of his burns were intradermal and shallow. Delayed resuscitation, together with an ill-advised dose of narcotic, led to his untimely and unnecessary death.)*[11]

It is important not to confuse the restlessness of burn shock with pain. A patient thrashing about in bed, without complaints of pain, may well be in burn shock. In this case, a narcotic is contraindicated; the physician usually orders an increased rate of parenteral fluid administration.

TOPICAL ANTIBACTERIAL AGENTS

Due to the disruption of skin integrity caused by burns, a loss of antibacterial protection occurs; as a result, burn wound infection poses a great threat to the thermally injured patient. A number of agents may be applied locally to burn wounds to prevent or control infection. Among those in use are the following:

Silver sulfadiazine (Silvadene)

Mafenide acetate (Sulfamylon)

0.5% Aqueous silver nitrate solution

While these local agents are often effective in controlling infection, they may, in varying degree, alter body electrolyte levels or cause other undesirable effects.

SILVER SULFADIAZINE Silver sulfadiazine (Silvadene) is a soft, white, water-miscible cream containing silver sulfadiazine in micronized form. This substance is bactericidal for a number of gram-negative and gram-positive bacteria, as well as for yeast. Application to a depth of ¹⁄₁₆ inch is recommended once to twice daily; dressings may or may not be used. Since silver sulfadiazine is not a carbonic anhydrase inhibitor, cases of acidosis have not been reported (a problem that can occur with mafenide therapy—see discussion following). Silver sulfadiazine is somewhat better tolerated by patients than mafenide and is associated with less pain during dressing changes. Any of the adverse reactions associated with sulfonamides may occur with the use of silver sulfadiazine (e.g., rash or leukopenia). Silver sulfadiazine is only fairly effective in penetration of eschar.

MAFENIDE Mafenide acetate (Sulfamylon) is a white, water-soluble cream that can diffuse through avascular burn tissue to help control infection. It is smeared on open burns with a gloved hand or a sterile tongue blade once or twice daily. The cream can be left on the wound uncovered or a light fine mesh gauze dressing can be used; dry cream and exudate must be removed before more cream is applied. Because it is water soluble, it can be washed off by tub bathing, shower, or bed bath. The antibacterial spectrum of mafenide includes the *Clostridia* and a variety of other gram-negative and gram-positive organisms commonly found in major burn wounds. However, the use of mafenide is associated with certain problems:

Metabolic acidosis. Mafenide is absorbed into the burn wound and is apparently eliminated principally in the urine. The drug is a potent carbonic anhydrase inhibitor; the tendency to metabolic acidosis apparently is due to carbonic anhydrase inhibition at the renal tubular level. Absorption of the drug causes urinary bicarbonate excretion to

increase and the urine pH to become alkaline, which results in a plasma base bicarbonate deficit (metabolic acidosis). Most patients with adequate respiratory function tolerate mild acidosis fairly well; however, at times, the acidosis requires treatment with bicarbonate solutions or discontinuing the use of mafenide.

Hyperventilation. The respiratory system becomes the only functional method for maintaining body pH in the normal range because the renal buffering mechanism is impaired. Hyperventilation is a compensatory action to lighten the P_{CO_2} and raise the plasma pH. (Usually the acidosis is only partially compensated for by the respiratory system.) Hyperventilation is particularly evident in patients with burns of 50% or more of the body surface. After absorption, mafenide is metabolized into an acid, placing further demands on the lungs to maintain a normal pH. As a result, respiratory compensation may overshoot and cause a high arterial pH and a low P_{CO_2} (respiratory alkalosis).

Pain on application. Application of the cream causes most patients to complain of stinging, especially in the first few weeks. The use of analgesics is sometimes necessary.

Allergic reaction. This is mild in most cases but occasionally may require that the drug be discontinued.

AQUEOUS SILVER NITRATE SOLUTION The application of an aqueous 0.5% silver nitrate solution to burns is still used occasionally to combat infection. It is an effective antibacterial agent against organisms that regularly colonize burn wounds. The silver nitrate dressings are at least 1 inch thick, are held in place with stockinette, and are wet every few hours with warmed 0.5% silver nitrate solution. Dry sheets and a cotton blanket are used to cover the patient; the covers minimize convection currents and the rate of heat loss. Silver nitrate is associated with certain problems:

Leaching of electrolytes (Na, Cl, K, Ca, and Mg) and vitamins from the burn wound. This is due to the hypotonicity of the silver nitrate solution as well as to the ion-binding potential of the silver ion. 0.5% Silver nitrate contains only 29.4 mEq/liter of $AgNO_3$, and the solute composition of the solution is much different from that of body fluids. *The patient's plasma electrolyte levels must be monitored closely so that adequate replacement therapy can be instituted; children should be monitored especially closely. Patients treated with topical 0.5% silver nitrate dressings should routinely receive supplementation of sodium chloride, potassium, calcium, and water-soluble vitamins, especially vitamin C.*

Staining. The staining property of silver nitrate is esthetically objectionable to the patient and to the staff. Silver nitrate stains everything it touches brown or black.

Pseudomonas superinfections.

Dressings must be rewet every few hours.

SYSTEMIC ANTIBACTERIAL AGENTS

Because infection commonly occurs in burned patients, systemic antibiotics are commonly administered; the specific agent required is determined by culturing the source of the infection (such as wounds, urinary tract, or respiratory tract). Penicillin is often given as a prophylaxis against streptococcal or staphylococcal infections.

PHYSIOLOGIC BASIS FOR TREATMENT AND NURSING CARE DURING THE FLUID REMOBILIZATION PHASE

Remobilization of edema fluid represents an interstitial fluid-to-plasma shift that begins on the second or third day after the patient has been burned. Its usual duration is from 24 to 72 hours.

OBSERVING URINARY OUTPUT

Reabsorption of edema fluid takes place about the second to fifth postburn day. The blood volume is greatly increased and large amounts of urine are excreted. The nurse should be alert for increasing urine volume and should report its presence to the physician.

If the expected diuresis does not occur, the possibility of renal damage must be considered.

OBSERVING FOR PULMONARY EDEMA

Fatal pulmonary edema may occur because the reno-cardiovascular system is not capable of handling the volume of water and electrolytes shifting from the interstitial fluid into the plasma. (Recall that the volume of edema fluid in a burn may equal the total normal plasma volume.)

The nurse should be alert for signs of circulatory overload and pulmonary edema:

Venous distention

Shortness of breath

Moist rales

Cyanosis

Coughing of frothy fluid

PARENTERAL FLUID THERAPY

Once the fluid remobilization phase is reached, parenteral fluids should be sharply curtailed or discontinued; infusion of large volumes of fluids could easily cause circulatory overload with pulmonary edema. Oral fluids and food may supply adequate fluid and nutrition during this phase of tolerated; if not, moderate quantities of intravenous fluids may be necessary to meet daily needs. If possible, a high protein and high caloric diet is started by the second or third day.

PHYSIOLOGIC BASIS FOR TREATMENT AND NURSING CARE DURING THE CONVALESCENT PERIOD

NUTRITION

Good nutrition is of first importance for burned patients, all of whom have great nutritional needs (as a result largely of increased catabolism), several times those of the healthy person. Unfortunately, burned patients frequently do not eat well because of pain, generalized discomfort (frequent dressing changes, grafts, and so forth), and depression. Without adequate nutritional support, the burn wound will not heal and defenses against invasive sepsis are compromised.

A number of formulas are in use to calculate the caloric need of burned patients. If the patient is unable to consume all of the desired calories orally, tube feedings may be used. (Tube feedings are discussed in Chapter 5.) If the patient is unable to tolerate oral or tube feedings, intravenous hyperalimentation may be required. (Hyperalimentation is discussed in Chapter 6.) Use of a central line should be undertaken cautiously since the burned patient has increased susceptibility to catheter sepsis. The nurse should monitor the patient's response to nutritional intake by measuring daily weights. In addition, each shift should estimate the patient's caloric intake. If body weights or caloric intakes are inadequate, additional nutritional supplements are required.

OBSERVING FOR SPECIFIC ELECTROLYTE IMBALANCES

The convalescent phase is often complicated by inadequate electrolyte intake from the diet; if supplemental replacements are not given, the patient may insidiously develop deficits of potassium, sodium, and calcium.

REFERENCES

1. Curreri W: Burns. In Schwarts S et al (eds): Principles of Surgery, 3rd edition, p 299. New York, McGraw-Hill, 1979
2. Munster A: Burn Care for the House Officer, p 5. Baltimore, Williams & Wilkins, 1980
3. Schwartz S: Consensus summary on fluid resuscitation. J Trauma 19, No. 11 (suppl): 876, 1979
4. Burke J: Fluid therapy to reduce mortality. J Trauma 19, No. 11 (suppl): 866, 1979
5. Jelenko C: Fluid therapy and the HALFD method. J Trauma 19, No. 11 (suppl): 866, 1979
6. Munster, p 16
7. Munster, p 23
8. Monafo W: The Treatment of Burns: Principles & Practice, p 66. St. Louis, Warren H. Green, 1971
9. Curreri, p 298
10. *Ibid*, p 289
11. Monafo, pp 54, 55

BIBLIOGRAPHY

Arturson G: Types of resuscitation therapy. J Trauma 19, No. 11 (suppl): 1979
Baxter C: Fluid resuscitation, burn percentage, and physiologic age. J Trauma 19, No. 11 (suppl): 1979
Monafo W: An overview of infection control. J Trauma 19, No. 11 (suppl): 1979
Pruitt B: The effectiveness of fluid resuscitation. J Trauma 19, No. 11 (suppl): 1979

9

FLUID BALANCE IN THE PATIENT WITH DIGESTIVE TRACT DISEASE

CHARACTER OF GASTROINTESTINAL SECRETIONS

The average daily volume of gastrointestinal secretions is approximately 8000 ml as compared with a plasma volume of 3500 ml. Most of these secretions are reabsorbed in the ileum and proximal colon; only about 150 ml of relatively electrolyte-free fluid is excreted daily in the feces.

Gastrointestinal secretions consist of saliva, gastric juice, bile, pancreatic juice, and intestinal secretions. Their average pH values are listed in Table 9-1. The electrolyte content of gastrointestinal secretions is summarized in Table 9-2.

With the exception of saliva, the gastrointestinal secretions are isotonic with the extracellular fluid (ECF). In addition, material entering the gastrointestinal tract tends to become isotonic during the course of its absorption. Many liters of ECF pass into the gastrointestinal tract, and back again, as part of the normal digestive process. This movement of water and eletrolytes is sometimes referred to as the "gastrointestinal circulation."

Loss of gastrointestinal fluids is the most common cause of water and electrolyte disturbances. This fact becomes evident when one considers the large volume of fluids in the gastrointestinal tract and the many ways in which they can be lost from the body. Vomiting, gastrointestinal suction, diarrhea, fistulas, and drainage tubes are some of the abnormal ways in which these fluids can be lost. Fluids trapped in the gastrointestinal tract, as in intestinal obstruction, are physiologically *outside* the body (third-space effect). Any condition that interferes with the absorption of fluids from the gastrointestinal tract can cause serious water and electrolyte disturbances.

VOMITING AND GASTRIC SUCTION

The absorption of gastric secretions and ingested fluids is hindered by vomiting and gastric suction. To understand the imbalances likely to occur with these conditions, it is helpful to review the normal characteristics of gastric juice, which is the most acid of the gastrointestinal secretions. Gastric juice has a pH of from 1 to 3.5; occasionally, the pH is higher than 3.5. The main electrolytes in gastric juice are hydrogen, chloride,

TABLE 9-1 *Gastrointestinal secretions and their usual pH*

Secretion	pH
Saliva	6–7
Gastric juice	1.0–3.5
Pancreatic juice	8.0–8.3
Bile	7.8
Small intestine	7.8–8.0
Large intestine	7.5–8.0

potassium, and sodium. Imbalances most often associated with the loss of gastric juice include the following:

> Fluid volume deficit
> Metabolic alkalosis (base bicarbonate excess)
> Potassium deficit

Other imbalances can include sodium deficit, magnesium deficit, and metabolic acidosis (if large amounts of intestinal secretions are vomited).

FLUID VOLUME DEFICIT When a large volume of water and eletrolytes is lost from the body, fluid volume deficit results. Since the gastric secretions are greatly reduced when the stomach is at rest, the patient should receive nothing by mouth during gastric suction or persistent vomiting. If the suction or vomiting is prolonged and fluid replacement therapy is inadequate, fluid volume deficit will result. Consequently, the nurse should be alert for the following symptoms indicating the presence of fluid volume deficit:

> Dryskin and mucous membranes
> Longitudinal wrinkling of the tongue
> Oliguria
> Acute weight loss—in excess of 5%
> Body temperature drop
> Exhaustion

METABOLIC ALKALOSIS Excessive loss of gastric juice by vomiting or gastric suction causes metabolic alkalosis due to loss of hydrogen and chloride ions from the body. Loss of chloride ions causes a compensatory increase in bicarbonate ions. The base bicarbonate side of the carbonic acid–base bicarbonate ratio is increased and the pH becomes alkaline. The nurse should be alert for symptoms of metabolic alkalosis when the patient has sustained a prolonged loss of gastric juice by vomiting or gastric suction (particularly if

TABLE 9-2 *Approximate electrolyte composition of gastrointestinal secretions*

Secretion	Usual maximum volume/day	Sodium	Chloride	Potassium
		(mEq/liter in adults)		
Normal				
Saliva	1000	100	75	5
Gastric juice (pH)<4.0)	2500*	60	100	10
Gastric juice (pH>4.0)	2000*	100	100	10
Bile	1500	140	100	10
Pancreatic juice	1000	140	75	10
Succus entericus (mixed small-bowel fluid)	3500	100	100	20
Abnormal				
New ileostomy	500–2000	130	110	20
Adapted ileostomy	400	50	60	10
New cecostomy	400	80	50	20
Colostomy (transverse loop)	300	50	40	10
Diarrhea	100–4000	60	45	30

*Nasogastric suction volume usually much less than this unless pyloric obstruction exists.
(Condon R, Nyhus L: *Manual of Surgical Therapeutics*, 4th ed, p 212, Boston, Little, Brown & Company, 1978)

the patient has been allowed to drink water or ingest ice chips in large amounts). These symptoms include the following:

1. Slow, shallow respiration (compensatory respiratory reaction to retain CO_2 and to correct alkalosis)
2. Muscle hypertonicity and tetany (caused by decreased calcium ionization in alkalosis)
3. Changes in sensorium
   a. Personality changes may be the first symptoms to appear
   b. Previously placid patient may become irritable and uncooperative
   c. Patient may be disoriented

It should be noted that vomiting of gastric contents alone is not as common as vomiting of fluids from lower in the gastrointestinal tract.

METABOLIC ACIDOSIS Vomiting of alkaline duodenal fluid in additon to gastric juice can cause plasma pH to remain essentially normal or even cause metabolic acidosis if the loss of alkaline fluid exceeds loss of acidic gastric fluid.

POTASSIUM DEFICIT Gastric juice is rich in potassium. A prolonged loss of this fluid frequently leads to potassium deficit, particularly if potassium replacement therapy is inadequate. The nurse should be alert for symptoms of potassium deficit, including the following:

Muscular weakness
Gaseous intestinal distention
Soft, flabby muscles
Muscle cramps (particularly in the legs)
Paresthesia and flaccid paralysis of extremities
Weak, irregular pulse
Respiratory failure
Heart block and cardiac arrest (as late symptoms)

SODIUM DEFICIT The nurse should be aware that gastric suction or prolonged vomiting can lead to sodium deficit, especially if plain water is drunk. Symptoms of sodium deficit include those listed below:

Anorexia, apathy, sometimes great apprehension
Abdominal cramps
Hypotension (declines further when sitting or standing)
Syncope with position change
Rapid, thready pulse

Oliguria, low specific gravity of urine
Hand veins fill slowly
Fingerprinting on sternum

MAGNESIUM DEFICIT Prolonged vomiting or gastric suction can result in magnesium deficit, an imbalance not as common as those listed above. The magnesium concentration in gastric juice is only 1.4 mEq/liter. In addition, the body conserves magnesium well. However, its continued loss by suction or vomiting, plus its dilution with magnesium-free replacement fluids, can result in symptoms of magnesium deficit. These include the following:

Hyperirritability
Gross tremors, which may occur in any extremity but are most common in the arms
Confusion and disorientation
Tachycardia
Hypertension
Hallucinations, usually visual but may be auditory
Abnormal sensitivity to sound (hyperacusis)
Convulsions (usually generalized and not preceded by an aura)

KETOSIS OF STARVATION Another imbalance is ketosis of starvation. Unless adequate parenteral nutrition is provided for the patient with prolonged vomiting or gastric suction, ketosis of starvation will occur. As a result of the absence of carbohydrate, the body must use fat for energy purposes. Because of increased fat utilization, ketone bodies accumulate in the blood. Since ketones are strong acids, they can convert metabolic alkalosis into metabolic acidosis or accentuate an already existing metabolic acidosis. The odor of acetone on the breath indicates starvation ketosis; other symptoms of metabolic acidosis include deep, rapid respiration and weakness.

NURSING IMPLICATIONS The nurse should be alert for symptoms of the imbalances described above, should discuss their occurrence with the physician, and, in addition, should attempt to minimize the loss of water and electrolytes by vomiting and gastric suction.

Vomiting Nursing actions for patient with vomiting are described below:

1. Discourage oral intake, particularly water, if vomiting is persistent and frequent. Ingested substances

stimulate gastric secretions. If the substance is hypotonic, electrolytes will move from the ECF into the stomach. When the stomach is emptied by vomiting, water and electrolytes from the gastric secretions and ECF are lost. Obviously, oral intake, while vomiting persists, promotes water and electrolyte depletion.

2. Report vomiting early so that appropriate treatment can be started before water and electrolyte losses become serious. Medications to relieve nausea may prevent further vomiting. Nutrition by the parenteral route allows the stomach to rest.

3. Administer p.r.n. medications, as prescribed, to relieve nausea.

4. Measure or estimate as accurately as possible the amount of vomitus lost from the body so that lost water and electrolytes can be replaced by parenteral fluids. All fluids lost and gained by the body should be recorded on the intake–output record.

5. Measure body weight daily to detect significant changes in fluid balance. Daily weights are helpful in detecting fluid volume deficit, particularly if vomitus has not been measured. A patient on a starvation diet should lose about 0.5 lb a day, whereas a loss in excess of this amount probably implies a fluid volume deficit. A weight gain implies fluid volume excess if the patient is on a starvation diet.

6. Report substantial improvement in the patient's condition early so that he can be returned to oral intake as soon as tolerated.

Gastric suction The following are important nursing actions in the care of the patient with gastric suction:

1. Irrigate the tube with isotonic sodium chloride solution, multiple electrolyte solution, or as prescribed by the physician. Plain water is dangerous as an irrigating solution because it can wash out electrolytes. (Some controversy exists over whether sodium losses are significant with irrigation with tap water.) Some physicians prefer hourly irrigation of the tube with 30 ml to 50 ml of air to prevent clogging; air does not interfere with electrolyte balance or with intake–output record keeping. However, when blood or thick secretions are occluding the tube, it may be necessary to use isotonic saline. In addition, the use of air may lead to uncomfortable abdominal distention and bloating.

2. Record the irrigating solution volume as intake and whatever is recovered as output.

3. Prohibit the intake of large quantities of water or ice chips by mouth, since water washes electrolytes from the stomach, causing metabolic alkalosis and loss of sodium and potassium. Sometimes the physician writes an order to permit the administration of ice chips sparingly; the term "sparingly" can readily be stretched by well-meaning but ill-advised persons to the point where the intake of plain ice is excessive. Some physicians allow approximately an ounce of plain ice chips per hour to relieve oral dryness.

4. Measure and record the amount of fluid lost by suction, as well as all other fluid gains and losses.

5. Measure daily weight variations to help detect early fluid volume deficit or excess.

DIARRHEA, INTESTINAL SUCTION, AND ILEOSTOMY

Abnormal loss of intestinal fluid occurs in the presence of diarrhea, intestinal suction, and ileostomy. Imbalances likely to occur in these situations are discussed below.

FLUID VOLUME DEFICIT Intestinal hypermotility shortens the opportunity for absorption of intestinal fluids and, thus, results in increased fluid loss in bowel movements. The hypermotility can be caused by a disease process, such as ulcerative colitis, or by the frequent use of an irritant carthartic. The liquid stools expelled as a result of hypermotility contain water and electrolytes derived from secretions, ingested food and fluids, and ECF brought into the bowel to render ingested substances isotonic. Obviously, prolonged diarrhea is a serious threat to water and electrolyte balance. The amount of fluid lost in intestinal suction averages around 3000 ml daily. Thus, the nurse should be alert for symptoms of fluid volume deficit when the patient has sustained large fluid losses from the intestinal tract. (Symptoms of fluid volume deficit are listed in the discussion of gastric suction and vomiting.)

METABOLIC ACIDOSIS Intestinal juice varies in composition according to the area of the intestine in which it was formed. However, the intestinal secretions are all alkaline, including pancreatic juice and bile, which are mixed with intestinal juices in the intestines. The chief electrolytes in intestinal juice include sodium, potassium, bicarbonate, and chloride.

The intestinal secretions are alkaline because of the preponderance of bicarbonate ions. Loss of bicarbonate results in a decreased pH. Symptoms of metabolic acidosis (primary base bicarbonate deficit) include the following:

Shortness of breath on exertion (mild deficit)

Deep, rapid breathing (moderate or severe deficit)

Weakness and general malaise

SODIUM DEFICIT Intestinal secretions have a relatively high concentration of sodium; excessive loss of these secretions results in sodium deficit. (Symptoms of sodium deficit are listed in the discussion of gastric suction and vomiting.)

POTASSIUM DEFICIT Relatively large amounts of potassium are contained in the intestinal fluid; therefore, potassium deficit occurs frequently with diarrhea, prolonged intestinal suction, and recent ileostomy. (Symptoms of potassium deficit are listed in the discussion of gastric suction and vomiting.)

NURSING IMPLICATIONS The nurse should be alert for symptoms of fluid imbalances likely to occur in the presence of diarrhea, ileostomy, or intestinal suction.

Diarrhea Important nursing actions in the care of the patient with diarrhea include the following:

1. Discourage oral intake, particularly irritating foods apt to stimulate peristalsis. Less fluid is formed when the intestinal tract is at rest.
2. Report diarrhea early so that appropriate treatment can be started before water and electrolyte losses are severe. Medications to reduce peristalsis may relieve diarrhea. Nutrition by the parenteral route allows the intestinal tract to rest.
3. Administer p.r.n. medications as prescribed to prevent diarrhea.
4. Measure, or estimate as accurately as possible, the amount of liquid feces lost from the body so that lost water and electrolytes can be replaced by parenteral fluids. All fluids lost and gained by the body should be recorded on the intake–output record.

5. Measure body weight daily to detect significant changes in fluid balance. Daily weights are helpful in detecting fluid volume deficit, particularly if the liquid stools have not been measured.
6. Report substantial improvement in the patient's condition early so that he can return to oral intake as soon as tolerated.

Ileostomy Described below are nursing considerations and actions important in the care of the patient with recent ileostomy:

1. Measure and record the fluid lost by ileostomy, as well as other fluid losses and gains by the body.
2. Be alert for symptoms of water and electrolyte disturbances in the immediate postoperative period; potassium deficit is the most frequent imbalance. Other imbalances may include sodium deficit and fluid volume deficit. Patients with ileostomies are more likely to develop water and electrolyte disturbances when their stomas first begin to function, as shown by a comparison of the amount of water and electrolytes lost in a 24-hour period in the early postoperative period with the amounts lost in a similar period after the ileostomy has adapted. Fluid loss from a recent ileostomy may be as high as 4000 ml in 24 hours; each liter of the fluid may contain 130 mEq of sodium and 20 mEq of potassium. An adapted ileostomy usually loses no more than 500 ml in 24 hours; each liter of fluid may contain 50 mEq of sodium and 10 mEq of potassium.

Intestinal suction Nursing responsibilities in the care of the patient with intestinal suction include the following:

1. Irrigate the tube with an isotonic or hypotonic electrolyte solution. Plain water should not be used to irrigate intestinal suction tubes, particularly those located low in the intestines, since plain water is injurious to the mucosa of the ileum. In addition, it promotes increased secretion of intestinal juice and causes electrolytes to be washed out.
2. Record the irrigating solution volume as intake and whatever is recovered as output.
3. Prohibit the intake of large quantities of water or ice by mouth since electrolytes may be washed out by the suction apparatus. Some physicians allow approximately an ounce of plain ice chips per hour to relieve oral dryness.

4. Measure and record the amount of fluid lost by suction, as well as all fluid gains and losses by the body.

PROLONGED USE OF LAXATIVES AND ENEMAS

ELECTROLYTE IMBALANCES The prolonged use of laxatives and enemas can result in serious water and electrolyte disturbances, particularly potassium deficit. Other possible deficits include sodium deficit and fluid volume deficit.

Cathartics increase the water and electrolyte output through the fecal route by hastening the excretion of fecal contents and, thus, reducing the absorption time. Irritant cathartics cause hypermotility of the bowel by irritating the bowel mucosa. Saline cathartics draw water from ECF into the bowel. The distended bowel produces mechanical stimulation and the large amount of fluid is propelled out of the bowel. A large fluid volume deficit can result from continued use of saline cathartics, which interfere with electrolyte absorption from the intestines.

Enemas can also affect electrolytes; repeated tap water enemas can result in hyponatremia. On the other hand, use of commercial hypertonic cleansing enemas (containing sodium biphosphate and sodium phosphate) can result in some degree of sodium absorption, and thus should be avoided in patients requiring sodium restriction. (Measures to help patients overcome the habitual use of laxatives and enemas are described in Chapter 5.)

NURSING IMPLICATIONS The nurse should teach patients to avoid the repeated use of cathartics and enemas; when frequent bowel irrigations are indicated, an isotonic electrolyte solution is probably indicated. A roughly isotonic sodium chloride enema solution can be easily made by adding 1 tsp of table salt to a liter of tap water (furnishes 120 mEq of Na and 120 mEq of Cl).

COLON PREPARATION FOR TESTS

Standard colon cleansing techniques for diagnostic studies usually include several days of dietary restrictions, such as low-residue and liquid diets; purgatives, such as castor oil and magnesium citrate; and numerous cleansing enemas, either isotonic saline or tap water.

Disadvantages of the standard method are that it is time-consuming, usually requiring several days; it is frequently unsuccessful in cleansing the bowel adequately; use of purgatives may cause nausea, cramping, and interference with sleep; normal dietary intake is disrupted for several days; and it is associated with potentially serious losses of salt and water from the body.

Burbige and associates studied the effect of bowel preparation for colonoscopy on fluid and electrolyte balance in 50 patients.[1] The preparation used in the study included a liquid diet for 2 days; magnesium citrate, 240 ml at 6 PM the evening before the test; a glass of water at 6 PM, 7 PM, 8 PM, and 9 PM the evening before the test; and 1-liter tap water enemas begun 2 hours before the examination until returned clear. A change in body weight occurred in 46 of the 50 patients with an average weight loss of 2.1 lb (range 0.2 lb–7 lb). Although there were some changes in electrolyte status before and after the preparation, the only statistically significant change was a drop in plasma sodium; however, the researchers did not regard the variations as clinically significant. According to the researchers, shifts of body fluid and electrolytes are the most potentially important result of rigorous catharsis. While loss of only 2 lb may seem small, one should remember that this loss occurs rather quickly; also, some patients lose more than 2 lb (one patient in the study lost 7 lb). Data indicate that there are significant shifts of fluid among body compartments during rigorous cathersis; if the shifts occur rapidly in an elderly person with cardiopulmonary disease, the results can be dangerous. (The literature reports a patient who apparently suffered a myocardial infarction due to hypovolemia following rigorous catharsis.)[2] The aged patient having gastrointestinal radiographs should probably receive intravenous fluids during the period of reduced oral intake and increased fluid loss by catharsis and enema. Glomerular filtration rate (GFR) is decreased in the aged person, making any further decrease in ECF volume potentially hazardous.

FISTULAS

Fistulas are abnormal communications between the intestine and the skin (external fistula) or another hollow viscus (internal fistula). Fistulas may result from trauma, or they can occur spontaneously as a complication of pancreatitis, inflammatory bowel disease, neoplasia, or other gastrointestinal disorders.

The amount and kind of fluid lost through a fistula depend on its location. An educated guess as to the content of the fluid and imbalances likely to accompany its loss can be made by reviewing the usual electrolyte content of the fluid in the region of the fistula (see Table 9-2). For example, one would expect metabolic alkalosis to occur if the fluid loss is mostly gastric; or, metabolic acidosis if the fluid loss is duodenal or pancreatic. Jejunal fistulas are not common and usually do not result in a serious acid–base problem. When in doubt as to the origin of the fistula, the fluid's pH and electrolyte content can be analyzed by the laboratory.

In addition to pH changes, fistulas can cause a serious contraction of ECF volume. A gastric, duodenal, or jejunal fistula may drain 3 liters to 6 liters daily; a pancreatic fistula may drain 2 liters daily. Pancreatic fistulas are particularly serious because pancreatic enzymes may digest the abdominal wall and cause peritonitis.

If possible, the nurse should attempt to measure the fluid lost by an external fistula by applying a stoma bag over its orifice. Sometimes a soft sump tube is used for drainage; however, a stoma bag is usually preferred, since a tube tends to keep the fistula open. If the drainage cannot be directly measured, an estimate should be made of the volume. Statements as to how much of a dressing is saturated, as well as extent of gown and linen saturation, help in planning of fluid replacement therapy.

Nutrition is extremely important in patients with gastrointestinal fistulas; elemental diets (Chapter 5) or intravenous hyperalimentation (Chapter 6) may be highly effective in promoting healing. Fistulas may close spontaneously with proper supportive care, or they may require surgical intervention.

INTESTINAL BYPASS SURGERY AND ELECTROLYTE BALANCE

A surgically created "short-bowel" syndrome has been used to control severe obesity in persons unable to lose weight by diet and exercise. The bypass operations are designed to produce a state of controlled malabsorption by shortening the small intestine in order to have less area for absorption of foods. The original surgery (anastamosis of the proximal jejunum to the transverse colon) has been largely abandoned due to the high incidence of complications; in many patients it was followed by unmanageable diarrhea and severe water and eletrolyte disorders. Less drastic surgery (jejunoileal bypass) is now used in some centers. The proximal 14 inches of jejunum may be anastomosed end-to-side to the ileum 4 inches above the ileocecal valve.

It should be remembered that surgery on very obese patients can result in appreciable postoperative complications, such as thromboembolism, dehiscence, pneumonia, atelectasis, wound infection, and incisional hernias.

Among the later complications that may be associated with jejunoileal bypass are the following:

Persistent, uncontrolled diarrhea (beyond the "usual" diarrhea occuring in all bypass patients)

Fatty infiltration of the liver; liver failure

Renal failure

Hyperoxaluria and calcium oxalate urinary calculi

Cholelithiasis (most common 2 to 3 years after surgery)

Malnutrition

Pancreatitis

Hypokalemia

Hypocalcemia

Hypomagnesemia

Hyperchloremic acidosis

ELECTROLYTE PROBLEMS

The major factor leading to electrolyte disorders in bypass patients is the diarrhea accompanying the surgical creation of a short bowel. All bypass patients have diarrhea, the amount varying with each patient. Diarrhea is due to the rapid transit effect of a short-bowel and perhaps to inadequate reabsorption of bile acids in the remaining ileum. (High concentrations of bile acids in the large intestine favor diarrhea.) Contributing to the diarrhea is bacterial overgrowth in the excluded segments of small bowel; refluxing of intestinal fluid in these loops occurs, leading to stagnation and eventually to evacuation at irregular intervals. Most patients have six to eight loose stools per day for several months following surgery; some may have as many as 24 loose stools daily.[3]

Up to 40% of bypass patients develop symptoms of electrolyte depletion during the first year; later the incidence decreases to 5%.[4]

The incidence of severe mineral depletion apparently diminishes after 2 years due to bowel adaptation; however, oral supplements should be continued. Patients should be instructed to watch for signs of electrolyte disturbances and to report them; the need for taking prescribed electrolyte supplements should be made clear.

POTASSIUM DEFICIT Hypokalemia is the most frequent electrolyte disorder in bypass patients. It is due primarily to loss of potassium in diarrheal stools. According to one study, hypokalemia (less than 3.5 mEq/liter) was reported in 27% of females and 19% of males despite potassium supplementation.[5]

Symptoms of hypokalemia can include weakness; muscle cramps, especially in the legs; and irregular pulse. Severe hypokalemia can lead to cardiac or respiratory arrest; cardiac deaths attributable to hypokalemia have been reported. Some patients require rehospitalization for intravenous potassium administration; others need only oral supplements.

CALCIUM DEFICIT Hypocalcemia in bypass patients is the result of calcium loss in feces; reasons for calcium loss include short bowel with rapid transit effect, steatorrhea, and low vitamin D levels. Despite low plasma calcium levels, clinical manifestations of hypocalcemia are relatively infrequent. They may include tetany, paresthesias, and muscle cramping. Intravenous calcium administration is required for treatment of severe hypocalcemia. Oral calcium supplements should be given prophylactically, since plasma calcium levels may fall to extremely low levels very quickly.

MAGNESIUM DEFICIT Hypomagnesemia is due to diarrhea and decreased magnesium absorption associated with reduction in bowel length. Despite frequent low plasma magnesium levels, clinical manifestations of hypomagnesemia are relatively infrequent. One should observe for neuromuscular irritability.

It has been noted that severe magnesium depletion in bypass patients may produce a failure of parathyroid hormone (PTH) release in reponse to hypocalcemia; intravenous administration of magnesium may be necessary before serum calcium levels will increase.

Maintenance of good nutritional patterns and avoidance of alcohol ingestion will help prevent magnesium deficit.

HYPERCHLOREMIC ACIDOSIS Mild hyperchloremic acidosis has been reported in some patients with bypass operation; the cause of the hyperchloremia is unclear. Oral sodium bicarbonate can be administered when necessary.

PROTEIN DEPLETION Malabsorption and decreased food intake result in significantly lowered serum albumin levels in some bypass patients. Levels usually reach their low point at about 6 and 12 months after operation, and more nearly normal levels are reached thereafter with greater food intake and improved absorption. Listed below are some of the clinical manifestations of hypoproteinemia:

- Muscular weakness
- Hair loss
- Mild edema (presumably due to hypoproteinemia and increased vascular permeability)
- Anemia
- Increased susceptibility to infection

Oral protein supplements are helpful but may exacerbate the diarrhea and produce greater losses.

HYPEROXALURIA AND STONE FORMATION Increased renal excretion of oxalate occurs in all bypass patients because oxalate is no longer normally excreted by the bowel. Increased oxalate absorption from the intestine may be due to unavailability of sufficient calcium to bind the oxalate for elimination. Abundant crystal formation is commonly observable in the urine of bypass patients, and calcium oxalate stones may form. Some physicians recommend calcium supplements to bind oxalate in the intestine for excretion, thus decreasing oxalate absorption and subsequent renal excretion. Dietary restriction of oxalates may also be recommended for stone formers. (Oxalate stones are discussed further in Chapter 10.)

CONCLUSION Approximately 80% of the patients with jejunoileal bypass have satisfactory results; however, in 10% to 20% of the patients it has been necessary to revise or reverse the shunt. Some of the principal reasons have included severe diarrhea, unmanageable electrolyte and metabolic problems, and liver failure.

Gastric bypass is advocated by some because it appears to have led to fewer late complications and less need for rehospitalization. A modification of the gastric bypass procedure, called gastric partitioning ("stomach stapling"), is particularly favored because it avoids opening the gastrointestinal tract or altering its continuity. Gastric partitioning surgery favors early satiety after the ingestion of small amounts of food and thus promotes weight loss.

TRAPPED GASTROINTESTINAL FLUIDS (THIRD-SPACE EFFECT)

Fluids trapped in an obstructed bowel or in the peritoneal cavity are lost because they are not available for use by the body (third-space effect). Such spaces collect at the expense of, and produce a deficit in, ECF volume. Third-space fluid losses cannot be measured directly as one measures fluid losses caused by vomiting or suction. Trapped fluids present a problem in planning fluid replacement therapy. Gastrointestinal conditions associated with fluid accumulation in the body include intestinal obstruction, peritonitis, and cirrhosis of the liver.

INTESTINAL OBSTRUCTION

It has been estimated that 20% of general surgical emergency admissions to hospitals are for operative or nonoperative treatment of intestinal obstruction.[6] Intestinal obstruction is said to exist when there is interference with the normal progression of intestinal contents. A mechanical obstruction is defined as an actual physical barrier, such as adhesions, hernia, tumor, or diverticulitis, blocking normal passage of intestinal contents. Functional obstruction, or paralytic ileus, refers to inability of the bowel to propel intestinal contents forward due to inhibition of the neuromuscular apparatus. Motor activity, though diminished, is never completely absent. Hypokalemia is capable of producing ileus, as is manipulation of the intestines during surgery.

FLUID AND ELECTROLYTE PROBLEMS The most obvious route of fluid and electrolyte loss in bowel obstruction is by vomiting or gastrointestinal intubation after treatment is initiated. The nature of the fluid and electrolyte disturbances accompanying intestinal obstruction depends largely on the site of the obstruction.

If the *upper small intestine* is obstructed, the patient will vomit intestinal juice and gastric juice. The loss of acid and alkaline fluids may be approximately equal, preventing serious disturbances in pH.

If the obstruction is in a *distal segment* of the small intestine, the patient may vomit larger quantities of alkaline fluids than of acid fluids. (Recall that secretions below the pylorus are mainly alkaline). Thus, metabolic acidosis can result from a low intestinal obstruction.

If the obstruction is below the *proximal colon*, most of the gastrointestinal fluids will have been absorbed before reaching the point of obstruction. Solid fecal matter accumulates until symptoms of discomfort develop. Accumulation of gas and fluid behind the obstruction can led to distention of the ileum if the ileocecal value is incompetent. Reverse peristalsis may cause severe vomiting of a fecal nature late in bowel obstruction.

Small intestinal obstruction traps gastrointestinal fluids and gas proximal to the obstruction. Large quantities of water and electrolytes continue to be secreted into the bowel lumen, even in the absence of oral intake.

Decreased selectivity of the intestinal membrane allows plasma proteins to enter both the intestinal lumen and the gut wall. Because of distention, the mucosa of the bowel above the obstruction is stimulated to secrete more fluid. The edematous bowel wall is not able to absorb the large fluid volume; thus, the distention becomes progressively greater.

Ten liters or more of fluid can collect in the bowel, resulting in severe ECF volume deficit. Plasma volume is substantially reduced and hypovolemic shock often ensues. Plasma concentrations of electrolytes are initially preserved, since the fluid lost is primarily isotonic; however, the patient usually becomes thirsty and drinks water, leading to hyponatremia. As stated above, acid–base disturbances are largely dependent on the site of the obstruction.

In addition to fluid, the gut is distended by gas, mostly from swallowed air. Stasis of the gut also causes accumulation of gas produced by bacterial action and diffusion from the bloodstream.

NURSING ASSESSMENT The nurse should frequently assess the vital signs of patients with intestinal obstruction. A fall in blood pressure with an increased pulse rate indicates further contraction of the plasma volume and oncoming circulatory collapse. Such findings should be quickly reported to the physician. Fluid administration before surgery aims at stabilizing the vital signs sufficiently to withstand the stress of surgery. Balanced electrolyte solutions and plasma fractions are frequently used as replacement fluids. It is generally not possible to totally correct the fluid deficit prior to surgery since time is limited. Frequently, however, there is sufficient time to administer 2 liters to 3 liters of an electrolyte solution (such as lactated Ringer's) over a 3- to 4-hour period to prepare the patient for operation. (The administration of only electrolyte-free solutions may easily induce water intoxication.) The rapidity of fluid replacement may compromise some patients' cardiovascular status; therefore, CVP should be monitored in these patients. A steadily rising CVP indicates too rapid fluid administration. (Nursing responsibilities in parenteral therapy are described in Chapter 6.) The hourly urinary output should be observed. Oliguria or anuria are indications of inadequate fluid replacement. Urinary specific gravity is high (above 1.025) when fluid replacement is inadequate. When possible, the urinary output should be at least 50 ml per hour before surgery.

Recall that the volume of fluid trapped in the intestine can only be estimated. Weight measurements are also valueless in detecting the amount of fluid trapped in the bowel. *Careful observation of vital signs, the patient's appearance, urinary volume, and specific gravity are significant.* Mild tachycardia may suggest an acute fluid loss of 5% of body weight; an acute fluid loss greater than 10% of body weight can cause hypovolemic shock.

The nurse should carefully observe and record the nature and frequency of vomiting (such observations are helpful in determining the level of obstruction). Brownish liquid vomitus with a fecal odor is characteristic of distal ileal obstruction. Greenish watery vomitus, without abdominal distention, may suggest obstruction of the proximal jejunum. Patients with colon obstruction do not usually vomit until advanced obstruction is present or unless the ileocecal valve is incompetent. (Recall that most fluids are absorbed before reaching the colon.) Distention is most marked in patients with obstruction of the distal ileum or of the colon. On the other hand, obstruction of the proximal gut causes minimal distention, since vomiting relieves the pressure.

PERITONITIS

Peritonitis involves the loss of ECF into the peritoneal cavity as an inflammatory exudate, causing fluid volume deficit. Calcium deficit can also occur in generalized peritonitis and acute pancreatitis because large amounts of calcium are immobilized in the diseased tissues and exudates, for reasons unclear.

Functional obstruction of the bowel (ileus) frequently occurs in association with peritonitis; imbalances most often seen in ileus include fluid volume deficit, hyponatremia, and hypochloremia. With extensive peritonitis, fluid translocation of 4 liters to 6 liters or more in a 24-hour period is not uncommon.[7] With profound hypovolemic shock, metabolic (lactic) acidosis occurs owing to tissue hypoxia.

CIRRHOSIS OF THE LIVER WITH ASCITES

Ascites represents the accumulation of a large quantity of water and electrolytes in the peritoneal cavity; sequestration of the ascitic fluid in this "third-space" renders it relatively unavailable to other physiological compartments. Ascites results from a combination of factors: mechanical obstruction to venous outflow from the cirrhotic liver, resulting in portal hypertension; hypoalbuminemia; and hyperaldosteronism. Decreased plasma albumin, combined with the increased capillary pressure produced by portal hypertension, allows albumin-rich fluid to shift out into the peritoneal cavity. It is thought that portal hypertension is more important in the pathogenesis of ascites than is hypoalbuminemia, since ascites disappears when portal hypertension is relieved by surgical shunting procedures. Ascites caused by liver disease is a transudate with a relatively high protein content (varying from approximately 20% to 50% of the protein level in plasma). Water and electrolyte disturbances that may occur in cirrhosis of the liver are summarized in Table 9-3.

CLINICAL MANIFESTATIONS Symptoms of cirrhosis are variable but may include the following:

Weakness and fatigue

Anorexia, nausea, vomiting, diarrhea, or constipation; blood from the portal vein backs into the gastrointestinal organs and interferes with their normal activity

Abdominal fullness, at first due to flatulence, later to ascites

Weight loss (may be masked by excessive fluid retention)

Enlarged liver; at first the liver may be enlarged with fatty tissue, later it becomes small, hard, and nodular

TABLE 9-3 *Water and electrolyte disturbances in cirrhosis of the liver with ascites*

Water and electrolyte disturbance	Cause
Sodium and water retention	Secondary hyperaldosteronism Recall that aldosterone causes Na retention, which in turn causes water retention. (The liver normally inactivates aldosterone and prevents its excessive buildup in the body; however, the cirrhotic liver does not perform this function well.)
Mild hyponatremia is quite common (A severe hyponatremia, less than 125 mEq/liter, is a serious prognostic sign)	Impaired renal water excretion (pathogenesis not clear); perhaps related to excess ADH secretion or inability of liver to deactivate ADH) Severe dietary Na restriction Frequent paracentesis Na moves into the cells to replace K loss (K loss is the result of diuresis, malnutrition, or hyperaldosteronism)
Decreased plasma protein level (hypoalbuminemia results in reduced plasma osmotic pull)	Albumin lost in ascites, (one liter of ascitic fluid contains as much albumin as 200 ml of whole blood. Normally, blood flowing into and out of the liver has the same volume. In cirrhosis, however, the volume of hepatic venous outflow is reduced by one half. This mechanical blockage leads to the formation of an ultrafiltrate of blood, which escapes into the abdominal cavity.) Decreased albumin synthesis resulting from liver dysfunction The normal albumin/globulin ratio of approximately 3:1 is disrupted; the ratio may be reversed so that globulin levels are higher than albumin. (Recall that globulin exerts a weaker osmotic pull than albumin, and that it is produced mainly by lymphatic tissues.)
Generalized edema	Reduced plasma protein level allows fluid to leave the plasma space and enter the tissue space Excessive aldosterone and ADH levels cause fluid retention Portal hypertension causes increased hydrostatic pressure and, thus, edema of the lower extremities
Potassium deficit (severe potassium deficit is often a late manifestation of liver disease)	Prolonged use of potassium-losing diuretics Hyperaldosteronism (recall that hyperaldosteronism causes K loss) Poor dietary intake because of anorexia and nausea Vomiting and diarrhea (increased loss of K)
Hyperventilation with respiratory alkalosis	May be related to hyperammonemia (A high ammonia level acts as a respiratory stimulant and may induce hyperventilation)
Elevated blood ammonia level	Under normal conditions, the large amounts of ammonia formed in the intestines by bacterial action are absorbed into the bloodstream and carried to the liver to be converted to urea for renal excretion. However, in cirrhosis, the liver cannot convert the ammonia to urea, causing the blood ammonia level to rise. Bleeding into the GI tract increases the production and absorption of ammonia.
Magnesium deficit (which occurs in some cases of alcoholic cirrhosis)	Poor dietary intake Alcohol produces magnesium diuresis
Calcium deficit may occur	Possibly a result of inadequate storage of vitamin D by the diseased liver
Renal tubular acidosis may occur in some patients (the mild metabolic acidosis is rarely symptomatic)	Apparently there is a relationship between liver disease and renal tubular acidosis; the reason for the relationship is obscure
Metabolic alkalosis is common in patients treated with thiazides, furosemide, ethacrynic acid, metazine, or bumetanide[8]	Related to the hypokalemia induced by these agents
Mild metabolic acidosis may occur in patients treated with spironolactone alone	Due to interference by spironolactone with Na^+/H^+ exchange in the distal tubules[8]

Enlarged spleen (spleen is engorged with venous blood from the portal system)

Low-grade fever

Jaundice

Edema due to hypoalbuminemia

Putrid, fecal order to the breath

Hypoglycemia (glycogen metabolism impaired)

Spider angiomata (small dilated superficial vessels resembling small bluish red spiders may appear in the skin of the face, forearms, and hands; the "spider" will disappear when pressure is applied to the central point with a pencil and reappear when the pressure is removed)

Palmar erythema (bright red color on the palms of the hands, due to increased circulation; redness disappears when pressure is applied)

Enlarged male breasts and atrophy of testicles, caused by the liver's inability to deactivate estrogen

Embarrassed respirations if the amount of ascites is large; however, if a high ammonia level is present, hyperventilation may occur

Enlarged abdominal veins, esophageal varices, and internal hemorrhoids caused by portal hypertension; these may be sites of profuse hemorrhage

Increased bleeding tendencies resulting from vitamin K deficiency and decreased prothrombin formation

Deficiency of vitamins A, C, and K caused by inadequate formation, utilization, and storage by the liver

Intensified reaction to certain medications, caused by the liver's inability to detoxify them; morphine and barbiturates should be avoided in patients with advanced liver disease

Greater susceptibility to infection, secondary to decreased formation of antibodies by the liver

Elevated ammonia level at first causes dullness, drowsiness, loss of memory, slow, slurred speech, and personality changes—later these may progress to confusion and disorientation and, finally, to stupor and coma

"Hepatic flap" is caused by a high ammonia level—this tremor is characterized by spasmodic flexion and extension at the wrists and fingers, as well as lateral finger twitching; hepatic flap can be elicited in susceptible patients by elevating the arms, hyperextending the hands, and spreading the fingers

NURSING ASSESSMENT OF THE PATIENT WITH ASCITES Measurements of abdominal girth are frequently used to monitor the degree of ascites. The abdomen should be measured at the same place each time; anatomic landmarks such as the iliac crests or umbilicus should be used or ink marks made on the skin to indicate where measurements should be taken. Remember, they give only an estimate of the degree of ascites, since gaseous distention of the intestines can vary from time to time. Measurements should be made with the patient in the same position each time. With treatment, one would expect to see a gradual decrease in abdominal girth to the level it was before ascites developed.

Daily weights should be measured, using the same scales each time. The same amount of clothing should be worn, and weights taken at the same time of the day (after voiding). Usually body weights are most accurate when obtained in the early morning, before breakfast. If treatment with sodium restriction and diuretics is effective, one would expect to see a 1- to 2-lb weight loss per day (more if generalized edema is present). (Evaluation of edema is described in Chapter 4.)

Accurate intake–output measurements are indicated for the cirrhotic patient being treated with sodium restriction and diuretics. During diuresis of excess fluid (ascites and generalized edema) one would expect urinary output to exceed fluid intake.

Vital signs should be observed periodically to detect signs of hypovolemia, especially when large doses of diuretics are used or severe dietary sodium restriction is employed.

With adequate treatment and subsequent removal of ascitic fluid, the patient should have less shortness of breath and improved respirations (due to decreased pressure on the diaphragm); moreover, he should be better able to tolerate lying flat. The ascitic patient frequently positions himself upright to allow for freer chest expansion.

TREATMENT

Sodium restriction Patients with ascites are first treated with low-sodium diets; accurate intake–output records, weight charts, and abdominal girth measurements are helpful in planning the degree of sodium restriction. Dietary restriction varies according to need; usually, daily limitations of 500 mg to 1000 mg are necessary. Restriction to less than 200 mg daily prevents additional fluid retention in most patients, but is not usually as well accepted. (Low-sodium diets are discussed in Chapter 5.) The patient/family should be taught how to manage a sodium-restricted diet at home; written information that the patient can take home is helpful, as is accessibility to a dietary consultant when questions arise at home.

Diuretic therapy If dietary sodium restriction alone does not suffice, diuretic therapy is necessary. Since furosemide, ethacrynic acid, and thiazide diuretics are potassium-losing agents, oral potassium supplements should be supplied to prevent hypokalemia. (Symptoms of hypokalemia include muscle weakness and flaccidity, sluggish bowel activity, and irregular pulse.) Hypokalemia is particularly dangerous in the cirrhotic patient, since it can contribute to the development of hepatic encephalopathy.

Another way to avoid hypokalemia in patients with ascites is the concomitant adminstration of a potassium-sparing and a potassium-losing diuretic; this not only helps to avoid hypokalemia, it also provides two agents to eliminate sodium. Furosemide, ethacrynic acid, and thiazides block sodium chloride reabsorption in the proximal nephron or the ascending limb of Henle's loop; spironolactone acts in the distal nephron. With the use of this combination of agents, the need for potassium supplementation is nil or greatly reduced. Routine potassium replacement in this instance is dangerous and can result in hyperkalemia.

In many cases effective diuresis can be induced by the use of a potassium-sparing diuretic (such as spironolactone) alone; potassium supplements can be lethal in this situation and should be avoided. Use of potassium-sparing diuretics is extremely dangerous in patients with renal damage, since hyperkalemia can easily be induced. (Symptoms of hyperkalemia include intestinal colic, diarrhea, and arrhythmias.)

If excessive diuresis is produced, hyponatremia, hepatic coma, or progressive renal insufficiency (hepatorenal syndrome) may develop.

Removal of ascitic fluid by sodium restriction and diuretic therapy should be achieved slowly (at a rate of about 500 ml–1000 ml/day) to allow for equilibration between peritoneal and intravascular spaces.[9] (Diuretics are discussed further in Chapter 5.)

Nursing implications in diuretic therapy The nurse should monitor the patient's response to diuretic therapy, being particularly alert for signs of hypovolemia (induced by excessive sodium and water loss), hypokalmeia (when potassium-losing diuretics are used), and hyperkalemia (when potassium-sparing diuretics are used).

A patient's understanding of his diuretic therapy should be promoted by the nurse; for example, the patient should be able to name the diuretic as well as its common side-effects. Desirability of taking the diuretic in the morning to avoid disturbing sleep at night should be pointed out. In addition, the discharged patient should be taught to monitor and record daily weights and to observe for worsening of edema. (Evaluation of edema is discussed in Chapter 4.)

If a potassium supplement is required, the nurse should determine if the patient understands the dosage, reason for taking it, and possible side-effects. Since some potassium supplements are unpleasant, the importance of taking the prescribed medication should be impressed upon the patient.

If diuretic therapy is successful, one would expect to see a gradual weight reduction (in the form of fluid loss); urinary output that gradually exceeds fluid intake until ascites subsides; and a gradual decrease in abdominal girth to the level it was at before ascites developed. Percussion of the abdomen should reveal that ascites is reduced or absent.

Water restriction A low plasma sodium level is an indication to moderately restrict the patient's water intake. It is unusual to find central nervous system symptoms attributable to hyponatremia (such as confusion and somnolence). However, an occasional patient may have an extremely low serum sodium; such a patient usually has severe liver disease and may also have hepatic encephalopathy. According to one source, ascites production ceases with a 200-mg sodium diet and a 500-ml fluid restriction are imposed.[10] Fluids must be evenly spaced over the waking hours to avoid excessive thirst and discomfort.

Limitation of activity Assumption of the supine position is often associated with a spontaneous diuresis, resulting in significant weight loss with mobilization of peripheral edema fluid and ascitic fluid.[11] The diuresis induced by bedrest results from improvement in cardiac output and effective circulating blood volume. The patient with severe ascites is usually better able to tolerate the lateral recumbent position.

Paracentesis Eventually, patients with ascites may become refractory to diuretics and require paracentesis to relieve severe symptoms of intra-abdominal pressure and respiratory distress. Since paracentesis can be associated with serious problems, it should be performed only when respiratory distress or patient discomfort is *severe.* As much as 20 liters of ascitic fluid may accumulate in one week; unfortunately, its removal only causes more to form. Since ascitic fluid forms faster than it is resorbed, hypovolemia may occur if too much fluid is removed at once. The mechanism involves a reaccumulation of ascitic fluid in the abdominal cavity resulting from flow of water and sodium from the interstitial fluid and plasma, causing a diminished plasma volume. In addition, abdominal blood vessels that have been under heavy external pressure and have carried little blood suddenly open with the removal of the fluid pressure. As a result, blood is suddenly rerouted and shocklike symptoms can develop. The nurse should be alert for pallor, weak and rapid pulse, and hypotension following the removal of a large amount of fluid by paracentesis. Removal of as little as 1000 ml to 1500 ml may precipitate hypotension in some patients. Following paracentesis the use of an abdominal binder might be considered to prevent rapid reaccumulation of ascites.[12]

As stated earlier, paracentesis should be avoided, if possible, because it results in protein loss, decreases effective blood volume, and increases the secretion of aldosterone and ADH. Existing hypoalbuminemia is made worse by paracentesis. Although salt-free albumin is sometimes administered intravenously to help correct hypoalbuminemia, it is expensive, and the results are negligible. (The amount that can be given is often ineffective in restoring a normal plasma albumin level.) The patient should be encouraged to eat enough protein for body needs without inviting hepatic encephalopathy. Acute sodium depletion with shock or renal failure may also follow paracentesis, particularly when it is performed repeatedly. In addition to sodium loss in ascites, body water is retained in excess of sodium and further dilutes the plasma. This dilutional hyponatremia appears to be due to stimulation of ADH secretion. Symptoms of hyponatremia can include nausea, vomiting, abdominal cramping, and decreased blood pressure.

Peritoneovenous shunt A peritoneovenous (LeVeen) shunt is an alternative to the treatment of cirrhosis that does not respond to sodium restriction and diuretic therapy. The shunt consists of a long tube (with openings along the sides) inserted into the abdominal cavity, running under subcutaneous tissue into the jugular vein. With the shunt, reinfusion of ascitic fluid into the venous system is allowed (a one-way valve pre-

vents backflow of blood and clotting of the shunt). Pressure changes of respiration permit the shunt to operate. Hemodilution by ascitic fluid is a potential complication of this procedure, sometimes requiring packed cell transfusions; the hematocrit level is used to monitor the extent of hemodilution. Other complications can include wound infection and peritonitis.

Preventing ammonia buildup Bleeding in the gastrointestinal tract increases ammonia formation, caused by digestion of blood proteins, and may precipitate hepatic coma. (Approximately 15 g–20 g of protein are added to the system with each 100 ml of blood digested.)[13] The nurse should be alert for gastrointestinal bleeding in the stool (melena) and in the vomitus (hematemesis). Sometimes enemas or cathartics are used to rid the intestine of blood, thus decreasing ammonia formation.

Ammonia formation caused by excessive bacterial growth in the small bowel can be decreased by the administration of bowel-sterilizing antibiotics, such as neomycin. Recall that neomycin (given orally or by enema) kills urease-producing bacteria, causing less urease to be produced, hence less urea to be broken down into ammonia. The nurse should remember that large doses of neomycin can damage both the kidneys and the ears; deafness may occur, particularly when neomycin is administered with furosemide (Lasix) or etharcrynic acid (Edecrin). One should be alert for early signs of hearing difficulties.

Since constipation contributes to the accumulation of ammonia, it must be prevented. Bowel evacuation removes blood from the intestine and thus decreases ammonia production; therefore, laxatives (such as milk of magnesia) may well be used to produce bowel movements. Enemas may also be used.

Ammonium chloride or carbonic anhydrase inhibitors (such as Diamox) should not be used because they may precipitate hepatic coma. As stated earlier, hypokalemia and metabolic alkalosis are associated with the use of potassium-losing diuretics; these imbalances should be prevented by the administration of potassium supplements as necessary. Among other effects, hypokalemia favors increased production of ammonia by the kidneys; metabolic alkalosis favors conversion of ammonium to ammonia.[13] Hypovolemia also predisposes to ammonia production by interfering with renal excretion of urea, causing more urea to enter the bowel to be converted into ammonia.

Impending hepatic coma is an indication to temporarily eliminate protein from the diet. A decrease in protein intake results in decreased ammonia formation; as the patient improves, more protein can be added. High ammonia levels must be discouraged since they contribute to hepatic encephalopathy. Symptoms of hyperammonemia include somnolence, lethargy, convulsive seizures, and coma.

IMBALANCES ASSOCIATED WITH ANTACIDS

Numerous antacid preparations are available commercially (Table 9-4) and are widely used (with or without the advice of a physician). Their purpose is to raise gastric pH to a level of at least 3.5 (to prevent pepsin activity). While effective in alleviating gastric distress, these agents are capable of disrupting electrolyte balance. Among the more common neutralizing agents are aluminum hydroxide, magnesium hydroxide, calcium carbonate, magnesium trisilicate, and sodium bicarbonate.

Aluminum hydroxide (Amphojel) frequently is used for its ability to relieve gastric acidity; however, this preparation can have a constipating effect. To counteract this effect, aluminum hydroxide is combined with magnesium hydroxide (which has a tendency to produce osmotic diarrhea). Examples of a combination of aluminum hydroxide and magnesium hydroxide include Aludrox, Maalox, and Mylanta. Aluminum hydroxide is sometimes used to bind phosphate in the intestines, causing the elimination of phosphate from the body. This effect is desirable in the treatment of certain cases of renal failure. However, excessive use may cause hypophosphatemia. Symptoms can include nausea, fatigue, muscle weakness, and osteomalacia.

Magnesium-containing antacids are contraindicated in patients with renal insufficiency because absorption of even small amounts of magnesium (coupled with impaired renal excretion) can result in hypermagnesemia. Some commercial antacids containing magnesium include Gelusil, Gaviscon, Camalox, Maalox, Mylanta, Aludrox, Kolantyl, among others. Early symptoms of hypermagnesemia include nausea and vomiting, a sense of heat and thirst, and hypotension (due to peripheral dilatation). Later symptoms can

TABLE 9-4 *Antacids*

Preparation	$Al(OH)_3$	$Mg(OH)_2$	$CaCO_3$	$MgCO_3$	$NaHCO_3$	Mg Trisilicate	$KHCO_3$
Alka-2			X				
Alka-Seltzer					X		X
Aludrox	X	X					
Alurex	X	X					
Amphojel	X						
Camalox	X	X	X				
Delcid	X	X					
Gaviscon	X				X	X	
Gelusil	X					X	
Kolantyl	X	X					
Maalox	X	X					
Marblen	X		X	X			
Mylanta	X	X					
Titralac			X				

include drowsiness, depressed deep tendon reflexes, depressed respirations, and cardiac manifestations. Cardiac effects of hypermagnesemia can include prolonged P-R interval, prolonged QRS intervals, tall T waves, premature ventricular contractions, and cardiac arrest.

Calcium carbonate is a very effective antacid, yet it should not be the sole agent for continued antacid therapy. Hypercalcemia may result, since some of the calcium is absorbed; renal stones may occur secondary to hypercalcemia. Also, an acid-rebound effect results as calcium stimulates the release of gastrin. The consumption of large amounts of calcium carbonate can cause the milk-alkali syndrome with nausea, vomiting, confusion, polydipsia, polyuria, and deterioration of renal function. Plasma calcium levels should be checked periodically when calcium carbonate is used. The dose of calcium carbonate should not exceed 8 g in 24 hours. It should be remembered that calcium preparations may also cause constipation. Examples of commercial calcium carbonate preparations include Titralac, Tums, and Alka-2.

Long-term use of sodium bicarbonate as an antacid is not recommended because it is absorbed into the bloodstream and can cause metabolic alkalosis. Patients requiring low sodium intake should avoid sodium bicarbonate preparations, since they have a very high sodium content (276 mg of Na in each Alka-Seltzer tablet and 1000 mg of Na in a teaspoon of baking soda). These preparations are frequently used as home remedies for "acid" indigestion even though the antacid effect is short-lived owing to their rapid emptying from the stomach.

A number of antacid preparations contain sizable amounts of sodium and, thus, should be used cautiously when sodium restriction is necessary. Magaldrate (Riopan) has a low-sodium content (0.7 mg/5 ml) and may be recommended for patients requiring a low sodium intake.

REFERENCES

1. Burbige E, Bourke E, Tarder G: Effect of preparation for colonoscopy on fluid and electrolyte balance. Gastrointest Endosc 24, No. 6:286, 1978
2. Rogers B, Silvis S, Nebel O et al: Complications of flexible fiberoptic colonoscopy and polypectomy. Gastrointest Endosc 22:73, 1975
3. Miller B: Jejunoileal bypass: A drastic weight control measure. Am J Nurs, March 1981, p 565
4. Maxwell M, Kleeman C (eds): Clinical Disorders of Fluid and Electrolyte Metabolism, 3rd ed, p 1509. New York, McGraw-Hill, 1980
5. DeWind L, Payne J: Intestinal bypass surgery for morbid obesity. JAMA 236:2298, 1976
6. Condon R, Nyhus L: Manual of Surgical Therapeutics, 4th ed, p 136. Boston, Little, Brown & Company, 1978
7. Schwartz, S (ed): Principles of Surgery, 3rd ed, p 1399. New York, McGraw-Hill, 1979
8. Maxwell & Kleeman, p 1256
9. *Ibid*, p 467
10. Krupp M, Chatton M: Current Medical Diagnosis and Treatment-1980, p 398. Los Altos, Lange Medical Publications, 1980
11. Maxwell & Kleeman, p 686
12. Spivak J, Barnes H: Manual of Clinical Problems in Internal Medicine, 2nd ed, p 223. Boston, Little, Brown & Company, 1978
13. Leonard B, Redland A: Process in Clinical Nursing, p 177. Englewood Cliffs, Prentice-Hall, 1980

Among their numerous cardinal functions, the kidneys excrete water, electrolytes, and organic materials and conserve whatever amounts of these substances the body requires. They act both autonomously and in response to blood-borne messengers, such as the mineralocorticoids and antidiuretic hormone (ADH). They excrete the breakdown products of protein metabolism, drugs, and toxins; produce the hormone erythropoietin, essential for red blood cell production; and convert an unusable form of vitamin D to one that the body can use. Failure of renal function causes a variety of water and electrolyte disturbances and other metabolic derangements.

Topics to be considered in this chapter include fluid and electrolyte disorders associated with acute renal failure, chronic renal failure, dialysis, nephrotic syndrome, ureteral transplants into the intestine, and urinary calculi.

Before continuing, the reader is encouraged to review the sections in Chapter 4 dealing with assessment of urinary specific gravity, urinary ph, urinary volume, and laboratory tests important in evaluating renal function.

10

Fluid Balance in the Patient with Urologic Disease

ACUTE RENAL FAILURE

Acute renal failure is of rapid onset (over days or weeks) and implies a pronounced reduction in urine flow in a previously healthy person. The condition can be caused by any of the following:

Severe and prolonged shock

Severe fluid volume deficit

Hemolytic blood transfusion reaction; lysis of red blood cells results in renal vasoconstriction and tubular blockage with hemoglobin

Severe crushing injuries; severely crushed muscles release large amounts of myoglobin into the bloodstream; myoglobin can block the tubules and might also produce vasoconstriction

Nephrotoxic chemicals such as lead, mercury, arsenic, and carbon tetrachloride

Drugs such as phenacetin, sulfonamides, streptomycin, kanamycin, radiographic contrast agents, tetracycline, and amphotericin

Endotoxemia

Renal vascular occlusion

Acute renal failure occurs in a variety of clinical settings; it can occur in obstetric patients as a result of prepartum and postpartum hemorrhages, toxemia, septic abortion, and the use of nephrotoxic abortifacients. Acute renal failure in postoperative patients has its highest incidence among those having cardiac, aortic aneurysm, biliary, and gastrointestinal surgical procedures. The use of radiocontrast agents in diagnostic radiology (particularly in large doses) has been incriminated as predisposing to acute renal failure; instances have been observed following gallbladder series, renal angiography, and intravenous pyelography. Indeed, clinical settings can be used to classify acute renal failure. An average breakdown is as follows: 40% surgery, 10% trauma, 10% nephrotoxins, 10% pregnancy, and 30% miscellaneous medical causes.[1] Patients with predisposing conditions should be observed for the possible development of acute renal failure. The nurse should measure the urinary output carefully, as well as all other fluid losses and gains. A reduced urinary output may be due to excessive fluid loss through another route, to inadequate intake, or to acute renal failure. It should be noted that although decreased urine volume (oliguria) is *most* characteristic of acute renal failure, urine output can vary from total anuria to polyuria. Categories have been defined as follows: total anuria (zero urine output), anuria (less than 50 ml to 100 ml/24 hours), oliguria (less than 400 ml/24 hours), and polyuria (greater than 1000 ml to 2000 ml/24 hours).[2] The incidence of nonoliguric acute renal failure, once thought to be rare, has been noted in 20% to 30% of the cases of acute renal failure.[3]

Acute renal failure may be classified as either reversible or irreversible. In irreversible renal failure, kidney function does not return and uremia progresses. Reversible acute renal failure can be divided into two phases—oliguria and diuresis.

OLIGURIC PHASE

The first manifestation of acute renal failure is usually decreased urinary output, appearing within a few hours after the causative event. Anuria is rare (except with urinary tract obstruction); instead, a 24-hour output of about 50 ml to 150 ml is the rule for the first few days. After this time, the urine output gradually increases. The oliguric phase may last one day or several weeks; the average duration is 10 to 12 days in severe cases.

The nurse should be alert for symptoms of the major problems of this phase. They include potassium excess, fluid volume excess, metabolic acidosis, and ure-

mia. Other problems include calcium deficit, sodium deficit, and anemia. Death usually is due to cardiac arrest caused by potassium excess or pulmonary edema caused by fluid volume excess.

POTASSIUM EXCESS Potassium excess usually occurs in the oliguric phase and can be life-threatening. Recall that the kidneys normally excrete 80% or more of the potassium lost daily from the body. When the kidneys are not functioning, potassium excretion is greatly reduced. This fact, in addition to the normal endogenous breakdown of muscle protein, causes an accumulation of potassium in the extracellular fluid (ECF). Factors that can further increase protein catabolism include inadequate caloric intake, acute stress or infection, massive crushing injuries, and the presence of large quantities of necrotic tissue. (Recall that the catabolism of necrotic tissue releases potassium into the ECF.)

The presence of acidosis augments the intracellular to extracellular shift of potassium, thereby hastening potassium buildup in the plasma. Recall that the normal plasma potassium level is 5 mEq/liter. Electrocardiogram (EKG) changes caused by hyperkalemia (6.5 mEq–8.0 mEq/liter) may include tall tented T waves, depressed ST segments, low amplitude P waves, prolonged P-R intervals, and widened QRS complexes with prolongation of the Q-T interval. A plasma potassium level above 9 mEq/liter may cause severe cardiac arrhythmias and cardiac standstill. Excessive potassium causes weakness, and dilatation of heart muscle and cardiac arrest in diastole.

Potassium intoxication is one of the main causes of death in acute renal failure. The immediate cause of death from hyperkalemia is disruption of cardiac conduction; the EKG is a better guide to therapy than any single electrolyte determination. Daily or twice daily EKGs should be obtained and kept in the patient's room for comparison with the previous tracings. In addition, daily serum potassium and sodium levels should be obtained.

The nurse should be alert for the following symptoms of potassium excess:

EKG changes (described above)

Anxiety and restlessness

Muscular weakness progressing to flaccid paralysis (primarily affects limbs and respiratory muscles)

Respiration decreased as a result of respiratory muscle weakness

Paresthesias or numbness, particularly of the mouth, hands, and feet

Decreased pulse rate, finally resulting in brady-cardia

Cardiac arrhythmias

Falling arterial blood pressure

Cardiac arrest and death

Treatment

Restricted intake Dietary potassium intake should be sharply restricted and potassium-containing medications should be discontinued. The potassium content in certain medications is often overlooked. For example, penicillin G contains 17 mEq of potassium per 10 million units. Salt substitutes should not be allowed, since they contain potassium also. Potassium-conserving diuretics (triamterene and spironolactone) are contraindicated because the additional potassium retention could prove lethal. Stored bank blood may contain as much as 30 mEq of potassium per liter and should not be used to transfuse patients with renal damage.

Cation exchange resins Cation exchange resins may be used to increase potassium excretion from the bowel. The resins may be taken by mouth, if tolerated, or may be instilled as a retention enema. When the resin is administered orally, the solution in which it is suspended should be recorded as part of the daily fluid allowance. While the cation exchange resins are given to remove potassium ions from the intestinal tract, they also remove other cations, such as sodium, calcium, and magnesium. It may be necessary to replace these ions if the resins are used for more than a few days. Such patients should have regular electrolyte studies.

Sodium polystyrene sulfonate (Kayexalate) is a sodium-potassium exchange resin. During its use, sodium ions are partially released by the resin in the intestine and replaced by potassium ions from the body. The bound potassium is then excreted in the feces. Hypokalemia may occur if the effective dose is exceeded; therefore, it is necessary to determine the plasma potassium level daily. Since sodium is released in the intestine during electrolyte exchange, the resin should be given with caution to patients with heart failure. Each gram of resin adds approximately 2 mEq (46 mg) of sodium. Signs of excessive sodium retention include hypertension and edema. The average adult oral dose of sodium polystyrene sulfonate is 15 g, one to four times daily in a small quantity of water or syrup

(3 ml–4 ml/g resin). Since obstruction has been reported with the use of exchange resins, the use of sorbitol, a liquid that promotes osmotic diarrhea, can be mixed with the resin to increase the excretion of the bound potassium. The patient should be told that mild diarrhea is desired when Kayexalate is used. The resin, mixed with sorbitol, may also be given as an enema; each adult dose, usually 30 g, is administered in 150 ml to 200 ml of water. Best results are achieved when the emulsion is warmed to body temperature before use. The enema should be retained for at least 1 to 2 hours (if back leakage occurs, the patient's hips should be elevated on pillows or a knee-chest position should be assumed temporarily). The enema vehicle administered with the cation exchange resin should be measured before instillation and after expulsion of the cation resin enema. A Harris flush may be given to facilitate removal of the previously administered resin prior to instillation of another retention enema. Oral administration of Kayexalate is preferred because it allows the resin to come into contact with a greater surface area of the gastrointestinal tract than is possible with rectal administration.

Cation exchange resins actually remove potassium from the body (approximately 1 mEq of potassium/g of resin). However, it is important to remember that cation exchange resins are of the most value as a *preventive* rather than as an emergency method to reduce severe potassium excess.

Emergency measures Emergency measures to temporarily relieve dangerously high hyperkalemia (>7 mEq/liter) may include the following:

1. The intravenous injection of $NaHCO_3$ solution (such as 44 mEq over a 5-minute period) causes rapid movement of potassium into the cells, temporarily lowering the plasma potassium level (1 to 2 hours). In addition, sodium bicarbonate provides sodium ion for antagonizing the cardiac effects of potassium. This procedure may have to be repeated if the EKG still shows remarkable abnormalities or if other symptoms persist. The patient should be observed for sodium overload and calcium deficit (tetany).
2. The intravenous infusion of hypertonic dextrose (usually 20% dextrose in water) and regular insulin temporarily reduces the plasma potassium level by inducing a temporary shift of extracellular potassium into the cell. These substances are usually given in the ratio of 1 unit of regular insulin to 5 g

of dextrose. (The regular insulin is sometimes given subcutaneously.) This form of treatment may take from half an hour to an hour; however, it usually is effective for several hours. Once begun, insulin–dextrose therapy should not be stopped until the total body potassium has been reduced by other means, since dangerous shifts of potassium from the cellular to the extracellular space may occur, with the return of hyperkalemia. Also, rebound hypoglycemia should be observed for after the administration of a concentrated carbohydrate solution, since the pancreas may continue to secrete extra insulin for a short period after the carbohydrate is discontinued.

3. The intravenous administration of calcium acts rapidly to provide temporary myocardial protection from the toxic effects of hyperkalemia without actually lowering the plasma potassium level. (Recall that calcium antagonizes the cardiac effects of potassium.) While the EKG is continuously monitored, 50 ml to 100 ml of calcium gluconate can be given at a rate of 2 ml/minute; this can be followed by the addition of calcium to the hypertonic dextrose and insulin solution mentioned above.[4]

It is important to remember that these emergency measures for treatment of hyperkalemia have only temporary effects. They serve to "buy time" for the institution of treatment measures to actually rid the body of excess potassium (such as cation exchange resins, hemodialysis, or peritoneal dialysis).

Dialysis Usually hyperkalemia can be controlled by conservative measures; however, dialysis (either hemodialysis or peritoneal dialysis) is indicated when conservative treatment is ineffective. Dialysis can also be indicated to correct serious metabolic acidosis, uremia, pulmonary edema, and deterioration in the patient's general condition. (Hemodialysis works more rapidly in relieving these disturbances.) Dialysis is discussed in detail later in this chapter.

Measures to reduce catabolism Factors contributing to catabolism include immobilization, infection, and fever. Moderate activity should be encouraged as soon as possible and infections should be treated promptly with appropriate antibiotics, usually in reduced doses owing to diminished renal function. If the patient has dirty wounds, undrained pus collections, or necrotic tissue, the plasma potassium concentration rises as a result of catabolism. To prevent severe potassium excess, it is important that necrotic tissue and pus be removed.

Although androgenic hormones have been demonstrated to decrease protein catabolism, their effect in patients with *high* rates of protein catabolism from trauma or infection seems negligible. The virilizing effects of testosterone may outweigh its anabolic effect.

ECF VOLUME EXCESS Fluid volume excess is a frequent problem during the oliguric phase when the body is unable to excrete excess fluid. Usually it is due to the excessive administration of fluids, either orally or intravenously.

Symptoms include elevated venous pressure, distention of neck veins, edema, puffy eyelids, bounding pulse, and shortness of breath. Hypertension, pulmonary edema, and congestive heart failure are complications of fluid volume excess. Pulmonary edema is manifested by severe dyspnea, moist rales, and frothy sputum.

Treatment The amount of fluid administered must be carefully planned to suit the patient's needs, with the body considered as a closed system. Excessive fluid intake should be avoided. Fluid intake should equal urinary output and other losses of water (such as occurs in diarrhea, vomiting, and fever) plus a basic ration of 400 ml/day for the average adult. Each degree of Centigrade temperature elevation/24 hours may necessitate adding 100 ml to the total daily fluid intake. The nurse must keep an accurate account of *all* routes of fluid gains and losses because an inaccurate record could lead to a fluid overdose and fluid volume excess with its dangerous sequelae. One should recall that the presence of excessive catabolism (as occurs in trauma and infection) increases endogenous water production and may necessitate decreasing the fluid intake. *If one is to err in calculating fluid intake, it should be in the direction of too little rather than too much.* Some authorities recommend that the electrolyte content of all fluids put out by the patient be measured. In general, patients with acute renal failure should not be given electrolytes in parenteral fluids except to replace documented losses.[5] (Nursing considerations in measuring and recording fluid intake and output are discussed in Chapter 4.)

The nurse should let the patient help decide how his limited fluid intake will be distributed over the 24-hour period.

Accurate daily body weight measurements are also

necessary for determining the desired fluid dose; they are probably the simplest and most accurate check on overall fluid balance. There should be a weight loss of 0.2 kg to 0.5 kg/day.[6] (Nursing considerations in obtaining accurate body weight measurements are discussed in Chapter 4.)

Sodium intake usually is restricted to help avoid fluid volume excess. It is sharply restricted when hypertension and congestive heart failure are present.

METABOLIC ACIDOSIS Metabolic bodily processes normally produce more acid wastes than alkaline wastes. Thus, when the kidneys fail to function, acid accumulation exceeds alkali accumulation. Metabolic acidosis is the result of the retention of these acid metabolites; their accumulation in the bloodstream causes the pH to drop. The hydrogen ions are retained to a large extent as sulfuric, phosphoric, and organic acids. The increased hydrogen ion is buffered by serum bicarbonate in an effort to maintain normal pH; the net result is a lower concentration of bicarbonate. Acidosis parallels uremia in severity. Decreased food intake causes increased utilization of body fats and the accumulation of ketones in the bloodstream; these acids further decrease the pH. Acidosis may develop more rapidly if marked tissue trauma or uncontrolled infection is present.

Vomiting commonly occurs with the development of uremia. If vomiting occurs at the time metabolic acidosis is developing, it is possible that the metabolic alkalosis accompanying vomiting may help to counteract acidosis. However, because gastric hypoacidity frequently occurs with uremia, vomiting may not have a significant effect on the pH. Extensive diarrhea contributes to the development of metabolic acidosis.

Metabolic acidosis causes a compensatory increase in pulmonary ventilation, which causes the elimination of large amounts of carbon dioxide from the lungs with a resultant decrease in the carbonic acid content of the blood. The pH is partially corrected by this mechanism. Anorexia, weakness, apathy, and coma may also be symptoms of metabolic acidosis.

Treatment The metabolic acidosis is tolerated fairly well and usually does not require treatment unless the plasma bicarbonate level is below 16 mEq/liter (normal 26 mEq/liter) or frank Kussmaul respiration, stupor, or coma occurs. Alkalinizing agents such as sodium bicarbonate or sodium lactate may be given orally or intravenously, as indicated, to elevate plasma pH. (So-

dium salts must be administered with caution to prevent fluid overload.)

Treatment by dialysis may be preferable in many cases, since the infusion of alkalinizing agents such as sodium bicarbonate or lactate may precipitate pulmonary edema or may cause clinical tetany in the presence of hypocalcemia.

Calcium may be indicated when alkalinizing agents are given to treat acidosis, since symptoms of calcium deficit (tetany and seizurelike movements) may be induced by an increase in plasma pH. (Calcium ionization is decreased when alkalinity of the ECF increases.) Restoration of a normal pH may disclose a calcium deficit that was not evident when the plasma pH was below normal. *Usually a depressed bicarbonate is restored only gradually and partially to normal to avoid this difficulty.* Since calcium acts synergistically with digitalis, it must be administered cautiously to digitalized patients.

SODIUM BALANCE The majority of patients with acute renal failure exhibit some degree of hyponatremia. Contributing to sodium deficit is the administration of excessive amounts of water, which dilute the plasma sodium. Sodium deficit may also be due to a shift of sodium into the cells, particularly if acidosis is present. Occasionally, sodium deficit becomes severe as a result of excessive loss of sodium in vomiting and diarrhea.

Moderate hyponatremia, *per se*, is usually asymptomatic. If the sodium level is markedly reduced (below 120 mEq/liter), enough water may be shifted into the cells to cause signs and symptoms of water intoxication (nausea, emesis, muscular twitching, grand mal seizures, and, finally, coma).

Sodium deficit usually is treated by limiting the water intake. Rarely, a hypertonic solution of sodium chloride (such as 3% or 5% NaCl) is administered *slowly* by the intravenous route in a *small* dose to correct a severe sodium deficit produced by excessive vomiting or diarrhea. Hypertonic solution of sodium chloride should be administered with caution because it can easily result in fluid volume excess with congestive heart failure and pulmonary edema.

It is possible that failure of the kidneys to excrete sufficient sodium, caused by reduced glomerular filtration rate (GFR), may result in sodium retention. Sodium retention, in turn, leads to water retention with resultant fluid volume excess (described earlier). Fluid volume excess is treated by limiting sodium and fluid intake and, when severe, by dialysis.

CALCIUM AND PHOSPHATE DERANGEMENTS The plasma calcium concentration is often below normal and may be related to the increased concentration of phosphate in the bloodstream. (Recall that phosphate is a constituent of one of the retained metabolic acids). A reciprocal relationship sometimes exists between calcium and phosphate so that an increase in one causes a decrease in the other. However, the elevated serum phosphate correlates poorly with the hypocalcemia of acute renal failure (which can go as low as 6.5 mg–8.5 mg/100 ml within 2 days of oliguria).[7] Decreased absorption of calcium from the gut occurs in uremia and contributes to the hypocalcemia. The cause of hypocalcemia in acute renal failure is not altogether clear; its severity is seldom so marked as occurs in chronic renal insufficiency.

The major significance of calcium deficit is that it enhances the toxic effects of potassium on the heart. (Recall that calcium and potassium have antagonistic actions on heart muscle.)

A plasma calcium level below 2 mEq/liter usually causes tetany; however, these symptoms rarely occur in renal failure, since the decreased blood pH of metabolic acidosis favors calcium ionization. (Calcium ionization increases in acidosis and decreases in alkalosis.) Chvostek's sign may be positive even though other symptoms of calcium deficit are not present. If metabolic alkalosis develops as a result of treatment with alkaline fluids, calcium deficit becomes manifest with the development of muscle twitching and convulsions. The EKG is usually normal but may show a prolonged Q-T interval.

Sustained hypocalcemia resulting from hyperphosphatemia causes the parathyroid glands to hypertrophy in an attempt to compensate for the low calcium level by production of parathormone. Calcium is withdrawn from the osteoid tissue and eventually results in bone changes if the calcium–phosphate derangement persists. (See discussion of bone changes under section on chronic renal failure.)

Calcium deficit can be treated with calcium gluconate by the oral or intravenous route—orally if nausea and vomiting are absent, intravenously in all other cases. The slow administration of 30 ml of calcium gluconate intravenously will temporarily ameliorate symptoms of hypocalcemia; once symptoms are relieved, calcium should be infused continuously (such as 100 ml of calcium gluconate per day).[8]

Calcium, as mentioned earlier, can cause pronounced improvement in EKG changes produced by potassium excess, even though the plasma potassium concentration is not changed.

Some physicians administer vitamin D in the active form (25-hydroxy-cholecalciferol) in an attempt to increase calcium absorption from the gut. Measures to reduce the high phosphate level may cause an increased calcium level. Therefore, aluminum hydroxide gels (Amphogel and Basalgel) may be administered to combine with phosphate in the gastrointestinal tract. Phosphate-binding gels may not be well accepted by patients since they cause constipation, nausea, and vomiting. (Excessive use of these gels can cause phosphate depletion and hypercalcemia.)

MAGNESIUM EXCESS The plasma magnesium level may increase when oliguria is present as a result of decreased renal excretion of this substance. Hypermagnesemia may be intensified to a serious level when magnesium-containing drugs are administered, such as magnesium sulfate, milk of magnesia, Mylanta, Gelusil, Maalox, and Creamalin. Symptoms include weakness, drowsiness, impaired respiration, orthostatic hypotension, and coma. Magnesium excess is primarily a disorder of acute and chronic renal disease and contributes to the central nervous system features associated with uremia.

URINE CHANGES The urine usually is bloody for the first few days, becoming clear about the end of the oliguric phase. If renal failure is due to hemolytic blood transfusion reaction, the urine has a port wine color. As renal failure advances, there is a tendency for the urinary specific gravity to become fixed at a low level (isosthenuria), similar to that of the glomerular filtrate (close to 1.010). A low specific gravity indicates that the renal tubules have lost their ability to concentrate urine.

ANEMIA Anemia may develop within 48 hours after the onset of acute renal failure. The red blood cells are normal in color (normochromic) and shape (normocytic) but are decreased in number. The hematocrit usually stabilizes at a level of 20% to 25%.

The inadequate production of a renal enzyme that stimulates release of erythropoietin is the probable cause of anemia. (Recall that erythropoietin normally stimulates the bone marrow to produce red blood cells—thus, a deficit of this substance results in anemia.) Other factors that may contribute to the occurrence of anemia include high plasma levels of urea, potassium, and hydrogen ions. It has been found that the life span of red blood cells is decreased when uremia is present, further contributing to the development of anemia. The low hemoglobin level predisposes to acidosis, since the buffering action of hemoglobin is diminished.

Mild bleeding, such as bruising or bleeding of gums, may occur early. Severe bleeding may occur into the gastrointestinal tract, lungs, or brain. The cause of the bleeding tendency in uremic patients is thought to be related to platelet defects resulting in imparied conversion of prothrombin to thrombin. Decreased platelet adhesiveness increases bleeding time.

The anemia of renal failure usually is tolerated well unless the hemoglobin level falls below 7 g/100 ml of blood. Severe anemia may be manifested by fatigue, dyspnea on exertion, palpitations, tachycardia, and angina.

Unless symptoms are present, anemia usually is left untreated. Sometimes transfusion is absolutely necessary owing to active bleeding, angina, or hypovolemia. If so, it is best to give fresh frozen packed red cells, since the risk of hyperkalemia is less than when bank blood is used; also, the risk of fluid volume overload is less. Unfortunately, the transfused cells have a shortened life span (probably a result of the abnormal constituents in the patient's blood). Transfusions depress normal bone marrow reticulocytosis, may cause hepatitis, and increase antibody levels that can later cause rejection of renal transplants. Washed cells keep the leukocyte antigenic exposure to a minimum, making rejection of a future transplant less likely to occur.

Androgenic steroids may be given for their erythropoietic effect. Administration of androgens to females causes masculinizing effects such as deepening of the voice, increased hair growth, and skin thickening.

INFECTION Patients with acute or chronic renal failure are highly susceptible to infection; in fact, infection is a major cause of mortality in such patients. Not only is the acutely uremic patient more susceptible to infections, but once started, the spread of bacteria is difficult to control. Exposure to others who are ill must be avoided; also, the patient should receive adequate rest periods and avoid chilling. Indwelling urinary catheters frequently cause urinary tract infections and should not be used in patients with oliguric acute renal failure once the diagnosis is established. If a catheter is required, it should be connected to a closed drainage system to minimize contamination. Scrupulous perineal and catheter care are indicated, along with the application of Betadine or a topical antibiotic at the catheter's point of entry into the urethra. Culture and sensitivity tests should be performed when infection occurs. Recall that fever, one of the cardinal symptoms of infection, may be absent in the acutely uremic patient with an infection. (Hypothermia is a well-recognized complication of renal failure.) The dosage of potentially nephrotoxic antibiotics (such as kanamycin, polymyxin, and gentamicin) must be adjusted to each patient's renal function and metabolic status. Sulfonamides, nitrofurantoin, and tetracycline should be avoided altogether. Chloramphenicol, erythromycin, penicillin, and penicillin derivatives can be used with only slight dosage reduction.[9]

UREMIA Uremia (azotemia) is a toxic condition caused by failure of the kidneys to excrete urea, potassium, organic acids, and other metabolic waste products. It progresses most rapidly during the first few days of oliguria. Uremia is described in the section on chronic renal failure.

DIETARY MANAGEMENT

Oral intake Protein intake is reduced to minimize urea formation, and a high caloric intake is encouraged to minimize protein catabolism and generally improve nutrition. Although protein must be limited to reduce urea formation, it is an important nutrient that must be supplied in the diet to prevent malnutrition. Proteins of high biologic value, such as eggs, milk, fowl, meat, and fish, are recommended for patients with renal failure in amounts of 0.25 g/kg body weight/day for adults and 0.5 g/kg body weight/day in the child.[10] Amin Aid (an oral commercial preparation of essential amino acids) may be used as a dietary supplement.

Giordano showed a decrease in azotemia (uremia) in chronic renal failure patients treated with a diet composed largely of essential amino acids and carbohydrates.[11] The Giordano diet primarily involves the ingestion of essential amino acids. Under these circumstances, the blood urea nitrogen (BUN) is split to ammonia, reabsorbed, and made available for synthesis by transamination into nonessential amino acids; the essential and nonessential amino acids are then synthesized into protein.

Caloric intake should be at least 35 calories to 45 calories/kg normal body weight in the adult and 50 calories to 75 calories/kg in the child; the precise desired caloric intake varies with age, sex, physical activity, and the degree of preexisting undernutrition. Examples of foods that may be used to increase caloric intake in the renal failure patient include butter, honey, hard candy, gumdrops, tapioca, sherbets. Some patients tolerate a mixture of equal parts of karo syrup and ginger ale, flavored with lemon juice and served chilled. Controlyte (Doyle Pharmaceutical) is a commercially available electrolyte-free powder that provides 1000 calories in 7 oz and is a useful source of calories for the renal failure patient.

Liberal carbohydrate and fat intake is indicated to decrease endogenous protein catabolism; such catabolism is harmful because it releases nitrogenous metabolites (the most likely cause of uremic symptoms) and potassium and hydrogen ions into the ECF. In addition, excessive protein catabolism may lead to malnutrition, which is an additional threat to recovery from acute renal failure. It is well recognized that nonprotein calories (carbohydrates and fats) have a protein-sparing action. To be most effective, these nonprotein calories should be evenly spaced over the 24-hour period.

As mentioned earlier in the chapter, the amount of fluid administered must be carefully computed with the body considered as a closed system. The actual amount of fluid allowed depends on the patient's dry weight (his tissue weight without excess fluids) and the point at which symptoms of fluid and electrolyte problems appear.

The physician calculates a fluid dose that will just replace sensible and insensible losses, relying heavily on the nurse to supply accurate intake and output records as well as weight charts. It is important that the correct amount of fluid be administered; *too much* fluid can cause hypertension and congestive heart failure; too little fluid decreases the GFR and further compromises renal function.

All substances that are liquid at room temperature should be counted as fluid intake, (e.g., ice cream, popsicles, sherbet, soups, beverages, liquid medicines). It is important that fluids be spaced carefully over the 24-hour period to avoid thirst. Cold liquids seem to be especially effective in quenching thirst. In monitoring daily weight changes, it is useful to remember that daily fluctuations reflect body fluid changes more than changes in actual weight. (Recall that 1 kg (2.2 lb) is equal to about 1000 ml of fluid.)

High-potassium foods should be excluded from the diet because the nonfunctioning kidneys are unable to excrete potassium. Examples of foods to avoid include bananas, citrus fruits, fruit juices, tea, coffee, legumes, and nuts. Since most salt substitutes contain potassium, they should not be used.

Tube feedings composed of essential amino acids, glucose, vitamins, and water have been administered to patients with acute renal failure to increase caloric intake. Use of tube feedings can be hazardous, however, when frank uremia is present because of the possibility of producing gastric bleeding. Antacid therapy by mouth or by nasogastric tube may be employed to decrease the risk of serious bleeding from stress ulcers; the antacids should be magnesium-free to avoid the danger of hypermagnesemia. The oligo-anuric patient should, if possible, receive no sodium.

Early dialysis is now done more frequently than in the past and allows for a more liberal diet. (Dietary considerations for the dialysis patient are discussed in the sections dealing with hemodialysis and peritoneal dialysis.)

Nausea and vomiting are frequent in the patient with acute renal failure and may require treatment with phenothiazines; however, repeated use leads to drug accumulation and may cause hypotension, making hemodialysis more difficult to accomplish.

Parenteral nutrition If oral intake or tube feedings are impossible, it becomes necessary to supply nutrients by the intravenous route. A 50% dextrose solution may be administered by slow drip through a central vein over the 24-hour period, or a solution of 10% dextrose, Intralipid (fat emulsion), and an essential amino acid preparation (such as Nephramine) may be given. Some physicians favor the use of total parenteral nutrition (TPN) for acute renal failure patients. TPN solutions employ hypertonic dextrose, essential amino acids, and vitamins. (Freamine E is a commercial parenteral preparation of essential amino acids.) Because the solution is very hypertonic, many calories are contained in a small volume of liquid; thus, even oliguric patients may receive a substantial number of calories. The use of TPN in acute renal failure has been reported to be associated with greater anabolism and shortening of the convalescent period. Abel and associates demonstrated that acute renal failure patients treated with essential L-amino acids and 57% dextrose solution had a significantly lower incidence of infectious complications and decreased overall mortality as compared with patients treated with intravenous glucose alone.[12] Potential complications of TPN include fluid overload, magnesium deficit, calcium deficit, phosphate deficit, potassium deficit, and metabolic acidosis. The TPN solution should be given cautiously at the prescribed rate to avoid fluid volume overload and hyperosmotic complications. Blood sugar levels should be monitored closely and insulin given as necessary. Electrolyte levels should also be closely monitored. (TPN is discussed more thoroughly in Chapter 6.) TPN in acute renal failure patients is most effective when daily hemodialysis is used to control fluid and electrolyte balance.

In uremic patients who are eating poorly and are not receiving TPN, amino acids and glucose may be infused into the venous line of the dialyzer as a nutritional supplement with each dialysis with little risk or discomfort.[13]

DIURETIC PHASE

The early diuretic phase of acute renal failure begins when the 24-hour urine volume exceeds 400 ml/24 hours, usually around the tenth day after onset; in some instances, it may not occur for 14 to 21 days. The amount of urinary output depends on the treatment the patient received during the oliguric phase. If fluid overloading was allowed, the urine volume may be more than 5000 ml daily. If the patient was well managed, the urine volume is not excessive. Urinary output usually increases in a stepwise manner, but occasionally increases rapidly.

During the early part of the diuretic phase, the partially regenerated tubules are unable to concentrate urine, and the glomerular filtrate is excreted virtually unchanged. Thus, the patient's condition does not improve in the first few days of the diuretic phase; indeed, uremia may be more severe during this period than at any other time. Convulsions, stupor, nausea, vomiting, hematemesis, bloody diarrhea, or hemorrhage may occur. The reason for the severity of the uremia lies in the rapid contraction of the total body fluid; early and prompt replacement of fluid is needed until the BUN level begins to fall. Dialysis is indicated if uremia is severe. Later in the diuretic phase, metabolic wastes begin to be cleared.

Hypokalemia and hyponatremia should be watched for during this phase, since potassium and sodium losses are increased when urine output is great. Urinary losses of these substances should be measured and replaced quantitatively, unless hyperkalemia is present. Body weight should be measured twice daily; if it decreases too rapidly, water and electrolytes must be supplied in sufficient amounts to prevent hypovolemic shock. As a rule, fluid intake is restricted to less than that excreted in the urine (since quantitative replacement may serve to perpetuate the obligatory water and salt loss). Urine and plasma electrolyte levels should be measured daily in addition to daily hematocrits. EKGs are of ancillary value and should be obtained periodically. Alkalinizing agents may be needed if acidosis is severe; however, they should be used cautiously to avoid precipitating symptoms of hypocalcemia.

Prognosis is favorable for patients with little underlying disease. On the average, the BUN becomes normal within approximately 15 to 20 days after the onset of the diuretic phase; the clinical signs and symptoms of uremia usually subside rapidly although it may be several months before the patient is well enough to resume full activities. The associated anemia improves gradually over a period of months.

Prognosis is unfavorable for patients with severe underlying renal damage. Significant permanent renal damage (following acute tubular necrosis) can occur from a variety of insults, such as heatstroke; severe or prolonged septic shock; severe fluid volume deficit, especially with the administration of diagnostic contrast media; and nephrotoxins, such as methoxyflurane and ethylene glycol.

Despite advances in treatment, the mortality rate remains at approximately 50%. Two thirds of the deaths occur during the oliguric phase and one third during the diuretic phase.[14] The mortality rate is related to the etiology, being highest in patients with marked tissue destruction (as in crush injuries, severe trauma, or major surgery).

CHRONIC RENAL FAILURE

Humans are normally born with 2 million nephrons; they can survive, although with difficulty, with as few as 20,000 nephrons. Chronic renal failure is the result of progressive irreversible loss of functional nephrons and occurs gradually over years or months. Because of the advent of long-term dialysis and renal transplantation, death is no longer the inevitable result when the nephron population falls below the minimal number compatible with unassisted life. Causes of chronic renal failure may include glomerulonephritis, pyelonephritis, polycystic kidneys, essential hypertension, or urinary obstruction. Chronic renal failure is sometimes classified into three stages:

1. Diminished renal reserve—renal function is somewhat diminished but the BUN level remains normal; symptoms may include nocturia and polyuria. One of the earliest regulatory functions of the kidney to fail is the maximal urine concentrating ability. It becomes necessary for the kidneys to excrete a large urinary volume (up to 3000 ml a day) to rid the body wastes.
2. Renal insufficiency—renal function is impaired to the point where metabolic wastes begin to accumulate in the bloodstream; homeostasis is usually maintained sluggishly.
3. Uremia—renal function is so markedly impaired that homeostasis can no longer be maintained; the BUN level rises sharply and serious fluid and electrolyte disturbances occur.

Prognosis is variable, depending on the degree of irreversible renal damage. Serious abnormalities in the body fluids may not appear until more than 75% of the nephron population has been destroyed, indicating the kidneys' remarkable homeostatic ability. When the GFR is as low as 15% to 20% of normal, the untoward consequences of nephron reduction begin to appear. The greater the fall in GFR, the more serious and extensive are the symptoms; virtually no organ system escapes damage from advanced renal insufficiency. Death is imminent when the GFR has reached 1% to 3%, unless the patient can be maintained on dialysis.

POTASSIUM IMBALANCES Plasma potassium concentrations vary widely in patients with chronic renal failure. Some have normal levels until oliguria and starvation occur, causing hyperkalemia. Other causes of hyperkalemia may include severe metabolic acidosis, catabolic illnesses, excessive protein intake, administration of stored blood, and gastrointestinal hemorrhage. Use of potassium-conserving diuretics (triamterene and spironolactone) may result in fatal hyperkalemia in renal failure. As a rule, hyperkalemia does not usually occur until late in chronic renal failure.

In some cases of chronic nephritis with polyuria, the plasma potassium may be low. Hypokalemia may also result from anorexia, vomiting, diarrhea, and excessive aldosterone production.

Since potassium metabolism varies from patient to patient, each patient must be evaluated individually to ascertain his need for potassium restriction or replacement.

METABOLIC ACIDOSIS In late chronic renal failure, metabolic acidosis occurs as a result of the impaired renal excretion of acid metabolites and the inability of the tubules to form ammonia. (The plasma bicarbonate level usually stabilizes at approximately 18 mEq to 20 mEq/liter) Although the respiratory mechanism partially corrects the decreased blood pH, it is sometimes necessary to treat the acidosis medically; sodium bicarbonate tablets may be given, in dosages of several grams a day, depending on the severity of the acidosis and on the sodium intake limitations. Acidosis usually is not treated unless the plasma bicarbonate is less than 16 mEq/liter or symptoms are present (such as anorexia, weakness, apathy, and coma). Drugs that impose an exogenous acid load (such as aspirin, ammonium chloride, or methionine) worsen acidosis and are contraindicated. Calcium administration may be necessary when alkalinizing agents are given to correct acidosis, since tetany may be induced by an increase in plasma pH.

SODIUM BALANCE Most patients with chronic renal failure maintain sodium balance until relatively late in the course of their disease, provided they are maintained on an average salt intake. Because the need for sodium varies with each patient, the desired level of intake is best gauged by observing the 24-hour urinary sodium output when the patient is on a known salt intake.

Excessive loss of sodium causes contraction of the ECF volume and constitutes one of the most common causes of an acute exacerbation of the uremic state. Excessive sodium loss can occur from the kidneys, vomiting, diarrhea, or profuse sweating. Salt depletion is treated by the oral or intravenous administration of sodium chloride, taking care to avoid overexpansion of the ECF; such treatment often results in a rise in GFR and striking clinical improvement.

If a patient exhibits sodium retention, the dietary intake of sodium must be reduced. (Sodium-restricted diets are discussed in Chapter 5.) Diuretics may be used as an alternative to dietary sodium restriction since they promote sodium excretion; however, high-dosage levels are often required at low GFRs, and the potential for toxicity is increased. A delicate balance must be struck between allowing enough salt to prevent sodium depletion, with its associated fluid volume deficit, and limiting salt intake to avoid fluid volume excess, with its associated hypertension, congestive heart failure, and pulmonary edema.

When the uremic patient presents with hyponatremia, the physician must determine if the problem is sodium loss or if it is water overload (as may occur when the patient has received excessive amounts of water or hypotonic fluids). Usually it is the former; if it is the latter, water restriction will usually serve to correct the hyponatremia. If dangerous water intoxication is present, it may be necessary to partially correct the low plasma sodium level by the cautious intravenous administration of 3% or 5% sodium chloride solution. (Dialysis, however, is frequently the treatment of choice for this complication). If a low plasma sodium level is due to actual sodium loss, it may be necessary to cautiously administer sodium-containing solutions.

CALCIUM AND PHOSPHATE DERANGEMENTS AND BONE CHANGES

Hypocalcemia und hyperphosphatemia occur in chronic renal failure when the GFR falls below 25 ml to 30 ml/minute (normal 100 ml–120m./min). Phosphates are retained in excess when renal function is impaired. Since a reciprocal relationship exists between phosphate and calcium, *hyperphosphatemia* causes *hypocalcemia*. Decreased absorption of calcium from the gut occurs in uremia and contributes to the hypocalcemia.

Efforts should be made to reduce the phosphate level in the bloodstream (to 5 mg/100 ml or less); dietary intake of phosphate is restricted; and phosphate-binding gels are given orally to bind phosphate in the intestine and thereby decrease the entry of phosphate into the ECF. (Hyperphosphatemia is thus minimized and its contribution to a drop in the ionized calcium level is minimized). Symptoms of hypocalcemia *rarely* occur in renal failure patients, since acidosis increases ionization of calcium. However, rapid correction (or overcorrection) of metabolic acidosis in a uremic, hypocalcemic patient can precipitate tetany and even grand mal seizures. Should hypocalcemic symptoms appear, calcium compounds may be administered orally or intravenously; severe symptoms require treatment by the intravenous route with either calcium gluconate or calcium chloride. (Caution should be exercised if the patient is receiving a digitalis preparation, since calcium ions enhance the action of digitalis.) Vitamin D may also be administered to increase calcium absorption from the intestine.

The low plasma calcium level stimulates the parathyroids, causing hyperplasia and increased secretion of parathyroid hormone. Excess of this hormone eventually causes dissolution of the bone and adds both calcium and phosphate to the ECF; decreased bone density and strength occur. Pruritus is often associated with calcium–phosphate derangement and may disappear after subtotal parathyroidectomy, although this procedure is performed less frequently now than in the past.

Normal serum levels of calcium (9 mg–10.5 mg/100 ml) and phosphate (3 mg–4 mg/100 ml) have a product of approximately 30 to 40. That is, the calcium level multiplied by the phosphate level equals approximately 30 to 40. Normal plasma levels of these electrolytes prevent their precipitation into soft tissues such as the blood vessels, skin, joints, and myocardium. An attempt is made to deep the Ca × P product below 70. Metastatic calcification is said to occur when calcium precipitates in such areas as the joints, cornea of the eye, heart, or lungs.

Chronic negative calcium balance produces renal osteodystrophy in patients with chronic renal disease. Usually the bone disease is not severe enough to cause symptoms, although pain and stiffness of the limbs and joints may occur. Demineralization of bone, most often in the hands, may be revealed by radiographs; sometimes it isn't discovered until autopsy. In childhood, chronic renal failure may cause bone lesions similar to those of vitamin D deficiency rickets; efforts are directed toward early renal transplantation.

One type of osteodystrophy (osteomalacia) is due to failure of calcium salts to be deposited in newly formed osteoid tissue. Another type of osteodystrophy (osteitis fibrosa) is due to reabsorption of calcium from the bone and replacement with fibrous tissue.

Treatment of renal osteodystrophy may include phosphate-binding gels, supplemental calcium, vitamin D, and, in extreme cases, parathyroidectomy.

ANEMIA Normocytic, normochromic anemia occurs in almost all patients with chronic renal failure. (See the discussion of anemia under the section on acute renal failure.)

CARDIOPULMONARY CHANGES Cardiopulmonary complications of chronic renal failure include congestive heart failure with pulmonary edema, pericarditis, pleurisy, and hypertension.

Congestive heart failure and pulmonary edema may occur when excessive salt and water are administered. Treatment consists of restricting salt and water intake, administration of diuretics (such as furosemide and ethacrynic acid, in modified doses), and administration of digitalis, also in a modified dose. (The digitalis dosage must be adjusted according to the patient's level of renal function to avoid digitalis toxicity.)

Pericarditis and pleurisy usually are late symptoms; they may be associated with myocarditis, cardiac tamponade, and pneumonitis. These conditions are most apt to occur when uremia is severe and are thought to be related to poor chemical control. Frequent hemodialysis is the treatment of choice for uncomplicated uremic pericarditis. Fluid may be aspirated from the pericardial sac when effusion is present, or a pericardiectomy may be preformed.

Patients with chronic renal failure usually have at least a mildly elevated diastolic blood pressure for reasons that are not always clear. Severe or symptomatic hypertension must be treated with antihypertensives such as methyldopa or with propranolol. In general, the pressure should be lowered gradually and great care should be taken to avoid hypotension. Moderate dietary sodium restriction with or without the addition of a diuretic may also facilitate the control of hypertension; but as emphasized earlier, care must be taken to avoid fluid volume depletion and associated hypotension. (Recall that severe uremia can be precipitated when renal perfusion is inadequate owing to volume depletion and hypotension.)

UREMIA Uremia is a toxic condition that may result from acute or chronic renal failure; it is caused by failure of the kidneys to excrete urea, creatinine, uric acid, organic acids, potassium, hydrogen ions, and other metabolic waste products. Symptoms of uremia may include the following:

Chronic fatigue

Nocturia

Insomnia

Anorexia (frequently caused by sight of food)

Intractable vomiting (often occurs in early morning)

Gastritis, hematemesis, and hiccoughs

Ammonia odor to breath and unpleasant metallic taste in mouth

Stomatitis; salivary urea is hydrolyzed by urease into ammonia, causing mucous membrane irritation

Pruritus—skin is dry and scaly, with excoriations caused by scratching (probably related to calcium and phosphate disturbances)

Pale sallow skin resulting from anemia and urochrome deposition in skin—discoloration most prominent on face and other body parts exposed to light

Increased bleeding tendency revealed by epistaxis, bleeding gums, easy bruising, petechiae, and conjunctival hemorrhage (probably caused by platelet defects)

Coarse muscular twitching, first occuring during sleep

Deep, rapid respirations (a respiratory attempt to compensate for metabolic acidosis)

Chest discomfort (may be caused by pericarditis or pleurisy)

Involuntary leg movements (restless leg syndrome)

Peripheral neuropathy (numbness, pain, and burning sensations in legs and arms; these changes begin distally and are symmetric—they may progress to motor weakness and paralysis)

Decreased libido and impotency in males; suppressed ovulation, libido, and menstruation in females

Hypertensive symptoms such as headache and visual difficulties (visual difficulties caused by retinal hemorrhages)

Delayed wound healing (sutures need to remain in place longer than usual) and increased susceptibility to infection

Hypothermia; patients with severe uremia may have body temperatures of 95°F (35°C) or less, even when infection is present

Decreased tolerance for all drugs that are excreted by the kidneys

Gradual diminution of mental acuity over a period of time, leading to coma (resulting from toxic substances in bloodstream and acidosis)

Generalized convulsions

Transient episodes of agitated psychotic behavior (hallucinations and paranoid delusions) interspersed with periods of lucidity

Uremic frost (urea crystal excreted through sweat glands, heaviest on nose, forehead, and neck) is rarely seen today because of improved management of uremic patients

Increased propensity for congestive heart failure (contributed to by the combination of hypertension, anemia, and acidosis)

Muscle wasting and weakness

Anemia (may be profound due to a combination of decreased erythropoietin, decreased red cell survival time, and blood loss)

Treatment Pruritis may be relieved by starch or vinegar baths. Nails should be kept short and clean to avoid skin trauma and infection caused by scratching.

Application of lotions or aquaphor ointments helps to relieve dry, cracked skin. Antipruritic medications should be given as necessary. In some patients with intractable itching, subtotal parathyroidectomy seems to provide relief (although this is done infrequently).

Good oral hygiene is essential to prevent oral infections and to alleviate oral irritation caused by ammonium hydroxide; a 0.25% acetic acid solution is beneficial as a mouthwash to help neutralize this substance. (Recall that ammonium hydroxide is formed in the mouth as a result of hydrolysis of oral urea.) Chewing gum, hard candy balls, and cold liquids are helpful in alleviating the unpleasant metallic taste caused by uremia. Gastrointestinal bleeding may necessitate the use of enemas to remove blood from the intestine and, thus, prevent elevation of potassium and BUN.

If diarrhea is a problem, administration of Lomotil may be indicated. If laxatives are necessary for the treatment of constipation, they should be *free of magnesium*, since patients with renal damage do not excrete magnesium normally and may thus develop serious hypermagnesemia.

Since the uremic patient is subject to disorientation and convulsions, his bed should remain in the low position with the siderails up. A padded tongue blade, suction apparatus, and oxygen equipment should be immediately at hand. Seizures should be treated with IV Valium, short-acting barbiturates, or Dilantin. The patient should be assessed frequently for changes in the level of consciousness and neuromuscular irritability. It is necessary to explain to the family that the patient's altered behavior is related to his toxic condition and not necessarily to anything they have done. The patient's family should be included in all explanations since the uremic patient frequently has some degree of impaired mental status. Seizures and other manifestations of uremic encephalopathy usually respond to control of uremia by dialysis.

DIETARY MANAGEMENT The level of renal impairment at which dietary protein restriction should begin is controversial. Some authorities recommend maintaining chronic renal failure patients on an unrestricted diet so long as the GFR is above 15 ml to 20 ml/minute and there are no special problems (such as severe acidosis, persistent hyperkalemia, or hypertension).[15] Once the GFR falls below 10 ml/minute, some degree of dietary protein intake is indicated to reduce potassium and phosphate intake and inhibit the development of acidosis. Substitution of essential amino acids may have the positive antiuremic effects of a low protein diet without the disadvantage (malnutrition) of a severely restricted protein diet. (See the discussion of dietary management of acute renal failure presented earlier in the chapter.)

Oral intake may be discouraged by the presence of nausea and vomiting; antiemetics can be helpful. If vomiting occurs, food can be offered again after an interval of 30 minutes. *It is important to keep the uremic patient's nonprotein caloric intake high to prevent endogenous protein catabolism.* Since the discomfort of stomatitis may decrease the patient's desire to eat, frequent oral hygiene measures should be performed to minimize oral discomfort (and prevent secondary infection). Attractive small frequent feedings may encourage eating. Encouraging activity may also stimulate appetite, provided congestive heart failure requiring bedrest is not present.

Sodium requirements must be individualized for each patient; soduim restriction may be necessary for some and harmful for others. Excessive sodium intake can result in edema, hypertension, and congestive heart failure; hence, patients retaining excessive sodium and water require a low-sodium diet. On the other hand, patients who excrete large quantities of sodium in the urine require a normal diet to prevent sodium deficit. The physician has to determine sodium needs according to the response to test diets and clinical observations such as excessive weight gain. Abnormal losses of sodium (as in vomiting, diarrhea, or excessive sweating) increase the need for sodium. (Low-sodium diets are described in Chapter 5.)

As a rule, potassium retention is not seen until the GFR falls below 5 ml/minute.[16] However, patients should be instructed to avoid high-potassium foods (such as bananas, oranges, and dried fruits). The potassium content of various foods may be reviewed in Tables 5-7 and 5-8.

Vitamin D and calcium supplements are given to treat the renal bone dystrophy of chronic renal failure.

Chronic renal failure patients who do not have oliguria or anuria should be urged to drink from 2000 ml to 3000 ml of water daily to aid in the elimination of urinary waste products. *If the patient is unable to excrete large volumes of water, however, fluid intake should be limited to his needs.* Fluid intake should be increased during periods of abnormal extrarenal fluid loss (such as occurs in vomiting, diarrhea, fever, and excessive sweating). In the final stages of chronic renal failure, patients may experience extreme thirst, probably a result of cellular dehydration produced by the increased osmotic effect of the high urea level.

Type IV hyperlipidemia wht elevated serum levels of triglycerides is often present in patients with chronic uremia and in those treated with dialysis; the hypertriglyceridemia may contribute to cardiovascular disease. The efficacy of dietary modification on the condition has not been documented.

DIALYSIS

Dialysis may be defined as the differential diffusion of water and solutes through a semipermeable membrane separating two solutions. The physical principles of diffusion, osmosis, and filtration are important to dialysis. The two major clinically effective dialysis techniques are hemodialysis (HD) and peritoneal dialysis (PD). Dialysis may be indicated in the treatment of patients with acute renal failure, chronic renal failure, acute poisonings, and certain severe body fluid disturbances. PD is a relatively inefficient technique compared with HD but, given enough time, PD is able to compare favorably with HD in the amelioration of the biochemical abnormalities of uremia. The choice of dialysis method is based on availability of facilities and experienced personnel, preference of the physician, and individual patient characteristics. These criteria will be discussed below.

HEMODIALYSIS

HD works by circulating blood extracorporeally through an artificial kidney (hemodialyzer). Hemodialyzers function by allowing arterial blood to pass through numerous tiny thin-membraned channels surrounded by a dialysate bath; substances then pass through the membrane from an area of greater concentration to an area of lesser concentration until an equilibrium is reached. The major advantage of HD is that it acts rapidly and efficiently (when managed properly) to rid the body of unwanted substances. It has proved successful in maintaining life in patients with otherwise fatal renal failure. However, HD is at best an imperfect substitute for normal kidney function, since it cannot completely reverse many of the sequelae of uremia (such as anemia, neuropathy, and osteodystrophy). Although usually performed in hospital settings,

it can be performed at home with special training. Patients selected for home HD must be medically and psychologically stable and must have the intellectual capacity to perform the procedure themselves. Home HD patients require extensive teaching and the help of a psychologically stable partner.

HD is used in the treatment of acute and chronic renal failure. Acute renal failure can usually be managed medically without the use of dialysis; nevertheless, dialysis simplifies treatment by improving patient comfort, facilitating the maintenance of better nutritional status, and reducing the risk of infection. In the past, dialysis was reserved for patients with severe fluid and electrolyte problems; now, some centers begin treatment early in order to *prevent* the development of uremic symptoms. The use of frequent HD not only prevents uremia with its complications but permits the oliguric patient to consume sufficient high-quality protein, fluid, and calories to avoid the complications of malnutrition. Some dialysis centers dialyze only for the control of uremic symptomatology; others choose to institute dialysis when the BUN reaches some arbitrary figure, usually 100 mg/100 ml.[17]

Access HD requires reliable long-term access to the circulation; this may be accomplished by an arteriovenous (AV) fistula, external arteriovenous (AV) cannula, direct cannulation, bovine vein grafts, or synthetic vascular prostheses. An AV fistula is a surgically created fistula between an artery and a vein. This method has several advantages over an external cannula: it allows the patient more freedom, it does not require maintenance care, it can be used for months or years, and it circumvents the problems of hemorrhage and clotting associated with a cannula. Anastamosis of a vein and artery results in an arterialized vein; when used for HD, two large needles must be inserted into the arterialized vein, one distally for the arterial line and one proximally for the venous line. A tourniquet is placed between the two needles to prevent blood from the venous line from entering the arterial line and thus entering dialyzer. The AV fistula is created weeks or months before it is needed; its average life is 3 to 4 years. Complications of an AV fistula can include hemorrhage, thrombosis, ischemia of the hand ("steal" syndrome), infection, aneurysm formation, and failure of the prominent vein to develop. Blood should not be drawn from the arm with the fistula, nor should blood pressure be checked on this arm.

When no suitable peripheral vessels are available for AV fistula or cannula, autogenous vein grafts or bovine grafts may be used (bovine grafts are antigenic and thus should not be used in patients awaiting transplantation).

HEMODIALYSATE The approximate composition of a hemodialysate developed for treatment of acute renal decompensation is presented in Table 10-1. Note that the dialysate approximates the normal ionic composition of normal ECF; the only nonelectrolyte solute is dextrose. The absence of other solutes allows the maximal removal of waste metabolites such as urea, creatinine, uric acid, and phosphate.

The relatively high concentration of acetate (a bicarbonate precursor) is provided to treat the metabolic acidosis that frequently accompanies uremia. (Substitution of bicarbonate for acetate in the dialysate has been recommended for critically ill patients with unstable cardiopulmonary status, since acetate has been described as having a myocardial depressant effect and a vasodilatory effect[18]).

Since hyperkalemia is frequently present in renal failure patients, the dialysate is free of, or low in, potassium. Potassium dialyzes readily and, in the presence of hyperkalemia, reductions in the serum potassium concentration can be as much as 4.0 mEq/liter[19]. If hypokalemia is present, extra potassium can easily be added to the solution to increase the plasma potassium concentration. Hypokalemia is particularly dangerous in patients receiving digitalis, since it may induce digitalis toxicity and potentially fatal arrhythmias. The optimum potassium concentration of dialysate is variable and dependent on a variety of clinical factors; generally, a concentration of less than 2 mEq/liter is being used for both acute and chronic HD.[19]

The calcium concentration in the dialysate is usually slightly higher than normal ionized serum calcium concentration (approximately 3 mEq/liter), allowing a small amount of calcium to dialyze into the patient. (Recall that hypocalcemia is common in renal failure patients).

The magnesium concentration of 1.5 mEq/liter in the dialysate is used by most dialysis centers and allows for correction of any magnesium abnormalities that might exist. (Most uremic patients have a mild degree of hypermagnesemia.) The use of a magnesium-free dialysate has been reported to be well tolerated and, possibly, to improve neuropathy.[20]

The sodium concentration of most HD solutions is slightly lower than the normal serum sodium level; this appears to date back to the early days of dialysis when its presumptive purpose was to create a concentration gradient for sodium, allowing it to dialyze out of the patient. Currently, sodium removal is achieved primarily by the convective transport that attends ultrafiltration.[21] The advisability of using a hemodialysate with a sodium concentration lower than the normal serum sodium level has been questioned by some authorities. Researchers have indicated that the use of a dialysate with a sodium concentration of 145 mEq/liter has lessened intradialytic symptoms of nausea and leg cramps without compromising maintenance of dry weight and control of hypertension.[22] In any event, choice of the desired dialysate sodium concentration must be made after considering the sodium content of the local water supply.

The glucose concentration in most dialysate solutions is between 100 mg and 200 mg/100 ml; this approximates the usual blood sugar level and thus does not allow any significant diffusion of glucose either into or out of the patient.

ULTRAFILTRATION Ultrafiltration is produced by creating a blood-to-dialysate hydrostatic pressure gradient and varies with the type of hemodialyzer used. The degree of ultrafiltration must be determined by the individual patient's need; factors to consider include the presence or absence of edema and hypertension and the amount of interdialytic weight gain. The goal of ultrafiltration is to maintain the total body sodium and water composition as close to normal as possible.

ADVANTAGES HD is preferable to PD under the following clinical circumstances:

HD is definitely indicated when it is important to

TABLE 10-1 *Composition of a hemodialysate*

Solute	Concentration Range
Sodium	130–135 mEq/liter
Potassium	0–2.0 mEq/liter
Magnesium	1.0–1.5 mEq/liter
Calcium	2.5–5.0 mEq/liter
Acetate	32–38 mEq/liter
Chloride	96–103 mEq/liter
Dextrose	0–200 mg/100 ml

achieve *rapid* removal of dialyzable fluid and toxic solutes. For example, a patient who does not comply with dietary restrictions and has grossly abnormal blood values may need HD to remove the excess fluid and metabolites rapidly. HD is especially superior to PD when the removal of exogenous poisons is required, since virtually all dialyzable drugs clear better with HD. Urea clearance in HD is approximately 80ml to 160 ml/minute as opposed to approximately 15 ml to 20 ml/minute in manual PD.

HD is generally preferable to PD when the patient has a *localized* intraperitoneal infection or abscess; the introduction of a dialysis catheter into the peritoneal cavity carries the risk of spreading the infection.

HD is the necessary choice when PD is technically impossible, as in the presence of dense intra-abdominal adhesions or distortion of the anatomic integrity of the peritoneal cavity. (The peritoneal cavity can be distorted by the inadvertent introduction of PD solution into the anterior abdominal wall or by extensive retroperitoneal hemorrhage.)

Some still view HD as the modality of choice for chronic dialysis; however, improvements in PD-automated equipment and methodologies have made PD a competitive alternative.

Complications The following complications may accompany HD:

Dialysis disequilibrium syndrome may occur as a result of the more rapid removal of urea from blood than from the brain, causing water to pass by osmosis into the brain, with resultant cerebral edema. Symptoms include headache, mental confusion, hallucinations, nausea, vomiting, visual blurring, leg cramps, peripheral paresthesias, anxiety, hyperventilation, and convulsions. (This disturbance has also occurred during peritoneal dialysis, although much less frequently, since PD reduces the BUN more slowly than does HD.) Thus, in patients with a very high BUN, the HD treatment must be modified to reduce dialysis efficiency.

Hypotension may result from ultrafiltration or when the blood flow from the patient to the dialyzer is rapid.

Heparinization may result in bleeding from a recent surgical site or from an area of ulceration in the gastrointestinal tract; it represents a serious risk to the patient with a hemorrhagic tendency. (The extracorporeal circulation of blood during HD requires that anticoagulation with heparin be achieved to prevent clotting in the artificial kidney.) This risk can be minimized either by regional heparinization or constant slow infusion of a minimal heparin dosage. (In regional heparinization, heparin is added to the blood as it enters the dia-

lyzer and is neutralized by protamine as the blood exits from the dialyzer and enters the patient.)

Blood reactions are always a possibility when blood is administered, particularly to patients who have received multiple transfusions in the past. (Blood transfusion reactions are described in Chapter 6.)

Hepatitis may occur from the administration of hepatitis-contaminated blood.

Cramping of muscles may result from rapid sodium and water removal.

The "hard water syndrome" can occur with the use of a dialysate in which calcium concentration exceeds 5 mEq/liter (usually seen in areas with a high calcium content in the local water supply).[23] The syndrome is manifested by a rise in blood pressure, nausea, vomiting, headache, somnolence, and intense burning of the skin. Both calcium and magnesium can contribute to this syndrome; the occurrence of nausea, vomiting, and hypertension indicates calcium as the causative agent, whereas the presence of central nervous system disturbances associated with hypotension suggests magnesium as the causative agent. The routine use of deionizers has decreased the occurrence of the hard water syndrome.

Hypermagnesemia can result from the accidental use of a dialysate containing a large amount of magnesium. Inadvertent use of a dialysate containing 15 mEq/liter of magnesium has been reported to cause flushing, hyperreflexia, muscular weakness, ataxia, and blurred vision.[24]

Respiratory alkalosis may be present after HD owing to a persistence of predialysis hyperventilation (which probably existed as a compensatory response to the metabolic acidosis or uremia).

Intradialytic symptoms such as nausea, vomiting, headache, and muscle cramps seem to be related to the use of a dialysate that is slightly low in sodium (these symptoms seem to be minimized by using a dialysate with a higher sodium concentration.)[25]

Metastatic calcification (calcium precipitation in such areas as the joints, cornea of the eye, heart, or lungs) has been seen in patients on chronic HD programs.

Eight to ten grams of essential and nonessential amino acids and 3g to 4g of peptides are lost in a single HD treatment in a nonfasting patient.[26] It is important to note, however, that the protein loss accompanying HD is not usually as great as that which occurs with chronic PD.

MONITORING THE HD PATIENT Prior to HD, it is important to measure and record weight, blood pressure (standing and lying), temperature, pulse, and respirations. During HD the patient should be observed for abnormal physiologic reactions such as fluid volume excess (elevated blood pressure, congestive heart failure, or pulmonary edema), hypovolemia (hypotension), transfusion reactions, disequilibrium syndrome, and other complications of HD. (See section dealing with complications of HD.) Vital signs (temperature, pulse, respiration, and blood pressure), weight, and clotting time should be monitored during dialysis and technical problems (such as blood leaks) should be observed for. The reader is referred to more specific texts on HD techniques if detailed information is desired.

DIET Patients requiring dialysis therapy are often already protein and calorie depleted. HD adds to the problem by causing some loss of protein and water-soluble vitamins. About 30% to 40% of the amino acids lost during HD are essential; therefore, the need for high-quality protein in the diet is increased. The actual amount of dietary protein required varies with the type of dialyzer used and the frequency of dialysis. Caloric intake should be optimal to prevent catabolism of protein for energy needs. Supplemental water-soluble vitamins should be administered daily.

PERITONEAL DIALYSIS

Because of its large surface area (22,000 cm^2), the peritoneum can be used as a dialyzing membrane for the removal of toxic substances, body wastes, water, and electrolytes by the processes of osmosis, filtration, and diffusion. As in HD, substances pass through the membrane from an area of greater concentration to an area of lesser concentration. PD is being performed with increasing frequency in hospitals, and many patients carry out their own treatment at home.

DIALYSATE A solution for PD must, in most instances, approach the electrolyte composition and tonicity of the plasma so that normal plasma constituents will not be altered (Table 10-2). It must be at least slightly hyperosmotic to plasma to prevent its absorption, with resultant development of fluid volume excess.

Various concentrations of dextrose may be used to render the solution hyperosmotic. Dialysis solutions with 1.5% dextrose are available commercially;* these

*Inpersol (Abbott Laboratories); Dianeal (Travenol Laboratories; Peridial (Cutter Laboratories).

TABLE 10-2 *Approximate electrolyte content of peritoneal dialysis solution*

Electrolytes	Usual ranges
Sodium	140–141 mEq/liter
Potassium	0–4 mEq/liter
Calcium	3.5–4.0 mEq/liter
Magnesium	1–1.5 mEq/liter
Acetate or lactate	43–45 mEq/liter
Chloride	101–102 mEq/liter

slightly hyperosmotic solutions are useful in removing abnormal plasma constituents, such as potassium or magnesium. Osmolality of a 1.5% dextrose dialysate is approximately 365 mOsm to 372 mOsm/liter; the typical uremic patient's serum osmolality usually ranges from 310 mOsm to 350 mOsm/liter.[27] Since the osmolality of the 1.5% dialysate is somewhat higher than the usual serum osmolality of uremic patients, dialysate is not absorbed from the abdomen. In nonuremic patients with a normal serum osmolality of 280 mOsm to 300 mOsm/liter, it is possible that a 1.5% dialysate could produce a degree of negative fluid balance (certainly not as much, however, as could a 4.25% solution).

A 1.5% solution does not usually affect fluid balance unless the dialysis rate is rapid. As just mentioned, a stronger hyperosmotic effect can be achieved with the use of a dialysate containing a 4.25% dextrose concentration, resulting in a more rapid removal of fluids. (A 4.25% dextrose dialysate has an osmolality of approximately 525 mOsm/liter.)

Potassium is omitted from commercial PD solutions to permit individualization for each patient; varying amounts of potassium chloride may be added to the dialysate as deemed necessary. A potassium-free dialysate is used to treat hyperkalemia; solutions containing about 4 mEq of potassium/liter are used for patients with normal plasma potassium levels. If the patient has hypokalemia, a solution with a greater than normal potassium concentration may be employed to allow potassium to dialyze into the patient. The addition of potassium to the dialysate is particularly important for digitalized patients, since the plasma potassium may fall rapidly during dialysis. (Recall that potassium deficit makes the patient more susceptible to digitalis toxicity.)

It is interesting to note that absorption of a large amount of glucose from the dialysate causes cellular incorporation of potassium (in the conversion of absorbed glucose to glycogen); this process lowers the serum potassium level until later when glycogenolysis occurs. (This process can take place in either HD or PD when a dialysate with a high glucose content is used.)

The sodium concentration of the typical dialysate (140 mEq/liter) generally results in a normal serum concentration of sodium after dialysis. However, hypernatremia has been observed after infusion of a hypertonic (4.25% dextrose) dialysate for the purpose of removing excess fluid.[28]

The presence in the dialysate of a bicarbonate precursor (either lactate or acetate) helps correct the metabolic acidosis associated with uremia. (Bicarbonate cannot be added to PD solutions, since the calcium in the dialysate causes the precipitation of calcium carbonate.)

The calcium content of the dialysate is slightly higher than the normal ionized serum calcium level; thus calcium tends to dialyze into the patient and helps correct the hypocalcemia that frequently accompanies renal insufficiency. (It is important to remember that calcium administration can contribute to the development of digitalis toxicity in the patient receiving a digitalis preparation.)

The magnesium content of the dialysate (1 to 1.5 mEq/liter helps correct any magnesium abnormalities. Most patients with renal insufficiency have a slightly elevated ionized serum magnesium level; if a large exogenous magnesium load has been imposed (as by the intake of magnesium-containing antacids or laxatives), the serum magnesium level can be markedly elevated.

It is desirable to warm the dialysis solution to body temperature before administration. Infusion of the dialysate at body temperature is more comfortable for the patient, increases peritoneal clearance by 35%, and prevents hypothermia. The bottles can be warmed by immersion in a warm water bath set at 37°C (98.6°F). Rubber bands hold the wet labels in place; a glass marking pencil can be used to indicate the contents of the bottle. To avoid contamination of the equipment, the outside of the bottles should be dried carefully before use. A warming cabinet is sometimes used to warm the dialysis solution; care should be taken to avoid overheating. Air should be cleared from the administration tubing, since air pockets in the peritoneal cavity impede drainage.

Medications should be added before the bottles are hung for infusion. Aqueous heparin in a dose of 1000 units per each 2-liter exchange may be used to prevent occlusion of the peritoneal catheter with clotted blood.[29] Antibiotics may be added to the dialysate if necessary; they are definitely recommended when there is any indication that there has been a break in sterile technique. Nephrotoxic or ototoxic antibiotics (such as kanamycin and neomycin) are contraindicated. Extreme care is required to prevent contamination when medications are added; recall that the most frequent serious complication of peritoneal dialysis is peritonitis. Some dialysates are available in single 2-liter bottles; because medications are added to only one container, the opportunity for contamination is reduced. If two separate 1-liter bottles are used, contamination risk can be minimized by adding medications to only one of the bottles. Solutions from two separate bottles mix quickly in the abdomen.

PREPARATION Prior to the initiation of PD the patient should be given a careful explanation of the treatment; reassurance and support are required throughout the lengthy procedure. The position of choice during PD is a modified semi-Fowler's. Meperidine (Demerol) may be given one half hour before the insertion of the catheter; the bladder and colon should be emptied to avoid injury. Inadvertent bowel perforation by the catheter results in pain and fecal drainage in the dialysate return, or copious watery diarrhea. If the catheter is accidentally introduced into the bladder, the patient may void dialysate solution. Accidental introduction of the catheter into a blood vessel causes blood to appear in the catheter or in the first dialysate return. The patient's weight and vital signs should be checked prior to the procedure for later reference. PD can be performed either manually or by machine.

MANUAL PROCEDURE After the abdomen is shaved and prepared, an abdominal paracentesis is performed and a catheter inserted into the peritoneal cavity. (Preparation of the dialysate solution is described in the preceding section.) Two liters of solution are allowed to enter in about 10 minutes. If the patient complains of discomfort when the fluid is instilled, the flow rate may have to be reduced; the physician should be notified if the pain does not subside. Analgesics may be administered if necessary; lidocaine may be added to the dialysate or injected through the dialysate tubing to alleviate pain. If the pain persists, the physician may need to change the location of the peritoneal catheter.

The fluid is left in the peritoneal cavity about 20 minutes and then is drained into a bottle or plastic drainage bag. Drainage of the fluid from the peritoneal cavity should not take much longer than 20 minutes. To facilitate drainage, it may be necessary to massage the abdomen gently or adjust the patient's position.

The dialysis solution is instilled and drained at regular intervals until the patient shows definite improvement. PD usually is performed for 18 to 48 hours in the treatment of the uremic patient. Studies have suggested that the optimum dialysis exchange rate is 3.5 liters/hour.[30] With the use of automated equipment, dialysate exchange rates as high as 12 liters/hour can be achieved with an increase in urea clearance from the usual 15 ml to 20 ml/minute up to 40 ml/minute.[31]

The nurse should use meticulous technique during PD to prevent *peritonitis*, the most common complication of peritoneal dialysis. It can develop from poor aseptic technique at almost any point in the dialysis procedure (e.g., when the catheter is inserted, when the bottles are changed, when the tubing is connected and disconnected, when medications are added to the bottles, during catheter care, or when taking a specimen of dialysate for culture and sensitivity). Bacteria commonly associated with peritonitis include *Escherichia coli*, *Staphylococcus aureus*, and *Staphylococcus epidermidis*. Treatment of peritonitis is imperative to prevent the dreaded septicemia and eventual scarring of the peritoneum, which makes it less permeable and thus hinders future dialysis treatments. The nurse should observe for symptoms of peritonitis, which include a cloudy dialysate (possibly containing shreds of fibrin or mucus), persistent abdominal cramps or pain, a feeling of fullness in the abdomen, and temperature elevation. (Recall that patients with chronic renal failure often have a subnormal body temperature; therefore, one should be alert for even a slight temperature elevation.) Peritonitis should also be suspected if the dialysate return decreases suddenly or comes in drips or spurts. When peritonitis is suspected, a sample of the dialysate should be taken for culture and sensitivity tests. Antibiotics may be added to the dailysate and given systemically (intravenously and orally).

Abdominal dressings should be checked frequently for bleeding or leakage of solution from around the catheter exit site. Wet dressings should be changed immediately to minimize the chance of wound infection and skin irritation. Vital signs should be observed periodically throughout the dialysis treatment for significant changes. The time at which each exchange begins and the amount of fluid instilled and removed should be recorded. When the total amount of drainage is more than the total amount of fluid infused, the patient's cumulative fluid balance is said to be negative. When the total amount of drainage is less than the total amount of fluid infused, the patient's cumulative fluid balance is said to be positive. The physician should stipulate the limits of positive and negative balance considered desirable for each patient.

A dialysis flowsheet is used to record the fluid exchange (Fig 10-1). For each exchange, the nurse should record the starting time, amount of fluid instilled, its concentration and any added medications, finishing time of inflow, starting time of outflow, color of outflow, finishing time of outflow, and the amount of positive or negative balance. (For example, if 2000 ml were instilled and only 1800 ml drained, the balance for this exchange would be +200 ml.) Vital signs should be checked frequently and the patient observed for signs of hypovolemia (particularly when a 4.25% dialysate is used), peritonitis, fluid overload with respiratory difficulties (caused by dialysate pressing against diaphragm), hyperglycemia, glycosuria, and other complications.

Fluid intake and output from all routes should be measured and recorded. Finally, body weight measurements are often taken twice daily, since these readings are invaluable in assessing the patient's state of fluid balance. Weight should be consistently measured at the same point during the dialysis procedure (such as immediately after the outflow is completed). Recall that 2 liters of dialysate fluid weigh approximately 2 kg (4.4 lb). An in-bed scale is of great value in monitoring weight changes during dialysis. (See the section on weight measurement in Chapter 4.)

ADVANTAGES Advantages of PD as compared with HD include the following:

All hospitals can use PD, while HD usually is available only in larger medical centers. PD techniques are far less complex and are more easily learned than those of HD. PD requires only a fraction of the personnel and equipment needed for HD.

Fluid and electrolyte shifts occur more gradually in PD than in HD, making hypotensive episodes and arrhythmias less apt to occur. Therfore, the risk to the hypotensive patient is lower in PD.

Immediate life-threatening events (such as hemorrhage and embolism) are not as apt to occur during

PERITONEAL DIALYSIS RECORD

ABBOTT

PATIENT_____ ROOM _____

ATTENDING PHYSICIAN _____

HOSPITAL NUMBER _____ AGE _____ SEX _____

DIAGNOSIS_____

_____

_____

WEIGHT PRIOR_____ WEIGHT ON COMPLETION_____

DATE	INPERSOL SOLUTION TYPE			MEDICATION ADDED TO SOLUTION	OTHER MEDICATION OR REMARKS	SOLUTION IN			SOLUTION OUT			DIFFER-ENCE PLUS OR MINUS	CUMULA-TIVE DIFFER-ENCE
	1.5%	1.5%K	4.25%			START-ING TIME	FINISH TIME	VOLUME	START-ING TIME	FINISH TIME	VOLUME		
	✔			KCl 8mEq		7$\frac{00}{A}$	7$\frac{10}{A}$	2000	7$\frac{30}{A}$	7$\frac{50}{A}$	2200	-200	-200
	✔			KCl 8mEq		7$\frac{50}{A}$	8$\frac{02}{A}$	2000	8$\frac{20}{A}$	8$\frac{40}{A}$	1900	+100	-100

97-0169/R1—1.25—APRIL, 1974 · LITHO IN U.S.A.

FIG. 10-1. Peritoneal dialysis flowsheet. (Courtesy of Abbott Laboratories, North Chicago, Illinois)

PD as in HD, making patients less fearful to use it, particularly in the home.

Patients with poor peripheral veins present an access route problem for HD and thus may be treated with PD. PD also is indicated for patients in shock whose arterial blood flow is insufficient for HD.

Infants and young children may be treated effectively with PD. HD is technically difficult to perform on young children.

Patients who refuse blood transfusions (such as Jehovah's Witnesses) and those in whom anticoagulation is considered dangerous (recent surgery, gastrointestinal bleeding, or bleeding defects) may be treated with PD.

It has been suggested that PD more efficiently removes yet-to-be characterized uremic toxins in the "middle molecular" weight range (400 daltons–5000 daltons) than does HD because the peritoneal membrane is more permeable.[32]

PD may be started quickly (within minutes) while HD requires a longer period to initiate (a minimum of 45 minutes and, more often, 1 to 3 hours). Time is a crucial factor in acute, life-threatening emergency situations such as severe hyperkalemia or acidosis.

Dialysis disequilibrium syndrome is less apt to occur in PD, since it is slower and less efficient than HD. Thus, when the BUN is significantly in excess of 100 mg/100 ml, it may be advisable to use PD to produce a more *gradual* reduction over a 24- to 48-hour period. (Remember, however, that disequilibrium syndrome *can* occur during PD, and that HD techniques can be modified to reduce dialyzing efficiency, thus minimizing the danger of disequilibrium syndrome.)

PD may have a therapeutic effect on generalized chemical or bacterial peritonitis by virtue of the dilution and removal of bacteria or chemical irritants and by the introduction of antibiotics into the peritoneal cavity. [33,34]

DISADVANTAGES The following are disadvantages of PD:

It is considerably slower than HD (urea clearance is only about 15 ml to 20 ml/minute with PD as compared with 80 ml to 160 ml/minute with HD). Patients on PD require approximately three to six times as much dialyzing time as those on HD to achieve essentially the same quantitative removal of solute and biochemical improvement.

Bacterial or chemical peritonitis is a potential complication of PD. It is evidenced by pain and tenderness of the abdomen, a sense of fullness, and the slow return of a cloudy dialysate fluid. Recall that the dialysate should normally be returned a pale yellow color. (Peritonitis is discussed more thoroughly in the section dealing with procedure above.)

Protein loss during PD is significant and may reach 1 to 2g during each 2-liter exchange.[26]

Certain conditions, such as gangrenous bowel, bowel perforation, localized peritonitis, ileus with abdominal distention, abdominal drains, extensive abdominal adhesions, or severe pulmonary insufficiency, preclude the use of PD.

Use of hypertonic dextrose solutions in PD may cause hyperglycemia and hyperosmotic coma, particularly in diabetic patients and in repeated dialyses with prolonged dwell times. Blood sugar and urinary sugar levels should be monitored at regular intervals and insulin given as necessary.

Infusion of 2 liters of hypertonic dialysate can cause as much as 1000 ml of ECF to move into the peritoneal cavity within 30 to 45 minutes.[27] In patients with cardiopulmonary disease, this volume may be sufficient to impair diaphragmatic movement and significantly increase dyspnea. (In such patients it may be necessary to limit the dialysate to 1 liter per each exchange.) If severe symptoms occur (dyspnea, tachycardia, tachypnea, rales, and orthopnea) the dialysate fluid should be withdrawn rapidly.

CHRONIC PD It has been estimated that the number of potential new patients requiring treatment for end-stage renal failure in the United States each year is at least 70 to 80 per one million population.[35] Any chronic renal failure patient is a potential candidate for PD. Patients who may be unable to learn home HD may be able to learn home PD; in addition, home PD does not require a partner to aid in the procedure. Although chemical control in PD is not as good as in HD, patients seem to do equally well clinically. Repetitive maintenance PD has been made more feasible by two technical advances: automated PD equipment and bacteriologically safer peritoneal catheters. Initially, long-term PD was reserved for patients not suited to long-term dialysis; however, it has become more competitive with HD as a treatment for *all* patients with end-stage renal disease.

Automated PD systems Closed automated PD systems have eliminated many of the problems associated with the manual procedure. For example, the automated systems are capable of warming the dialysate and regulating inflow and outflow for the entire treatment, making it unnecessary to repeatedly enter the system to connect and disconnect containers and drainage bags. Thus, the incidence of peritonitis is less with automated PD than with the manual method. Adjustable timers control the duration of each step, and alarms are activated when preset limits are exceeded. The simplicity and safety of automated PD techniques make it particularly suitable for home dialysis and overnight use. Many patients can sleep through their automated PD treatment and thus start dialysis at bedtime and disconnect themselves in the morning, allowing more free time during waking hours.

Usually PD is performed three or four times a week, for 8-to 12-hour periods. The average time required for PD is approximately 40 hours/week.

Patients who manage their own PD treatments are taught the importance of maintaining sterile technique. In addition, they are taught how to operate their specific dialysis machinery and how to do simple maintenance and troubleshooting. Patients are shown how to record vital signs and weight before each dialysis treatment and to record the percent of the dialysate solution and the number of containers used. The data are reviewed at each clinic visit.

Peritoneal catheter The Tenckhoff peritoneal catheter is made of silastic tubing and has two attached Dacron felt cuffs; one to fit in the subcutaneous tissue above the peritoneum and the other below the skin exit site. The tip of the catheter has many openings and rests in the pelvic gutter. Tissue grows into the cuffs in 10 to 14 days with complete healing in 2 to 4 weeks; the cuffs anchor the catheter, acting as infection barriers. Once healed, the patient may shower and carry on normal day-to-day activities. Self-examination for skin exit infection should be taught. Daily catheter care may include the application of Betadine and a light sterile dressing. The external end of the catheter is covered with a removable rubber cap when not in use. Several hundred milliliters of dialysis solution may be left in the peritoneal cavity after dialysis to prevent catheter plugging. Poor drainage from the catheter may be caused by constipation, malposition, or obstruction from infection. The catheter is removed if an infection develops in the subcutaneous tunnel or if other problems develop. Although pain may occur during the first few weeks after insertion, there should be no discomfort after this period. Permanent dialysis catheters can last as long as 5 years when aseptic routines for catheter care are followed.[36]

Continuous ambulatory peritoneal dialysis Continuous ambulatory peritoneal dialysis (CAPD) is an alter-

native to intermittent PD for renal failure patients. The only equipment needed for this method is a plastic bag and tubing; the plastic bag (full or empty) is connected to a permanent peritoneal catheter at all times. Two liters of dialysate are instilled into the peritoneal cavity by gravity and then the bag and tubing are folded up; after the specified dwell time, the solution is drained and a new bag of dialysate is hung. The process is repeated four to five times daily (the final exchange dwells uninterrupted during the sleeping hours). Because repeated manual connections of the dialysate bags must be made, the possibility of peritonitis is appreciable. Besides increased patient freedom, advantages of CAPD include the simplicity of equipment and decreased cost (about one third as much as intermittent PD). Some researchers state that CAPD is superior to intermittent PD in its ability to control biochemical abnormalities.[37]

Diet Patients treated with CAPD require greater protein intake than those maintained with HD, since a greater loss of protein occurs during PD.[38] Caloric intake should be optimal to prevent protein catabolism for energy needs. Supplemental water-soluble vitamins should be administered daily.

NEPHROTIC SYNDROME

The nephrotic syndrome is a clinical entity of diverse etiologies that is characterized by the concurrent occurrence of marked proteinuria, hypoalbuminemia, and generalized edema. Causes of nephrotic syndrome may include inflammatory renal disease (e.g., acute glomerulonephritis), glomerular disease associated with another systemic disease (e.g., lupus erythematosus), mechanical disorders (e.g., thrombosis of a renal vein), poisons (e.g., mercury), and miscellaneous causes (e.g., renal transplant.). The nephrotic syndrome can also be associated with carcinomas and lymphomas. Sometimes the nephrotic syndrome has no apparent cause and is said to be idiopathic. (About 80% of the cases of nephrotic syndrome in children are of the idiopathic variety.) The central pathophysiologic lesion common to all the causes of the nephrotic syndrome is an increase in glomerular capillary permeability to serum albumin.

Owing to increased glomerular permeability, large amounts of albumin are lost through the kidney, resulting in decreased plasma osmotic pressure. (Recall that albumin provides about 90% of the plasma colloid oncotic pressure.) Hypoalbuminemia causes fluid to be lost from the plasma to the interstitial space and also into the potential spaces of the body (such as the abdominal, pleural, and pericardial cavities). The resultant contraction of the plasma volume activates humoral and neurogenic reflexes, which then conserve water and sodium in an attempt to repair the volume deficit; however, the retained fluid only further dilutes the plasma proteins, causing more fluid to shift to the interstitial space. Edema develops first in dependent parts and later becomes generalized; as fluid collects in the serous cavities, the patient may become anorexic and short of breath. The edema may appear insidiously and increase slowly; however, it often appears suddenly and accumulates rapidly.

Care should be taken to avoid traumatizing the edematous skin. Since edema is often gravitational, elevation of the lower extremities may be helpful, as may the use of support stockings to reduce venous pooling and fluid accumulation. Diuretic therapy is indicated only for the relief of *symptomatic* edema; overuse of diuretics leads to contraction of the effective circulating blood volume and increases the risk of electrolyte imbalances, hypotension, and impairment of renal function. Salt-free albumin, dextran, and other oncotic agents have transient effects and are of little value. The administration of salt-free albumin is sometimes used for temporary relief in patients who are incapacitated by anasarca (generalized edema).

The urine contains large amounts of protein, 4 g to 10 g/day or more; the high protein losses result in malnutrition, and this, in turn, is responsible for muscle wasting. (It should be noted that muscle wasting may be disguised by the presence of edema.) Most of the excreted protein is albumin and the albumin blood level is usually reduced to about 2.5 g/100 ml (normal is 3.5 g–5.5 g/100 ml). For reasons that are not clear, severe hypoalbuminemia (*i.e.*, less than 2 g/100 ml) appears to be associated with an increases risk of thromboembolic events.[39] Hyperlipidemia may also be present and appears to be related to the lowered albumin level; the urine is foamy, deeper in color than usual, and contains oval fat bodies.

Renal biopsy can identify the glomerular lesion and thus serve as a basis for selecting therapy and estimating prognosis. Adrenocorticosteroids have proved useful in treating nephrotic syndrome in children and in some adults (*e.g.*, those with lipoid nephrosis, sys-

temic lupus erythematosus, proliferative glomerulonephritis, or idiosyncrasy to toxin). Cyclophosphamide (Cytoxan) has also been used to treat patients with nephrotic syndrome.

In summary, the primary goals of dietary management of the nephrotic syndrome are sodium restriction, adequate protein intake, and abundant caloric intake. A restricted dietary sodium intake (500 mg–1000 mg) may be indicated to help control edema. Protein intake of 100 g to 150 g/day for the average adult and 2 to 3 g/kg for the child should be encouraged to replace the urinary protein losses.[40] Urinary protein losses are best replaced with dietary protein of high biologic value (such as meat, milk, fish, fowl, soybean, and egg protein). A high caloric intake must be maintained to prevent catabolism of protein for energy needs. Some nephrologists recommend measures to reduce lipidemia (such as the use of hypocholesterolemic agents and unsaturated or polyunsaturated fats in the diet); however, the benefit of this approach has yet to be demonstrated. Since anorexia is a problem, dietary compliance is often less than desirable.

Because patients with the nephrotic syndrome are unduly susceptible to infection, it is necessary to protect them from others with infections. Bedrest is indicated during a bout of infection.

Prognosis depends upon the basic disease responsible for the nephrotic syndrome. About half of the cases of childhood nephrosis appear to run a benign course when managed properly and leave insignificant sequelae; adults with nephrosis fare less well (particularly when the underlying pathology is glomerulonephritis, systemic lupus erythematosus, diabetic nephropathy, amyloidosis, and renal vein thrombosis).

URETERAL TRANSPLANT INTO THE INTESTINAL TRACT

ELECTROLYTE IMBALANCES

Ureteral transplants can be made into the sigmoid colon (ureterosigmoidostomy), terminal end of the ileum (ureteroileostomy), or into a segment of the ileum isolated from the intestinal tract (ilealconduit). Patients with ureteral transplants into the intestinal tract (*particularly ureterosigmoidostomy*) may develop metabolic acidosis and eventually potassium deficit.

Metabolic acidosis occurs when urine is reabsorbed from the intestinal tract. Recall that urine is usually acid (average pH of 6) and has a high chloride content. Absorption of urinary chloride into the bloodstream causes a compensatory decrease in bicarbonate. The decrease in the bicarbonate side of the carbonic acid–base bicarbonate balance causes the blood pH to drop.

Potassium deficit occurs largely as the result of excessive renal chloride excretion. This can be explained in the following way: urinary chloride is absorbed from the intestine into the bloodstream and must eventually be reexcreted by the kidneys. A cation, such as potassium or sodium, is excreted with chloride. The continued absorption of urinary chloride by the intestine can deplete the body of potassium.

The patient with electrolyte disturbances following ureteral transplants may have weakness, intense fatigue, anorexia, nausea, and vomiting. When acidosis is severe, hyperpnea is noted.

CONTRIBUTING FACTORS TO DEVELOPMENT OF ELECTROLYTE IMBALANCES *Electrolyte disturbances do not occur in all patients with ureteralintestinal transplants.* Persons with normal renal function can withstand the absorption of urine without changes in pH or electrolyte levels. Some degree of renal damage must be present for patients with such implants to develop electrolyte imbalances. Unfortunately, many patients with good kidney function develop renal damage as a result of urinary tract infections caused by intestinal organisms and *then* begin to develop electrolyte imbalances.

Absorption of urine is increased when the urine remains in the intestine for a prolonged period. Thus, the likelihood of metabolic acidosis is greater in patients with ureteral-sigmoid transplants than in those with implants into the intact ileum, since the sigmoid is larger than the ileum and can accommodate larger volumes of urine before peristalsis is initiated. Since exposure of a large area of the intestinal mucosa is associated with a higher urinary absorption than is the exposure of a small area, it is often better to transplant the ureters into an ileal conduit. *It has been demonstrated that urinary diversion using the ileal conduit is associated with a much lower incidence of metabolic acidosis;* unless stomal obstruction occurs, external drainage of the urine is usually sufficiently prompt to avoid significant bicarbonate losses.

Prolonged periods of inactivity, as when the patient is bedfast during an illness, favor unsatisfactory urinary drainage. Stricture of the ileal conduit stoma can also cause urinary retention.

TREATMENT Treatment of the electrolyte disturbances usually consists of the administration of potassium and sodium, as gluconate, lactate, or citrate salts. Patients with ureteral transplants into the intact bowel (such as ureterosigmoidostomy) may require the insertion of a rectal catheter to drain urine when they are confined to bed for a prolonged period. The success of treatment is inversely proportional to the degree of renal damage present.

NURSING IMPLICATIONS To prevent or minimize electrolyte disturbances in patients with ureteral transplants into the intestine, the nurse should take the following measures:

1. Encourage the patient to drink approximately 3000 ml of fluids daily, unless intake is restricted. This amount of fluids ensures frequent emptying of the intestine.
2. Encourage the patient to walk; activity favors emptying of the intestines.
3. Encourage the patient with ureteral transplant into the intact bowel to evacuate every few hours. This practice limits the absorption time of urine. The use of a rectal tube helps reduce the contact time between urine and the colonic mucosa.
4. If the patient has an ileal conduit, see that the ileostomy bag is emptied before it is completely full so that urine drainage will not be hindered or back up into the ureters.
5. Look for and report symptoms of electrolyte disturbances, such as weakness, intestinal distention, soft flabby muscles, deep rapid respiration, and disorientation.

URINARY CALCULI

The most common type of urinary stone in North America is calcium oxalate (with or without calcium phosphate); next, in descending frequency, are magnesium ammonium phosphate (struvite) stones, uric acid stones, and cystine stones.[41]

PATHOGENESIS

Etiologic factors are varied and can be defined in only about one half of the cases. Thirty to forty percent of stone formers have a condition called *idiopathic hypercalciuria* (excessive excretion of calcium in the urine, of unknown etiology); these individuals have a *normal* serum calcium level. (Idiopathic stone disease is five times more common in males than in females.)

Other causes of calcium stones include primary hyperparathyroidism (5% to 7%), high vitamin D intake, excessive intake of milk or alkali, destructive bone disease due to neoplasm or metabolic problems (such as thyrotoxicosis or corticosteroid excess), sarcoidosis, and prolonged immobilization.

Causes of oxalate stones may include familial oxaluria, ileal disease (ileal resection or bypass), and high dietary oxalate intake (nuts, spinach, cocoa, beets, tea, and rhubarb).

Some uric acid stone formers have an idiopathic condition called *hyperuricosuria* (excessive uric acid in the urine); these persons have a *normal* serum uric acid level. Uric acid stones may also result from treatment of gout with large doses of uricosuric agents (such as probenecid), which increase urinary excretion of uric acid. Other causes include hyperuricosuria due to high purine intake, increased excretion of uric acid secondary to therapy of neoplastic disease with agents that cause rapid destruction of cells, and myeloproliferative disease (such as lymphoma, leukemia, and polycythemia vera).

Cystine stones are the result of an inherited disorder of amino acid metabolism.

Calculi of all types are most likely to occur in persons having a *low urine volume secondary to low fluid intake.*

Alterations in urinary pH favor certain stones; for example, a low (acidic) urinary pH predisposes to uric acid and cystine stones; a high (alkaline) urinary pH predisposes to inorganic stones.

Urinary infections with urea-splitting organisms (particularly *Proteus Mirabilis*) cause urea to convert to ammonia, which in turn increases urinary pH; magnesium ammonium phosphate (struvite) can then precipitate to form stones. Clumps of bacteria, bits of necrotic tissue, and blood clots can serve as a nucleus (nidus) for stone formation, particularly in the presence of urinary stasis.

TREATMENT AND NURSING CONSIDERATIONS

Before appropriate therapy can be instituted, the nature of the stone must be discovered. Diuretics, acidifying or alkalinizing agents, and various specific drugs may be indicated, depending (when possible) on laboratory analysis of the stone. Urine of stone formers should be carefully strained; any stones found should be saved for analysis. Other laboratory tests of value in the evaluation of patients with renal calculi include serum calcium, phosphate, uric acid, and creatinine levels; urinalysis, urine culture, 24-hour calcium and uric acid excretion levels, and urinary pH in the fasting state.

The mainstay of treatment of all renal calculi is the maintenance of a dilute urine to avoid supersaturation of precipitates. To maintain a continuously dilute urine (more dilute than the ECF) an adult must excrete approximately 4 quarts of urine in 24 hours. Enough fluid must be given to maintain this urine volume; in many cases, this method alone, if followed conscientiously, keeps stones from forming or at least decreases the rate of formation.[42]

IDIOPATHIC HYPERCALCIURIA Idiopathic hypercalciuria is present in 35% of all stone formers. A thiazide diuretic (such as hydrochlorothiazide, 50 mg twice daily) reduces urinary calcium excretion and thus may be used to treat hypercalciuric stone formers. By the fourth day of hydrochlorothiazide therapy, urinary calcium should decrease to about half the pretreatment level. Inorganic orthophosphates have proved helpful in reducing hypercalciuria when used alone or with the thiazides; phosphates seem to act by binding calcium within the intestine and by reducing excretion of calcium within the urine.

CALCIUM PHOSPHATE AND MAGNESIUM AMMONIUM PHOSPHATE STONES Excessive alkali ingestion is often noted in patients who produce calcium phosphate stones. Struvite (magnesium ammonium phosphate) stones commonly occur in patients with alkaline urines secondary to infection from urea-splitting organisms. Since both calcium phosphate and struvite salts are soluble in an acid urine, acidifying agents are employed.[43] Among these are potassium acid phosphate (3 g to 6 g daily), ascorbic acid (3 gm daily), or methionine (8 g to 12 g daily).[43,44] A urinary pH below 6.5 has been recommended for persons with struvite stones; they should be treated with appropriate antibacterial therapy of a long-term basis. Recall that urinary salts are more likely to precipitate when urinary stasis and infection are present.

URIC ACID STONES Patients with gout are hyperuricosuric, particularly if treated with pharmacologic agents that increase uric acid excretion in the urine (such as probenecid). Prevention of uric acid stones by inhibiting formation of uric acid is possible by blocking the conversion of xanthine to uric acid with allopurinol in doses of 200 mg to 300 mg/day. The alternative to the use of allopurinol is the reduction of excess dietary purine (almost always due to excessive meat). Allopurinol is sometimes used in association with antileukemic and anticancer agents to prevent uric acid stones (recall that these agents cause rapid cell destruction and thus increase uric acid formation). Again, patients with uric acid stones should be encouraged to maintain a large urine volume!

OXALATE STONES Patients who have had ileal resection or small-bowel bypass for treatment of obesity may form oxalate stones because they tend to overabsorb dietary oxalate and become hyperoxaluric. The resultant high oxalate excretion encourages stone formation by directly oversaturating urine with calcium oxalate. One form of treatment for these patients consists of the ingestion of calcium in the form of calcium carbonate (Titralac) to bind the free oxalate so that it cannot be absorbed.

Excessive ascorbic acid ingestion has been linked to the development of oxalate stones. It appears reasonable to limit vitamin supplements to a tolerable level in oxalate stone formers, since the ingestion of large amounts of ascorbic acid increases oxalate production.

According to one source, the solubility of oxalate salts in the urine is not pH-dependent; thus, manipulation of urinary pH is not useful for this variety of stone.[45]

CYSTINE STONES Cystine stones are the result of an inherited disorder of amino acid metabolism. Cystinuria is treated by a large fluid intake, preferably enough to produce a urine volume of 4 liters/24 hours, and by alkalinization of the urine to a pH of 7.6 to 8.0.

Patients unable to tolerate large doses of bicarbonate may be treated with D-penicillamine.

URINARY *pH* A urinary pH above 6.5 is recommended for uric acid stone formers and a pH above 7.5 is recommended for cystine stone formers. Calcium phosphate and magnesium ammonium phosphate (struvite) salts tend to become soluble in an acid urine; hence, a urinary pH below 6.5 is recommended for persons with struvite stones. Oxalate stones do not appear to be pH-dependent.

Diet alone does not sufficiently change urinary pH; instead, the physician relies on acidifying or alkalinizing pharmacologic agents to achieve a more uniform and sure control of urinary pH.

Some persons rely on cranberry juice to assist in acidification of urine. Nickey found a reduction in urinary baseline pH when cranberry juice was used alone, but a greater decrease when it was used in conjunction with ascorbic acid.[46] Kinney and Blount studied the effect on urinary pH of sweetened 80% cranberry juice (either 150 ml, 180 ml, 210 ml, or 240 ml) with each meal and found significant decreases in urinary pH between each control group and its corresponding experimental group.[47] The authors state, however, that the concentration and volumes of juice that produced these effects may not be practical for clinical use. (Note that commercially prepared cranberry juice is only approximately 26% cranberry juice!)

PREVENTIVE NURSING MEASURES

Nursing measures to help prevent the development or recurrence of urinary calculi include the following:

1. Forcing fluids, unless contraindicated, to increase urinary volume to 4 liters daily in adults susceptible to stones (remember that crystals have less tendency to accumulate mass when urine flow is great and continuous.) As stated earlier, enough fluid must be given to maintain this urine volume; in many cases, this method alone, if followed conscientiously, keeps stones from forming or at least decreases the rate of formation. Patients should be encouraged to drink large quantities of fluids not only during the day but before retiring and during the night if awakened to urinate. (Some physicians encourage awakening the patient during the night

to force fluids.) Fluid intake should be increased when there is an abnormal route of fluid loss (such as in diarrhea, vomiting, or excessive sweating). Also, the ingested fluids should not contain substances conducive to the particular person's type of stone. (For example, calcium stone formers should not drink large quantities of milk; oxalate stone formers should not drink large quantities of tea or cocoa.)

2. Assisting in the manipulation of urinary pH to favor solution of the components likely to develop into stones:

    Explaining to the patient the need for taking the prescribed acidifying or alkalinizing agents

    Teaching the patient to test and record his own urinary pH to aid in regulation of the prescribed agent

    Teaching the patient to recognize untoward effects associated with the prescribed agent

3. Assisting the patient and family with understanding of dietary restrictions and allowances

4. Assisting the patient with understanding of other prescribed medications (such as diuretics, allopurinol, antibiotics)

5. Preventing urinary stasis by turning the immobilized patient frequently, elevating the head of the bed, and having the patient sit up, if allowed

6. Encouraging weight bearing and ambulation as soon as possible to minimize the hypercalciuria accompanying prolonged immobilization

7. Preventing urinary tract infections, whenever possible, by thorough patient teaching and meticulous sterile technique in handling of catheters and drainage apparatus

8. Checking drainage tubes frequently for patency to avoid urinary stasis

REFERENCES

1. Maxwell M, Kleeman C (eds): Clinical Disorders of Fluid and Electrolyte Metabolism, 3rd ed. p 768. New York, McGraw-Hill 1980
2. *Ibid*, p 745
3. Anderson R, Linas S, Berns A et al: Nonoliguric acute renal failure. N Engl J Med 296 (20):1134, 1977
4. Maxwell & Kleeman, p 782
5. Sanders J, Gordon L: *Handbook of Medical Emergencies*, 2nd ed, p 267. Garden City, NY, Medical Examination Publishing Company, 1978
6. Maxwell & Kleeman, p 780
7. *Ibid*, p 776
8. *Ibid*, p 781
9. *Ibid*, p 785
10. Hodges R: Nutrition in Medical Practice, p 246. Philadelphia, Saunders, 1980
11. Giordano C: Use of exogenous and endogenous urea for protein synthesis in normal and uremic subjects. J Lab Clin Med 62:231, 1963
12. Abel R, Beck C, Abbott W et al: Improved survival from acute renal failure after treatment with intravenous essential L-amino acids and glucose: Results of a prospective blind study. N Engl J Med 288:695, 1973
13. Halpern S: Quick Reference to Clinical Nutrition, p 233, Philadelphia, JB Lippincott, 1979
14. Maxwell & Kleeman, p 778
15. *Ibid*, 821
16. Hodges, p 244
17. Maxwell & Kleeman, p 834
18. *Ibid*, p 854
19. *Ibid*, p 848
20. Stewart W, Fleming L, Anderson D et al: Changes in plasma electrolytes and nerve conduction velocities during hemodialysis without magnesium. Proc Eur Dial Transplant Assoc 4:285, 1967
21. Maxwell & Kleeman, p 843
22. Stewart W, Fleming L, Manuel M: Benefits obtained by the use of high sodium dialysate during maintence hemodialysis. Proc Eur Dial Transplant Assoc 9:111, 1972
23. Drukker W: The hard water syndrome: A potential hazard during regular dialysis treatment. Proc Eur Dial Transplant Assoc 5:283, 1968
24. Govan J, Porter C, Cook J et al: Acute magnesium poisoning as a complication of chronic intermittent hemodialysis. Br Med J 2:279, 1968
25. Maxwell & Kleeman, p 857
26. Halpern, p 223
27. Maxwell & Kleeman, p 867
28. *Ibid*, p 864
29. *Ibid*, p 868
30. Penzotti S, Mattocks A: Effects of dwell time, volume of dialysis fluid, and added accelerators on peritoneal dialysis of urea. J Pharm Sci 60:1520, 1971
31. Tenckhoff H, Ward G, Boen S: The influence of dialysate volume and flow rate on peritoneal dialysate volume and flow rate on peritoneal clearance. Proc Eur Dial Transplant Assoc 2:113, 1965
32. Babb A, Johansen P, Strand M et al: Bi-directional permeability of the human peritoneum to middle molecules. Proc Eur Dial Transplant Assoc 19:247, 1973
33. Burns R, Henderson L, Hager E, Merrill J: Peritoneal dialysis, clinical experience. N Engl J Med 267:1060, 1962
34. Burnett W, Brown G, Rosemond G, Caswell H: Treatment of peritonitis using peritoneal lavage. Ann Surg 145:675, 1957
35. Blumenkrantz M: Maintenance peritoneal dialysis as an alternative for the patient with end-stage failure. *Clinical Digest* 6, March 1977, p 1
36. Denniston D, & Burns K: Home peritoneal dialysis. Am J Nurs, Nov 1980, p 2022
37. Oreopolous D, Robson M, Iza S et al: A simple and safe technique for continuous ambulatory peritoneal dialysis. Trans Am Soc Artif Intern Organs 24:484, 1978
38. Halpern, p 227
39. Glassock R: The nephrotic syndrome. Hosp Pract, Nov 1979, p 128
40. Hodges, p 238

41. Nemoy N: Axioms of renal calculi. Hosp Med, Feb 1979, p 8
42. Derrick F & Carter W: Kidney stone disease: Evaluation and medical management. Postgrad Med 66:119, Oct 1979
43. *Ibid*, p 1687
44. Krupp M & Chatton M (eds): Current Medical Diagnosis and Treatment. Los Altos, CA, Longe Medical Publications, 1980, p 567
45. Frank L, McDonald D: Urology. In Schwartz S (ed): *Principles of Surgery*, 3rd ed, p 1687, New York, McGraw-Hill, 1979
46. Nickey K: Effect of prescribed regimes of cranberry juice and ascorbic acid (abstr). Arch Phys Med Rehabil 56, Dec 1975, p 556
47. Kinney A, Blount M: Effect of cranberry juice on urinary pH. Nursing Res 28, No. 5: 287, 1979

ALTERATIONS IN ELECTROLYTE CONCENTRATIONS: EFFECTS ON THE MYOCARDIUM

Alterations in electrolyte concentrations affect the initiation and conduction of myocardial impulses, myocardial repolarization, and may result in cardiac arrhythmias. In addition alterations in electrolyte concentrations affect myocardial contractility and function. Clinically, potassium and calcium imbalances are the most common alterations affecting the myocardium, although almost every electrolyte has some effect on the heart. The effects of these alterations are usually reversible if detected early; therefore, nurses must monitor persons at risk for the early signs and symptoms of these imbalances so that appropriate therapy can be instituted. Although cardiac disease itself causes few electrolyte imbalances, the patient is frequently at risk because of therapies administered to treat his disease or because of a coexisting noncardiac condition.

The electrocardiogram (EKG) reflects the sum of all ionic influences on the myocardium. Therefore, disturbances in the concentrations of electrolytes alter EKG tracings. The EKG provides important diagnostic help in detecting disturbances in electrolyte concentration, especially of potassium and calcium. Specific changes in the EKG are often associated with particular electrolyte disturbances, but these changes may also be produced by cardiac disease or certain medications. Also, it must be remembered that not all patients with electrolyte disturbances display the characteristic EKG changes. When these changes do occur, they do not necessarily occur at the expected level of concentration. Evidence of an electrolyte disturbance on the EKG should be confirmed by plasma electrolyte studies. Changes resulting from myocardial damage tend to be localized, while changes caused by electrolyte disturbances often appear in all leads. When EKG changes are seen, the possibility of medication effect must also be considered (e.g., digitalis or quinidine).

POTASSIUM Potassium influences both impulse conduction and muscle contractility; alterations in its concentration may change myocardial irritability and rhythm. The normal plasma potassium concentration

11
Fluid Balance in the Patient with Cardiac Disease*

*Revised by Martha Spies, R.N., M.S.N., Assistant Professor of Nursing, St. Louis University.

is from 4.0 mEq to 5.6 mEq/liter. EKG changes produced by plasma potassium level variations are illustrated in Figure 11-1.

Hyperkalemia Severity of cardiac symptoms from hyperkalemia depends on the rate of increase in the serum potassium, with the most severe manifestations being associated with a rapid rise in serum levels. Also influencing the effects of hyperkalemia on the myocardium is the serum sodium level (recall the relationship between potassium and sodium on the resting membrane potential of the cell). Hyponatremia potentiates the action of hyperkalemia on the heart. Acidosis and hypocalcemia can also make the effects of hyperkalemia on the myocardium more severe.

The cardiac effects are usually not seen until the serum potassium level reaches 7 mEq/liter, but they may occur at levels as low as 6 mEq/liter Myocardial effects of hyperkalemia are slowed conduction speed of impulses and increased speed of repolarization. In profound hyperkalemia, the heart becomes dilated and flaccid owing to a decreased strength of contraction resulting from a decrease in the number of active muscle units. In addition to disordered rhythm and conduction, there is an increased sensitivity to the therapeutic and toxic effects of lidocaine, procainamide, and quinidine. At high serum potassium levels, artificial pacemakers may fail because of lack of conduction from the pacing site.

The sequence of EKG changes in hyperkalemia includes the following:

1. Tall, peaked, tented T wave with shortened Q-T interval
2. Widened QRS complex
3. Decreased P wave amplitude, prolonged P-R interval, and depressed ST segment
4. Merger of QRS complex with T wave
5. Arrhythmias, especially supraventricular tachycardia, premature ventricular contractions, atrial or ventricular fibrillation, or ventricular arrest in diastole, which can occur at any time

Interventions to prevent or treat the cardiac effects of hyperkalemia are described below:

1. Restriction of potassium intake in all forms (*i.e.,* diet, medications, intravenous fluids) to about 40 mEq daily
2. Facilitation of potassium excretion through the use of diuretics, sodium chloride infusions, cation exchange resins, or dialysis

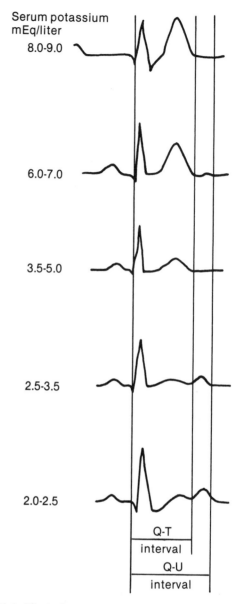

FIG. 11-1. Effect of serum potassium levels on EKG. Hyperkalemia initially causes the T wave to increase in magnitude. With increasing serum potassium levels, the ST segment becomes depressed, the U wave disappears, and the QRS duration and P-R interval increase. With serum potassium levels greater than 10 mEq/liter, ventricular fibrillation often ensues. Hypokalemia broadens and lowers the magnitude of the T wave, increases the magnitude of the U wave, and causes the T and U waves to fuse. It should be noted that, with respect to the heart rate, the Q-T interval and Q-U interval do not increase. With severe hypokalemia, ST segment depression is observed. (Wilson R (ed): Principles and Techniques of Critical Care, p 29. Kalamazoo, The Upjohn Company, 1976)

3. Prevention of tissue catabolism
4. Insertion of a pacemaker if the patient has brady-cardia or heart block (most likely to occur when the serum potassium is greater than 7 mEq/liter and the QRS complex is widened)
5. Utilization of emergency measures to temporarily reduce the serum potassium level if the patient has a tachyarrhythmia (most likely to occur if the serum potassium is greater than 7.0 mEq/liter and the QRS complex is widened):
   - Ten percent calcium gluconate, 10 ml to 20 ml, may be given intravenously; this treatment produces a response in approximately 2 minutes with a duration of 15 to 120 minutes. Calcium gluconate changes the membrane potential to more normal levels so that conduction becomes more normal. The patient must be continuously monitored for bradycardia; if it occurs, the infusion must be stopped.
   - An intravenous bolus of 45 mEq sodium bicarbonate or an intravenous infusion of 90 mEq in 500 ml $D_{10}W$ may be given. The effect of this treatment begins in 5 to 60 minutes and can last 2 to 4 hours. Administration of sodium bicarbonate affects the membrane potential so that conduction is more normal; it also temporarily shifts potassium into the cells.
   - An IV bolus of 50 ml of 50% dextrose with 10 units to 25 units of regular insulin may be given; this form of treatment begins to act in 15 to 60 minutes and the effect lasts 2 to 4 hours. Dextrose and insulin administration induces a temporary shift of extracellular potassium into the cell.

None of the last three therapies will decrease *total* body potassium; they are attempts to delay or block the effects of hyperkalemia on the myocardial cell membrane until more definitive means to decrease total body potassium content can be instituted. (A thorough discussion of the treatment of hyperkalemia is presented in Chapter 10.)

Hypokalemia The major cardiac effects of hypokalemia are prolonged cardiac repolarization and decreased strength of myocardial contraction. Patients with hypokalemia have an increased sensitivity to digitalis and can be resistant to antiarrhythmics such as lidocaine, procainamide, and quinidine. Hypernatremia and alkalosis can make the effects of hypokalemia more severe.

The sequence of EKG changes in hypokalemia includes the following:

1. ST segment depression
2. Broad, sometimes inverted, progressively flatter T waves
3. Enlarging U wave that sometimes becomes superimposed on the T wave to give the appearance of prolonged Q-T interval
4. Prolonged QRS and P-R intervals (at very low serum potassium levels)
5. Conduction disturbances and arrhythmias; prolonged repolarization associated with ectopic beats

Treatment of hypokalemia aims at correcting the basic cause for the imbalance and providing potassium replacement. Oral potassium supplements and foods high in potassium may be given. If arrhythmias or conduction disturbances are present, parenteral replacement therapy is required. Potassium replacement must take place *slowly* because a rapid rise in the serum potassium level predisposes the patient to the effects of hyperkalemia (even when the serum level remains less than 5.3 mEq/liter). Acute, severe hypokalemia is usually treated by administering 10 mEq to 20 mEq of KCl per hour (usually 50 ml intravenous piggyback) by constant infusion.[1] This should be given only when close monitoring is available. It generally takes 40 mEq to 60 mEq to raise the serum potassium level by 1 mEq/liter.

CALCIUM The cardiac effects of calcium include cardiac excitability and myocardial contractility. Alterations in serum calcium levels primarily affect repolarization. Arrhythmias and conduction disturbances are uncommon with calcium imbalances when serum potassium and magnesium levels are normal. The EKG changes produced by serum calcium level variations are illustrated in Figure 11-2.

Hypercalcemia The most common cause of hypercalcemia is metastatic bone disease. Other causes include an increased serum protein level, hyperparathyroidism, hypervitaminosis, and immobility. Cardiac manifestations of hypercalcemia include the following:

1. Shortened ST segment resulting in shortened Q-T interval
2. Increased contractility leading to cardiac arrest in systole
3. Potentiation of digitalis toxicity
4. Widened and rounded T wave
5. Slightly widened QRS and P-R intervals

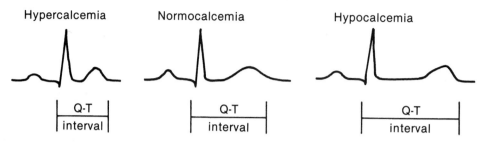

FIG. 11-2. EKG changes produced by serum calcium level variations. (Wilson R (ed): Principles and Techniques of Critical Care, p 26. Kalamazoo, The Upjohn Company, 1976)

Treatment of hypercalcemia frequently begins with the administration of saline infusions to dilute the extracellular calcium concentration and increase the renal excretion of calcium.

Hypocalcemia A decreased ionized calcium level can be caused by hypoparathyroidism, alkalosis, and acute pancreatitis, as well as by other factors. The cardiac manifestations of hypocalcemia include those listed below:

1. Prolonged ST segment and Q-T interval
2. Prolonged ventricular systole
3. Decreased contractility (especially associated with acute losses) leading to congestive heart failure or death in diastole
4. Reduction of the heart's sensitivity to digitalis

Treatment of hypocalcemia includes correction of the underlying cause and calcium replacement. *Caution is necessary when calcium is given intravenously to a digitalized patient because calcium increases the sensitivity of the heart to digitalis and may precipitate toxicity.*

MAGNESIUM The cardiac effects of magnesium are not fully known although cardiac muscle contains high concentrations of magnesium. Hypermagnesemia suppresses ventricular extrasystoles and arrhythmias, including those related to digitalis toxicity. At serum magnesium levels of 7.1 mEq to 10.0 mEq/liter, the patient can have sinus bradycardia, prolonged P-R, QRS, and Q-T intervals. Heart block and cardiac arrest in diastole can occur at levels above 15 mEq/liter.

Hypomagnesemia can produce arrhythmias such as ventricular premature contractions and ventricular fibrillation; it also potentiates the arrhythmias associated with digitalis toxicity. In mild hypomagnesemia, the EKG will reveal tall, peaked T waves, slight widening of the QRS complex, and occasional mild ST

segment depression. In severe hypomagnesemia (0.2 mEq–0.6mEq/liter), there is secondary hypokalemia and hypocalcemia. At this level, the EKG shows prolonged P-R interval, wide QRS complex, slight ST depression, and broad and flat (or inverted) T waves. These changes have been shown to be associated with a prolonged relative refractory period, which may produce the ventricular arrhythmias.[2]

SODIUM Extracellular sodium ion concentrations at levels compatible with life have little effect on the activity of the heart. It should be noted that hyponatremia potentiates the cardiac effects of hyperkalemia.

Acid–base alterations Acid–base alterations primarily affect cardiac function by changing the potassium and calcium ion concentrations.

Acidosis causes the heart rate to increase because it stimulates the cardio-accelerator center. Acidosis depresses myocardial contractility, decreases the responsiveness of the heart to epinephrine, and lowers the threshold for ventricular fibrillation; in addition, acidosis is associated with an increased T wave amplitude.

Alkalosis stimulates the vagus nerve and causes a slowed heart rate. In alkalosis, there may be a decrease in T wave amplitude. In sinus tachycardia, the T-P phenomenon may be seen, in which the Q-T interval is prolonged enough so that the succeeding P wave falls close to or on the downstroke of the previous T wave.

ELECTROLYTES IN CARDIAC RESUSCITATION

SODIUM BICARBONATE Metabolic acidosis occurs in cardiac arrest; the lack of available oxygen causes cellular processes to automatically shift to anaerobic metabolism, which yields lactic acid as the end product. The drop in plasma pH interferes with the normal functioning of many enzyme systems. Acidosis inhibits resuscitation, making the heart unresponsive to treatment such as defibrillation.

Sodium bicarbonate, an alkali, is often used to correct metabolic acidosis; 50 ml of a 7.5% solution (containing 3.75 g or 44.6 mEq) may be administered by direct venous injection. This amount can be repeated every 5 to 10 minutes, as indicated, until spontaneous cardiac function is restored and blood gas studies indicate the pH is normal. The administration of a large amount of sodium bicarbonate could conceivably cause metabolic alkalosis and fluid volume excess as a result of sodium loading; thus, the arterial pH should be monitored and the patient observed closely for fluid retention. Alkalosis aggravates tissue hypoxia by causing a firmer bond between oxygen and hemoglobin, decreasing the amount of oxygen available to cells.

Lack of respiration causes respiratory acidosis to be superimposed on the metabolic acidosis; the excess CO_2 must be removed by effective ventilation.

CALCIUM CHLORIDE Calcium chloride may be given intravenously or intracardially to strengthen the cardiac contraction; the dose is usually 10 ml of a 10% solution. Calcium is especially helpful in the treatment of patients with potassium excess. Since calcium and potassium have antagonistic effects, raising the plasma calcium level decreases the cardiotoxic effects of hyperkalemia. (Calcium should not be administered intravenously to the digitalized patient because calcium ions enhance the cardiotoxic action of digitalis.)

POTASSIUM CHLORIDE Potassium chloride may be given intravenously to correct arrhythmias caused by digitalis. (See the section dealing with potassium administration earlier in this chapter and in Chapter 6.)

FLUID BALANCE IN MYOCARDIAL FAILURE AND CARDIOGENIC SHOCK

DETERMINANTS OF CARDIAC OUTPUT

In the cardiac patient, fluid balance both affects and is affected by cardiac output, which is the product of the heart's stroke volume and heart rate. These factors are interrelated, and changes in one will cause compensatory changes in others.

Stroke volume is determined by four factors. The first is *preload*, which refers to the pressure or quantity of blood in the ventricle at the end of diastole (ventricular end-diastolic pressure [VEDP]). When the quantity of blood at the end of diastole increases, the ventricle is compliant and can adjust by increasing the strength of contraction (Starling's law). Preload can be measured in the right ventricle by central venous pressure (CVP) monitoring and in the left ventricle by pulmonary wedge pressures. Preload is affected by blood volume, ventricular function, and systemic vasoconstriction.

The second factor affecting stroke volume is *afterload*, which refers to vascular resistance, or the pressure that the heart pumps against. Increases in afterload (e.g., increased arterial blood pressure) stimulate the baroreceptor reflex, which leads to decreased ventricular contractility and decreased cardiac output.

The remaining factors related to stroke volume are *cardiac contractility* and the *coordinated contraction of the myocardium*. These factors affect the ability of the heart to increase cardiac output in the face of a constant amount of resistance.

Heart rate is the other determinant of cardiac output. Extreme bradycardia (20–40 beats/min) can lead to a decreased cardiac output because the ventricle does not eject blood often enough to maintain blood flow. The highest cardiac output is achieved at heart rates of 100 to 120/minute. When the heart rate exceeds 120/minute, cardiac output may fall as diastolic filling time is shortened.

HEART FAILURE*

Heart failure refers to an inability of the heart to pump enough blood to meet the metabolic demands of the body. When the ventricles fail, they lose their compliance and cannot adjust to changes in venous return; preload increases. Causes of heart failure include factors within or affecting the heart itself, such as cardiomyopathy, infarction, arrhythmias, valvular disease, cardiac tamponade, electrolyte imbalances, and factors affecting the circulation, including inadequate venous return and circulatory overload.

Ventricular failure can cause symptoms primarily

---

*This section on congestive heart failure and the section on hemodynamic monitoring were prepared with the assistance of Catherine A. Smith, R.N., M.S.N., Clinical Specialist in Cardiovascular Nursing, St. Louis, Missouri.

associated with pulmonary congestion (left-sided heart failure), venous congestion (right-sided heart failure), or decreased cardiac output (cardiogenic shock). Failure of one side of the heart eventually leads to failure of the other. Heart failure can be acute or chronic.

LEFT-SIDED HEART FAILURE Left-sided heart failure is primarily a backward failure that causes damming of blood back from the left side of the heart to the pulmonary vessels. Symptoms of pulmonary vascular congestion include dyspnea, orthopnea, dry cough, and pulmonary edema. Pulmonary capillary pressure above 30 mm Hg causes transudation of fluid into the alveoli and diminished oxygen–carbon dioxide exchange.[3] Acute pulmonary edema is characterized by severe dyspnea, profound anxiety, cyanosis, noisy respirations, and pink frothy sputum.

Physical signs associated with left-sided heart failure may include third and fourth heart sounds (ventricular gallop and atrial gallop respectively) and fine moist rales in the lungs. Ventricular gallop in adults is almost never present in the absence of significant heart disease. Rhythms associated with left-sided failure may include sinus tachycardia, atrial premature contractions, paroxysmal atrial tachycardia, and ventricular premature beats. Other signs of left ventricular failure may include pulsus alternans (an alternating greater and lesser arterial pulse volume), expiratory wheezing breath sounds, and Cheyne-Stokes respirations.

RIGHT-SIDED HEART FAILURE In unilateral right-sided heart failure, blood is not pumped adequately from the systemic circulation into the lungs; back pressure of the right heart causes systemic edema. Right-sided heart failure may result from stenosis of the pulmonary or tricuspid valve, cor pulmonale (chronic bronchitis, emphysema, bronchiectasis), or massive pulmonary embolism. Acute right-sided failure rarely occurs alone; when it does, symptoms are related to a low cardiac output. In rare conditions where the right heart fails acutely, cardiac output may be so low that death occurs rapidly. Chronic right-sided failure leads to progressive development of peripheral congestion, hepatosplenomegaly, and ascites. Failure of the right ventricle causes a rise in the right atrial and vena caval pressures. Neck veins will appear distended when the patient is lying in bed with the head of the bed elevated between 30 and 60 degrees

(see Fig. 4-1). As with left-sided failure, sinus tachycardia and other rhythms associated with pump failure may be present. Right ventricular third and fourth heart sounds may be heard.

Usually, right-sided heart failure is the result of left-sided failure; in this case, symptoms of both right- and left-sided failure are present.

ADAPTIVE MECHANISMS IN DECREASED LEFT VENTRICULAR FUNCTION

HEMODYNAMIC MECHANISMS

When cardiac output falls, the first response is stimulation of the sympathetic nervous system and inhibition of the parasympathetic nervous system. This tends to support arterial pressure even when the stroke volume is decreased. Blood is preferentially shunted to the brain and heart while being directed away from the skin, kidneys, and skeletal muscles; this shunting increases cellular oxygen extraction from the blood from 25% to 60%. If the shunting is unsuccessful in preventing cellular hypoxia, the cells shift to anaerobic metabolism, which can lead to metabolic acidosis. Venous tone is increased to maintain venous return and ventricular filling. There is also an increased heart rate, probably caused by increased release of norepinephrine. In chronic heart failure, the ventricular myocardium will hypertrophy, increasing the heart's ability to maintain adequate stroke volume when tissue needs increase.

SALT AND WATER RETENTION BY THE KIDNEYS

Decreased cardiac output causes a decrease in glomerular filtration rate. This, along with the shunting of blood away from the kidney, can result in almost complete anuria when the cardiac output decreases by one third.[4] The kidneys respond by retaining sodium and water. If the cardiac output remains low, the renin-angiotensin system will be activated to increase secretion of aldosterone from the adrenal cortex. The effect of aldosterone is to cause further sodium and water retention while increasing potassium excretion.

When cardiac output remains low, the release of

antidiuretic hormone (ADH) from the posterior pituitary is stimulated. ADH acts on the distal tubules to cause water retention; patients may retain water in excess of sodium.

The effects of aldosterone and ADH may be further increased if inactivation of these hormones by the liver is inhibited because of liver congestion or cirrhosis. Circulating blood volume may also be increased by an increase in the production of red blood cells. This is accomplished as hypoxia stimulates the release of erythropoietin, which then causes the bone marrow to increase the production of erythrocytes.

All of these mechanisms increase the blood volume 5% to 15% in mild heart failure and 30% to 50% in severe heart failure. These adaptive mechanisms aim at preserving stroke volume in compensated heart failure. However, blood volume and venous return can increase preload beyond the ability of the heart to adapt. This leads to decompensated heart failure, resulting in even further fluid retention.

SUMMARY OF WATER AND ELECTROLYTE DISTURBANCES ACCOMPANYING HEART FAILURE

The adaptive mechanisms described above, in addition to some of the therapies for heart failure, can lead to a number of fluid and electrolyte alterations. These imbalances are summarized in Table 11-1.

NURSING ASSESSMENT OF PATIENTS WITH HEART FAILURE

SIGNS AND SYMPTOMS OF HEART FAILURE

The signs and symptoms of chronic ventricular failure are presented in Table 11-2, which can be used to provide an understanding of the factors to assess in patients at risk for developing heart failure. Knowledge

TABLE 11-1 *Water and electrolyte disturbances in congestive heart failure*

Cause	Water and electrolyte disturbance
Excessive aldosterone secretion	Increased retention of sodium and water by the kidneys resulting in:
Decreased renal blood flow secondary to cardiac failure and vasoconstriction	Increase in total sodium content of body Increase in total extracellular water volume
	Relatively greater retention of water than sodium
Excessive secretion of ADH causes increased retention of water	May depress serum sodium to abnormally low levels, even though the total body sodium is above normal
Hydrostatic pressure is increased by the excessive venous blood volume	Shift of fluid from the intravascular compartment to the interstitial compartment with edema
Excessive aldosterone secretion promotes potassium excretion	Potassium deficit
Excessive use of potassium-losing diuretics and prolonged loss of potassium by vomiting or diarrhea represent typical causes of potassium deficit	
Slowing of the circulation interferes with the excretion of metabolic acids	Mild metabolic acidosis
Increased liberation of lactic acid from anoxic tissues and failure of the body to metabolize it rapidly	
Pulmonary congestion interferes with the elimination of carbon dioxide from the lungs	Respiratory acidosis
Mercurial and thiazide diuretics cause a greater excretion of chloride ions than sodium ions; loss of chloride ions causes a compensatory increase in bicarbonate ions, hence alkalosis	Metabolic alkalosis if mercurial or thiazide diuretics are used extensively
Extensive use of potent diuretics plus severely restricted sodium intake	Sodium deficit
Excessive loss of sodium from other routes, such as repeated paracentesis, vomiting, or diarrhea	

TABLE 11-2 *Signs and symptoms of congestive heart failure and their causes*

Symptom or sign	Cause
Fatigue with little exertion or at rest	Tissue anoxia due to decreased cardiac output
Dyspnea on exertion	Cardiac output is inadequate to provide for the increased oxygen required by exertion
Increased respiratory difficulty Dyspnea even at rest Orthopnea	Increased tissue hypoxia caused by progressive failure of the heart as a pump. Increased interstitial edema decreases the lung compliance and increases the work of breathing
Paroxysmal nocturnal dyspnea (with left heart failure)	When recumbent, edema fluid from the dependent parts returns to the bloodstream, increasing preload and causing decompensation
Troublesome cough producing noncharacteristic sputum, although it may at times be brownish or blood-tinged Pulmonary rales and wheezing	Transudation of serum with hemosiderin-filled macrophages into the alveoli causes pulmonary congestion
Tachycardia	Effort to compensate for decreased cardiac output
Abnormal third heart sound	Associated with rapid ventricular filling in noncompliant ventricle
Elevated pulmonary capillary wedge pressure	The left ventricle cannot maintain stroke volume in the face of increased venous return
Elevated venous pressure	Increase in total blood volume Accumulation of blood in the venous system resulting from incomplete emptying of the heart
Cardiomegaly	Hypertrophy of myocardium helps to maintain stroke volume
Cyanosis, particularly of lips and nail beds	Venous distention Inadequate oxygenation of blood
Visible distention of peripheral veins, most noticeable on face, neck, and hands	Elevated venous pressure
Decreased urinary output	Decreased cardiac output and renal blood flow Sodium and water retention caused by excess aldosterone secretion Increased water retention caused by excess ADH secretion
Edema first appears in dependent parts	Hydrostatic pressure is greatest in dependent parts of the body
Edema later becomes generalized	Progressive cardiac failure causes substantial increase in hydrostatic pressure in all parts of the body
Engorgement of the liver and other organs	Decreased cardiac output causes damming of venous blood Increase in total blood volume and interstitial fluid volume
Nausea and vomiting	Edema of the liver and intestines Impulses arising from the dilated myocardium in acute congestive heart failure Digitalis toxicity
Anorexia	Potassium deficit Digitalis toxicity
Constipation	Poor nourishment and inadequate bulk in diet Lack of activity Depression of motor activity by hypoxia
Pulmonary edema with severe dyspnea, coughing of pink frothy fluid, cyanosis, shock, and death	Increased venous pressure may cause serum and blood cells to transudate into the alveoli

of the underlying pathophysiology also allows the nurse to provide effective nursing therapy in this complex situation.

HEMODYNAMIC MONITORING

The use of hemodynamic monitoring in heart failure has made it possible to measure pressures on a continuing basis directly within the heart chambers and great vessels and to monitor cardiac output. Changes in these pressure measurements can indicate early left ventricular failure before other signs and symptoms develop; thus, they have become an important part of nursing assessment for left ventricular failure.

Left ventricular function is monitored indirectly by measuring pulmonary artery and pulmonary capillary wedge pressures with a flow-directed catheter. These pressures provide an indication of the left ventricular end-diastolic pressure (LVEDP). During left ventricular diastole, the mitral valve is open and the pressures become equalized in the left ventricle and left atrium. This pressure is reflected retrogradely into the pulmonary veins, pulmonary capillary bed, and pulmonary artery. Thus, the pulmonary artery end-diastolic pressure (PAEDP) is a reflection of the (LVEDP) (Fig. 11-3). Monitoring PAEDP provides a guide to early changes in LVEDP—also known as left ventricular filling pressure. Increase in LVEDP in acute cardiac disease indicates failure of the left ventricle as a pump. The increased pressure is due to the diminished ability of the left ventricle to empty its contents. Because any ob-

struction between the left ventricle and the pulmonary artery will distort these pressures, the relationship of PAEDP to LVEDP will not be close in persons with mitral disease, chronic obstructive pulmonary disease, or pulmonary hypertension. In addition to assessing left ventricular function, the PAEDP is useful in monitoring the effect of therapies aimed at restoring left ventricular function. These therapies include inotropic agents, vasodilator medications, diuretics, and dietary/fluid interventions.

SWAN-GANZ CATHETER One type of flow-directed catheter (Swan-Ganz) is illustrated in Figure 11-4. The catheter is used as a diagnostic tool to obtain hemodynamic pressures and to determine cardiac output by means of a cardiac output computer. The 7F thermodilution catheter body is a quadruple-lumen design with a balloon at the distal end.

Lumen1—terminates at the tip of the catheter. Chamber pressures, pulmonary artery pressure, capillary wedge pressures, as well as blood samples, can be obtained through this lumen.

Lumen 2—terminates 30 cm from the catheter tip, placing it in the right atrium when the distal lumen opening is in the pulmonary artery (allowing simultaneous measurement of CVP and pulmonary artery pressure). This lumen carries the injectate necessary for cardiac output computation. An exact amount of dextrose solution of known temperature (such as 0 to 5° C [32° to 41° F] is injected into the right atrium or superior vena cava and the resultant change in blood temperature is detected by the thermistor in the pulmonary artery. Cardiac output is inversely proportional to the integral of temper-

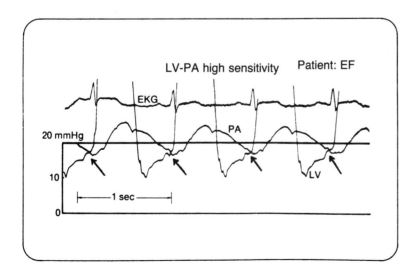

FIG. 11-3.
Simultaneous pressures are recorded from the pulmonary artery *(PA)* and left ventricle *(LV)* in a patient with acute myocardial infarction. At end-diastole the PAEDP and LVEDP are quite similar. Note arrows. (Reprinted with permission of the American Heart Association: Coronary Care—Invasive Techniques for Hemodynamic Measurement, p 20, 1973. Arrows have been added by the author.)

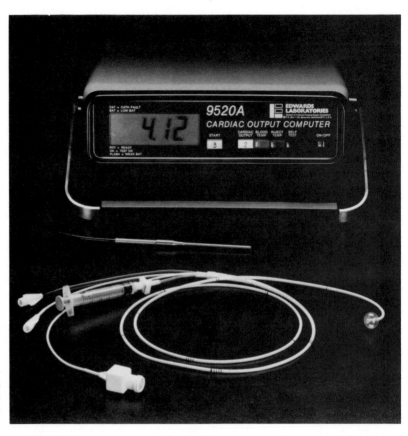

FIG. 11-4.
Swan-Ganz thermodilution catheter and cardiac output computer. (Courtesy of American Edwards Laboratories, Division of American Hospital Supply Corporation, Santa Ana, California)

ature change (the cooler the blood, the less is cardiac output, and vice versa).

Lumen 3—contains the electrical leads for the thermistor, which is positioned at the catheter surface 4 cm proximal to its tip.

Lumen 4—used to inflate and deflate the 1.5-cc capacity balloon.

A newer Swan–Ganz catheter has been developed with three lumens; right atrial and right ventricular electrodes are combined in one lumen. In addition to providing for the determination of intracardiac pressures and cardiac output, right atrial and right ventricular electrocardiograms can be recorded for the diagnosis of complex arrhythmias; the right atrial and right ventricular electrodes can also be used for temporary pacing.

The flow-directed catheter is inserted by means of a cutdown or by percutaneous technique through a suitable needle or sheath. The catheter, connected to a transducer and monitoring system, is advanced into the vena cava near the right atrium; at this point, the balloon is inflated to the recommended volume. Fil-

tered carbon dioxide is the inflation medium of choice since it is rapidly absorbed should the balloon accidentally rupture; however, air is sometimes used for reasons of convenience. The risk of balloon rupture and the possibility of air embolus entering the arterial system (as in the presence of intracardiac shunts) must be assessed by the physician prior to substituting air for carbon dioxide. The balloon must never be inflated with liquid (fluid interferes with balloon deflation and "flotability" of the catheter). The inflated balloon serves two purposes in the insertion procedure. First, it allows the catheter to be moved with the flow of blood through the heart chambers; secondly, the inflated balloon covers the catheter tip, minimizing the occurrence of premature ventricular contractions during passage of the catheter. Care should be taken not to overinflate the balloon, since rupture may occur.

The catheter is carefully advanced under continuous pressure and EKG monitoring. Usually it will pass within 10 to 20 seconds through the right atrium, through the right ventricle, into the pulmonary artery, and into the pulmonary wedge position (Fig. 11-5).

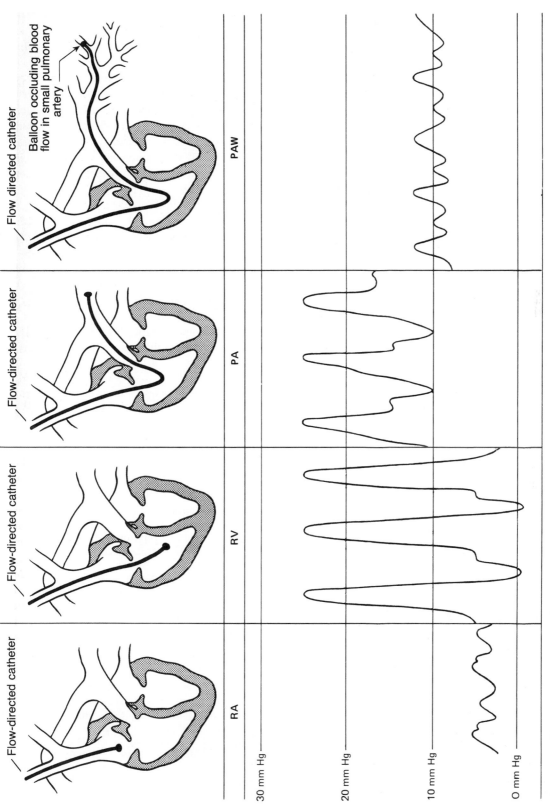

FIG. 11-5. Flow-directed catheter positions with corresponding pressure tracings. (*RA*, right atrium; *RV*, right ventricle; *PA*, pulmonary artery; *PAW*, pulmonary artery wedge)

Pulmonary artery pressure is observed as soon as the balloon passes through the pulmonary valve. Once the balloon becomes lodged in the wedge position, as noted on the pressure monitor, the balloon is quickly deflated. (Wedge position refers to lodging of the catheter in a small branch of the pulmonary artery.)

Lengthy balloon inflations during pulmonary wedge pressure recordings should be avoided, since this is an occlusive maneuver and may cause infarction in the area of the lung supplied by the involved branch of the pulmonary artery. The balloon should be deflated as soon as the wedge pressure is recorded; it should never remain inflated more than 30 to 60 seconds.[5] After balloon deflation, the pulmonary artery contour pressure wave should return. Pulmonary infarction may be manifested by the sudden occurrence of hemoptysis.

Because the possibility of thromboembolic and infectious complications increases with the length of time of catheterization, the duration of catheterization should be limited to the minimum required by the patient's condition. Prophylactic antibiotics and anticoagulation protection should be considered when the catheter remains in place for more than 48 hours. Thrombus formation at the tip of the catheter can be prevented by the continuous infusion of a heparinized intravenous fluid. Care should be taken to maintain proper aseptic technique.

Pulmonary artery perforation may rarely occur; it can be prevented by keeping the balloon inflation time to a minimum and never using fluid as the inflation medium. Also, reinflation of the balloon should be done very slowly and stopped as soon as the wedge pressure is recorded. (The balloon should never be inflated beyond the capacity recommended by the manufacturer of the catheter.)

PRESSURE PARAMETERS The right atrial pressure is normally between 1 mm and 6 mm Hg; right ventricular pressure is 20 mm to 30 mm Hg systolic and 0 to 5 mm Hg diastolic. Causes of increased right atrial pressure include hypervolemia and increased vascular tone; conversely, decreased right atrial pressure is caused by hypovolemia and by loss of systemic venous tone. Failure of the right ventricle causes a rise in right atrial pressure, as does tricuspid stenosis and pericardial tamponade. Abnormal right ventricular pressures are caused by right ventricular failure, pulmonary hypertension, pulmonary stenosis, pulmonary insufficiency, and untreated left ventricular failure. These right-sided pressures provide information similar to that furnished by CVP measurement (see Chapter 6). CVP readings are not reliable indicators of left ventricular performance.

The pulmonary artery pressure is normally 20 mm to 30 mm Hg systolic and 8 mm to 12 mm Hg diastolic. The pulmonary artery wedge pressure is normally between 4 mm and 12 mm Hg; it reflects left ventricular preload. Elevation above normal is seen in mitral stenosis, mitral insufficiency, and left ventricular failure. The pressures obtained by a flow-directed catheter are illustrated in Figure 11-5.

Cardiac output, which refers to the quantity of blood pumped each minute, can be obtained with the use of a four-lumen catheter. The comparison of cardiac output with body surface area is called the *cardiac index*; the normal is 2.8 liters to 3.2 liters/min/m.[2]

The relationship between pulmonary artery wedge pressure readings and cardiac index can be used to identify the presence and type of left ventricular failure, as well as to indicate the type of medical therapy needed and to evaluate the therapy's effectiveness.

TYPICAL HEMODYNAMIC PATTERNS AFTER MYOCARDIAL INFARCTION The following are examples of typical patterns of hemodynamic measurements seen in a patient after a myocardial infarction:

Normal pulmonary artery wedge pressure and normal cardiac index. This indicates that the patient is not experiencing left ventricular failure at the moment. The patient's activity can increase as long as he remains stable clinically.

Increased pulmonary artery wedge pressure and increased cardiac index. This indicates pulmonary congestion, such as pulmonary edema. The pressures reflect LVEDP and adaptive mechanisms in response to the increased preload. Therapy may include vasodilators to decrease the work of the heart or diuretics to decrease preload.

Decreased pulmonary artery wedge pressure and decreased cardiac output. This indicates systemic hypoperfusion, which can lead to tissue hypoxia. It also may indicate hypovolemia (which occurs in 20% to 40% of patients who have a myocardial infarction) caused by one or more of the following: inadequate fluid intake, excessive sweating, long-term use of diuretics, salt-restricted diet, vomiting, and persistent vasoconstriction. This hypovolemia may cause a low pulmonary artery diastolic pressure, as well as a low cardiac index. The patient with cardiogenic shock caused by hypovolemia has a good prognosis if the fluid volume is replaced rapidly (but cautiously) using the hemodynamic pressure changes to evaluate the therapy.

Increased pulmonary artery wedge pressure and decreased cardiac output. This indicates systemic hypoperfusion and pulmonary congestion. In this situation, the patient may receive cardiotonics, vasodilators, or intra-aortic balloon pumping (IABP). Cuff blood pressures may be difficult to obtain because of the low cardiac output and intense vasoconstriction. Useful information can be obtained by calculating the mean arterial pressure (MAP), which is the systolic pressure plus twice the diastolic pressure divided by three

$$\text{(for example: } \frac{120 + (2 \times 80)}{3} = 93\text{)}.$$

Any value less than 75 indicates poor perfusion of the coronary arteries, which will cause further deterioration of left ventricular pressure.

TREATMENT OF CONGESTIVE HEART FAILURE: NURSING IMPLICATIONS

When possible, treatment involves elimination of the underlying disease producing the heart failure. For example, surgical correction of a valvular disorder or removal of a calcified pericardium may restore cardiac function to normal and produce a spontaneous diuresis. Unfortunately, most persons with congestive heart failure have irreversible cardiac damage, such as that caused by myocardial infarction. When the primary disease cannot be eliminated, the only alternative is to make the most efficient use possible of remaining cardiac function.

DECREASING MYOCARDIAL WORKLOAD

Rest causes a reduction in the tissue's oxygen need and decreases the workload of the heart. It also produces a physiologic diuresis. Sometimes rest alone is sufficient to alleviate the symptoms of congestive heart failure. The amount of rest required varies with the person and may range from complete bedrest to only slight restriction of activity.

The prescription of physical rest by the physician must be specific enough to have meaning to the patient. Too often, patients are given ambiguous directions to "rest" or to "take it easy." Such vague statements are not only useless; they may actually be harmful because each person interprets rest differently. The nurse can help by encouraging the patient to ask the physician specific questions regarding the activity permitted.

Another important nursing responsibility consists in observing the patient's responses to exercise, including pulse and respiratory rate changes. An increase in heart rate greater than 10% over the baseline rate indicates that an activity may exceed the capacity of the failing heart to respond. Careful reporting of these observations helps the physician determine the desired amount of activity. Because the patient's condition may fluctuate widely from day to day, the nurse must often use her own judgment in controlling his activity. For example, assume that a patient has been allowed up in a chair for 30 minutes in the morning. Even though this period is permitted, the appearance of dyspnea, chest pain, or a substantially increased pulse rate before the 30 minutes are up indicates that the patient should be put back to bed.

Emotional rest is also important in the management of congestive heart failure. Periods of tension are associated with increased sodium and water retention, while periods of emotional relaxation are associated with diuresis. Nursing efforts should, therefore, be directed toward avoiding emotional problems and achieving a relaxing environment. A major nursing responsibility is emphasizing the importance of emotional rest to the patient's family and to nonprofessional personnel giving direct patient care. At times, judicious use of sedatives may help promote needed rest and relaxation.

Other interventions to ensure decreased myocardial workload include relief of any existing pain and reduction of fever, if present. Also, patients with heart failure may receive oxygen therapy to maintain an arterial PO_2 of at least 80 mm Hg. This reduces hypoxia and the work of breathing. Situating the patient in a semi-Fowler's position facilitates ventilation by allowing full chest expansion and diaphragmatic excursion. The upright position may also decrease venous return.

DECREASING VENOUS RETURN

There are three aspects involved in the long-term management of venous return: low-sodium diet, diuretics, and fluid restriction.

LOW-SODIUM DIET Restriction of sodium ions in the diet is a valuable aid in the management of congestive heart failure. In general, the fewer number of sodium ions in the body, the less water is retained.

The degree of sodium restriction necessary to control edema varies with the severity of the heart failure. Many patients can achieve a sufficiently low intake of sodium simply by not adding salt in cooking or at the table and by avoiding high-sodium foods such as salted crackers, bacon, ham, salted nuts, foods with sodium salt preservatives, and so on. As a rule, restriction of ths intake of salt to 2 g to 5 g daily instead of the usual 10 g or more in an average diet is sufficient to control edema. However, more drastic sodium restriction to less than 1 g/a day—even to 250 mg/day or less—may be required for some patients. A reduction of 1 g of salt/day is equivalent to a decrease of 0.4 g of sodium. It is important that diet orders specify the sodium amounts accurately and that the patient understand how to use the nutrition information provided on food labels.

The degree of sodium restriction necessary to control edema also varies with other facets of treatment, such as rest and the use of diuretics. For example, an ambulatory patient requires more severe sodium restriction than a patient at bedrest, because rest in itself encourages diuresis. A patient receiving potent diuretics has much less need of severe sodium restriction than a patient not receiving diuretics. Indeed, drastic restriction of sodium intake can be dangerous in the patient receiving a potent diuretic.

Although dietary sodium restriction is simple in theory, it frequently is difficult to achieve. Many patients fail to adhere to low-sodium diets because they lack an understanding of the foods allowed and to be avoided. All too often, the only diet instruction given to the patient consists of handing him a copy of his diet on the day of his discharge from the hospital.

The nurse should make every effort to make the diet acceptable to the patient. First of all, she should give him an explanation of why he must be on the diet. Secondly, the dietitian should discuss the diet with the patient and learn what his food preferences are. The possible use of salt substitutes should be discussed with the physician. All of these contain potassium and, thus, should not be used in the presence of oliguria and severe kidney disease. Additional sessions should be held while the patient is in the hospital in order to increase his knowledge and acceptance of the diet. The nurse should support the dietitian's efforts. In instances where a dietitian is not available, the nurse should carry the full responsibility of diet instruction; for this reason, she should have a working knowledge of low-sodium diets. (See Chapter 5 for a more thorough discussion of sodium-restricted diets.) Literature concerning these diets is available from the American Heart Association at the request of the patient's physician. The diets have been arranged in exchange lists to facilitate meal planning with a variety of foods. Excellent books on the preparation of attractive low-sodium diets are available.

Patients on severe sodium restriction should be observed for symptoms of sodium deficit, especially if they are receiving potent diuretics or if they are losing sodium through such routes as vomiting, diarrhea, excessive perspiration, or repeated paracentesis.

DIURETICS Diuretics are a valuable aid in the symptomatic treatment of congestive heart failure. Their primary purpose is to promote the excretion of sodium and water from the body. If hemodynamic monitoring is being used, it may be noted that diuretics produce a decrease in pulmonary artery wedge pressure (preload). In varying degrees, most diuretics tend also to promote the excretion of potassium. Diuretics that are associated with hypokalemia include the thiazides, the mercurials, chlorthalidone (Hygroton), furosemide (Lasix), ethacrynic acid (Edecrin), and the carbonic anhydrase inhibitors, such as acetazolamide (Diamox). Mercurial diuretics are used less frequently than the other diuretics because many of them require intramuscular administration and they have more side-effects, including cramps, diarrhea, skin rashes, and local pain at the injection site. (See Chapter 5 for a more thorough discussion of diuretics.)

Excessive loss of potassium ions during diuretic therapy can be either prevented or corrected by the administration of a suitable potassium supplement, such as K-Lyte, Kaon Elixir, Potassium Triplex, or K-Lor. The use of diuretics should be decreased when sodium loss is occurring from another route; a low-sodium diet can take the place of diuretic therapy in many persons.

Edecrin and Lasix are potent diuretics that are effective even after their action has produced hypochloremic alkalosis. They are unusually potent and have a rapid onset. Patients receiving these diuretics should be observed closely for signs of too vigorous diuresis, such as lethargy, weakness, dizziness, anorexia, vomiting, leg cramps, mental confusion, and circulatory collapse.

Potassium-conserving diuretics are capable of producing diuresis in congestive heart failure. Spironolactone (Aldactone) is an aldosterone antagonist, which acts by blocking the potent sodium-retaining effect of aldosterone on the renal tubules. Aldactone should not be given in conjunction with a potassium supplement because of the danger of potassium excess. (Recall that aldosterone causes potassium loss; therefore, its antagonist permits potassium retention.) Triamterene (Dyrenium) also promotes sodium loss and potassium retention although it has a different mode of action. (It interferes with the exchange of sodium ions for potassium and hydrogen ions.) Again, potassium supplements should not be given because hyperkalemia may result; symptoms of hyperkalemia include paresthesia of the extremities, nausea, weakness, and intestinal cramping with diarrhea. If hyperkalemia is severe, cardiac arrest can occur. Only patients with adequate renal reserve should receive potassium-conserving diuretics.

Primary nursing responsibilities in the care of heart failure patients who are receiving diuretics and are on sodium-restriction include keeping an accurate account of fluid intake and output and measuring the weight daily (see the discussion of both of these procedures in Chapter 4). The data obtained form these measurements are of inestimable use to the physician in regulating the dose of diuretics and the degree of dietary sodium restriction for each patient.

FLUID ADMINISTRATION Water intake is usually not restricted in the long-term management of congestive heart failure unless there is a body sodium deficit or dilution of the serum sodium by the excessive retention of water.

Undue loss of sodium may be caused by the excessive use of diuretics, vomiting, diarrhea, severe diaphoresis, or repeated paracentesis. A drastically reduced sodium intake may predispose to sodium depletion, although persons on low sodium intake for prolonged periods usually develop remarkable sodium conservation, something that does not happen in the case of patients on low potassium intake, since there is no true body conservation of potassium. If the water intake of patients in a mild state of sodium depletion is not reduced, the depressed serum sodium level may become further depressed; a frank state of sodium deficit may then develop.

The operation of abnormal routes of sodium loss should alert the nurse to search for hyponatremia, especially if the sodium intake is low and diuretics are being given. Although the total sodium content of the body is elevated in congestive heart failure, water retention caused by excessive ADH hormone secretion may dilute the serum sodium concentration to below normal levels. Moreover, part of the sodium in the extracellular fluid moves into the cells to replace the potassium loss which so often occurs with congestive heart failure. There is no characteristic clinical picture accompanying this state. When it is well developed, however, the usual therapeutic measures fail to reduce the edema that accompanies it.

One of the chief features of intractable heart failure is the inability of the kidney to respond to the usual diuretics. Treatment may consist of sodium restriction, restriction of water intake to 1000 ml per day, and the use of more potent diuretics. It has been found that patients refractory to other diuretics often respond successfully to Edecrin and Lasix.

Intravenous fluids The intravenous route for fluid administration may be necessary in critically ill patients with congestive heart failure. Many physicians are hesitant to administer fluids to such patients for fear of causing circulatory overload and pulmonary edema. While there is little doubt that intravenous administration of fluids to a cardiac patient carries some risk of causing circulatory overload, the fear of this complication has been exaggerated to the point that many cardiac patients receive inadequate fluid therapy. The recent increase in the use of hemodynamic monitoring devices has done much to alleviate this problem. Frequent checks of venous pressure and pulmonary artery wedge pressure during fluid administration give early warning of circulatory overload and serve as guides to the safe administration of needed water and electrolytes.

The nurse should pay careful attention to the volume, speed, and composition of fluids administered to the patient with congestive heart failure. The response to fluids should be observed frequently and the flow rate adjusted accordingly. (See Chapter 6 for a more detailed discussion of venous pressure monitoring and nursing responsibilities in intravenous fluid administration.)

INCREASING MYOCARDIAL CONTRACTILITY AND CARDIAC OUTPUT

DIGITALIS Digitalis preparations are given to patients with decreased left ventricular function because of their inotropic action—that is, digitalis increases the force of myocardial contraction. The mechanism of action for this inotropic effect is not completely understood; it may act by increasing the availability of free calcium ions to the contracting sites of the heart muscle.[6] As the heart contracts more forcefully, tissue perfusion increases and the compensatory responses caused by hypoxia decrease, allowing a corresponding increase in renal function and diuresis. Inotropic agents are generally given to increase cardiac output and decrease pulmonary artery wedge pressure through increasing the strength of myocardial contraction.

Excessive doses of digitalis may result in the following toxic symptoms:

Aversion to food (which usually precedes other symptoms by 1 or 2 days)

Nausea

Excessive salivation

Vomiting

Abdominal pain

Diarrhea

Urticaria

Gynecomastia

Headache, fatigue

Confusion, particularly in elderly patients with arteriosclerosis; central nervous system symptoms are often late effects

Blurred vision, diplopia, optic neuritis

Yellowish green "halo," vision or presence of white dots (frost) on objects

Bradycardia (caused by atrioventricular block)

Variety of arrhythmias, including ventricular tachycardia, in which the heart beats rapidly and irregularly (arrhythmias may precede extracardiac manifestations)

Bigeminal pulse

It is important to differentiate between the combined anorexia and nausea of heart failure and that produced by digitalis toxicity. Patients receiving digitalis should have periodic EKGs to detect early the development of digitalis toxicity. Serum digitalis levels also provide important information. Before administering the drug, the nurse should check the apical-radial pulse for a full minute, noting rate, rhythm, volume, and pulse deficit. It is important for the nurse to know which patient is receiving a digitalis preparation as well as the patient's baseline data, especially heart rate and rhythm, so that any changes can be evaluated accurately in terms of representing expected results or

serious arrhythmias. Unless otherwise ordered, the drug should be withheld and the physician notified when the following appear:

Apical pulse below 60

Occurrence of a marked change in rate or regularity (premature ventricular contractions); bigeminy; atrial fibrillation; sudden marked change from a tachycardia (e.g., a change in heart rate from 160 to 80 may indicate that a 2:1 heart block has developed); sudden spurts of a rapid pulse (e.g., paroxysmal atrial tachycardia with varying block)

An overdose of digitalis may have a depressant action, causing conduction disturbance and excessive slowing of the heart. It may also cause increased myocardial irritability, producing extrasystoles or tachycardias. The nurse should be alert for a coupled pulse beat (bigeminy) in which the regular beat is followed almost immediately by a weak beat and a pause. A bigeminal pulse is a common sign of digitalis toxicity in adults; bigeminal pulse and other irregularities should be reported to the physician. The nurse should also report pulse deficit, caused by failure of the extrasystoles to produce a pulse at the wrist. If digitalis is not discontinued, the premature beats can take on the rhythm of ventricular tachycardia and progress to ventricular fibrillation.

Calcium ions enhance the action of digitalis; thus, decreasing the plasma calcium concentration is helpful in counteracting the cardiotoxic effects produced by digitalis. The plasma calcium level can be reduced by the intravenous infusion of disodium edetate (Endrate); this drug ties up excess calcium and removes it from the body. Conversely, calcium should never be administered intravenously to a digitalized patient.

Symptoms of digitalis toxicity may be induced by potassium deficit, since this deficit sensitizes the heart to digitalis. The patient maintained on digitalis without toxicity can, in the presence of potassium deficit, exhibit arrhythmias typical of digitalis intoxication. An irregular pulse caused by digitalis toxicity can sometimes be corrected by the administration of a potassium salt, either by mouth or, if necessary, parenterally. Magnesium has also been reported to correct the toxic rhythms produced by digoxin.

Patients prone to develop potassium deficit (such as those receiving furosemide, mercurial, or thiazide diuretics, or those with vomiting, diarrhea, or poor food intake) should be observed with special care for signs of digitalis toxicity.

PULMONARY EDEMA

Pulmonary edema is a manifestation of severe left ventricular failure. The signs and symptoms of pulmonary edema include those listed below:

Restlessness

Severe dyspnea

Gurgling respirations

Cyanosis

Coughing up of frothy fluid

Elevated pulmonary artery wedge pressure

The nurse must identify patients at risk for pulmonary edema and provide nursing care that decreases the heart's workload. This will include monitoring activity levels, careful patient teaching, and emotional support so that anxiety does not stimulate the sympathetic nervous system, causing increased cardiac output. Preventive interventions also include careful monitoring of intravenous fluids, especially when sodium chloride must be given to correct a profound sodium deficit.

The occurrence of pulmonary edema is an emergency situation. The patient should be quickly placed in a high-Fowler's position, and oxygen started. Often, positive pressure breathing devices are used for assisting ventilation; this increases intra-alveolar pressure so that fluid does not continue to move into the alveoli. Positive pressure ventilation also increases the intrathoracic pressure so that venous return and, therefore, preload are reduced.

Intravenous administration of morphine sulfate may be used to achieve multiple effects in the patient with pulmonary edema. It acts in the following ways:

Decreases preload through peripheral venous vasodilation, which decreases venous return

Decreases afterload by decreasing arterial blood pressure

Decreases the myocardial workload

Decreases anxiety, thus decreasing sympathetic nervous system stimulation

Morphine must be given cautiously to patients with increased intracranial pressure, severe pulmonary disease, or decreased level of consciousness.

Alternating tourniquets may be used to obstruct venous return to the heart; they can remove up to 700 ml of blood from the circulating volume. The removal of 200 ml to 500 ml of blood by phlebotomy may be tried to relieve the workload on the heart and to reduce venous pressure. Diuretics may be given intravenously to produce a rapid diuresis. All of these measures help to decrease preload, which will be reflected in a decreased pulmonary artery wedge pressure.

An intravenous vasodilator such as sodium nitroprusside may be given to decrease afterload. This decreases the workload of the heart by lowering the peripheral resistance to left ventricular output. This can cause an increased cardiac output and decreased pulmonary artery wedge pressure. Pulmonary venous pressure usually decreases significantly because of the peripheral venous pooling effect caused by vasodilators. Myocardial oxygen consumption is decreased.

Careful monitoring of arterial pressure, cardiac output, and pulmonary artery wedge pressure should be done when parenteral vasodilator therapy is employed in the management of such patients. A major possible complication of vasodilator therapy is a pronounced drop in arterial pressure, which could increase myocardial ischemia by restricting coronary blood flow.

An inotropic agent such as digitalis may be given. In the undigitalized patient, 0.5 mg to 1.0 mg of digoxin may be given intravenously. This can be followed by additional doses as indicated.

One should remember that acidosis, hypoxia, and electrolyte imbalance can precipitate or prolong cardiac failure. Because of this, arterial blood gases and electrolyte levels should be monitored closely in the acutely ill patient. Any chronic abnormality must not be corrected too abruptly. Fluid volume deficit, if present, is corrected first; then pH alterations are treated, along with imbalances of potassium and calcium. It should be noted that myocardial contractility may be decreased when the pH is greater than 7.55 or less than 7.20, the PCO_2 is less than 25 mm Hg; the serum potassium level is elevated or the serum calcium level is decreased. Sodium bicarbonate should be given as necessary to correct metabolic acidosis. Adequate ventilation is mandatory to prevent hypoxia and respiratory acidosis; if necessary, intubation and artificial ventilation should be employed.

CARDIOGENIC SHOCK

CHARACTERISTICS

Cardiogenic shock occurs in approximately 15% of patients who have acute myocardial infarctions. It occurs when 40% of the left ventricular myocardium has decreased function (caused by infarction or ischemia) and is associated with an 80% mortality rate. Other causes of cardiogenic shock include myocardial ischemia, papillary muscle rupture, and severe cardiomyopathy.

Cardiogenic shock is characterized by left ventricular failure, low cardiac output, arterial hypotension, and peripheral vasoconstriction. The patient will usually demonstrate an arterial systolic blood pressure of less than 90 mm Hg. When the mean arterial pressure falls below 75 mm to 85 mm Hg, there is danger that coronary blood flow will be inadequate, leading to further damage to the myocardium. The patient can also demonstrate a decreased urinary output of less than 20 ml/hour, which is related to decreased glomerular filtration rate. An altered mental status may occur due to cerebral hypoxia. Poor skin perfusion can cause cold, clammy skin (due to peripheral vasoconstriction).

TREATMENT AND NURSING IMPLICATIONS

The treatment of cardiogenic shock depends on the alterations that the patient is experiencing. One basic problem that many patients will encounter is cellular hypoxia (related to decreased cardiac output). It is therefore important to implement interventions aimed at treating the hypoxia. Oxygen may be given by cannula, mask, or mechanical ventilation with intubation. It is important to prevent atelectasis and retention of CO_2 by suctioning the patient as necessary and promoting adequate coughing. The goal of these interventions is to prevent acidosis, which decreases myocardial contractility, and to maintain an arterial PO_2 greater than 80 mm Hg.

Inotropic agents such as digoxin, dopamine, and dobutamine may be given to strengthen the myocardial contractility. Vasopressors such as norepinephrine may be given if the mean arterial pressure is less than 75 mm Hg. Sometimes a combination of medications is given. Dopamine, 5 mcg to 15 mcg/kg/min, is often given first to increase myocardial contractility and renal perfusion. A therapeutic response may not be achieved before dopamine causes an increased heart rate, which increases the heart's need for oxygen. When this occurs, dobutamine, 2.5 mcg to 15 mcg/kg/min, is added. Dobutamine increases myocardial contractility while decreasing vascular resistance and afterload. Nitroprusside may also be given to decrease

afterload in the presence of severe, persistent vasoconstriction. During therapy with any of these medications, particularly with vasodilators, intra-arterial and pulmonary artery wedge pressures must be monitored closely. If the patient's mean arterial blood pressure is less than 75 mm to 85 mm levarterenol bitartrate (Levophed) may be given to increase vascular resistance, arterial pressure, and cellular oxygenation. Levophed also has a slight inotropic effect.

Intravenous fluids should be given to correct any hypovolemia. However, fluids must be given cautiously to prevent overload, particularly pulmonary congestion. The pulmonary artery wedge pressure should be maintained between 18 mm to 22 mm Hg to ensure adequate left ventricular filling.

If the cardiogenic shock persists longer than two hours, the patient may be a candidate for IABP. In this procedure, a catheter with a balloon is placed in the aorta. The pump is set so that the balloon inflates during ventricular diastole and deflates during systole. This cycle provides for better coronary artery and systemic perfusion with no increase in peripheral resistance to left ventricular output. The IABP does not correct the cause of cardiogenic shock but does allow time for diagnostic evaluation. The IABP does not decrease mortality in patients who do not have a surgically correctable condition; its use requires skilled nursing and medical care. Complications associated with IABP are bleeding at the site of insertion, thrombosis of the distal femoral artery, aortic dissection, rupture of the balloon, and sepsis associated with the indwelling catheter.

CALCIUM ANTAGONISTS

Calcium antagonists (such as nefedipine, verapamil, and diltiazem) are medications whose primary effect is through their action on vascular smooth muscle contraction. These calcium blockers inhibit the influx of extracellular calcium during the slow phase of muscular contraction (which is the second part of the contraction). This phase occurs as calcium moves to the intracellular space through pores in the cell membrane, which allows the slow movement of calcium. Research indicates that calcium blockers can prevent vasospasm in a vessel wall, arteriolar constriction, and platelet aggregation. They also lessen oxygen demand by decreasing the strength of myocardial contraction.

Calcium antagonists are being used in the treatment of rest angina, especially when coronary spasm has been demonstrated. The therapeutic effect in these clients is probably due to coronary vasodilatation rather than a decrease in heart rate, systemic vasodilatation, or decrease in myocardial contractility. Calcium blockers are also being used in effort angina, especially when unrelieved by nitrates or β-adrenergic blockers. The therapeutic effect in these clients is probably due to systemic vasodilatation rather than coronary dilatation. Calcium antagonists may also be used to treat paroxysmal supraventricular tachycardia and atrial fibrillation of flutter. This effect may be due to a decrease in calcium influx through slow channels in the myocardial conduction system. Other uses include myocardial infarction, systemic hypertension, pulmonary hypertension, chronic heart failure, peripheral vasospastic disorders, and cardiopulmonary bypass.

The side-effects of calcium antagonists are dose-related and are caused by vasodilatation of regional vascular beds. The most common are dizziness, light-headedness, and hypotension.

REFERENCES

1. Wilson R (ed): *Principles and Technique of Critical Care*, p 24. Kalamazoo, The Upjohn Company, 1976
2. Maxwell M, Kleeman C (eds): Clinical Disorders of Fluid and Electrolyte Metabolism, 3rd ed, p 168. New York, McGraw-Hill, 1980
3. Guyton A: Textbook of Medical Physiology, 6th ed, p 295. Philadelphia, WB Saunders, 1981
4. Wilson, p 8
5. Sampline J, Pitluk H: Hemodynamics and respiratory monitoring. In: Berk J, Sampliner J (eds): Handbook of Critical care, 2nd ed, p 70. Boston, Little, Brown & Co, 1982
6. Horvath P, DePew C: Toward Preventing Digitalis Toxicity, Nurses' Drug Alert 4, No. 4:25, 1980

BIBLIOGRAPHY

Leonard B, Redland A: Process in Clinical Nursing. Englewood Cliffs, Prentice-Hall, 1981

Edwards Laboratories Product Manual. Santa Ana, Edwards Laboratories, 1977

Forrester J, Diamond B, Chatterjee K, Swan H: Medical therapy of acute myocardial infarction by application of hemodynamic subsets. N Engl J Med 295:1356, 1404, 1976

Hurst J, Logue R, Schlant R, Wenger N (eds): The Heart, Arteries and Veins, 3rd ed. New York, McGraw-Hill 1974

Krupp M, Chatton M: Current Medical Diagnosis and Treatment. Los Altos, Lange Medical Publications, 1980

Mantle J, Rogers W, Papapietros S, Russell S, Rackley C: Cardiovascular evaluation and therapy of unstable patients. In Kinney M, Dean C, Packa D, Voorman D: AACN's Clinical Reference for Critical Care Nursing. New York, McGraw-Hill 1981

Urrows S: Fluid and Electrolyte Balance in the Patient with a Myocardial Infarction. Nurs Clin North Am 15, No. 3, 1980

12

Fluid Balance in the Patient with Endocrine Disease

INFLUENCES OF HORMONES ON ELECTROLYTES

ADRENOCORTICAL HORMONES

The adrenal mechanism is closely associated with the regulation of sodium and potassium balance. The primary adrenal cortex secretions are mineralocorticoids and glucocorticoids. Aldosterone is the most important mineralocorticoid; cortisol is the chief glucocorticoid.

MINERALOCORTICOIDS Aldosterone causes sodium retention and simultaneous loss of potassium in the urine; it acts on the distal tubule, causing sodium to be exchanged for hydrogen and potassium ions (Fig. 12-1.) This same exchange takes place in the sweat glands in response to aldosterone. The effect of aldosterone on sodium and potassium is many times greater than cortisol's effect on these electrolytes. It should be noted that increased secretion of aldosterone promotes metabolic alkalosis (base bicarbonate excess). Conversely, decreased aldosterone secretion promotes acidosis. These changes in body pH are not excessive, however, and are usually adequately controlled by normal acid–base regulatory mechanisms.[1]

GLUCOCORTICOIDS The glucocorticoids increase the blood glucose level by encouraging gluconeogenesis in the liver. While their activity in the liver is anabolic, their overall action is catabolic, since they increase protein breakdown and nitrogen excretion. Other effects of the glucocorticoids include an anti-inflammatory effect and a mild influence on sodium and potassium concentrations (sodium retention and potassium loss). Glucocorticoids cause a decrease in the serum calcium level, probably by inhibiting the absorption of calcium from the gastrointestinal tract. Lastly, the glucocorticoids enhance water diuresis by preventing the movement of water into the cells; they may also directly inhibit the action of antidiuretic hormone (ADH).[2]

PARATHYROID HORMONE

Parathyroid hormone (PTH) has a major effect on calcium and phosphate and hence on bone metabolism. PTH causes an *increase* in plasma calcium concentration primarily by increasing the rate of bone resorp-

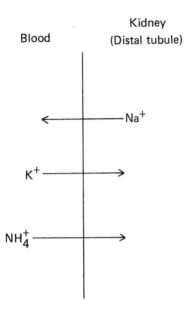

FIG. 12-1. Effects of aldosterone on kidney. Aldosterone causes sodium to be exchanged for hydrogen and potassium ions. (Ezrin C et al: Systematic Endocrinology, p 164. Hagerstown, Harper & Row, 1973)

tion; it also raises plasma calcium by increasing the rate of calcium reabsorption from the gastrointestinal tract and by increasing calcium reabsorption by the kidneys. When the blood calcium drops below the normal level, secretion of PTH increases; conversely, when the blood calcium exceeds the normal level, PTH secretion decreases.

Another function of PTH is to increase the renal excretion of phosphate and, thus, to lower the plasma phosphate concentration. Plasma phosphate level has no direct effect on PTH secretion, but hyperphosphatemia will indirectly stimulate PTH secretion because it suppresses bone resorption, thereby lowering serum calcium levels. Bicarbonate diuresis, which occurs in response to PTH action on the renal proximal tubule, may produce metabolic acidosis in patients with hyperparathyroidism.[3]

CALCITONIN

Calcitonin (thyrocalcitonin) is a weak hypocalcemic hormone secreted by the thyroid in response to hypercalcemia. It *lowers* the serum calcium level by inhibiting bone resorption and perhaps also by limiting calcium absorption from the gastrointestinal tract and from the renal tubules.

THYROID HORMONES

Thyroid hormones are not usually considered to be important in normal regulation of water and electrolyte metabolism; however, at extremes of hypofunction and hyperfunction of the thyroid gland, there may be profound disturbances. These effects are largely indirect. (See the sections dealing with thyroid storm and myxedema coma later in this chapter).

ANTIDIURETIC HORMONE

The posterior pituitary gland releases a water-conserving hormone referred to as the antidiuretic hormone, or ADH. As the name implies, it *inhibits diuresis*. It seems more direct, however, to think of it as *conserving water*. The release of ADH is influenced by changes in both extracellular solute concentration and volume. High solute concentration causes stimulation of ADH secretion with resultant water conservation, diluting the solute load. An increase in blood volume causes a drop in ADH secretion, which in turn causes increased renal water excretion; conversely, a decreased blood volume causes a rise in ADH secretion, which results in renal water conservation.

INSULIN

Although insulin probably has no direct effect on water and electrolyte metabolism, its lack influences intermediary metabolism and can produce life-threatening disturbances in water, electrolyte, and acid–base homeostasis. (See the discussion of water and electrolyte abnormalities associated with uncontrolled diabetes mellitus later in this chapter.)

ENDOCRINE DISORDERS THAT CAUSE FLUID BALANCE DISTURBANCES

ADRENOCORTICAL HORMONE EXCESS

CUSHING'S SYNDROME The clinical condition brought about by excesses of hormones the principal action of which is glucocorticoid is usually referred to as *Cushing's syndrome*. Causes of Cushing's syndrome may include adrenocorticoid tumor (adenoma or carcinoma), excessive pituitary secretion of adrenocorticotropic hormone (ACTH) (Cushing's disease), secretion of an ACTH-like substance by a nonpituitary tumor (the ectopic ACTH syndrome), or administration of excessive quantities of glucocorticoids in the treatment of nonendocrine disease. Cortisol overproduction is the hallmark of Cushing's syndrome. Excessive mineralocorticoid and androgen production may occur with excessive pituitary secretion of ACTH (Cushing's disease) and with the ectopic ACTH syndrome.[4]

Electrolyte disturbances Although the glucocorticoids have only slight mineralocorticoid activity, they may be present in large enough amounts in Cushing's syndrome to produce abnormalities in electrolyte metabolism. These abnormalities may include the following:

Increased loss of potassium and hydrogen ions in the urine due to mineralocorticoid effect on the distal renal tubule

Depletion of total body potassium due to loss of cellular potassium as a result of increased catabolism

Some degree of sodium retention with hypervolemia, which may lead to hypertension and congestive heart failure

Hypokalemic alkalosis (recall that hypokalemia is frequently associated with metabolic alkalosis); this is an ominous sign because it indicates the presence of a large amount of glucocorticoid hormone.

Also, excess glucocorticoid hormones can cause demineralization of bone with resultant increased calcium excretion in the urine, leading to the development of renal calculi. Additional electrolyte problems may be caused indirectly by the effect glucocorticoid hormones have on the cardiovascular system and on the kidney. For example, the hypertension frequently associated with Cushing's syndrome can eventually lead to heart failure and renal failure, with all their sequelae.

Clinical manifestations Clinical findings in Cushing's syndrome are listed below:

Moon facies (edematous facial appearance)

Buffalo hump (extra deposition of fat in the thoracic region)

Obesity with protuberant abdomen, thin extremities

Muscle wasting and weakness (negative nitrogen balance)

Hypertension due to sodium retention with hypervolemia

Gastric ulcers (may occur secondary to increased production of pepsin and hydrochloric acid)

Capillary fragility and bruising (e.g., hematoma may occur following venipuncture)

Purple striae

Thinning of skin

Mental changes (severe depression may be present; suicide is an important consideration)

Low glucose tolerance, often with glycosuria (elevated blood sugar is probably a result of excessive gluconeogenesis)

Loss of bone matrix and demineralization (osteoporosis of skull, spine, and ribs is common)

Hypercalciuria and renal calculi

Cardiac enlargement (seen on radiograph)

In women with Cushing's syndrome, the clinical picture may be complicated by androgen excess, causing hirsutism, acne, oligomenorrhea, and infertility.

Treatment Treatment of Cushing's syndrome is directed at the cause of adrenocortical hyperfunction; adrenal tumors are removed surgically, as are pituitary adenomas.

For persons who are taking exogenous corticosteroids for non-endocrine diseases, the incidence of Cushingoid features can be minimized by administering the drug when serum cortisol levels are normally highest. (Recall that the serum cortisol level fluctuates according to sleep-awake patterns [diurnal cycles]; it is higher in the morning and lower in the evening.[5]) Administration of corticosteroids when the cortisol level is normally low (as in the late evening) increases the likelihood of Cushingoid features developing. Many physicians prefer to administer corticosteroids twice a day—a larger dose in the morning and a smaller dose in the afternoon.

HYPERALDOSTERONISM States of mineralocorticoid excess are usually characterized by the singular effects of these hormones on blood pressure and electrolyte metabolism.

Primary aldosteronism Primary aldosteronism is caused by excessive aldosterone secretion from an adrenal adenoma arising from the zona glomerulosa. It is a relatively rare condition that is more common in females. Aldosterone acts on the renal tubule to cause sodium retention and excretion of potassium and hydrogen. Electrolyte abnormalities that may result include those listed below:

Hypokalemia

High serum bicarbonate concentration (metabolic alkalosis)

Elevation of serum sodium concentration into the high-normal or high range

Decreased calcium ionization, caused by metabolic alkalosis

Although the patient with primary aldosteronism has a higher than normal total body sodium content, there is characteristically *no edema* (unless some other problem is present, such as heart failure).

Clinical manifestations of primary aldosteronism may include the following:

Moderately elevated blood pressure, probably resulting from slight increase in body sodium content

Polyuria, especially nocturnal, and polydipsia caused by hypokalemia and impaired renal tubular reabsorption of water

Paresthesias with frank tetanic manifestations caused by decreased calcium ionization associated with alkalosis

Treatment for primary aldosteronism is surgical removal of adenomas; nodular bilateral hyperplasia is better treated with spironolactone and antihypertensive agents.

Secondary aldosteronism Secondary aldosteronism is a state of increased secretion of aldosterone associated with certain diseases such as cirrhosis, congestive heart failure, and the nephrotic syndrome (these conditions are discussed in their respective chapters). Secondary aldosteronism is a normal physiologic response to fluid volume deficit and hemorrhage; in these situations, it acts to expand the plasma volume. The hypokalemic alkalosis so typical of primary aldosteronism does *not* occur in secondary aldosteronism.

ADRENOCORTICAL INSUFFICIENCY

Adrenocortical insufficiency may be due to primary adrenocortical disease (Addison's disease) or to a deficiency of pituitary ACTH secretion (secondary adrenocortical insufficiency) due to pituitary atrophy, necrosis, or tumor. Also, some of the features of adrenal

insufficiency might be caused by insufficient cortisol administration during a period of stress in patients taking this drug for any of the numerous steroid-responsive illnesses.

PRIMARY ADRENOCORTICAL INSUFFICIENCY Primary adrenocortical insufficiency, or Addison's disease, may be due to autoimmune destruction of the adrenals (cause unknown), infectious destruction of the glands, bilateral adrenal hemorrhage (from anticoagulation therapy), metastasis to adrenals, or adrenalectomy. Persons with primary adrenocortical insufficiency have deficiencies of both glucocorticoids and mineralocorticoids. In mild states of deficiency, and without stress on the person, the condition may be fairly well tolerated provided adequate dietary salt is provided. Addison's disease typically has a slowly progressive course; almost total destruction of the adrenal cortex is necessary before symptoms of insufficiency appear.[6] However, it must be remembered that the chronic course of the disease can be abruptly interrupted if a stressful event intervenes that precipitates adrenal crisis (described later in this chapter).

Electrolyte disturbances Electrolyte disturbances associated with lack of adrenocortical hormones include the following:

Decreased extracellular fluid volume due to large losses of sodium, chloride, and water in urine

Hyponatremia due to urinary sodium wasting

Hyperkalemia due to renal retention of potassium

Mild metabolic acidosis due to failure of potassium and hydrogen ions to be secreted in exchange for sodium reabsorption

Decreased aldosterone secretion is largely responsible for the increased urinary excretion of sodium and retention of potassium, although decreased cortisol secretion undoubtedly contributes to these changes. Water loss accompanies the increased urinary secretion of sodium and results in extracellular fluid volume deficit.

Clinical manifestations Clinical manifestations of primary adrenocortical insufficiency may include those listed below:

Weakness, easy fatigability out of proportion to activity

Anorexia, nausea, vomiting and diarrhea, manifestations of steroid deficiency that hasten the development of volume contraction and vascular collapse

Orthostatic hypotension with narrow pulse pressure; giddiness accompanying change from supine to upright position (it is unusual for the untreated Addisonian patient to have a systolic pressure above 110 mm Hg[7])

Diffuse tanning over nonexposed as well as exposed body parts or accentuation of pigment over nipples, scars, or pressure areas such as the knuckles and elbows (*Note:* Lack of cortisol enhances ACTH and melanocyte-stimulating hormone secretion. Hyperpigmentation may precede other symptoms by months or even years.[8])

Weight loss due to fluid volume deficit and negative caloric balance

Hypoglycemia, noticed several hours after meals; symptoms include hunger, nervousness, sweating, headache, and confusion

Mental changes; paranoid psychoses may be a presenting symptom in chronic adrenal insufficiency

Decrease in heart size noticed on radiograph

Poor resistance to stress; mild adrenal insufficiency may become severe under stress and can lead to vascular collapse

Treatment The patient with Addison's disease has a lifelong disorder and is committed to lifelong steroid therapy. Approximately 20 mg of cortisone by mouth in the morning and 10 mg in the afternoon are typically prescribed; however, the dosage in each patient must be individually adjusted on the basis of symptoms, blood pressure, and serum electrolytes.[9] Some patients do not obtain sufficient salt-retaining effect from cortisone and require either extra dietary salt or the administration of a mineralocorticoid, such as fludrocortisone, 0.1 mg each day or every other day.

Nursing interventions The patient with adrenal insufficiency should be instructed to wear an identification bracelet or necklace indicating his disability so that prompt therapy can be instituted if he is rendered unconscious for any reason. Also, the patient should be instructed to consult his physician during stressful periods, since extra steroid replacement therapy may be needed during these times. Some physicians favor teaching the patient to double his normal dose of cortisol for several days when stressful events (such as infection) occur; with more prolonged illness, the patient is encouraged to consult the physician for further directions.

When hormonal replacement therapy is adequate,

dietary salt intake should be consistent with the patient's taste and usually differs little from the average salt content of a normal diet. However, the patient should be instructed that extra salt may be needed in hot weather. In the presence of severe diarrhea or profuse perspiration, the great losses of salt and fluid warrant an increased salt intake and an increase in steroid therapy (at the physician's discretion). Regular meals should be scheduled to avoid hypoglycemia (recall that cortisol deficiency interferes with the liver's ability to maintain adequate glucose production).

The nurse should be aware that overtreatment with cortisone may produce unfavorable symptoms such as acne, moon facies, elevated blood sugar, peptic ulcer, bleeding tendencies, and hypertension. Overtreatment with mineralocorticoids may cause excessive fluid retention, with weight gain and hypertension; hypokalemia, with arrhythmias and fatigue, and metabolic alkalosis, perhaps with tetany. The nurse should be alert for these symptoms and should make the patient aware of them. Of course, the patient should also be aware of the symptoms of *undertreatment*, which are mainly those of the primary disease. The patient should be instructed to weigh himself daily and be alert for significant weight changes from day to day, indicating either excess fluid retention or loss. Also, he should be made aware of the need for systematic medical follow-up in the control of his disease.

ADRENAL CRISIS Acute adrenal insufficiency is a rare condition referred to as *adrenal crisis*. It is a true medical emergency caused by sudden marked deprivation or insufficient supply of adrenocortical hormones, both cortisol and aldosterone. Adrenal crisis may occur following sudden withdrawal of adrenocortical hormone in a patient with chronic insufficiency or in a patient with normal adrenals, but with temporary insufficiency due to suppression by exogenous glucocorticoid administration; following injury to both adrenals (e.g., trauma or surgery); following stress (e.g., trauma or surgery) in a patient with latent insufficiency, and following bilateral adrenalectomy.

Nursing interventions The nurse should be alert for acute adrenocortical insufficiency (adrenal crisis) when patients with decreased adrenal function are exposed to stress, such as surgery, trauma, emotional upset, excessive heat, or prolonged medical illness. (Adrenal crisis as a result of surgical procedures in patients on prolonged adrenocortical hormone therapy is discussed in Chapter 7). Adrenal crisis may also occur when a patient with chronic adrenocortical insufficiency fails to take his prescribed hormones. To prevent this disorder, the patient should be educated as to the need for adhering to his prescribed regimen and should be aware of the danger of stopping corticosteroid therapy abruptly. The need to keep an adequate supply of the medication *on hand* should be made clear, particularly when traveling. Corticosteroids should never be packed in luggage that could be lost or delayed; instead, they should be carried on the person or in a purse or tote bag. Also, the patient should be instructed to have his prescription filled before all of the medication is gone. Lastly, it is important that the patient understand his illness and the reason for corticosteroid therapy; an informed patient is often more cooperative in maintaining close control of his illness.

Clinical manifestations Clinical manifestations of acute adrenocortical insufficiency include the following:

> Extreme weakness
>
> Acute onset of nausea, vomiting, and epigastric pain
>
> Hyperthermia often present (fever may be 105°F [40.5°C] or higher)
>
> Hypotension (moderate to severe)
>
> Confusion or coma
>
> Shock, if treatment is not begun early

Laboratory data usually reveal hyperkalemia, hyponatremia, an elevated blood urea nitrogen (BUN) due to fluid volume deficit, and hypoglycemia. Hypercalcemia may also be present. The patient with adrenal crisis might appear hopelessly ill, yet regain his strength and well-being within hours of effective treatment with fluids, electrolytes, and steroids.

Treatment Initial treatment consists of both fluid and cortisol replacement. A suggested fluid for intravenous administration is 5% dextrose in isotonic saline (0.9% NaCl). This fluid expands the blood volume and supplies glucose to correct hypoglycemia. The actual amount of fluid replacement varies with the degree of fluid deficit, but it is usual to require 2 or 3 liters of saline in the first 8 hours of treatment. Rapid hydration is important in lowering the dangerous effect of hyperkalemia on the heart. Fluid administration promotes dilution of extracellular potassium levels and pro-

motes renal excretion of potassium by increasing urinary volume. If the initial plasma potassium level is greater than 6.5 mEq/liter, it may be necessary to administer 50 mEq to 100 mEq of sodium bicarbonate intravenously to temporarily force potassium into the cells, allowing time for other therapies to lower the actual potassium level.[10] As a rule, however, hyperkalemia is quickly corrected by fluid replacement and corticosteroid therapy, making other measures to lower serum potassium rarely necessary. Metabolic acidosis is also usually corrected by control of hypovolemia and should not be treated with bicarbonate unless arterial pH is less than 7.1.[11]

Hydrocortisone succinate (Solu-Cortef) is given stat through the intravenous line and is followed by 50 mg to 100 mg of the same agent every 2 to 4 hours thereafter.[12] Large doses of cortisol are required because of its short biological half-life. Hydrocortisone has sufficient mineralocorticoid properties to influence the renal tubules to retain sodium and water and to excrete potassium. During this acute phase, the nurse should monitor the vital signs closely. A fall in blood pressure and a rapid, thready pulse may indicate inadequate hydrocortisone and fluid replacement therapy. Once the patient becomes stable and the precipitating cause is controlled, the corticosteroid dosage is *gradually* reduced daily until maintenance levels are reached.

The nurse should be aware that overtreatment of adrenal crisis can cause edema, hypertension, hypokalemia with flaccid paralysis, and perhaps even psychotic reactions.

PTH EXCESS (HYPERPARATHYROIDISM)

The exact incidence of hyperparathyroidism is not known, but it may be as frequent as 28/100,000 population.[13] Hyperparathyroidism is usually classified as either primary or secondary. Primary hyperparathyroidism is usually caused by a single parathyroid adenoma; other causes can include primary hyperplasia and hypertrophy of all four glands and rarely, carcinoma of one gland. Secondary hyperparathyroidism is almost always associated with hyperplasia of all four glands; it is most commonly seen in chronic renal disease but is also seen in rickets, osteomalacia, and acromegaly. In primary hyperparathyroidism, the normal feedback control of PTH secretion is lost, and excessive hormone production continues despite elevated plasma calcium levels. In secondary hyperparathyroidism, as seen in chronic renal failure, chronic hypocalcemia is a stimulus to PTH secretion.

Electrolyte disturbances The most consistent biochemical abnormality in primary hyperparathyroidism is hypercalcemia. The degree of hypercalcemia, however, may be slight in some patients. Hypophosphatemia accompanies the hypercalcemia in about 50% of the patients; in others, the serum phosphate levels may be normal.[14] Hypomagnesemia and hypokalemia may also be present and require correction. The plasma bicarbonate is normal or low (metabolic acidosis) in primary hyperparathyroidism, whereas it is frequently elevated in patients with other types of hypercalcemia.

In secondary hyperparathyroidism due to renal failure, serum phosphate is high owing to renal retention, and serum calcium is low or normal.

Clinical manifestations The clinical manifestations of primary hyperparathyroidism are directly or indirectly related to hypercalcemia. They may include the following:

Hypercalcemic manifestations (in general, these symptoms are proportional to the degree of hypercalcemia)

Thirst is prominent.

Anorexia and vomiting are prominent.

Abdominal pain secondary to peptic ulcer or pancreatitis may be the presenting symptom.

Constipation may occur.

Paresthesias may occur.

Hypertension is common.

Tiredness, listlessness, lethergy, apathy, and depression are frequently observed; other patients may display agitation and insomnia.

Decreased memory, poor calculation, and decreased attention span are seen.

Delirium, confusion, somnolence, and even coma may occur.

Changes progressing from neurotic behavior to frank psychoses have been documented and shown to be reversible when hypercalcemia was controlled.

Occipital headache occurs frequently.

Fingernails and toenails may be unusually strong and thick.

Band keratopathy is seen on slitlamp examination (calcium may precipitate in the eyes, most prominent at the medial and lateral margins of the limbus of the cornea). *Note:* In secondary (renal) hyperparathyroidism, calcium may precipitate in soft tissues.

Urinary calcium is often high.

Shortened Q-T interval noted on EKG.

Manifestations related to urinary system

Polyuria is an early symptom, followed by polydipsia; this may be severe enough to mimic diabetes insipidus.

Calcium oxalate or phosphate stones may form; over 75% of patients with primary hyperparathyroidism seek medical attention owing to urinary stones or renal calcifications.[15]

Renal damage caused by obstruction or secondary infection may occur.

Calcification in the region of the kidneys occurs.

Note: These symptoms of hyperparathyroidism are produced by the excess filtration of calcium through the glomeruli. Calcium sediment deposits in the kidneys and produces tubular damage. Polyuria occurs, owing to the increased renal solute load and to damaged renal tubules. Polydipsia (excessive thirst) follows excessive water loss through the kidneys. Uremia may eventually follow the renal damage imposed by calcium excess.

Manifestations related to skeletal system (due to bone demineralization)

Bone pain, local or diffuse

Pathologic fractures occurring through areas weakened by cysts, tumors, or generalized resorption

Evidence of bone resorption seen on radiograph, particularly in the phalanges

Treatment Treatment should be directed at controlling the cause of hyperparathyroidism; until this can be accomplished, a large fluid intake (orally and intravenously) is mandatory to promote renal excretion of calcium and thus decrease the serum calcium level. A dilute urine minimizes the formation of renal calcium stones.

In severe cases, infusion of as much as 100 ml to 200 ml/hour of isotonic saline (0.9% NaCl) may be necessary; of course, careful consideration must be paid to the patient's cardiovascular status. (Recall that urinary calcium excretion is enhanced by saline infusion, since sodium competitively inhibits tubular reabsorption of calcium.) Inducing sodium diuresis by infusing sodium salts intravenously and administering diuretics (such as furosemide) is the emergency treatment of choice. Thiazide diuretics are contraindicated, since they actually may decrease urinary excretion of calcium, adding to the hypercalcemia.

Other measures to control hypercalcemia may include reduced calcium intake, increased oral intake of phosphate, and mobilization of the patient. Some authorities recommend the oral administration of phosphate, as phospho-soda, to lower the serum calcium level; others feel phosphates should be used only when saline therapy is ineffective. Phosphates are contraindicated in patients with renal impairment and high serum phosphate levels because of the potential for irreversible soft-tissue deposition of calcium-phosphate products in the kidney, heart, lungs, and blood vessels.[16] It should be remembered that immobilization causes increased bone resorption, thus worsening the hypercalcemic state; this is why mobilization should be encouraged at the earliest possible time.

Accurate measurement of fluid intake and output is essential for maintenance of adequate fluid replacement. The serum calcium level must be monitored frequently. Cardiac monitoring is necessary to determine the effects of hypercalcemia on cardiac conduction and irritability. The patient taking digitalis is in special danger because hypercalcemia renders him more sensitive to the toxic effects of digitalis. Propranolol (Inderal) may be helpful in preventing the adverse cardiac effects of hypercalcemia. In life-threatening situations due to hypercalcemia, in which pharmacologic therapy does not adequately lower the serum calcium level, either peritoneal dialysis or hemodialysis can efficiently remove calcium and lower the serum calcium level.

PTH INSUFFICIENCY (HYPOPARATHYROIDISM)

Underproduction of PTH occurs in primary hypoparathyroidism and in the accidental removal of parathyroid tissue during thyroidectomy. Renal tubular damage can interfere with the action of PTH and produce symptoms of hypoparathyroidism (pseudohypoparathyroidism). Decreased parathyroid activity results in the following:

Decreased plasma calcium concentration
Increased plasma phosphate concentration

Symptoms of hypoparathyroidism are primarily those of neuromuscular irritability produced by a decrease in ths serum concentration of ionized calcium. Included in the symptoms are those listed below:

Numbness of extremities
Tingling of hands, feet, and circumoral region
Mood changes
Voice changes caused by spasms of vocal cords
Muscular spasm, induced by compressing blood supply to area
Abdominal cramps
Diarrhea
Carpopedal attitude of hands (Fig. 12-2)
Facial spasm, induced by tapping over nerve course in front of the ear (Chvostek's sign)
Laryngeal spasms
Convulsions

Other symptoms of hypoparathyroidism are influenced by the duration of the PTH deficiency and the age at which it developed. For example, cataracts or calcifi-

FIG. 12-2. Manifestations of hypocalcemia. Trousseau's sign is elicited in a patient with hypoparathyroid tetany. (Ezrin C, Godden J, Volpe R, Wilson R (eds): Systematic Endocrinology, p 151, 2nd ed. Hagerstown, Harper & Row, 1979)

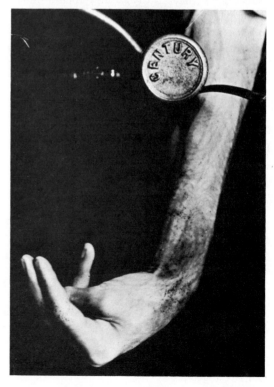

cation of various body parts such as the basal ganglia of the brain may occur when hypoparathyroidism has long been present. Formation of new teeth is restricted when hypoparathyroidism occurs in a child, although the degree of hypoplasia depends upon the age at which hypoparathyroidism began. Stunting of growth and mental retardation may also be present if the onset of disease occurs in childhood.

The danger of the accidental removal of parathyroid tissue during thyroidectomy is always present because the parathyroids are small and resemble fatty tissue. Removal of half of the parathyroid glands usually does not present symptoms. (Most persons have four parathyroid glands; some have fewer, and some have as many as seven). However, removal of three of four causes symptoms of hypoparathyroidism until the fourth gland is able to hypertrophy sufficiently to fulfill the function of all the glands. Tetany may also be produced by temporary interference with the parathyroid blood supply following thyroidectomy.

The nurse should be alert for symptoms of deficit of ionized calcium during the postoperative care of patients who have undergone thyroidectomy. Such symptoms usually appear a few days after the operation. Early complaints are numbness and tingling in the hands and feet. Compression of circulation to the hand while checking the blood pressure may cause spasm of the forearm muscles and palmar flexion of the hand. Any sustained form of pressure on a motor nerve (as when the arm is allowed to rest over a chair or when the patient sleeps on an arm or croses the legs) can elicit symptoms of tetany. Other symptoms of deficit, such as general irritability or "jumpiness," may also be noted. It is crucial to detect calcium deficit early so that appropriate hormonal therapy, calcium replacement, or both can be started before the onset of laryngeal spasms and convulsions.

Hypocalcemia caused by hypoparathyroidism may be treated by the administration of calcium salts. Calcium gluconate given intravenously may control tetany; calcium gluconate or lactate may be administered orally for the same purpose. An increased dietary intake of high-calcium, low-phosphate foods will also be beneficial. Milk and cheese, high in phosphate, should be omitted.

Dihydrotachysterol is a substance having an action similar to that of PTH. It increases calcium absorption from bone and thus causes an increased plasma calcium concentration. Calciferol, 1 mg to 5 mg daily, is frequently the drug of choice.[17]

PROFOUND THYROID HORMONE EXCESS (THYROID STORM)

The term *thyroid storm* describes the crisis that can result from severe hyperthyroidism. Fortunately, it is a rare condition. Thyroid storm is seen most often in patients with diffuse toxic goiter; however, it may occasionally occur in patients with multinodular toxic goiter. Thyroid storm may be precipitated by thyroid surgery in an inadequately prepared patient or by surgery for other conditions when the presence of underlying hyperthyroidism was not suspected. Other forms of stress (such as trauma, burns, infections, and severe emotional stress) may also precipitate thyroid storm in the patient with underlying hyperthyroidism.

Water and electrolyte disturbances in thyroid storm may include the following:

Dehydration (increased evaporative water loss occurs from the skin and lungs in order to aid in dissipation of excess body heat)

Hypernatremia

Hypercalcemia

Clinically, thyroid storm is manifested by the following:

Marked hyperpyrexia (sometimes reaching or exceeding 106°F [41.1°C])

Palpitations

Tachycardia, 160 to 200 beats/min

Skin warm, profuse sweating

Nausea, vomiting, and diarrhea

Tremors (usually mild; however, may become severe enough to interfere with normal activities)

Extreme weakness

Thyroid enlargement

Mental changes; chronic state of agitation, intensely fearful, even paranoid; in contrast, elderly patients more commonly display depression[18]

High-output cardiac failure

Treatment of thyroid crisis includes a number of important therapeutic considerations. The precipitating factor should, of course, be identified and treated. Thyroid-blocking drugs are given by mouth or by nasogastric tube to reduce thyroid hormone synthesis; iodine is administered by mouth or intravenously to inhibit the secretion of stored hormone. Lithium blocks the release of hormone from the thyroid gland and can thus be used to treat hyperthyroidism; however, its toxic side-effects make it less desirable for this purpose than the thiourea derivatives.

The β-adrenergic agent, propranolol (Inderal), is a drug of choice in the treatment of thyrotoxic crisis, since it antagonizes the peripheral effects of thyroid hormones.[19] Propranolol often improves or abolishes tachycardia, tremor, nervousness, and excess sweating and is used to control symptoms of hyperthyroidism until antithyroid drugs can take effect. Propranolol may also produce improvement in the high-output cardiac failure associated with hyperthyroidism. Hydrocortisone may be helpful, particularly for patients in shock.

Intravenous fluid replacement is indicated to correct the large insensible fluid loss; if congestive heart failure is present, careful monitoring of central venous pressure is necessary. The fluids should contain glucose and vitamins to meet the increased metabolic needs. Fever should be treated promptly with cooling blankets; the use of aspirin is discouraged because it displaces triiodothyronine from its carrier and increases its free level.[20]

PROFOUND THYROID HORMONE DEFICIENCY (MYXEDEMA COMA)

Myxedema coma is a medical emergency that, although uncommon, is being recognized more frequently today perhaps because of more widespread use of radioiodine therapy for hyperthyroidism. The disease is more common in females and is most prevalent in winter months. Mortality may be as high as 50% for reasons that are unclear.

Long untreated hypothyroidism may eventually terminate in myxedema coma; it usually occurs in *elderly* people who have been without medical care. Precipitating events may include infection, trauma, sedative drugs, and exposure to a cold environment. Virtually every organ in the body is affected by thyroid hormone deficiency.

Clinical features of myxedema coma may include the following:

Hypothermia (occurs in more than 80% of patients; temperatures as low as 75.2°F [24°C] have been seen; a temperature below 89.6°F [32°C] has been associated with a grave prognosis)[21]

Hypotension

Bradycardia, EKG changes of low voltage

Hypoventilation

Stupor or frank coma

Coarse, rough voice

Dry skin, puffy edema

Hypoglycemia

Hyponatremia

Pericardial and pleural effusions

In review, myxedemic patients characteristically have dry, flaky skin. In addition, the skin is coarse and there is subcutaneous swelling due to fluid and hyaluronic acid accumulation, most obvious in the eyelids, the periorbital tissues, and the dorsa of the feet and hands. The hair becomes coarse and dry; there is slowing of intellectual functioning and speech.

Fluid and electrolyte abnormalities of myxedema coma commonly include the following:

Respiratory acidosis (elevated PCO_2 due to alveolar hypoventilation)

Hyponatremia associated with inappropriate secretion of ADH and inability to excrete water normally; serum sodium level may approach 115 mEq/liter

Hypokalemia due to dilution phenomena

Lactic acidosis due to hypoxia

Treatment of myxedema coma includes the correction of hypometabolism by the intravenous administration of thyroxin (T_4). Hydrocortisone is also given intravenously, since patients often have relative adrenal insufficiency once the metabolic rate begins to increase following thyroid hormone therapy.

Hyponatremia is usually treated by water restriction; however, if the serum sodium level is less than 115 mEq/liter, a small amount of hypertonic sodium chloride solution may be administered cautiously intravenously. (See Chapter 2 for a more thorough discussion of treatment of SIADH.)

Fluid replacement is usually accomplished by the administration of 5% dextrose in isotonic saline (0.9% NaCl). Dextrose helps combat the hypoglycemia associated with myxedema coma, and sodium chloride helps correct the hyponatremia. Fluid therapy must be monitored carefully since myxedema coma patients generally have lowered insensible fluid losses and are more readily overloaded with fluids.

Assisted ventilation is frequently needed to treat hypoventilation. Depressant drugs must be avoided because they contribute to respiratory depression. Pulmonary infections should be vigorously treated.

Although hypothermia is present, active warming is contraindicated because rapid warming precipitates arrhythmias, increases the caloric requirements, and may induce peripheral vasodilation, thus aggravating hypotension. Most patients can be adequately treated by covering with blankets to conserve body heat.

ADH EXCESS

Secretion of inappropriate antidiuretic hormone (SIADH) is characterized by hyponatremia and a low serum osmolality resulting from decreased water excretion. (The reader is referred to Chapters 2 and 13 where the causes, recognition, and treatment of SIADH are discussed in detail.)

ADH DEFICIENCY (DIABETES INSIPIDUS)

ADH (vasopressin) plays a major role in aiding the body to conserve water to prevent dehydration. This hormone acts by increasing renal reabsorption of water. Without ADH, the ingestion of approximately 10 liters to 20 liters of water daily would be necessary to match urinary losses.[22] However, with ADH, urine output can be reduced to as little as 500 ml/24 hours. There are two types of diabetes insipidus (DI): central and nephrogenic.

Central (or neurogenic) DI is due to relative lack of ADH; it is sometimes referred to as *vasopressin-sensitive DI.* This form of DI may occur after head trauma or surgical procedures near the pituitary or as a result of infection, primary tumor, or metastatic tumor; it may also be idiopathic.

Nephrogenic DI is due not to lack of ADH, but to failure of the kidney to *respond* to the hormone; it is sometimes referred to as *vasopressin-resistant DI.* It may occur as an X-linked recessive trait; be acquired after primary renal disease or electrolyte disorders, such as hypokalemia or hypercalcemia, or be pharmacologically induced by certain drugs, such as demeclocycline (Declomycin), lithium, and methoxyflurane (Penthrane).

DI may be complete or partial, permanent or temporary. Prognosis varies with the cause of the condition.

Clinical manifestations Outstanding signs and symptoms of DI include the following:

Polyuria in central DI ranging from 3 liters to more than 10 liters daily (polyuria in nephrogenic DI is usually much less severe, with a urine output of 2 liters to 3 liters in 24 hours)[23]

Intense thirst, especially with craving for ice water (fluid intake volume corresponds to urinary volume)

Urinary specific gravity (SG) less than 1.005 (urine osmolality less than 200 mOsm/liter)

Increased *plasma* osmolality due to water loss

Inability to concentrate urine by fluid restriction

Central DI responds to ADH administration; nephrogenic DI does not.

Water and electrolyte disturbances As a rule, the patient with an intact thirst mechanism will drink sufficient fluids (if they are readily available) and will maintain balance. If water is not readily available, the polyuria will lead to severe dehydration with weight loss, tachycardia, and even shock. After appropriate treatment has been instituted, it is possible that the patient may continue his large fluid intake from habit and conceivably induce water intoxication.

Treatment and related nursing interventions The standard treatment of central DI has been ADH replacement by means of vasopressin administration by either the intramuscular route or the nasal route. During the first few weeks of treatment of DI, the nurse should caution the patient to avoid the continued excessive habitual ingestion of water because of the danger of inducing water intoxication.

Two nonhormonal forms of therapy may be used in the treatment of DI. Thiazide diuretic therapy is the specific treatment for nephrogenic DI, no other treatment has been found for this disorder. Paradoxically, the chronic administration of thiazide diuretics to the patient with nephrogenic DI or central DI can produce approximately a 50% reduction in the 24-hour urine volume and a comparable decrease in thirst.[24] The thiazide diuretic apparently acts by causing a sodium chloride diuresis, decrease in extra-cellular fluid volume, and thus, a decrease in the distal tubular delivery of fluid to be excreted.[25] It is important to teach the patient to restrict oral sodium intake while on diuretics, since the effectiveness of these agents is markedly diminished by *ad libitum* sodium intake.[26]

Some patients with central DI have some residual capacity to secrete ADH; in this situation, drugs that increase the release of ADH or enhance its action on the kidney may be used instead of hormonal therapy.

Such drugs include chlorpropamide (Diabinese), an oral hypoglycemic agent; clofibrate (Atromid-S), a hypolipidemic drug; and carbamazepine (Tegretol), an anticonvulsant. When these agents are used, the nurse must be alert for drug reactions. For example, chlorpropamide can cause hypoglycemic reactions, making it necessary to caution the patient to eat regular meals. Clofibrate can cause muscle aches and carbamazepine is limited in use because of a number of associated adverse reactions.[27] Unfortunately, nephrogenic DI will not respond to any of these agents.

INSULIN DEFICIENCY

Deficiency of insulin can lead to diabetic ketoacidosis (DKA) or nonketotic hyperosmolar coma (NKHC). These conditions are discussed below.

DIABETIC KETOACIDOSIS

DKA is a medical emergency that accounts for approximately 14% of all hospital admissions for diabetics; it is the most common cause of hospital admission of diabetics under 20 years of age.[28] A 1000-bed community hospital may average one case of DKA per week.[29] Although improved treatment of DKA has significantly reduced the mortality associated with this disease, there is still at least a 5% to 10% mortality rate reported from most medical centers.[30]

DKA may be precipitated by a number of factors, most of which are associated with the diabetic's poor understanding of his disease. For example, the patient may stop taking insulin during an illness when food intake is diminished because he fears a hypoglycemic reaction. Or he may be unaware of the need for extra insulin during an acute infection or severe emotional stress.

Health care personnel should remember that diabetics frequently need extra insulin during acute stressful situations such as surgical emergencies, severe trauma, or myocardial infarction. DKA can also be precipitated in the latent diabetic by hyperalimentation exceeding insulin secretory capacity. The abrupt onset of diabetes in the juvenile can lead to DKA as the initial harbinger of the disease.

A *major* nursing function involves education of diabetics to help them learn intelligent management of

their disease. A well-educated diabetic will often give himself additional amounts of insulin during an obvious infectious process. Patient education is discussed later in the chapter.

METABOLIC CONSEQUENCES OF INSULIN LACK
The metabolic consequences of insulin deficiency are described below.

Hyperglycemia and hyperosmolality Lack of insulin leads to hyperglycemia primarily by promoting underutilization of glucose (recall that cells are relatively impermeable to glucose in the absence of insulin) and excessive production of glucose from fats and amino acids by the liver (gluconeogenesis). Because of these two processes, the blood glucose concentration rises markedly and increases plasma osmolality. One becomes aware of the influence of glucose on plasma osmolality when considering the formula for calculating plasma osmolality:

$$pOsm = 2\,Na + \frac{glucose}{18} + \frac{BUN}{2.8}$$

Normally, the plasma osmolality ranges from 280 mOsm to 300 mOsm/liter. The elevated osmolality of the extracellular fluid produces cellular dehydration by causing water to shift from the cells to the extracellular fluid (water moves from the area of lesser solute concentration to that of higher solute concentration).

Osmotic diuresis and fluid volume deficit When the blood glucose level exceeds the renal threshold, glucose is spilled out into the urine. (One should recall that the renal threshold in the diabetic may be higher than normal, particularly in the older patient.) As glucose spills out into the urine, it takes water and electrolytes with it, increasing the urine volume. Thus, one would initially expect to see polyuria and a high urinary SG owing to the high glucose content. The polyuria eventually leads to fluid volume deficit. As the fluid volume deficit increases, glomerular filtration rate and renal blood flow progressively diminish, causing the patient to become oliguric or even anuric in spite of marked hyperglycemia. A potential danger of profound fluid volume deficit is renal tubular damage and resultant acute renal failure.

Electrolytes In addition to water loss in osmotic diuresis, there is urinary loss of sodium, chloride, and potassium.

Sodium Despite losses of water in excess of solute in DKA, plasma sodium is usually below normal. In this instance, a plasma sodium level below 120 mEq/liter usually is a sign of severe electrolyte loss. If vomiting is present, the magnitude of hyponatremia is increased. Also, sodium moves into the cells as they become depleted of potassium, further lowering the plasma sodium level.

Potassium Probably the most important electrolyte disturbance that occurs in DKA is the marked deficit in total body potassium. Causes of potassium depletion can include starvation effect with lean tissue breakdown, loss of intracellular potassium, potassium-losing effect of aldosterone (stimulation of aldosterone secretion is produced by fluid volume deficit), loss of potassium with osmotic diuresis, and the presence of severe anorexia (no intake) and vomiting (increased loss).

In the untreated patient with DKA, the plasma potassium level may be normal or elevated even though there is a marked deficit of total body potassium. Factors that tend to elevate the plasma potassium level in the untreated patient include plasma volume contraction with oliguria, interfering with renal excretion of potassium; and metabolic acidosis, which causes potassium to shift into the extracellular compartment as hydrogen ion is buffered intracellularly. The hyperkalemia is usually quickly alleviated by fluid replacement therapy and reestablishment of urine output.

The plasma potassium falls rapidly and usually reaches its lowest point within 1 to 4 hours after institution of therapy. Reasons for the decreased plasma potassium level at this time include dilution by the intravenous fluids; increased urinary potassium excretion due to plasma volume expansion; formation of glycogen within the cells, involving utilization of potassium, glucose, and water, (which causes further withdrawal of potassium from the extracellular fluid); and correction of the acidotic state with reentry of potassium into the cells.

Phosphate Hypophosphatemia almost invariably occurs during the treatment of the patient with DKA. One potentially serious consequence of phosphate deficiency is decreased erythrocyte 2,3-diphosphoglycerate (DPG); a low level of 2,3-DPG may result in decreased peripheral oxygen delivery. Also, depressed myocardial function has been observed when the serum phosphate level is less than 2 mg/dl.[31] When the plasma phosphate concentration goes below 0.5 mg/dl,

serious disturbances in metabolism can result; seizures, respiratory failure, impaired leukocyte and platelet function, and gastrointestinal bleeding have been reported.[32]

Ketosis Insulin lack allows greater release of free fatty acids from peripheral fat stores and also may activate ketogenic pathways in the liver; the excess fatty acids are converted by the liver to ketones, resulting in ketosis. Contributing to the excessive buildup of ketones in the bloodstream is the decreased utilization of ketones by peripheral tissues (again, due to insulin lack). This impaired ketone metabolism, coupled with ketone overproduction by the liver, overloads the body's buffers, and metabolic acidosis ensues. (Recall that ketones are strong acids.) The anionic charge of bicarbonate is replaced by the negatively charged ketones.

The type of metabolic acidosis that occurs in DKA is manifested by a fall in bicarbonate with a reciprocal rise in the anion gap. Anion gap, also referred to as Delta, can be calculated by the following formula:

$$\text{anion gap} = Na^+ - (HCO_3^- + Cl^-) = 8 \text{ to } 16 \text{ mEq}$$

The elevated anion gap in ketoacidosis is due to flooding of the extracellular fluid with ketones. Anion gap is very important in diagnosing acid–base disturbances. For example, the vomiting that frequently accompanies DKA can superimpose a metabolic alkalosis on the preexisting ketoacidosis, sometimes normalizing the pH. However, the ketone anions remain elevated, disrupting metabolism. Measurement of anion gap reveals the abnormal level of ketones despite the pH.

Ketones are made up of acetoacetic acid, β-hydroxybutyric acid, and acetone. Excessive ketosis leads to ketonuria and excretion of volatile acetone from the lungs, resulting in the classic "fruity" odor of the breath associated with DKA.

It is not unusual for the plasma pH to drop to 7.25 or below and for the bicarbonate level to drop to a level of 12 mEq/liter or less. Possibly the greatest risks of prolonged uncorrected acidosis are decreased cardiac function, arrhythmia, and impaired hepatic handling of lactate.[33]

The body attempts to compensate for the metabolic acidosis associated with DKA by means of the kidneys and lungs. The kidneys eliminate hydrogen ions and conserve bicarbonate ions, resulting in decreased urinary pH. The lungs attempt to lighten the acid load by blowing off extra carbon dioxide by deep, relatively rapid respirations, thus decreasing the PCO_2. The expected fall in PCO_2 for compensation of metabolic acidosis can be calculated by the following formula:

$$PCO_2 \text{ (mm Hg)} = 1.5 \text{ (HCO}_3\text{)} + 8 \pm 2$$

A fall below the calculated amount indicates a superimposed respiratory alkalosis; the failure of the PCO_2 to decrease to the expected level indicates a complicating respiratory acidosis, which is a dangerous combination.

TABLE 12-1 *Clinical manifestations of DKA and their probable causes*

Clinical manifestations	Probable cause
Glycosuria	Blood glucose level exceeds renal threshold, causing glucose to spill out into the urine
Polyuria (initially) with high specific gravity	Osmotic diuretic effect of hyperglycemia; high renal solute load
Polydipsia	Thirst due to cellular dehydration (cells become dehydrated when water is drawn from them by the hypertonic extracellular fluid)
Ketonemia and ketonuria	Excessive accumulation of ketones in the bloodstream causes them to spill out into the urine
Tiredness, muscular weakness	Lack of carbohydrate utilization; hypokalemia
Increased longitudinal furrows in tongue (poor tongue turgor)	Fluid volume deficit
Poor skin turgor, dry mucous membranes	Fluid volume deficit
Anorexia, nausea, and vomiting	Follows onset of DKA; interferes with fluid intake (hastening the development of fluid volume deficit)
Gastric dilatation	Neuropathy? Water and electrolyte loss?
Abdominal pain (can simulate appendicitis, pancreatitis, or other acute abdominal problems)	Apparently is secondary to the DKA *per se*[34] Note: Anorexia, nausea, and vomiting precede the abdominal pain when it is due to DKA—this is in contrast to most surgical emergencies where the pain usually occurs first
Deep "air hunger" respirations (Kussmaul)	Compensatory mechanism to increase plasma *p*H by the elimination of large amounts of CO_2 from the lungs
Cherry-red skin and mucous membranes	Marked peripheral vasodilatation associated with ketosis
Postural hypotension	Changing the patient's position from supine to sitting causes more than a 10-mm drop in the systolic blood pressure; eventually the blood pressure becomes low even in the supine position
No fever; in fact, hypothermia is often present (if fever is present, it is almost always associated with the precipitating factor of the DKA)	Fluid volume deficit
Acetone breath odor (similar to that of overripe apples)	Acetone content of body is increased; since acetone is volatile, some of it is vaporized in the expired air Note: Vomitus odor to the breath frequently obscures this finding
Oliguria or anuria (late)	Fluid volume deficit causes decreased renal blood flow and decreased glomerular filtration rate Note: Before assuming that urine formation is scant, check for a distended, atonic bladder
Acute weight loss	Fluid volume deficit (the amount of acute weight loss is a good approximation of the magnitude of the fluid loss)
Decreased level of consciousness; patients with DKA are frequently stuporous, unresponsive, and, at times, in frank coma	Level of consciousness correlates best with the level of hyperglycemia (and plasma osmolality) at the time of admission

CLINICAL MANIFESTATIONS Commonly seen clinical manifestations of DKA are summarized in Table 12-1. Laboratory findings are summarized in Table 12-2.

DIFFERENTIATION BETWEEN DIABETIC COMA AND HYPOGLYCEMIC COMA It is imperative that the presence of hypoglycemia be ruled out before insulin is given. Hypoglycemia is characterized by anxiety, sweating, hunger, headache, dizziness, double vision, twitching, convulsions, nausea, pale wet skin, dilated pupils, normal breathing, and normal blood pressure.

When in doubt, it is advisable to administer 50 ml of 50% dextrose/water intravenously. If hypoglycemia is the problem, the patient's condition will quickly improve; if ketoacidosis is the problem, this small amount of dextrose will do no harm. (See Table 12-3 for a comparison of symptoms of DKA with those of hypoglycemia.)

TREATMENT

Fluid replacement Adequate and prompt rehydration is vital, particularly if the patient is in shock. The initial fluid of choice in most instances is isotonic saline (0.9% NaCl), since hypotonic fluids will not expand the extracellular fluid as quickly. Isotonic saline is usually infused at the rate of approximately 1 liter/hour in adults without cardiac failure until the blood pressure is stabilized and the urine output is 1 ml to 2 ml/minute.[35] If there is a question as to the adequacy of the patient's cardiovascular status, a central venous pressure (CVP) line may be necessary to help gauge the best rate of fluid replacement. (Nursing responsibilities associated with monitoring CVP are discussed in Chapter 6.) After the first 2 liters to 3 liters of isotonic saline have been administered, the intravenous fluid may be changed to half-normal saline (0.45% NaCl) to dilute the hyperosomolar plasma and provide free water for renal excretion.

When the blood pressure is stabilized and the urine output is adequate, the rate of fluid administration can be reduced to 1 liter every 2 hours and then to 1 liter every 3 to 4 hours. As soon as oral intake is adequate, intravenous fluids can be discontinued.

Adrenergic substances such as norepinephrine are not effective in correcting oligemic shock, and severe

TABLE 12-2 *Laboratory findings in DKA and their probable causes*

Laboratory finding	*Probably cause*
Blood sugar: (Normal = 80–120 mg/dl) Elevated above normal, often 400–600 mg/dl; may be as high as 2000 mg/dl	Faulty glucose metabolism causes glucose to accumulate in the bloodstream (lack of insulin decreases glucose uptake by most cells and also increases gluconeogenesis in the liver)
Plasma osmolality: (Normal = 280–300 mOsm/liter) Elevated above normal	Due primarily to hyperglycemia
Bicarbonate: (Normal plasma HCO_3 = 24 mEq/liter) Usually decreased below 15 mEq/liter; may be as low as 5 mEq/liter (or even unmeasurable) in severe DKA	Excessive ketonic anions in the bloodstream cause a compensatory drop in bicarbonate anions
Arterial pH: (Normal = 7.38–7.42) Usually less than 7.25, may be as low as 6.8 in severe uncompensated states	Associated with the metabolic acidosis caused by ketosis
High anion gap acidosis (Normal anion gap is less than 8–16 mEq/liter) Elevated	Due to excessive ketones in bloodstream
BUN: (Normal = 20 mg/dl) Elevated to an average of around 40 mg/dl	Partially due to increased protein metabolism with increased hepatic production of urea (because of insulin lack) Also related to fluid volume deficit and diminished glomerular filtration rate
Polymorphonuclear leukocytosis: Usually 15,000–20,000/mm^3 in DKA; may be as high as 90,000/mm^3 (with a shift to the left)	May be secondary to fluid volume deficit, acidosis, and adrenocortical stimulation

TABLE 12-3   *Differential diagnosis of diabetic coma and hypoglycemic reactions*

	Diabetic coma	Hypoglycemic reactions	
		Regular insulin	Modified insulin or oral agents
Clinical Features			
Onset	Slow—days	Sudden—minutes	Gradual—hours
Causes	Ignorance	Overdosage	
	Neglect of therapy	Omission or delay of meals	
	Intercurrent disease or infection	Excessive exercise before meals	
Symptoms	Thirst	"Inward nervousness"	Fatigue
	Headache		Headache
	Nausea		Nausea
	Vomiting	Hunger	Sweating (sometimes absent)
	Abdominal pain	Sweating	
	Dim vision		Dizziness
	Constipation	Weakness	
	Dyspnea	Diplopia	
		Blurred vision	
		Paresthesia	
		Psychopathic behavior	
		Stupor	
		Convulsions	
Signs	Florid face		
	Air hunger (Kussmaul's respiration)	Pallor	
	Finally, respiratory paralysis	Shallow respiration	
	Dehydration (dry skin)	Sweating	Skin may be dry
	Rapid pulse	Normal pulse	Pulse not characteristic
	Soft eyeballs		
	Normal or absent reflexes	Eyeballs normal	
		Babinski's reflex often present	
	Acetone breath		
Chemical Features			
Urine			
Glucose	Positive	Usually absent, especially in second voided specimen	
Acetone	Positive	Negative	
Diacetic acid	Positive	Negative	
Blood			
Glucose	>250 mg/dl ordinarily	60 mg or less/dl	
CO_2 combining power	<20 volumes/dl	Usually normal	
Leukocytosis	Present; may be very high		
Response to treatment	Slow	Rapid; occasionally delayed	May be delayed

(Diabetes Mellitus, 8th ed, p 168. Indianapolis, Eli Lilly & Co, 1980) acidosis must be corrected before the patient's circulatory system can respond to endogenously produced catecholamines.[36]

If the patient is stuporous, an indwelling urinary catherter must be inserted until lucidity is regained; it is imperative that hourly urine volume be closely monitored in the early phase of fluid replacement therapy. (Of course, a catheter should not be used unless absolutely necessary because of the danger of urinary tract infection.)

Glucose is not administered in initial intravenous fluids when the blood sugar is already markedly elevated; to do so would only potentiate the existing hyperglycemia and enhances glycosuria with renal salt and water loss. However, carbohydrate replacement is important in the overall treatment of DKA. It should be remembered that the total body stores of carbohydrates are greatly diminished in spite of hyperglycemia. Glucose solutions should be started when the blood sugar has fallen to approximately 250 mg/100 ml. A typical fluid might be 5% dextrose in 0.45% NaCl, or 5% dextrose in water. It is important that the patient *not* be allowed to become hypoglycemic. A precipitous fall in blood sugar might produce cerebral edema, since the glucose osmols in the brain decrease at a slower rate than those of the plasma, allowing fluid to be drawn into brain tissue. Not allowing the blood sugar to drop rapidly below 250 mg/dl should prevent the problem. Most reported cases of brain edema associated with DKA treatment have been associated with a rapid decrease in blood sugar to levels close to 100 mg/dl.

Insulin administration Rapidly acting (regular) insulin may be given intravenously (IV), intramuscularly (IM), or subcutaneously (SQ) in the treatment of DKA. The aim of insulin therapy is to give enough insulin to correct the problem without subjecting the patient to the risk of hypoglycemia. Studies have demonstrated that large insulin doses are generally no more effective in correcting DKA than are small doses.

The insulin dose for continuous IV infusion is usually 4 units to 8 units/hour; for the IM or SQ routes the dose is usually 5 units to 10 units of insulin/hour. These doses have been shown to lower blood sugar smoothly, improve ketosis, and repair acidosis at rates indistinguishable from those obtained by higher-dose regimens.[37] In addition, patients receiving low-dose insulin therapy are less likely to develop hypoglycemia

and hypokalemia than are patients receiving large doses.

Of course, all patients must be closely monitored for lack of response to insulin therapy; if necessary, more aggressive insulin therapy should be instituted. Some patients may be resistant to insulin and may require hundreds of units of insulin before a response can be demonstrated.

Continuous IV therapy As stated above, the insulin dose for continuous IV infusion is usually 4 units to 8 units/hour. This route is best for the hypotensive patient, since absorption from poorly perfused muscle and fat depots may be erratic.

There has been some concern that with the low-dose constant infusion of insulin there would be substantial loss of insulin by adsorption to containers or tubing. This should not significantly affect delivery, however, if sufficiently high concentrations of regular insulin are added to the container and the tubing is adequately rinsed with the solution. For example, the insulin-binding effect can be offset almost entirely by letting 50 ml to 100 ml of a solution containing 100 units of regular insulin/liter run through the tubing before administering it to the patient. Apparently this high concentration and "washout procedure" allows for saturation of the insulin-binding sites.

A dose sufficient for most patients can be achieved by administering 1 ml/minute of a liter of solution containing 100 units of regular insulin (after 50 ml have been run through the tubing). More accurate delivery of the IV insulin solution can be assured by infusion through an infusion pump; if one is not available, the solution should be administered through a volume-controlled administration set.

IM and SQ insulin therapy The IM and SQ routes are suitable only for patients with normal blood pressure and tissue perfusion. Sometimes patients with DKA are initially given regular insulin (0.2 units/kg body weight) by deep IM injection.[38] Thereafter, a dose of 5 units to 10 units may be given each hour until the blood sugar drops to 250 mg/dl and the DKA clears. At this time, insulin is administered SQ as indicated by blood glucose levels or the glucose content of double-voided urine specimens.

Electrolyte replacement Potassium Recall that hypokalemia is probably the most important electrolyte disturbance that occurs in diabetic ketoacidosis; therefore, potassium replacement during therapy is imperative. Paradoxically, however, elevation of plasma potassium may be present before fluid therapy is instituted and adequate urine flow is established. It is important to remember that plasma potassium concentration does not accurately reflect total body potassium in the presence of acidosis; this does not hold as acidosis is corrected.

As stated earlier in the chapter, the first 2 liters to 3 liters of fluid administered for the treatment of DKA usually consist of plain isotonic saline; ordinarily, potassium should not be added during the first 2 hours of fluid therapy.

Once good urine flow is established and the BUN is near normal, it is safe to add 20 mEq to 40 mEq of potassium to half-strength saline (0.45% NaCl). In the presence of a low plasma potassium level, potassium is usually administered at a rate of 20 mEq/hour intravenously if the urinary flow is adequate. This amount of potassium will usually prevent clinical manifestations of hypokalemia. If the urine flow is questionable, potassium is infused at somewhat less than 20 mEq/hour while the EKG is carefully monitored. (Since the IV administration of potassium is always associated with the risk of hyperkalemia, it is wise to monitor the serum potassium level at 1- or 2-hour intervals and utilize serial EKG tracings.)

When signs of severe hypokalemia are present, such as acute respiratory paralysis, muscle weakness, cardiac arrhythmias, or ileus, it may be necessary to administer potassium at a rate of 40 mEq/hour.[39] Great care must be exercised in administering such large doses; clinical response must be continuously assessed by means of the EKG.

Potassium may be administered as potassium chloride, potassium phosphate, or potassium acetate. Some authorities prefer the potassium phosphate salt, since phosphate is also depleted in DKA; administration of the phosphate ion allows it to move into the cells with potassium to replace intracellular deficits of both electrolytes.

Potassium supplementation is continued IV until the plasma potassium is within the normal range. Oral potassium-containing fluids such as orange juice may be administered once the patient is lucid, and free of nausea, vomiting, and gastric dilatation.

Sodium As mentioned earlier, the initial IV fluid used for the treatment of the hypotensive patient with DKA is isotonic saline (0.9% NaCl). This solution not only expands the extracellular fluid, it also supplies extra sodium to correct any sodium deficit.

It should be remembered, however, that the patient with DKA has lost relatively greater amounts of water than of sodium. Thus, eventually, hypotonic fluids such as half-strength saline (0.45% NaCl) must be administered. Hypotonic solutions provide "free-water" to correct cellular dehydration.

Phosphate As stated above, there are significant losses of phosphate with DKA. Some authorities favor replacing at least a part of the lost potassium with the buffered phosphate salt. To do so replaces both potassium and phosphate and allows for correction of intracellular deficits of both. It is imperative that significant renal failure be ruled out before phosphate is administered IV; serum phosphate levels should be monitored to prevent possible hyperphosphatemia. When oral intake is tolerated, skim milk is a good source of phosphate. Occasionally phosphate is replaced orally in the form of Phospho-Soda, 5 ml.

It is thought that phosphate replacement accelerates the recovery of reduced red blood cell 2,3-DPG levels, thereby decreasing hemoglobin-oxygen affinity and improving tissue oxygenation.

Serum phosphate and calcium concentrations should be monitored daily during phosphate therapy, since there is a remote chance severe hypocalcemia may occur.[40]

Bicarbonate There is disagreement as to the necessity for alkali therapy in the treatment of DKA. Some feel that bicarbonate replacement is unnecessary because insulin therapy reverses the biochemical abnormalities of DKA, including the bicarbonate deficit.

Those who favor bicarbonate replacement feel it is necessary when severe acidosis is present (*i.e.*, an arterial pH less than 7.15 and bicarbonate less than 8 mEq/liter). Profound acidosis may lead to life-threatening pulmonary edema, hypotension, and shock.[41] Therapy might be begun with the administration of 44 mEq to 88mEq of sodium bicarbonate in the adult and 0.5 mEq to 1.0 mEq/kg in children.[42] Bicarbonate should not be given to completely restore the plasma bicarbonate to normal levels because too rapid correction of acidosis can induce hypokalemia. Some authorities theorize that not only is bicarbonate replacement unnecessary, it may actually be detrimental in that rapid correction of acidosis might contribute to impaired tissue oxygenation. Hemoglobin's affinity for oxygen is *decreased* in acidosis, allowing hemoglobin to "unload" more oxygen at the tissue level; conversely, hemoglobin's affinity for oxygen is *increased* in alkalosis, interfering with oxygen release.

NURSING ASSESSMENT OF RESPONSE TO THERAPY During treatment the patient with DKA should be closely monitored by the physician and the nurse. Both medical and nursing interventions must be based on the patient's response to therapy. Continuous vigilance over the clinical and the laboratory status of the patient is of extreme importance in the eventual outcome; only constant observation will permit the critical judgments necessary to provide optimal therapy. During the acute phase, a diabetic flowsheet should be used to record sequentially vital signs, laboratory data, and treatments. Data on the flowsheet is usually recorded hourly at first and may include the following:

Temperature, pulse, and respirations

Blood pressure, supine and sitting

Hourly urine volume and SG

Glycosuria and ketonuria

Blood glucose level

Plasma ketone level

Plasma pH

Electrolyte levels

Fluid intake, amount and kind

Electrolyte intake

Level of consciousness

EKG findings

Deep tendon reflexes

Degree of fluid volume deficit The degree of fluid volume deficit can be monitored by assessing the blood pressure, neck veins, hourly urine volume, skin and tongue turgor, and body weight.

Most patients with DKA are hypotensive before treatment; the hypotension is most pronounced when the patient is changed from the supine to a sitting position. A drop in systolic pressure by more than 10 mm Hg on position change is indicative of fluid volume deficit. Eventually the volume depleted patient is hypotensive even in the supine position. Elevation of the systolic pressure towards normal is indicative of fluid volume expansion and is, of course, desirable. Collapse of the neck veins with raising the head off the bed is also a sign of fluid volume deficit for which the patient should be observed.

Fluid volume deficit is associated with tachycardia as the heart attempts to compensate for a diminished plasma volume by pumping faster. Thus, a decrease in pulse rate, associated with fluid volume replacement, is a favorable sign. Pulse volume should change from "weak and thready" toward normal with plasma volume expansion.

Failure of the DKA patient to excrete a normal urine volume is related to the fall in glomerular filtration rate and renal blood flow associated with fluid volume deficit. It is important that fluid volume replacement be prompt and of sufficient magnitude to prevent renal tubular damage, a potential result of profound fluid volume deficit. Hourly urine volume is one of the best parameters for gauging adequacy of fluid volume replacement. The desired urinary volume per hour for an adult is 30 ml to 50 ml. Urine should be measured in a device calibrated for accurate readings of small volumes. After institution of appropriate fluid replacement therapy, a rise in urinary volume should be noted. Failure of urine volume to increase can indicate either inadequate rate and volume of fluid replacement or renal tubular damage. If the problem is the latter, the patient must be treated for acute renal failure (see Chapter 10). Urine of the patient with severe renal damage will usually be of fixed low SG regardless of the small urinary volume, as opposed to the high SG associated with scanty urine volume in the patient *without* renal damage.

An indwelling urinary catheter is indicated for stuporous patients to allow for accurate urine measurement. It is imperative to maintain meticulous aseptic technique to prevent secondary urinary tract infection.

It has been noted that young adults may appear much less dehydrated than they actually are. Therefore, it is important to assess fluid volume deficit by several methods. On admission, the patient with DKA should be weighed and his preillness weight ascertained in the history. It is important to know the patient's preillness weight in order to approximate the degree of fluid volume deficit. For example, one might expect, at most, a 1 lb/day tissue weight loss in the patient who is not eating and, at most, a 0.5 lb/day tissue weight loss in the patient who is eating poorly. Any *acute* weight loss over these amounts represents a fluid volume loss. (Recall that 2.2 lb is equivalent to 1 liter of fluid.)

The degree of fluid volume deficit can be approximated by the following formula:

fluid volume
  deficit (liters)

$$= \frac{\text{patient's pOsm} - \text{normal pOsm}}{\text{normal pOsm}}$$
$$\times \text{ liters of body fluid}$$
$$(0.6 \times \text{body wt in kg})$$

Assume an adult patient has a plasma osmolality of 340 mOsm/liter and weighs 70 kg:

fluid volume
  deficit (liters)

$$= \frac{340 - 280}{280} \times (0.6 \times 70) = 9 \text{ liters}$$

It is obvious that fluid intake must exceed fluid output until the deficit is corrected.

Tongue turgor is one of the best gauges of fluid volume deficit because it is not influenced by age, as is skin turgor. The tongue should be assessed for excessive longitudinal furrows; as fluid volume replacement progresses, tongue turgor should improve. Skin turgor is usually a valid measure of hydration in all but the elderly and should be assessed periodically throughout the treatment phase.

Also associated with severe fluid volume deficit are "soft eyeballs," indicating a deficit of intraocular fluid. This situation should improve with fluid volume replacement.

Degree of hyperglycemia and hyperosmolality The degree of hyperglycemia is best monitored by directly measuring the blood sugar level in either venous or capillary blood rather than by arriving at it indirectly by urine specimens; renal threshold for glucose often increases with age, and renal disease, which is common in diabetics, may alter the renal threshold for glucose. A number of other factors make urinary glucose levels less reliable than blood sugar levels in the regulation of insulin dosage. A study by Feldman and Lebovitz using usual urine glucose testing methods reported a 23% incidence of falsely high results and a 33% incidence of falsely low results.[43] In addition, a number of medications can interfere with urinary glucose results.

It is particularly important that blood sugar levels be closely monitored during insulin therapy to prevent a precipitous drop in the blood sugar that can lead to cerebral edema. The blood sugar should not be allowed to drop rapidly below 250 mg/dl.

Sensorium DKA is associated with a variety of neurologic findings, the most frequent of which is depressed sensorium. According to one study, depression of sensorium occurred in 90% of the cases.[29]

While acidosis may undoubtedly affect the brain, diabetic coma seems to correlate best with the degree of hyperosmolality of body fluids.

The level of consciousness in DKA can range from alertness to frank coma. The disturbances of consciousness may be graded as follows: Grade I, awake and fully responsive; Grade II, disorientation or somnolence; Grade III, stupor with motor reactions to pain but without appropriate verbal responses; and Grade IV, deep coma without response to external stimuli.[44] Depression of sensorium in DKA is usually readily responsive to corrective therapy and is one gauge of effectiveness of treatment.

Besides assessing the patient's response to therapy, it is important to be aware of the level of consciousness in relation to maintenance of an airway. An adequate airway must be maintained in the patient with a depressed sensorium, particularly when gastric dilatation is present, making aspiration of regurgitated stomach contents likely. A nasogastric tube should be inserted when gastric dilatation is present to minimize the danger of aspiration pneumonia. If the patient is comatose and there is evidence of airway obstruction, endotracheal intubation is indicated.

Degree of ketoacidosis The degree of ketoacidosis is monitored by assessing respirations, level of consciousness, skin color, acetone breath odor, and, of course, by determinations of plasma pH, plasma ketone levels, and ketonuria.

Labored, deep, costal and abdominal respirations reflect the lungs' attempt at compensation for metabolic acidosis by breathing off more carbon dioxide than normal. More simply stated, respiration is stimulated by the acidemia, and the consequent decrease in PCO_2 helps bring the arterial pH back toward normal. (See the section of this chapter on metabolic consequences of insulin lack for a discussion of the expected fall in PCO_2.) At a pH of 7.2, pulmonary ventilation increases four times; at a pH of 7.0, about eight times. When the acidosis is extremely severe, the central nervous system function is disrupted to such a degree that the respiratory center becomes depressed. A change from deep, rapid respiration to rapid, shallow, gasping respiration may indicate a severe drop in blood pH (below 7.0) or impaired blood flow to the respiratory center because of fluid volume deficit and circulatory collapse. When this occurs, the patient loses the compensatory action of Kussmaul breathing, and acidosis worsens. It is important to monitor the quality of respirations and the arterial pH at frequent intervals during the acute phase of DKA.

Consciousness may be depressed with progressive acidosis and lethargy; disorientation and stupor may appear. These changes should be reversed with clearing of the ketoacidosis and hyperosmolality.

A cherry-red color to the skin and mucous membranes is characteristic of DKA and is due to marked peripheral vasodilatation. An improvement in skin color is indicative of response to therapy. The classic "fruity" breath odor ascribed to DKA may be present but is unreliable in measuring the level of ketosis, since it is frequently masked by the odor of vomitus.

Many feel that the abdominal pain commonly seen in DKA patients is secondary to the ketoacidosis *per se*. With clearing of the ketoacidosis, one would expect abdominal pain to subside. It is important to differentiate between the abdominal pain of DKA and that of an acute surgical emergency. As stated earlier in the section on clinical manifestations, the abdominal pain in DKA is always preceded by anorexia, nausea, and vomiting. The patient with a surgical emergency frequently experiences pain *first*. One must always keep in mind that an acute abdominal emergency such as pancreatitis can be present and, indeed, may have triggered the DKA. An intra-abdominal emergency may be more difficult to detect if ileus or gastric distention, commonly seen in DKA, is present.

Electrolyte derangements *Potassium alterations* As stated earlier, the plasma potassium level may be elevated in the patient with untreated DKA. For this reason, the plasma potassium level should be closely monitored and the EKG observed for peaked T waves. Hyperkalemia usually responds rapidly to fluid replacement and insulin therapy.

Since the potassium level falls rapidly with treatment, it is imperative that the patient be closely monitored for signs of hypokalemia:

Decreased plasma potassium level
Decreased deep tendon reflexes (may also be due to peripheral neuropathy)
Abdominal distention
Muscle weakness
Cardiac arrhythmia
Ileus

Flat or inverted T waves; depressed ST segments; Q-T intervals often appear prolonged because of U waves superimposed on T waves
Acute respiratory paralysis in severe hypokalemia

One should remember that hypokalemia is most likely to occur 1 to 4 hours into therapy.

Phosphate deficiency Phosphate deficiency is common in DKA, and hypophosphatemia usually becomes manifest within 4 to 12 hours of institution of therapy. The clinical consequences of hypophosphatemia are not clear. Arthralgias, myopathies, seizures, acute respiratory failure, impaired leukocyte and platelet function, and gastrointestinal bleeding have been ascribed to hypophosphatemia. When the plasma phosphate concentration falls below 0.5 mg/dl, serious disturbances in metabolism and organ function may ensue. Depletion of phosphate has deleterious effects on the oxyhemoglobin dissociation curve and on cardiac output; therefore, phosphate depletion has the potential for interfering with tissue oxygenation and initiating lactic acidosis.

PATIENT EDUCATION Development of ketoacidosis in a previously diagnosed patient should not occur; when it does, it is frequently due to inadequate patient education. Ketoacidosis is not a sudden complication; it takes several hours to several days to develop. Because of this, there is sufficient time to take preventive measures provided the condition is recognized. Thus, it behooves the nurse to be sure the patient and his family are familiar with the signs of ketoacidosis, and it behooves the physician to furnish specific directions to follow once early symptoms occur. An example of a patient instruction sheet to prevent ketoacidosis during intercurrent illness follows.

NONKETOTIC HYPEROSMOLAR COMA

NKHC primarily occurs in middle-aged to elderly diabetics, with the mean age reported in one study to be 62 years. Often the occurrence of NKHC is the initial harbinger of diabetes mellitus; in fact, as many as two thirds of the patients who develop NKHC have not yet been diagnosed as having diabetes. The typical patient has a mild enough form of diabetes to be managed by oral hypoglycemic agents or diet alone. NKHC is found equally in males and females; as many as 85% have

PATIENT INSTRUCTION SHEET: PREVENTION OF KETOACIDOSIS DURING INTERCURRENT ILLNESS

This plan should be used *only during illness,* not for day-to-day regulation of diabetes.

1 **At the first sign of illness, check second-voided urine for glucose before each meal and at bedtime**

Frequent urine testing will give clear warning of diabetic ketoacidosis, which usually takes several days to develop. Clinitest[R] is the preferred method for testing urine glucose during illness.

If your urine test shows a glucose level of 2 percent or higher, then urinary ketones should be measured with Ketostix[R].

2 **Continue to take your usual insulin injection each day**

Your need for insulin usually is increased by another illness. If your usual insulin dose is not taken, your blood sugar level will rise quickly, even if you do not eat anything. So, even if you are nauseated and vomiting and cannot eat, *never omit your usual dose of insulin.*

3 **Take extra insulin if necessary**

If you feel *ill* and your urine test shows a 2 percent level of glucose or higher, then you need more insulin than usual.

When to take extra insulin
Before each meal and at bedtime:
• Check your second-voided urine test.
• Each time your glucose level is 2 percent or higher, you should take extra insulin at that time.

How much extra insulin to take
The amount of *extra* insulin you should take is calculated as follows:
• Add up the total number of units you usually take each day (be sure to include all mixtures or split dosages in the sum).
• Divide the total by 4, and administer this number of extra units of *regular* insulin, repeated before each meal and at bedtime, whenever your urine glucose level at that time is 2 percent or higher. (See example.)

4 **If you are nauseated, substitute bland or liquid foods for your usual diet**

To avoid ketoacidosis or possible hypoglycemia, patients who take insulin must consume at least 150 grams of carbohydrate during each day.

What will satisfy this requirement?
1½ quarts of a fluid containing sugar (such as gingerale, cola, or fruit juice) taken in small amounts over a 24-hour period.

If vomiting prevents even this intake, you will need hospital care

This is an example only. You will need to substitute your own numbers, according to your usual daily dose of insulin and your glucose levels when you test your urine.

If your usual daily injection is 28 units of NPH plus 4 units of regular insulin, your total daily dose is 32 units of insulin.

Divide 32 by 4:

$$\frac{8\text{ units}}{4\overline{)32}}$$

SAMPLE daily administration of insulin when you are sick:

Before breakfast
   Always take your usual injection.   28 units NPH
                                        4 units regular
   Urine glucose level is 2%. Add   8 units regular

Before lunch
   Urine glucose level is 3%. Take   8 units regular

Before supper
   Urine glucose level is 2%. Take   8 units regular

At bedtime
   Urine glucose level is down to 1%.
   Do not take extra insulin this time   0 units
   ──────────
   TOTAL FOR DAY   56 units

5 Notify your physician if:

- **Your condition gets worse**

- **Your illness lasts longer than 24 hours**

- **You continue vomiting**

- **Ketones persist in your urine, or**

- **You have any questions**

When you are ill, call your physician to let him know the results of your urine tests. It is far better to call early in the course of an illness than to wait until you become more seriously ill.

Your physician's telephone number:

_____

(Benson E, Metz R: Diabetic ketoacidosis. Hosp Med p 34, May 1979)

---

underlying renal or cardiovascular impairment or both.[45] The reported mortality rate ranges from 50% to 60%. While the literature reports that NKHC occurs only one sixth as frequently as DKA, it is by no means uncommon. In fact, with the rising percentage of elderly patients receiving more aggressive forms of medical and surgical therapies, NKHC is becoming almost commonplace.

PATHOPHYSIOLOGY It has been postulated that in NKHC there is sufficient insulin to prevent ketosis but not enough to prevent hyperglycemia. The hyperglycemia results in hyperosmolality of the extracellular fluid, causing water to diffuse from the cellular to the extracellular space. The condition is made worse by the inability of the kidney to properly adjust water and electrolyte balance.

Severe hyperglycemia causes osmotic diuresis with large water loss and sodium and potassium depletion. Dehydration, hyperosmolality, and hyperglycemia are responsible for many of the clinical manifestations of NKHC.

PREDISPOSING FACTORS Many of the patients who develop NKHC have concurrent serious illnesses; in fact, mortality in such patients is often attributed to the concurrent illness. *Conditions* that have precipitated NKHC in susceptible patients include the following:[45]

Gram-negative infections, particularly pneumonia

Myocardial infarction

Uremia

Acute pyelonephritis

Lactic acidosis

Pancreatitis

Heatstroke

Gastrointestinal hemorrhage

Severe burns

Alcoholism

Intracranial disorders

Inadequate fluid intake for any reason (e.g., withholding fluids before surgery or for tests)

Medical treatments that have been reported to precipitate NKHC in susceptible patients include those listed below:

Hyperosmolar tube feedings

Hyperalimentation

Peritoneal dialysis

Hemodialysis

Medications

Steroids (inhibit peripheral insulin activity)

IV Dilantin ⎫

Thiazide diuretics ⎬ impair insulin release

Other diuretics

Cimetidine (Tagamet)

Chlorpromazine

Propranolol (Inderal)

Immunosuppressive agents

Mafenide (for burn therapy)

CLINICAL FINDINGS The prodromal period is longer in NKHC (several days to several weeks) than in DKA (several hours to several days).

The blood sugar tends to be higher than that seen in DKA; it is typically greater than 600 mg/dl and possibly as high as 4800 mg/dl. Serum osmolality is high. Ketones in plasma and urine are absent or minimal (recall that patients with NKHC have sufficient insulin to prevent ketosis but not enough to prevent hyperglycemia). The plasma bicarbonate is normal or only slightly reduced; plasma pH is usually normal or only

slightly low. Most striking in NKHC are symptoms of fluid volume deficit:

Dry skin and mucous membranes

Poor tongue turgor (increased longitudinal furrows)

Postural hypotension

Acute weight loss

Soft, sunken eyeballs

Other symptoms of NKHC may be of a neurologic nature:

Depressed sensorium; serum osmolalities in excess of 350 mOsm/liter are generally associated with a decreased level of consciousness

Motor seizures occur in about 15% of patients, usually focal but may be grand mal

Hemiparesis

Aphasia

Sensory defects

Hyperreflexia

Autonomic changes

The most common misdiagnosis of NKHC is cerebrovascular accident or an organic brain syndrome. Fortunately, most of the above changes revert to normal after correction of the plasma hyperosmolality.

Kussmaul respiration is not a feature of NKHC as it is of DKA; instead, patients with NKHC usually have rapid, shallow respirations.

It has been reported that the average BUN is between 60 mg and 90 mg/dl in NKHC, as opposed to an average of 40 mg/dl in DKA, reflecting a more severe fluid volume deficit in NKHC. The elevated BUN is also indicative of the catabolic state associated with starvation. In addition, serum creatinine is often elevated in NKHC, reflecting the common association of NKHC with renal impairment.

The plasma sodium level in patients with NKHC is usually above normal (actually, there is a *deficit* of total body sodium). Osmotic diuresis also leads to potassium depletion. Phosphate depletion syndromes are seen in NKHC, just as in DKA.

TREATMENT *Fluid replacement* Since the major problem in NKHC is severe fluid volume deficit, a primary treatment concern is fluid replacement. In fact, it is impossible to develop NKHC if adequate hydration is maintained.

A hypotonic electrolyte solution (such as 0.45% NaCl) is a logical fluid choice, since the patient with NKHC has lost far more body water than electrolytes. If signs of hypovolemic shock are present, however, the initial fluid should be normal saline (0.9% NaCl) to expand plasma volume; after correction of shock, hypotonic fluids are administered.

Electrolytes Potassium replacement should be begun as soon as adequate renal function is confirmed. As a rule, 20 mEq to 40 mEq of potassium chloride can be added to each liter of fluid during the period of fluid replacement and insulin administration.

Severe phosphate deficiency states should be treated by the administration of potassium phosphate; caution should be exercised because of the risk of causing hyperphosphatemia in patients with renal damage. (Recall that a substantial number of patients with NKHC have at least some degree of renal impairment.)

Insulin Administration of insulin is of secondary importance to fluid replacement. Good responses have been demonstrated by the administration of low doses of IV regular insulin. (See the discussion of IV insulin administration in the treatment of DKA.) As in DKA, care must be taken to avoid a precipitous fall in the blood sugar level below 250 mg/dl because of the danger of cerebral edema.

PREVENTION Vigorous efforts must be made to limit the increasing occurrence of NKHC by proper patient education and awareness by health care professionals of common precipitating factors associated with NKHC. The diagnosis must be made quickly, and appropriate and individualized treatment must be instituted to reduce the morbidity and mortality of NKHC.

LACTIC ACIDOSIS

Lactic acidosis can occur in both diabetics and non-diabetics. It is associated with serious illnesses in which tissue hypoxia exists, such as acute hemorrhage, hypotension, and heart failure. Anaerobic metabolism causes excessive lactic acid formation and a drop in plasma pH.

Since coma in lactic acidosis may be confused with the coma of ketoacidosis, it is important to be able to differentiate the two. Ketoacidosis is characterized by the following laboratory findings:

High blood sugar

High plasma acetone level

High urinary sugar content

High urinary ketone content

Low plasma bicarbonate level

Lactic acidosis is characterized by the following laboratory findings:

Blood sugar is variable (depending upon state of diabetic control)

Negative or slight elevation of plasma acetone level

Negative or slight urinary sugar content

Negative urinary ketone content

Decreased plasma bicarbonate level

The presence of an unexplained fall in pH associated with a condition producing hypoxia should lead one to suspect lactic acidosis.

Generally, lactic acidosis can be distinguished from ketoacidosis by the absence of severe ketosis and hyperglycemia. Treatment is directed at correction of the cause of hypoxia and administration of sufficient alkaline fluids (usually sodium bicarbonate) to correct the acidosis.

REFERENCES

1. Guyton A: Textbook of Medical Physiology, 6th ed, p 946. Philadelphia, WB Saunders, 1981
2. Ezrin C, Godden J, Volpe R, Wilson, R (eds): Systematic Endocrinology, p 163. Harper & Row, 1973
3. Maxwell M, Kleeman C (eds): Clinical Disorders of Fluid & Electrolyte Metabolism, 3rd ed, p 1307. New York, McGraw-Hill, 1980
4. Cryer P: Diagnostic Endocrinology, 2nd ed, p 69. New York, Oxford University Press, 1979
5. Gotch P: Teaching patients about adrenal corticosteroids. Am J Nurs, Jan 1981, p 80
6. Tzagrouris M: Acute adrenal insufficiency. Heart Lung 7: No. 4:605, 1978
7. Schimke R: Acute adrenal insufficiency. Crit Care Q, Sept 1980, p 19
8. *Ibid*, p 23
9. *Ibid*, p 25
10. Bacchus H: Metabolic and Endocrine Emergencies, p 114. Baltimore, University Park Press, 1977
11. Sanders J, Gardner L: Handbook of Medical Emergencies, 2nd ed, p 63. Garden City, NY, Medical Examination Publishing Co, 1978
12. Bacchus, p 113
13. Heath H et al: Primary hyperparathyroidism. N Engl J Med 302:189, 1980
14. Lukert B: Hypercalcemia, Crit Care Q, Sept 1980, p 13
15. Ezrin et al, p 125
16. Rush D, Hamburger S: Drugs used in endocrine metabolic emergencies. Crit Care Q, Sept 1980, p 8
17. Krupp M, Chatton M: Current Medical Diagnosis & Treatment, p 698. Los Altos, Lange Medical Publications, 1980

18. Hellman R: The evaluation and management of hyperthyroid crisis. Crit Care Q, Sept 1980, p 79
19. Rush & Hamburger, p 5
20. Larsen P: Salicylate-induced increase in free triiodothyronine in human serum. J Clin Invest 51:1125, 1972
21. Meek J: Myxedema coma. Crit Care Q, Sept 1980, p 134
22. Fairchild R: A review of diabetes insipidus. Crit Care Q, Sept 1980, p 111
23. *Ibid*, p 115
24. Crawford J, Kennedy G: Chlorothiazide in diabetes insipidus. Nature 182:891, 1959
25. Maxwell & Kleeman, p 1461
26. *Ibid*, p 605
27. Fairchild, p 118
28. Report on the National Commission on Diabetes to the Congress of the United States. DHEW Publ No. (NIH) 76-1022, 3:88, 1975
29. Skillman T: Diabetic ketoacidosis. Heart Lung 7, No.4:594, 1978
30. Powers D, Hamburger S, Rush D: Nursing management of diabetic ketoacidosis. Crit Care Q, Sept 1980
31. Martin D et al: Effect of hypophosphatemia on myocardial performance in man. N Engl J Med 297:901, 1977
32. Kyner J: Diabetic ketoacidosis. Crit Care Q, Sept 1980, p 73
33. Park R, Arieff A: Lactic acidosis. Ann Intern Med 23:33, 1980
34. Maxwell & Kleeman, p 1349
35. Kyner p 70
36. Maxwell & Kleeman, p 1359
37. Kitabechi A, Ayyaguri V, Guerra S: The efficacy of low-dose versus conventional therapy of insulin for treatment of diabetic ketoacidosis. Ann Intern Med 84:633, 1976
38. Maxwell & Kleeman, p 1354
39. Hockaday T, Alberti K: Diabetic coma. Clin Endocrinol Metab 1 (3):751, 1972
40. Freitag J, Miller L: Manual of Medical Therapeutics, 23rd ed, p 365. Boston, Little, Brown & Co, 1980
41. Wildenthal K: The effects of acid–base disturbances on cardiovascular and pulmonary function. Kidney Int 1:375, 1972
42. Maxwell & Kleeman, p 1357
43. Feldman J, Lebovitz F: Tests for glycosuria: An analysis of factors that cause misleading results. Diabetes 22:115, 1973
44. Keller U, Berger W: Prevention of hypophosphatemia by phosphate infusion during treatment of diabetic ketoacidosis and hyperosmolar coma. Diabetes 29:87–95, 1980
45. Sneid D: Hyperosmolar hyperglycemic nonketotic coma. Crit Care Q, Sept 1980, p 30

13

Fluid Balance in the Patient with Neurologic Disease

REGULATION OF FLUID BALANCE BY THE CENTRAL NERVOUS SYSTEM

RESPIRATORY CENTER AND pH REGULATION The amount of carbon dioxide given off by the lungs is controlled by the respiratory center in the medulla. Recall that carbon dioxide is crucial in the carbonic acid-base bicarbonate buffer system. Pathologic conditions can cause depression or stimulation of the medullary respiratory neurons, and, thus, can alter plasma pH.

INFLUENCE OF CENTRAL NERVOUS SYSTEM ON OSMOTIC BALANCE The central nervous system influences water loss from the kidneys, skin, and lungs. It controls the desire to drink as well as the motor ability to do so. Emotions influence water balance by affecting the drinking pattern; for example, neuroses sometimes cause compulsive water drinking.

Hyperosmolarity stimulates the hypothalamus to release antidiuretic hormone (ADH), the secretion of which is also influenced by volume receptors in the left atrium and pulmonary veins.

CLINICAL CONDITIONS

Brain infections, tumors, or trauma can cause a diversity of fluid balance problems, depending on the area of the brain involved. Injury to the hypothalamus and the brain stem presents the most problems because many metabolic functions are controlled in these areas.

Often, water and electrolyte balance is the most critical factor in determining the survival of the neurologic patient. For this reason, the nurse should become acutely aware of common problems so that their early detection and correction can be facilitated.

HYPERTHERMIA The heat control center is located in the hypothalamus. Direct injury to this area or pressure exerted by edema or masses in other areas of the brain can cause a body temperature elevation. A rise in temperature in a neurologic patient often is an ominous sign. However, fever may be due to other causes such as pneumonia, urinary tract infection, or dehydration (sodium excess).

Patients who have recovered from encephalitis or those with cerebral damage from birth injuries may have faulty temperature regulation. Sometimes multiple sclerosis is associated with low, irregular fever. In addition, an injury to the upper cervical spinal cord often results in high, irregular fever. Finally, surgical operations in the area of the third ventricle sometimes cause pronounced hyperthermia.

Fever should be reduced by pharmacologic or physical means. An automatically regulated hypothermic blanket can be most helpful in reducing temperature; sponging with ice water or alcohol is less effective. Fever speeds up body metabolism. Because of the increased energy expenditure, the patient needs more calories and water. Yet, as mentioned above, a neurologic patient is often unable to recognize thirst and hunger. Moreover, impaired motor function may interfere with the mechanics of drinking and eating. Thus, at a time when the need for food and fluids is great, the patient may be unable to respond with increased intake. Because fever resulting from brain lesions may continue for weeks or even months, careful attention must be paid to meeting the patient's need for food and fluids.

CONFUSED OR UNCONSCIOUS PATIENTS One should remember that a patient with a neurologic deficit may not have the capacity to respond to thirst; the sensorium may be clouded, or the patient may be unable to respond to thirst due to weakness or paralysis, or he may be unable to communicate his needs owing to aphasia or unconsciousness. Thus, the nurse must help assess and meet the fluid needs of such patients.

If the patient is unable to take oral fluids, other routes of intake are available. For example, a patient with difficulty in swallowing may not be able to drink sufficient fluids, yet his fluid requirement can easily be met with tube feedings. Because of the danger of tracheal aspiration of gastric contents, gastric tube feedings are contraindicated in unconscious patients; jejunostomy tube feedings are considered safer. (See Chapter 5 for a detailed discussion of tube feeding routes and methods.) A patient with nausea and vomiting can neither take fluids orally nor by gastric tube, but he can be provided with parenteral fluids. Keen observations by the nurse help the physician determine which replacement route is best and, thus, minimize the duration of the period of inadequate intake.

Fluid output All confused or unconscious patients should be placed on careful intake–output measurements. Fluid loss by all routes, such as sweating, vomiting, or diarrhea, should be carefully recorded. Body temperature checks should be made every 4 hours—oftener if indicated—to detect elevations. Evaluation of fluids lost from the body serves as a basis for fluid replacement; inadequate fluid replacement eventually results in fluid volume deficit.

It is often difficult to determine the urinary output of neurologic patients, since many of them are incontinent. Because of time-consuming bed changes, the nursing staff is often misled into thinking that incontinent patients have large urinary outputs. The amount of incontinent voidings should be estimated as closely as possible. Failure to assess the urinary output accurately may lead to inadequate fluid replacements. To avoid guesswork, seriously ill patients should have indwelling urinary catheters or external devices to catch urine. The insensible water loss can be assessed by accurate daily weight measurements.

Ventilation Efforts should be made to ensure maximal ventilation in order to prevent hypoxia and hypercapnia (elevated PCO_2); recall that severe hypoxia and hypercapnia cause a rise in intracranial pressure (ICP.) Frequently, an airway is necessary to prevent the tongue of the comatose patient from obstructing air flow; dentures and dental plates should be removed, since they can obstruct the airway. The patient with a prognosis of prolonged complete unconsciousness usually requires intubation and mechanical ventilation. Great care should be taken to prevent hypoxia during suctioning (that is, avoid tracheal suctioning longer than 15 seconds; provide extra oxygen, as allowed, immediately prior to tracheal suctioning).

PATHOLOGIC CONDITIONS AFFECTING THE RESPIRATORY CENTER Pathologic conditions of the central nervous system can cause stimulation or depression of the respiratory neurons, thus altering plasma pH.

Hyperventilation Neurologic conditions associated with overstimulation of the respiratory center include the following:

Meningitis

Encephalitis

Brain tumor

Fever

Hyperventilation results from the overstimulation of respiratory neurons. Increased pulmonary ventilation causes excessive elimination of CO_2, resulting in respiratory alkalosis. The nurse should watch for symptoms of respiratory alkalosis when the rate and the depth of respiration are increased. These symptoms include the following:

Light-headedness

Numbness, tingling of fingers and toes

Circumoral paresthesia

Tinnitis

Blurring of vision

Convulsions

Unconsciousness

(See the discussion of hyperventilation as a treatment for patients with elevated ICP later in this chapter.)

Hypoventilation Neurologic conditions associated with depression of the respiratory center include those listed below:

Direct trauma to the respiratory neurons in the medulla

Pressure on the respiratory neurons secondary to tumor, hemorrhage, or brain abscess

Bulbar poliomyelitis

Hypoventilation results from the depression of respiratory neurons. Retention of excessive amounts of CO_2 causes respiratory acidosis. Hypoventilation, which is far more dangerous then hyperventilation, presents these hazards:

Hypoxia with dilation of cerebral blood vessels and subsequent increased ICP (particularly with a profoundly low arterial PO_2 of less than 50 mm Hg)[1]

Increased CO_2 retention (hypercapnia) with dilation of the cerebral blood vessels and subsequent increased ICP (particularly when arterial PCO_2 is greater than 60 mm Hg)

Respiratory acidosis

Hypoventilation due to medullary depression may necessitate the use of a mechanical respirator. Treatment should also be aimed at eliminating the cause of respiratory center depression.

Nursing implications The nurse should be alert for changes in respiration. One should check the respiratory rate, depth, and rhythm and should watch for symptoms of respiratory acidosis when breathing is suppressed. These include the following:

Disorientation

Weakness

Coma, if acidosis is severe

Commentary Some of the symptoms described above, including disorientation, weakness, convulsions, coma, and blurred vision, may be caused by other conditions, such as brain tumors or cerebral vascular accidents. Such symptoms should be evaluated in the light of the patient's history and neurologic status.

Frequent blood gas analyses may be needed to help evaluate the patient's acid–base status.

SYNDROME OF SECRETION OF INAPPROPRIATE ANTIDIURETIC HORMONE Central nervous system disease may be followed by a condition referred to as *syndrome of secretion of inappropriate antidiuretic hormone (SIADH).* Recall that ADH causes water conservation. Neurologic conditions associated with SIADH may include head injury, stroke, brain tumors, infection, and hydrocephalus. Patients with SIADH develop *severe hyponatremia* when exposed to a high fluid intake, owing to abnormal water retention. In SIADH, the water retention is sufficient to expand extracellular fluid (ECF) volume; this inhibits the secretion of aldosterone, allowing sodium to be lost in the urine. Because sodium is lost in the urine, urine specific gravity (SG) is inappropriately high in relation to the plasma osmolality. Most of the retained water in SIADH patients moves into the cells; therefore, there is *no visible peripheral edema.* Although there is no peripheral edema, one can expect to see an increase in body weight because there is abnormal water retention, although usually no more than 3 liters to 5 liters. (Recall that 1 liter of water weighs approximately 2.2 lb.) Accurate measurement and recording of the patient's intake and output is essential in the early detection of SIADH; a fluid intake greatly exceeding fluid output will be noted. *One must remember that SIADH can cause cerebral edema, irreversible brain damage, and death if not detected and treated early!* The nurse should be alert for symptoms of SIADH in patients with brain injury, particularly when fluid intake is excessive. See Table 13-1 for a summary of symptoms and findings in SIADH.

Treatment In addition to any possible treatment of the precipitating condition, the following measures have been employed:

1. Restriction of water to the extent that negative water balance is induced. This may require as little as 400 ml to 700 ml/24 hr.

2. When there are symptoms of severe water intoxication, intravenous administration of 3% or 5% so-

TABLE 13-1 *Symptoms and findings in SIADH*

Decreased serum sodium concentration below 130 mEq/liter (even less than 115 mEq/liter). Symptoms usually appear at a plasma sodium level below 120 mEq/liter and can include the following:

- Lethargy
- Headache
- Loss of appetite
- Nausea and vomiting
- Abdominal cramping
- Irritability
- Personality change (uncooperative, hostile, and confused)
- Muscular twitching
- Deep tendon reflexes disappear (particularly if Na below 110 mEq/liter)
- Positive Babinski's sign (particularly if Na level below 100 mEq/liter)
- Hemiparesis or aphasia
- Stupor and convulsions
- Renal sodium wasting—urinary excretion of a significant amount of sodium in spite of the low serum sodium concentration. (Usually in sodium deficit [hyponatremia] one would expect a low SG of urine, perhaps 1.002 to 1.004, but in SIADH the SG may rise above 1.012)
- Absence of *peripheral* edema, but *cerebral* edema may occur, leading to irreversible brain damage if not detected early and treated
- Weight gain (usually in SIADH, 3 liters to 5 liters of water are abnormally retained; recall that 1 liter of fluid weighs approximately 2.2 lb)

lution of sodium chloride may be successful in increasing the osmolality of the body fluids and in controlling the central nervous system disturbance.

3. Administration of furosemide (Lasix) plus salt replacement has accomplished an increase in the plasma concentration of sodium.

4. If SIADH is chronic, many patients find the long-term restriction of water extremely unpleasant. Certain drugs (such as demeclocycline and lithium carbonate) inhibit the action of ADH on the kidneys, allowing increased urinary output, and thus liberalization of fluid intake.

5. Phenytoin (Dilantin) administration has been used in patients with SIADH to inhibit ADH release.

Essential to effective therapy are the frequent measurement of serum and urine electrolytes and osmolality and careful monitoring of fluid intake and output plus daily body weights.

SODIUM EXCESS FOLLOWING BRAIN INJURY Sodium excess following brain injury is usually caused by inadequate water intake. Recall that thirst and the motor activities necessary to respond to it are often affected by neurologic trauma. Since elderly persons, young children, and unconscious patients are least able to make thirst known, hypernatremia occurs most frequently among these groups.

High-protein tube feedings (which are average in sodium content) are contraindicated during the first week after trauma to the brain. Because of the stress response that follows trauma, the patient is unable to anabolize protein normally. The renal solute load is increased, resulting in excessive water loss, which eventually produces sodium excess. Fever and hyperventilation cause additional water loss and further contribute to sodium excess.

The neurologic patient with respiratory depression may require prolonged mechanical ventilation. Even with humidification, there may be a tendency for the ventilator to remove water and produce hypernatremia. If this occurs, the increased water loss must be replaced; the actual amount required is determined by the clinical condition, urine volume, urine SG, and accurate body weight measurement. Sometimes the converse will occur; the common use of nebulizers with respirators has increased the hazard of water retention. The nurse should keep accurate intake-output records (including the water contribution from nebulizers). Accurate daily weight measurements should also be made to help detect water deficit or water overloading.

Nursing implications The nurse should be alert for symptoms of sodium excess when the water intake is deficient or when excessive water loss occurs. Symptoms of sodium excess (water deficit) include the following:

Severe thirst (if the patient is conscious)

Dry, sticky mucous membranes

Flushed skin

Oliguria

Irritability (particularly in children)

Urine SG above 1.030

Fever

Plasma sodium above 147 mEq/liter

Delirium and hallucinations

Depressed level of consciousness

Fever after neurologic trauma calls for a check of the serum sodium level. The central nervous system symptoms may be caused by altered cerebral electrolyte concentrations and changed volume (shrinkage) of the cells. Hypernatremia may cause cerebral vascular damage (petechial and subarachnoid hemorrhages).

Efforts should be made to supply adequate water, by mouth if possible. If tube feedings are necessary, the nurse should carefully record the amount of water given, plus the total volume of the tube feeding mixture. The physician may order a specific volume for the 24-hour period in order to prevent or minimize cerebral edema and elevation of the ICP. If a specific fluid intake is not established, the nurse should give fluids in amounts sufficient to keep the tongue moist and the skin turgor normal.

It may be necessary to use the intravenous route to supply water if the patient is vomiting or has diarrhea.

Too rapid rehydration may cause further damage to a brain rapidly swelling because of osmosis. Usually several days are necessary for safe rehydration.

DIABETES INSIPIDUS Diabetes insipidus results from interruption of the hypothalamohypophyseal tract by a variety of lesions, so that the posterior pituitary gland no longer secretes ADH adequately. Causes include brain tumors, head injuries, encephalitis, and vascular disease. It may occur as a complication of surgery in the region of the third ventricle. Sometimes the disease is of an idiopathic origin (no demonstrable etiologic factors). The type of diabetes insipidus associated with neurologic disease should not be confused with *nephrogenic* diabetes insipidus, a condition in which there is an adequate supply of ADH but failure of the kidneys to respond ot it.

Symptoms of diabetes insipidus include polyuria and polydipsia; polyuria is caused by the decreased secretion of ADH. (Recall that ADH causes the body to conserve water; a deficiency of this hormone results in an abnormally high water loss into the urine.) Urinary output is usually 4 liters to 6 liters per day, although it may be as great as 12 liters to 15 liters. The large urinary loss of water causes the body fluid osmolarity to rise and produces thirst; as a result, a large volume of water is consumed. Urinary SG is low, remaining between 1.002 and 1.006.

Several tests have been devised to help diagnose diabetes insipidus. For example, if water can be withheld long enough to cause concentration of the urine to a SG of 1.010, the patient probably does not have diabetes insipidus. However, it is difficult to withhold fluids from the patient with diabetes insipidus, even temporarily, without causing a water deficit or even peripheral vascular collapse. Another test consists of the administration of hypertonic solution of sodium chloride. In the normal person, this procedure stimulates ADH secretion and, thus, decreases urinary volume. The person with diabetes insipidus is unaffected by extracellular osmolar changes; the only mechanism capable of decreasing urinary volume in such patients is the administration of pitressin (vasopressin). The administration of pitressin produces a more concentrated urine in the patient with diabetes insipidus than can be achieved with water restriction.

Posterior pituitary extracts or vasopressin are used to conserve about 90% of the water that would otherwise be lost by way of the kidneys of the patient with diabetes insipidus. A potential complication of the administration of pitressin, either diagnostically or therapeutically, is water intoxication (excessive retention of water with subsequent sodium dilution). Urinary volume and fluid intake should be measured and compared to detect excessive water retention. Early symptoms of water intoxication include listlessness, drowsiness, and headache; convulsions and coma may develop later.

Nursing implication The nurse should be alert for polyuria and polydipsia in patients with brain tumor, head injury, vascular disease, or cerebral infection. An accurate intake-output record is helpful in detecting these symptoms. Sometimes the neurosurgeon will leave orders to report large hourly urine outputs in the postoperative period (such as more than 200 ml in each of 2 consecutive hours or more than 500 ml in any 2-hour period).[2] Also, a low urinary SG (below 1.002) should be reported.

If diabetes insipidus is present, provision should be made for an easily accessible water supply. In addition, the patient should be as close to the bathroom as possible. Fluid balance is remarkably well maintained in the patient with diabetes insipidua, as long as his thirst mechanism is intact and he has access to as much water as he wants. Hypernatremia will result if the output of water is not balanced by intake of water.

NONKETOTIC HYPEROSMOLAR COMA (NKHC) NKHC may result from massive osmotic diuresis in the diabetic neurosurgical patient (particularly in the older, mild or undiagnosed diabetic). Medications frequently used in neurosurgical patients that may precipitate this disorder include hyperosmotic agents (such as urea and mannitol), corticosteroids, and phenytoin. Clinical signs may include the following:

Visual hallucinations

Mental confusion

Hyporeflexia

Unilateral Babinski's sign

Nystagmus

Focal motor seizures

Aphasia

Hemiparesis

Coma

This condition may be overlooked because the symptoms may be mistaken for a worsening of the original neurologic disorder. Treatment is aimed at the intravenous replacement of the fluid loss and administration of insulin. (See the discussion of NKHC in Chapter 12.)

ELEVATED INTRACRANIAL PRESSURE

Causes Elevated ICP occurs when the rate of cerebrospinal fluid formation is increased or when the rate of absorption of cerebrospinal fluid is decreased. Brain tumors may produce either or both of these effects. Any irritation to the meninges, such as an infection or tumor, causes large quantities of fluid and protein to pass into the cerebrospinal fluid system; the added volume causes a rise in ICP. A large hemorrhage can increase ICP directly by compressing the brain. The formation of arachnoidal granulation, caused by hemorrhage or infection, can interfere seriously with cerebrospinal fluid absorption and, thereby, increase ICP.

Cerebral blood flow is affected by carbon dioxide concentration, hydrogen ion concentration, and oxygen concentration. It has been shown that cerebral blood flow increases as the arterial PCO_2 increases; in fact, doubling the arterial PCO_2 by breathing carbon dioxide causes blood flow in the cerebrum to approximately double.[3] Danger of high ICP thus is greatest when the PCO_1 is high (e.g., an arterial PCO_2 greater than 60 mm Hg). The mechanism involved is the increased formation of carbonic acid and dissociation of hydrogen ions; hydrogen ions then cause vasodilatation of the cerebral vessels.[4] Any substance that increases the acidity of brain tissue (such as lactic acid or pyruvic acid) can also cause increased ICP. Hypoxia can cause increased cerebral blood flow and, thus, elevated ICP (particularly when the PO_2 is less than 50 mm Hg).

Symptoms The nurse should be alert for the symptoms of elevated ICP in all patients with cerebral abnormalities. Symptoms of elevated ICP may include those listed below:

Changes in level of responsiveness:

Lethargy (an early sign)

Increasing drowsiness

Confusion

Shift from quietness to restlessness

Less responsive to external stimuli

Headache

Pupillary changes

Vital sign changes (terminal events):

Slowed pulse rate

Slowed, irregular respiratory rate

Widened pulse pressure

Increased temperature

Treatment Ideally, treatment of elevated ICP is directed at removing the basic cause of the pressure increase (such as removing a tumor or draining a hematoma). However, the ideal treatment might not be possible, necessitating other modes of therapy.

Hyperventilation Hyperventilation with resultant decrease in PCO_2 (to a level of 25 mm–30 mm Hg) acts immediately to decrease cerebral blood flow and thus reduce ICP. Respiratory alkalosis has been induced in some emergency situations for short periods of time to reduce ICP until agents such as mannitol or urea can take effect (and neurosurgical treatment can be instituted if necessary). Hyperventilation may also be used in longer term situations in which ICP is elevated. The PCO_2 should not be lowered below 25 mm Hg, since this may be harmful.

Mannitol Mannitol is an osmotic diuretic capable of temporarily relieving elevated ICP when given intravenously. It may be given intravenously at a dose of 1 g/kg body weight over a 10- to 30-minute period, depending on the severity of the situation.[4] By elevating the osmolality of plasma, mannitol promotes diffusion of water from cerebrospinal fluid and brain tissue. It may take 1 to 2 hours for effects to occur; if necessary, the dose may be repeated once or twice every 4 to 6 hours.[5]

Although mannitol is chemically stable, it may crystallize when cooled excessively. If this occurs, the bottle should be warmed to 50°C (122°F) in a water bath, then cooled to body temperature before being administered. Mannitol should be given through a blood infusion set to filter any undissolved crystals; it should not be administered through the same infusion set with blood. The nurse should obtain a specific order concerning the rate of administration of the solution; it must be administered slowly enough to prevent overexpansion of the intravascular space.

Initially with the use of mannitol, there is an expansion of plasma volume caused by diffusion of water into the bloodstream; pulmonary edema or water intoxication (sodium deficit) is a possible complication. (Symptoms of pulmonary edema are described in Chapter 11; those of water intoxication are described in Chapter 17.) According to one source, signs of water intoxication have occurred only when the dosage has exceeded 200 g/8 hr.[6]

Because of the osmotic diuretic effect of mannitol, a catheter should be inserted into the bladder. An indwelling catheter ensures bladder emptying and facilitates measurement of urinary output. Urine output should be observed frequently and should exceed 30 ml to 50 ml/hr. If hourly output is less than this amount, the drug should be stopped and the physician notified. (Among the contraindications to the administration of mannitol are fluid volume deficit, renal disease of sufficient magnitude to produce anuria, and marked pulmonary congestion.)

If prolonged osmotic diuresis occurs with the use of mannitol, fluid volume deficit may be induced. Because of the potential fluid balance problems associated with the use of mannitol, it may be necessary to monitor central venous pressure (CVP) during its use. Certainly vital signs should be evaluated frequently. Most authorities recommend that plasma osmolality be monitored when osmotic diuretics are used. In addition, plasma electrolyte levels (especially sodium) should be carefully monitored. The infusion of mannitol should be terminated if the patient develops signs of pulmonary edema, heart failure, progressive renal dysfunction, or an excessively elevated plasma osmolality.

Another potential problem associated with the use of mannitol is rebound increase in ICP. The rebound may result from retention of mannitol in the brain tissue as the blood mannitol level is dropping; this situation reverses the pressure gradient and allows water to diffuse back into the brain tissue, increasing ICP.

The nurse should closely observe the patient for possible complications during the administration of mannitol. One would expect to see lessening of symptoms of elevated ICP with its use.

Glycerol Glycerol, an osmotic dehydrating agent, may be used to lower ICP. Although it is slower acting than mannitol or urea (duration of action is approximately 12 hours), it can be used for a longer period.[7] Glycerol may be given by mouth or by a nasogastric tube in a dosage of 1 g to 3 g/kg body weight per day in four to six divided doses.[5] One should be alert for complications such as a rebound increase in ICP, hyperglycemia, hemolysis, and acute tubular necrosis.[5]

Steroids Dexamethasone (Decadron) may be used both acutely and chronically in the treatment of cerebral edema. The initial dose is usually 10 mg intravenously followed by 4 mg (intravenously, intramuscularly, or orally) every 6 hours.[7] The effects begin in 4 to 6 hours and peak at 24 hours.[4] Its effects are slower acting than mannitol and thus are not as effective in an extreme emergency.[8] Although dexamethasone is known to be a potent anti-inflammatory agent, its direct effect on the cerebrovasculature is not known. With the use of steroids it is important to be alert for signs of gastrointestinal bleeding; antacids are frequently used prophylactically. (Other general nursing responsibilities related to the administration of steroids are discussed in Chapter 5.)

Fluids Most physicians avoid the administration of free water (such as D_5W) and prefer the administration of Ringer's lactate or 0.9% sodium chloride in any condition likely to be associated with cerebral edema and increased ICP. In general, hypotonic IV fluids should be avoided. Fluid restriction is believed to be important in the control of increased ICP. It is important that the nurse carefully monitor the volume and rate of administration of parenteral fluids given to patients with elevated ICP; an excessive volume or accidental rapid administration of the correct volume can cause a rise in ICP.

Position Unless contraindicated, a body position with 30 degree head elevation is recommended to decrease ICP. Flexion of the neck should be avoided; sharp flexion of the neck can increase ICP by interfering with venous outflow and, thus, increasing the amount of blood in the cranial cavity. Extreme flexion of the hips should also be avoided, since it has been found that this position can cause a rise in ICP; the prone position during cranial surgery consistently increases baseline pressure and thus should be avoided in patients with known or highly potential increased ICP.[1] Although intermittent use of the Trendelenberg position is useful in draining the tracheobronchial tree, it should not be used when the patient has an elevated ICP.

REFERENCES

1. Mitchell P, Mauss N: Intracranial pressure: fact and fancy. Nursing 76, June 1976, p 57
2. Howe J: Patient Care in Neurosurgery, p 206. Boston, Little, Brown & Co, 1977
3. Guyton A: Textbook of Medical Physiology, 6th ed, p 347 Philadelphia, WB Saunders, 1981
4. Samuels M: Manual of Neurologic Therapeutics, p 11. Boston, Little, Brown & Co, 1978
5. Freitag J, Miller L (eds): Manual of Medical Therapeutics, 23rd ed, p 403. Boston, Little, Brown & Co, 1980
6. Davis J, Mason C: Neurological Critical Care, p 266. Baltimore, VanNostrand Reinhold, 1979
7. Weiner H, Levitt L: Neurology for the House Officer, 2nd ed, p 154. Baltimore, Williams & Wilkins, 1978
8. Swift N, Mabel R: Manual of Neurological Nursing, p 32. Boston, Little, Brown & Co, 1978

14

FLUID BALANCE IN THE PATIENT WITH RESPIRATORY DISEASE*

The primary function of the lungs is to provide the body tissues with adequate amounts of oxygen (O_2) and to excrete appropriate quantities of carbon dioxide (CO_2), thereby maintaining normal arterial concentrations of O_2 and CO_2. This exchange of respiratory gases occurs by way of diffusion from areas of high to areas of low partial pressures of gas. The adequacy of the diffusion of gases is dependent upon the quality and quantity of the inspired gas, the functioning of the alveoli, the blood supply to the lungs, and the hemoglobin.

Through excretion of CO_2 the lungs also participate in acid–base balance by means of hydrogen ion (H^+; pH) regulation. Alveolar ventilation is responsible for the daily elimination of about 13,000 mEq of H^+, as opposed to the 40 mEq to 80 mEq excreted by the kidney. Therefore, it becomes obvious that pulmonary dysfunction can provide a rapid change in pH within a few minutes, while the H^+ excess of renal failure may not be evident until several hours or days after onset.

The lungs can be regarded as organs of physiologic adaptation, since they regulate the carbonic acid level of extracellular fluid (ECF) through exhalation or retention of CO_2. Under the control of the medulla, the lungs act promptly to correct systemic H^+ changes that are synonymous with acid–base disturbances. When metabolic acidosis occurs, the medulla signals the lungs to exhale CO_2 by means of deep, rapid respiration; when metabolic alkalosis occurs, the medulla orders the lungs to retain CO_2 by means of slow, shallow respirations. The lungs also participate in water and electrolyte balance; thus, disruption of normal pulmonary functioning can produce water and electrolyte imbalances. For example, the blockage of the bronchi and of the alveolar capillary membrane in bronchiectasis results in the inadequate elimination of CO_2 from the lungs with increased carbonic acid concentration in the ECF; respiratory acidosis results. High fever, with its associated hyperventilation, causes an excessive elimination of CO_2 from the lungs and a decreased carbonic acid concentration in the ECF, which results in respiratory alkalosis. In the adult respiratory distress syndrome (ARDS) there is refractory hypoxemia, tachypnea with hypocapnia (early), decreased pulmonary compliance, and evidence of diffuse pulmonary injury (*i.e.*, interstitial edema).[1]

*Revised by Anne Griffin Perry, R.N., M.S.N., Assistant Professor of Nursing; Coordinator, Respiratory Option Medical-Surgical Nursing Major, St. Louis University.

The lungs remove large quantities of water from the body as water vapor. The daily insensible water loss in the adult is approximately 400 ml, varying with the environmental humidity and the depth and rate of respiration.[2] Abnormal factors such as fever, sustained hyperpnea, and copious secretions greatly increase insensible water loss. Conversely, prolonged mechanical ventilation can decrease the amount of water excreted by the lungs. Finally, pulmonary infections and neoplasms can cause abnormal retention of water by the mechanism of secretion of inappropriate antidiuretic hormone (SIADH).

EFFECTS OF PCO_2, *pH*, AND PO_2 ON ALVEOLAR VENTILATION

The alveolar ventilation is the amount of fresh gas entering the respiratory zone each minute. To calculate this value, the anatomic or physiologic deadspace volume must be accounted for; then this value is multiplied by the frequency of respiration:

$$(V_T - V_D) \times f = V_A$$

| V_T tidal volume | V_D dead space volume | f frequency | V_A alveolar ventilation |

In the adult male with healthy lungs the physiologic and anatomic deadspace volumes are equal to 150 ml.[3] Therefore, the healthy male's alveolar ventilation is calculated as such:

$$(500 \text{ ml} - 150 \text{ ml}) \times 15/\text{min} = 5250 \text{ ml/min}$$

The alveolar ventilation is sensitive to changes in PCO_2, pH, and PO_2. Normally the sensitivity of alveolar ventilation to an increase in PCO_2 is rapid and occurs continually throughout daily activities such as eating, exercising, and resting (during which the PCO_2 is probably held to within 3 mm Hg).[4] The normal ventilatory response to an increased PCO_2 is an increased

rate and depth of ventilation, thereby increasing excretion of CO_2 by the lungs. For each mm Hg rise in PCO_2 over normal, the total ventilation increases about 21 ml/minute. The maximal effect of PCO_2 elevation on ventilation is reached at approximately 15% CO_2 concentration.[4] Whenever the PCO_2 is chronically above 50 mm Hg, the sensitivity of the respiratory center is reduced. A sudden further increase in PCO_2 can render the respiratory center insensitive to this stimulus, leaving hypoxemia as the primary respiratory stimulus. It is for this reason that patients with chronic CO_2 retention are placed on low O_2 flow rates by nasal cannulae or Venti-masks.

Plasma pH also affects ventilation. Acidosis stimulates ventilation, and increased amounts of CO_2 are excreted. Alkalosis has the opposite effect on ventilation, and CO_2 is retained. This regulation of pH is an adaptive mechanism; therefore, its value is short-term.

Arterial O_2 also influences ventilation, but the effect of a decreased PO_2 is not as rapid as that produced by PCO_2 and pH. The PO_2 must fall to 50 mm Hg before any increase in ventilation takes place.[5]

STUDY OF ARTERIAL BLOOD GASES

Blood gas analysis includes measurement of pH, PO_2, and PCO_2 (Table 14-1). In addition, it is possible to obtain alveolar-arterial oxygen gradient, $P(A-a)O_2$, from these values (Table 14-2). The advantage of the $P(A-a)O_2$ is that it provides an indication of potentially reversible lung disease when values are obtained serially. In addition, this value provides assessment of the degree of ventilation-perfusion inequalities.

PO_2 AND $P(A-a)O_2$

It is incorrect to consider normal PO_2 to be between 80 mm and 100 mm Hg for all persons, since one must account for changes associated with age. The PO_2 de-

TABLE 14-1 *Arterial blood gas values*

pH	PCO_2	PO_2	HCO_3	Base excess
7.38–7.42	38–42 mm Hg	80–100 mm Hg (decreases with age)	22–26 mEq/liter	−2–+2

TABLE 14-2 *Calculation of P(A − a)O$_2$*

P(A − a)O$_2$ is calculated in two steps

Step 1: Obtain alveolar oxygen tension (PAO$_2$)

$$PAO_2 = (PB - PH_2O)FIO_2 - \frac{PCO_2}{R}$$

PB = barometric pressure, 760 mm Hg at sea level

PH$_2$O = water vapor pressure at body temperature, 47 mm Hg at 37°C

FIO$_2$ = inspired oxygen concentration

PCO$_2$ = partial pressure of carbon dioxide obtained from blood gas measurement

R = respiratory quotient; usually measured by obtaining the ratio of CO$_2$ to produced O$_2$ consumed; usually given a constant value of .8

Step 2: Once the PAO$_2$ is calculated, the PA − aO$_2$ is obtained by the following formula:

P(A − a)O$_2$ = PAO$_2$ − PaO$_2$

Example: Arterial blood gases are drawn on Mr. Baines, a 50-year-old man who has been mechanically ventilated with 100% O$_2$ for 15 minutes. His body temperature is 37°C.

PaO$_2$ = 220 mm Hg   FIO$_2$ = 1 (100%)

PCO$_2$ = 64 mm Hg

pH = 7.31

$$PAO_2 = (760 - 47) \times 1 - \frac{64}{.8}$$

= 713 − 80

= 633 mm Hg

P(A − a)O$_2$ = 633 − 220

= 413 mm Hg

Since the normal P(A − a)O$_2$ corresponds to the physiologic shunt, the normal value on room air should be 10 mm Hg. However, this normal value will increase as the FIO$_2$ increases. Therefore a normal P(A − a)O$_2$ for this patient should be 90 mm to 110 mm Hg. This value of 413 mm Hg indicates a large degree of intrapulmonary shunting.

---

creases and the P(A−a)O$_2$ increases with age (Table 14-3).[6] However, a PO$_2$ less than 70 mm Hg requires the nurse to be alert for signs of developing hypoxemia. Hypoxemia leads to hypoxia and anaerobic metabolism; lactic acid formation results, and metabolic acidosis develops.

An acceptable PO$_2$ for the newborn breathing room air varies between 40 mm and 70 mm Hg.[7] Aging leads to degenerative lung changes in which the alveoli lose their elasticity; thus, the available surface area for the exchange of gases and the PO$_2$ decline with age.

PCO$_2$

The normal partial pressure of CO$_2$ in arterial blood varies between 38 mm and 42 mm Hg. Acute hyperventilation causes the PCO$_2$ to decrease and the pH to increase (e.g., PCO$_2$ 30 mm Hg ↓ and pH 7.5 ↑). This

TABLE 14-3 *Normal, upright PaO$_2$ and P(A − a)O$_2$ values*

Age	PaO$_2$ (mm Hg)	P (A − a)O$_2$ (mm Hg)
10	95–103	< 4
20	91– 99	< 7
30	87– 95	< 11
40	82– 90	< 16
50	78– 86	< 20
60	74– 82	< 24
70	70– 78	< 28
80	66– 74	< 32
90	61– 69	< 37

(Hodgkin JE: Blood gas and analysis and acid–base physiology. In Burton GG, Gee GN, Hodgkin JE (eds): Respiratory Care: A Guide to Clinical Practice, p 236. Philadelphia, JB Lippincott, 1977)

occurs because the rapid rate of ventilation increases the amount of CO_2 that is expired. Conversely, acute hypoventilation causes the PCO_2 to increase and the pH to decrease (e.g., PCO_2 60 mm Hg ↑ and pH 7.2 ↓).

pH

As mentioned earlier, normal arterial pH varies between 7.38 and 7.42. This value represents the amount of H^+ in the arterial blood, and it is an inverse relationship. In acidemia, there is an increased number of hydrogen ions, and this is represented by a decreased pH value. The opposite occurs with alkalemia.

ASTRUP METHOD AND SIGGARD–ANDERSON NOMOGRAM

The method used by Astrup and Siggard–Anderson to analyze acid–base disturbances is ingenious but complicated. Indeed, most American authorities regard the system as needlessly complicated and as presenting what would happen if acid–base disturbances occurred in a test tube rather than in the human body. Therefore, most centers have abandoned use of the system. For those who have interest in the terms that the system has introduced, the following information is provided.

STANDARD BICARBONATE Standard bicarbonate (HCO_3) is the bicarbonate concentration in the plasma of blood that has been adjusted so that its PCO_2 is 40 mm Hg. Moreover, its hemoglobin has been fully saturated with O_2. The normal value of HCO_3 is 24 mEq (22 mEq–26 mEq)/liter, the same as the figure usually regarded as normal for the CO_2 combining power, which is a measure of bicarbonate, not of CO_2 as the name implies. The reason is that as HCO_3 is broken down by addition of a strong acid, the CO_2 gas emitted is a measure of the HCO_3 originally present.

BASE EXCESS Base excess indicates the presence in the blood of an excess of alkali or a deficit of fixed acid (which does not include carbonic acid), or a deficit of alkali or an excess of fixed acid. Positive values indicate alkalinity of the body fluids; negative values indicate acidity. The normal value of base excess is 0 ($+2--2$) mEq/liter. It would appear more forthright to call an excess of base "base excess" and a deficit "base deficit" rather than speaking of a positive base excess and a negative base excess. Base excess cannot be derived from the bicarbonate value alone. Rather, it is arrived at by multiplying the deviation of HCO_3 from normal by the factor of 1.2, which represents the buffer action of red blood corpuscles.

METHODS FOR OBTAINING ARTERIAL BLOOD GAS SAMPLES

Arterial blood gas samples are usually obtained by repeated single arterial punctures. In the critically ill patient, an arterial catheter may be inserted to provide immediate access to arterial blood for blood gas analysis.

ARTERIAL PUNCTURE PROCEDURE The following is the procedure used in arterial puncture.

1. *Explanation to patient.* The procedure for obtaining blood for analysis should be explained to the patient. It is important to prevent unnecessary pain and anxiety, which may cause hyperventilation, resulting in a temporary change in blood gases.

2. *Selection of site.* The radial artery (at the wrist) is frequently used for arterial puncture since it is superficially located, has collateral circulation, and is not adjacent to a large vein. Some authorities feel that arterial punctures by nonphysicians should be done only in radial arteries. The brachial artery (in the antecubital fossa) is also used for arterial puncture since it has appreciable collateral circulation. The femoral artery is not used as frequently, since it has certain disadvantages. For example, there is not adequate collateral flow if the femoral artery becomes obstructed. Obstruction, therefore, threatens the entire lower limb.[8] Also, it is more difficult to stop bleeding in the large femoral artery than in the smaller radial and brachial arteries. Sometimes the femoral vein is mistaken for the femoral artery, since it has a high pressure. Sites for pediatric use include the radial, temporal, brachial, and femoral arteries. The radial artery often is the preferred site, since there is no associated vein (may be used in the newborn as well as in older children). If frequent arterial samples are required, an indwelling catheter may be inserted percutaneously in children over 5 years of age; the umbilical artery can be catheterized in the newborn.

3. *Positioning.* If the radial site is to be used, a rolled

towel should be placed under the wrist and the hand should be pushed down to obtain wrist extension. If the brachial site is to be used, the arm should be extended and supported with a towel under the elbow. The maximum point of pulsation in the artery should be identified.

4. *Cleansing the site.* Usually the area is cleansed with either Betadine or alcohol wipes. Some operators wear sterile gloves to perform an arterial puncture; the hands, at least, should be thoroughly washed.

5. *Local anesthetic.* Some physicians favor the use of a local anesthetic prior to arterial punch; others do not. Usually the local anesthetic is not necessary if the operator is skilled in obtaining arterial samples and can do so on the first attempt with minimal pain to the patient. A strong argument for the use of a local anesthetic is that it reduces pain, particularly when more than one attempt is necessary. Also, infiltration of the anesthetic around the arterial wall reduces the likelihood of arterial spasm.

6. *Heparinized glass syringe.* A 5-ml glass syringe (with a 19 or 20 gauge needle) should be flushed with 0.5 ml of 1:1000 heparin solution and then emptied, leaving heparin in the needle to prevent clotting of the blood sample. (For small pediatric patients, a heparinized glass tuberculin syringe with a 25 gauge needle may be used.) It should be remembered that too much heparin left in the syringe will affect the pH of the blood sample and, thus, cause false results. The syringe should be free of air bubbles. A glass syringe is preferred to a plastic syringe, since a glass syringe has freer movement and, thus, fills more easily. Use of a plastic syringe may necessitate "pulling back" on the plunger to obtain the blood sample. Also, air bubbles adhere more tenaciously to plastic syringes, making them less desirable in this situation.

7. *Obtaining the sample.* The needle should be inserted at an angle (to enter the artery obliquely) while the artery is stabilized with the free hand. An oblique entrance into the artery minimizes the formation of a hematoma, since the hole in the artery seals better when the needle is removed. Usually the initial thrust of the needle will bring a pulsating flow of blood into the syringe without pulling out on the plunger; this spontaneous pumping of blood into the syringe is proof of entry into an artery. If blood does not enter the syringe, the needle must be redirected; if the needle has gone through the artery, it must be slightly withdrawn until blood flows again. From 2 ml to 5 ml of blood should be obtained (depending of individual laboratory requirements).

8. *Aftercare of the site and preparation of the specimen.* After the sample is obtained, pressure should be applied over the site with a gauze sponge and the needle removed. (Pressure should be continuously applied to the puncture site for a period of 5 minutes; a longer time may be necessary if the patient is on anticoagulants.) After the needle is withdrawn from the artery, the syringe should be handed immediately to a second person, who at once plunges the needle into a cork or rubber stopper to seal the syringe from air. Should air accidentally enter the syringe, the specimen should be discarded. Gentle rotation of the syringe is required to mix the blood with the heparin to prevent clotting. The syringe should quickly be placed in a basin of ice slush to slow down O_2 metabolism. Failure to do so causes the gas analysis to yield inaccurate results since O_2 metabolism of blood continues even after it is drawn from the body. The specimen should be taken immediately to the lab after it is properly labeled. The label on the specimen should state the patient's name, the time of the puncture, body temperature at time of puncture, and the inspired O_2level.

9. *Body temperature.* It is important that the patient's body temperature be recorded accurately, since arterial PO_2 and PCO_2 increase as body temperature increases; conversely, they decrease as body temperature decreases. A 6% change in arterial PO_2 occurs with each degree of centigrade body temperature change.[9] Blood gas analyzers are controlled for normal body temperature; thus, the technician must correct the results to the patient's actual temperature (by use of the appropriate nomogram) if severe hypothermia or hyperthermia is present.

10. *Inspired O_2 level.* It is important to state on the requisition the patient's inspired O_2 level. For example: Is the patient breathing room air? Is he breathing 40% O_2? Or, is he receiving O_2 by nasal prongs at a rate of 2 liters/minute? This information is important in interpreting blood gases. If blood gases are obtained to measure the effectiveness of a new O_2 level, the patient should receive that level of O_2 continuously for at least 20 minutes prior to obtaining the arterial sample.

SAMPLING FROM AN INDWELLING ARTERIAL CATHETER In order for a specimen to be drawn safely and correctly from an indwelling arterial-line catheter, the nurse must institute measures that will maintain the patency and integrity of the system as well as ensure correct sampling techniques.

1. Preparation. Prior to obtaining the specimen, the nurse should assess the arterial-line system and waveform pattern to ensure correct functioning. In addition, the monitoring alarms should be turned off.

2. Discard specimen. Using proper technique for cleansing the "exit" port, the nurse must aseptically withdraw 3 ml to 5 ml of blood from the exit port and discard the specimen. This is important because the specimen is diluted with intravenous solution, and analysis would produce inaccurate results.

3. Arterial blood gas specimen. Aseptically withdraw the desired amount of blood into a heparinized syringe. Observe the syringe for the presence of air in the syringe; if this occurs, obtain a new specimen. Send the specimen to the laboratory, and record the body temperature as noted earlier.

4. Aftercare. Close the stopcock to the exit port and flush the arterial-line system. Assess the system for correct functioning and accurate stopcock position. Observe arterial waveforms, and reset alarms.

While the arterial-line provides a ready access to arterial blood specimens, it is not without hazards. Complications of cannulated arteries include tissue necrosis due to diminished arterial blood flow and infection. Tissue necrosis is most likely to occur when the involved artery has inadequate collateral circulation to supply distal tissue. Infection is a risk with any indwelling catheter. This risk can be reduced by aseptic handling of the dressing, changing the dressing and intravenous tubing every 24 hours, and using an antibiotic ointment over the puncture site.

MIXED VENOUS SAMPLING

The increased use of balloon-tipped flow-directed catheters has resulted in increased frequency of mixed venous blood sampling. These samples are withdrawn from the pulmonary artery (PA) port on the Swan-Ganz catheter. The mixed venous oxygen($P_{\bar{v}}O_2$) gives the best overall indication of the adequacy of tissue oxy-

genation.[10] The normal value at rest is 35 mm to 40 mm Hg. A value below 35 mm Hg indicates that tissue O_2 delivery is inadequate. It is important to realize that this indicates only that an inadequacy is present; it does not *define* the cause of the deficiency. Venous O_2 determinants from a peripheral line will indicate the adequacy of oxygenation in the tissue that is drained by that venous system. Samples from a central venous catheter have a slightly higher value than those from the PA line and do not accurately reflect tissue oxygenation.[10]

CONCLUSIONS TO BE DRAWN FROM BLOOD GAS LEVELS

The levels of PCO_2, PO_2, and $P_{\bar{v}}O_2$ provide insight into what is happening to the patient. Theoretically, PCO_2, PO_2, and $P_{\bar{v}}O_2$ can be high, normal, or low (Fig. 14-1). Initially, let's consider combinations of high, normal, or low PCO_2 and PO_2.

First, of course, the PO_2 and PCO_2 can fall within the limits of normal, in which case no deduction concerning disease can be made.

What if the PO_2 is normal and PCO_2 is elevated? This is impossible unless O_2 is being given to the patient.

Suppose the PO_2 is normal and the PCO_2 is depressed. Then the patient must be compensating successfully for a lung disorder. We say he is compensating because the low PCO_2 indicates hyperventilation and the normal PO_2 indicates a healthy state of oxygenation of the blood.

Now suppose the PO_2 is high and the PCO_2 is normal. We can only deduce from this that the patient is receiving O_2—indeed, he must be.

Now let us move on to a high PO_2 and a high PCO_2. Such a situation is clearly impossible.

How about an elevated PO_2 and a low PCO_2? This indicates that hyperventilation is going on.

What if the PO_2 is low and the PCO_2 is normal? Here we have poor oxygenation of the blood but without compensatory activity on the part of the lungs. (There is no overbreathing, otherwise the PCO_2 would be depressed, not normal.)

Suppose the PO_2 is low and the PCO_2 is high. Here we have a condition in which pulmonary ventilation is inadequate. This might occur with any obstructive lung disease. It is not unusual for patients with chronic obstructive pulmonary disease (COPD) to have

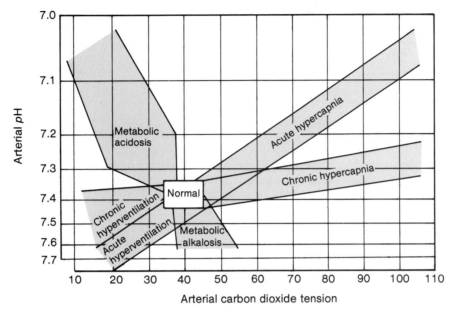

FIG. 14-1. Blood gas interpretation graph. (Burrows B, Knudson RJ, Kettell LJ: Respiratory Insufficiency. Chicago, Year Book Medical Publishers, 1975)

a "sixty-sixty" reading; that is, an arterial PO_2 of 60 mm Hg (normal 95 mm–100 mm) and an arterial PCO_2 of 60 mm Hg (normal 40 mm). Posttraumatic injuries of the chest commonly show a low PO_2 and a mild to moderate elevation of PCO_2.

What if the PO_2 is low and the PCO_2 is low? Here we have apparent efforts on the part of the lungs to compensate for a low O_2 saturation of the blood, but the efforts of the lungs are unavailing. We can see this in the syndrome known as "stiff lung" or in pulmonary embolism. Although acute pulmonary embolism usually is accompanied by hypoxemia and a decreased pH, blood gases may be within an acceptable range. Hypoxemia occurs in pulmonary edema as O_2 diffusion is impaired by transudation of fluid into the alveolar spaces. Thus, arterial PO_2 is below normal. The PCO_2 may range from low to high, depending on the severity of pulmonary edema and the patient's ability to breathe. Hyperventilation caused by hypoxemia can lower the PCO_2, and profound ventilatory failure can cause the PCO_2 to rise.

Expected blood gas changes for acute, partially compensated, and compensated acid–base disturbances are listed in Table 14-4.

The $P_{\bar{v}}O_2$ measurements are useful in determining the adequacy of tissue oxygenation in persons with primary respiratory disorders as well as secondary disorders and other acute alterations. Patients with severe respiratory and physiologic alterations are frequently placed on mechanical ventilation with positive end-expiratory pressure (PEEP). While PEEP can be beneficial, it can also reduce cardiac output, resulting in further diminished tissue oxygenation. In this instance, the PO_2 critically rises but the $P_{\bar{v}}O_2$ declines. It is perhaps of benefit to use the $P_{\bar{v}}O_2$ to determine individualized PEEP levels or supplemental inspiratory oxygen (FIO_2) levels.[11]

In addition to the arterial and mixed venous blood sampling, expired gases may also be analyzed. The mass spectrometer was developed to analyze expired gases. This equipment (Fig. 14-2) is capable of analyzing samples of O_2, CO_2, and other gases at the mouth. Continuous readings as well as computer printout for the patient's permanent medical record can be obtained. The expired gases analyses, along with arterial and venous samplings, can provide more complete information of the patient's cardiopulmonary status.

HYPOVENTILATION

Alveolar hypoventilation always results in hypercapnia (PCO_2 greater than 45 mm Hg) and hypoxemia (PO_2 less than predicted for the patient's age) when breath-

TABLE 14-4 *Blood gas disturbances*

Imbalance	Primary disturbance	pH	PCO₂	HCO₃⁻
Acute respiratory acidosis (early)	Hypoventilation	↓	↑	N
Partially compensated respiratory acidosis	"	↓	↑	↑
Compensated respiratory acidosis	"	N	↑	↑
Acute respiratory alkalosis (early)	Hyperventilation	↑	↓	N
Partially compensated respiratory alkalosis	"	↑	↓	↓
Compensated respiratory alkalosis	"	N	↓	↓
Metabolic acidosis (early)	Gain of acid or loss of base (causing base deficit)	↓	N	↓
Partially compensated metabolic acidosis	"	↓	↓	↓
Compensated metabolic acidosis	"	N	↓	↓
Metabolic alkalosis (early	Gain of base or loss of acid (causing base excess)	↑	N	↑
Partially compensated metabolic alkalosis	"	↑	↑	↑
Compensated metabolic alkalosis	"	N	↑	↑
Mixed respiratory acidosis and metabolic acidosis	Hypoventilation plus gain of acid or loss of base	↓	↑	↓
Mixed respiratory acidosis and metabolic alkalosis	Hypoventilation plus gain of base or loss of acid	? (depends on which is more severe)	↑	↑
Mixed respiratory alkalosis and metabolic acidosis	Hyperventilation plus gain of acid or loss of base	? (depends on which is more severe)	↓	↓
Mixed respiratory alkalosis and metabolic alkalosis	Hyperventilation plus gain of base or loss of acid	↑	↓	↑

ing room air. Alveolar hypoventilation may be caused by a variety of factors (Table 14-5). Each of these factors results in decreased gas exchange and CO_2 excretion.

HYPERCAPNIA

Hypercapnia, which is increased CO_2 concentration in arterial blood, signals the medulla to increase respirations. Symptoms that may occur with the sudden development of hypercapnia include increased pulse and respiratory rate, increased blood pressure, dyspnea, dizziness, sweating, feeling of fullness in the head, palpitation, mental clouding, muscle twitching, convulsions, and unconsciousness. However, patients with chronic pulmonary disease who gradually accumulate CO_2 over a prolonged period (days to months) may not develop these symptoms because compensatory changes have time to occur. An example is an emphysematous patient, kept alive with O_2 therapy for more than a year, who was mentally alert even when his arterial PCO_2 was 140 mm Hg (recall that the normal arterial PCO_2 is 38 mm–42mm Hg). Yet a rapid rise of arterial PCO_2 to 140 mm Hg would surely produce unconsciousness.

An elevated PCO_2 increases cardiac rate and force of contraction, whereas sudden reduction of PCO_2

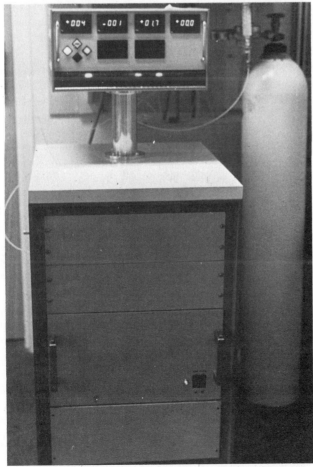

FIG. 14-2.
Mass spectrometer. (Harper RA: A Guide to Respiratory Care: Physiology and Clinical Applications. Philadelphia, JB Lippincott, 1982)

TABLE 14-5. *Cause of alveolar hypoventilation*

A. Pharmacologic
  1. Analgesics
  2. Drug overdose—central nervous system depressants
  3. Anesthesia effects during early postoperative period

B. Chest wall obstruction
  1. Anatomic—kyphoscoliosis, spondylitis, obesity, pregnancy
  2. Traumatic—flail chest

C. Pain—thoracic or abdominal incision
D. Neuromuscular
  1. Guillian-Barré
  2. Myasthenia gravis
  3. Poliomyelitis

E. Inappropriate mechanical ventilator settings

F. Space-occupying factors
  1. Atelectasis/pneumonia
  2. Pneumothorax or hemothorax

from very high levels may lead to ventricular fibrillation. CO_2 excess (particularly a PCO_2 greater than 60 mm Hg) can produce cerebral vasodilation and increased cerebral blood flow. The increased blood volume in the rigid cranium may produce increased cerebrospinal fluid pressure and papilledema. Extremely high PCO_2 levels result in total anesthesia and death.

Excessive CO_2 retention, of course, produces respiratory acidosis. The plasma pH drops below normal because the increase in CO_2 causes the carbonic acid content of the blood to increase. (See the discussion of respiratory acidosis in the next section of this chapter.)

HYPOXEMIA

Acute hypoxemia, reduces O_2 content in the blood, has varying causes and physiological effects. It should be noted that hypoxemia and tissue hypoxia are not synonymous, since tissue hypoxia may be present with a normal PO_2. Obviously, a decrease in inspired oxygen (FIO_2) results in decreased PO_2. This decreased FIO_2 can be the result of high altitudes, airway obstruction, or inappropriate mechanical ventilator settings. In addition, hypoxemia can be the result of overall hypoventilation, ventilation/perfusion (V/Q) mismatch, diffusion defect, and shunts.[11]

Overall hypoventilation The patient's minute ventilation may decline as a result of a reduced tidal volume or slowing of respiratory rate. As a result, the alveolar ventilation drops, producing hypoxemia.

Ventilation/perfusion mismatch V/Q mismatch occurs when the lungs are underventilated with respect to their circulation (perfusion). In this situation, the pulmonary blood is underoxygenated, resulting in hypoxemia.

Diffusion defect If the alveolar capillary membrane becomes thickened as a result of a disease process, the diffusion of O_2 can be reduced to the point that hypoxemia results. However, this is rare, since the hemoglobin in the pulmonary capillaries is fully saturated by the time the blood is one third of the way past the alveolus.[12]

Shunts Pathological shunts can occur from either anatomic or physiological sources. For example, an anotomic abnormality (such as tetralogy of Fallot) and a physiological alteration (such as atelectasis) result in hypoxemia. Hypoxemia is due to the decreased O_2 supply being delivered to the systematic circulation.

Clinical picture The hypoxemic patient has an increased pulse rate as the heart attempts to compensate for inadequate tissue oxygenation by supplying more blood. Other signs of hypoxemia include confusion, agitation, and an anxious facial expression.

Some observers regard cyanosis as the most characteristic clinical sign of hypoxemia. The term *cyanosis* means blueness of the skin; it is caused by an excessive amount of deoxygenated hemoglobin (Hb) in the skin blood vessels. Deoxygenated Hb has a strong blue color that is readily visible when present in a large concentration. Cyanosis becomes discernible only when the arterial O_2 saturation drops below 80% (normal arterial O_2 saturation is 95%). A number of factors influence the degree of cyanosis: one such factor is the thickness of the skin; for example, cyanosis is readily observable in newborn babies because they have thin skin. Cyanosis is often noted first in the lips and fingernails where the capillaries are numerous and the tissues over them are thin and transparent. Another factor influencing the ability to observe cyanosis is the rate of blood flow through the skin. The presence of cyanosis demands careful clinical evaluation for the possibility of tissue hypoxia; however, the absence of cyanosis does not mean than adequate tissue oxygenation is present. In fact, a severe state of tissue hypoxia may be present without cyanosis.

In shock, cyanosis may not be noted because the surface vessels are constricted and contain little blood. It is difficult to detect cyanosis in patients with severe anemia because there is sufficient O_2 available to saturate the small amount of Hb present. Patients with severe anemia and shock may suffer fatal tissue anoxia without the warning of cyanosis; such patients may display only skin pallor. Recognition of cyanosis also involves a subjective component of color preception. There is apt to be inconsistency among observers describing the same patient, or even in a single person watching the same patient over a period of hours.

RESPIRATORY ACIDOSIS

PATHOLOGIC MECHANISMS AND SYMPTOMS

Respiratory acidosis is caused by any clinical situation that interferes with pulmonary gas exchange, such as hypoventilation (Table 14-6.) This disorder is due to hydrogen ion accumulaiton as a result of hypercapnea. The increased PCO_2 leads to an increased carbonic acid concentration of the ECF.

Respiratory acidosis may be associated with no obvious clinical sign except dyspnea (difficult breathing) out of proportion with effort. Patients with pulmonary disease often complain of shortness of breath or of being unable to get their breath. It is a subjective symptom that cannot be measured objectively. Several factors enter into the development of dyspnea. One is an abnormality of the respiratory gases in the body fluids, particularly hypercapnea, and, to a lesser extent, hypoxemia; other factors include the patient's state of mind and the degree to which the respiratory muscles must work to achieve adequate ventilation. When one consciously controls breathing rate and depth, the sensation of dyspnea is likely to occur.

In addition to dyspnea, hyperpnea at rest may be another clinical assessment. Other findings associated with inadequate pulmonary ventilation include cyanosis and tachycardia.

The hydrogen ion excess of acidosis results in the loss of cellular potassium, resulting in hyperkalemia. This occurs due to the buffering of the cells. As the extracellular H^+ rises, the cells begin to buffer by re-

TABLE 14-6 *Conditions likely to be associated with respiratory acidosis*

Emphysema

Pneumonia

Asthma

Cardiac failure with pulmonary edema

Partial airway obstruction

Partial respiratory paralysis

Opiates or sedatives in excessive doses

Tight abdominal binders or dressings

Abdominal distention from ascites and bowel obstruction

Pain in the chest or upper abdomen, resulting in splinting of the diaphragm

Semicomatose states, as in cerebrovascular accidents

Improperly regulated respirator, causing too shallow or too slow breathing

Pneumothorax
Prolonged open-chest and open-heart operations

moving H^+ from the ECF and transferring it to the cell. For every H^+ taken in, one K^+ must be returned to the ECF. While this ion exchange maintains the electroneutrality of the cell, it results in hyperkalemia. As the serum potassium level increases, conduction blocks and ventricular fibrillation may follow.

The symptoms of respiratory acidosis may be difficult to detect. The nurse should be alert for their occurrence when the patient has a condition prone to be associated with respiratory acidosis.

CHRONIC PULMONARY DISEASES ASSOCIATED WITH RESPIRATORY ACIDOSIS

Chronic pulmonary diseases such as bronchiectasis, asthma, pulmonary fibrosis, or emphysema may cause respiratory acidosis. A factor common to all these conditions is chronic hypoventilation, resulting in primary retention of CO_2 and thus an increase in the carbonic acid content of the ECF.

BRONCHITIS–EMPHYSEMA SYNDROME

Pathologic mechanism and symptoms Chronic bronchitis and emphysema so frequently occur in the same patient that the term *bronchitis–emphysema syndrome* has been coined. Chronic bronchitis causes mucous gland hyperplasia and mucosal inflammation with the production of abnormal amounts of mucus. Emphysema involves chronic obstruction to the flow of air into and—even more important—out of the lungs. The chronic airway obstruction causes overdistention of the lungs with air. As a result, the alveoli become enlarged and eventually rupture and coalesce. There may be no symptoms until 20% to 30% of the lung tissue has been destroyed.

Conditions that may contribute to airway obstruction include respiratory infections, smoking, and breathing polluted air. The disease is most common in older persons, particularly men who have done manual labor. Poor ventilation and interference with gaseous exchange at the alveolar level produce hypoxia and hypercabia.

Retention of CO_2 causes a weighting of the carbonic acid side of the carbonic acid-base bicarbonate balance. As a result, this ratio becomes more than 1 to 20, and the balance is tipped in favor of acidosis. The pH of the blood is more acid than normal; the bicarbonate level is increased, since the body retains bicarbonate ions to balance the excessive quantity of carbonic acid. Thus, both the carbonic acid content and the bicarbonate content of the blood increase. If the ratio of carbonic acid to base bicarbonate becomes stabilized at 1 to 20, the pH of the ECF will be normal. This condition is sometimes referred to as *compensated respiratory acidosis*. However, if the body compensatory mechanisms fail and the 1 to 20 ratio is upset, the ECF pH will drop below normal. The condition is then referred to as *uncompensated respiratory acidosis*. In compensated respiratory acidosis, the plasma pH will be normal; in uncompensated respiratory acidosis, the plasma pH will drop below normal.

The emphysematous patient may fluctuate between compensated and uncompensated acidosis. For example, a respiratory infection may tip his delicate state of balance and precipitate uncompensated respiratory acidosis.

The symptoms of bronchitis–emphysema syndrome with respiratory acidosis may include the following:

Chronic fatigue

Dyspnea, first noted on exertion

Moderate cyanosis, early

Respiration with a prolonged expiratory phase accompanied by wheezing or a blowing sound (decrease in elasticity of lung tissue causes terminal bronchioles to collapse when the patient exhales at normal speed)

Large barrel-shaped chest

Chest that appears to be held in permanent inspiration, so that the shoulders appear elevated and the neck shortened

Use of accessory respiratory muscles in breathing

Chronic productive cough

Dull headache

Severe cyanosis in the terminal stages

Coma, if acidosis is severe

Treatment Ideally, the treatment of respiratory acidosis should consist of eliminationg the underlying pulmonary disease. Unfortunately, this is not possible in emphysema. For this reason, treatment is directed toward maximal relief of pulmonary obstruction and the improvement of pulmonary ventilation. Treatment may involve pharmacologic agents, pulmonary hygiene, supplemental oxygen, and symptomatic therapy.

Pharmacologic Agents. Bronchodilators help to reduce bronchial spasm and thus improve pulmonary ventilation. Bronchodilators belong to two classes: methylxanthines (theophylline type) and sympathomimetic (adrenergic) agents. They may be administered parenterally, orally, and through nebulizers. In using a nebulizer, best results are obtained when the patient first exhales completely and then inhales the medication directly into the respiratory tract.

Sputum may be thick in the emphysematous patient and difficult to expectorate. For this reason, an expectorant may be given to liquefy the sputum and make it easier to cough up. In addition, adequate hydration keeps the mucous membranes moist and thereby facilitates the removal of secretions.

Antibiotics may be necessary when a respiratory infection occurs. Unfortunately, the patient with chronic obstructive pulmonary disease is highly susceptible to such infections, particularly during the winter. Although ideally the patient should move to a mild climate during the fall and winter seasons, this is rarely possible. Hence, he should be taught to avoid exposure to respiratory infections, sudden chilling, and unnecessary exposure in damp or cold weather. If, in spite of these precautions, he develops a respiratory infection, he should immediately report that fact to his physician. The secretions of respiratory infections cause further obstruction to pulmonary ventilation in the emphysematous patient. Respiratory acidosis can worsen, and the patient may become seriously ill in a short period of time.

Pulmonary hygiene Patients with excessive respiratory secretions may be helped by postural drainage, which brings secretions up high enough so that they can be eliminated by coughing. Breathing exercises that utilize the abdominal muscles help the lungs empty and aid in the elimination of CO_2. The nurse should supplement the educational efforts of the physical therapist and the physician and encourage the patient to practice these exercises.

Supplemental O_2 administration Although O_2 is required to correct hypoxemia, only as much as is needed should be given. Various methods are available for O_2 administration. Nasal prongs supply an O_2 concentration of 28% at a flow rate of 2 liters/minute and a concentration of 40% at 6 liters/minute. When nasal prongs are used for patients with chronic alveolar hypoventilation, the rate of O_2 flow should not exceed 1 liter or 2 liters/minute until blood gases are available. A nonbreathing oxygen mask supplies an O_2 concentration of up to 90% to 95%. The Venti-mask is a specially constructed device that delivers a low concentration of O_2 (24%, 28%, 35%, or 40%) and is used for administering O_2 to emphysema patients.

The use of low-flow O_2 has been widely accepted. It has been recommended that the lowest flow of O_2 necessary to provide a PO_2 of 60 mm to 80 mm Hg be used. When PO_2 cannot be maintained with supplemental O_2, it may become necessary to institute mechanical ventilation. However, even on mechanical ventilation, care should be taken to use the lowest O_2 to avoid O_2 toxicity. A later section in this chapter will focus on this area.

Prevention of CO_2 narcosis during O_2 therapy Administration of a high O_2 concentration to a patient with an elevated PCO_2 may induce CO_2 narcosis, still a common complication encountered in clinical practice.

Chronic elevation of the CO_2 content of the ECF causes the respiratory center to become insensitive to CO_2. (Recall that while the respiratory center normally is extremely sensitive to changes in CO_2 concentration and that a slight elevation causes respiratory stimula-

tion, arterial CO_2 concentration of over 9% causes respiratory depression.) Hypoxia becomes the main stimulus to respiration when the CO_2 mechanism is not functioning. A reduction in arterial oxygenation stimulates respiration, and an elevation of arterial oxygenation removes the stimulus. Thus, if O_2 is administered in sufficient quantities to raise arterial oxygenation significantly, respiration will decrease. Decreased respiration favors CO_2 retention. Eventually, CO_2 narcosis will result unless the situation is reversed.

The nurse should be alert for the occurrence of CO_2 narcosis when O_2 is administered to a patient with respiratory acidosis. The following are symptoms:

Drowsiness
Irritability, depression, or euphoria
Warm, flushed skin
Respiratory depression
Tachycardia (arrhythmias may develop)
Hallucinations
Muscular tremors of face or extremities
Normal or elevated blood pressure
Convulsions
Paralysis of extremities
Deep coma

O_2 therapy should be used cautiously in patients with chronic respiratory acidosis. It is important to give no more than a 30% or 40% concentration of O_2 in air; a higher concentration may produce serious respiratory depression.

Additional therapy An absolute increase in the number of circulating red blood cells occurs in emphysematous patients as a result of hypoxia. Total blood volume also increases. For this reason, phlebotomy may be useful as a therapeutic measure.

ACUTE RESPIRATORY ACIDOSIS

The development of acute respiratory acidosis demands special measures. Bronchial aspiration may be necessary to rid the respiratory tract of mucus and purulent secretions. A mechanical respirator, used cautiously, may improve pulmonary ventilation. Overzealous use of a mechanical respirator may cause such rapid excretion of CO_2 that the kidneys will be unable to eliminate the excess bicarbonate ions with sufficient rapidity to prevent alkalosis and convulsions. For this reason, the elevated CO_2 concentration should be decreased slowly.

Fluid volume deficit accompanying respiratory acidosis may be treated by the intravenous administration of a Butler-type solution containing balanced quantities of extracellular and cellular electrolytes, plus carbohydrate, or with sixth molar lactate. O_2 can be administered cautiously to relieve severe hypoxia. (Frequent arterial blood gas studies are needed to guide O_2 therapy.) It should be remembered that patients with hypoxemia ($\downarrow PO_2$) and hypercarbia ($\uparrow PCO_2$) may display confusion, anger, and noisy behavior (caused by neurologic manifestations of the abnormal blood gases). It would be *dangerous* to sedate these patients to quiet their behavior; instead, methods should be instituted to improve ventilation.

ACUTE PULMONARY CONDITIONS ASSOCIATED WITH RESPIRATORY ACIDOSIS

Mechanical obstruction of ths respiratory tract with a foreign object prevents air flow into the lungs and results in severe anoxia. In addition, air flow from the lungs is interrupted. The sudden retention of CO_2 causes acute respiratory acidosis; it can also cause a mild rise in blood pressure.

Ventricular fibrillation and potassium excess are common causes of death in patients with acute respiratory acidosis. Treatment consists of the intravenous administration of a sixth molar sodium lactate solution.

Sudden relief of the obstruction, such as may be produced by tracheotomy, causes hyperventilation and may result in alkalosis and tetany as CO_2 is rapidly eliminated from the lungs. The bicarbonate level remains temporarily high. Rapid correction of acidosis may cause ventricular fibrillation.

Apnea also may follow the sudden release of a respiratory obstruction through tracheotomy, possibly because of hypotension and decreased blood flow to the respiratory center. The blood pressure should be checked immediately before and also after tracheotomy to detect hypotension. If hypotension occurs, the physician may request that a vasopressor be given.

Other acute pulmonary conditions that may be associated with respiratory acidosis include pulmonary edema, atelectasis, open chest wounds, and severe pulmonary infections.

OTHER CONDITIONS ASSOCIATED WITH RESPIRATORY ACIDOSIS

Overdoses of drugs Overdoses of morphine, meperidine, or a barbiturate result in depression of respiration and increased retention of CO_2. Before administering a drug of this class, the nurse should check the dose carefully, as well as the time when the drug was last given. In addition, the rate and depth of respiration should be observed before and after administration of the drug.

Pain Severe pain, particularly in the abdomen or thorax, results in splinting of the chest and shallow respiration. CO_2 is retained and respiratory acidosis may develop. Judicious use of analgesics is indicated to relieve pain and to allow the patient to breathe more efficiently.

Weak respiratory muscles Weakening of the respiratory muscles may be caused by such conditions as poliomyelitis or spinal cord injuries. Adequate pulmonary ventilation is not possible; an excessive amount of CO_2 is retained by the lungs, and respiratory acidosis may develop.

Inaccurate regulation of mechanical respirators Inaccurate regulation of a mechanical respirator may result in excessively shallow and slow respiration. Excessive CO_2 is retained by the lungs, causing respiratory acidosis. The use of mechanical ventilators to correct respiratory alterations will be described in a later section of this chapter.

Inhalation anesthesia The use of inhalation anesthetics, such as cyclopropane or ether, may be associated with hypoventilation and CO_2 retention. Mild CO_2 retention may be well tolerated for a while, particularly if hypoxia is not present. However, respiratory acidosis may develop as soon as 15 minutes after the start of inhalation anesthesia; it is most likely to occur in patients with chronic pulmonary disease, such as emphysema.

A patient may have normal color and still develop respiratory acidosis, particularly during an operation when the anesthetized patient is given O_2. While O_2 therapy to produce tissue oxygenation is valuable, if measures are not taken to increase the exhalation of CO_2 (such as with a positive pressure breathing device), an excessive amount of carbonic acid may form in the ECF and cause respiratory acidosis. The first indication of acidosis may be the development of ventricular fibrillation, probably caused by potassium excess. CO_2 retention potentiates vagus nerve activity so that minor stimuli, such as tracheal suction, may cause cardiac arrhythmias. Positioning the patient on the operating table in such a way that normal respiratory excursions are prevented contributes to the development of respiratory acidosis.

Orthopedic deformities Restriction of respiratory excursions by spinal deformities may result in CO_2 retention and acidosis, even though the lungs are normal.

HYPERVENTILATION AND RESPIRATORY ALKALOSIS

Hyperventilation causes an excessive loss of CO_2 and, thus, a decrease in the carbonic acid content of the blood. Unless the kidneys can eliminate bicarbonate (HCO_3) sufficiently to maintain a normal carbonic acid–base bicarbonate ratio, respiratory alkalosis will result. In addition to an excessive loss of CO_2, hyperventilation causes an increased insensible water loss—a fact that must be considered in supplying an adequate fluid replacement.

PATHOLOGIC MECHANISM AND SYMPTOMS

Respiratory alkalosis results from conditions associated with hyperventilation (see Table 14-7). Hyperventilation results in decreased PCO_2 and increases pH.

Symptoms vary in respiratory alkalosis. They may be only those of the underlying disease process, or they may be absent. Sometimes the patient may appear to be short of breath. He may use his upper chest muscles and accessory respiratory muscles during respiration; he may complain of pain and tenderness of the left side of his chest. Alkalosis may cause increased neuromuscular excitability because of the decreased ionization of calcium. (Recall that calcium ionization is decreased in alkalosis.) Anoxia may occur because alkalosis inhibits the release of O_2 from oxyhemoglobin. The most characteristic clinical picture of respiratory alkalosis is represented by the hyperventilation syndrome:

Dizziness or light-headedness

Inability to concentrate

Numbness and tingling of hands, feet, mouth, and tongue

Tinnitus

Blurred vision

Palpitation of the heart

Sweating

Dry mouth

Stiffness, aches, and cramps of muscles

Positive Chvostek's sign (tapping the facial nerve in front of the ear causes the facial muscles about the mouth to contract)

Positive Trousseau's sign (compression of the brachial artery for 1 to 5 minutes causes the muscles of the hand and wrist to go into spasm)

Twitching and convulsions

Loss of consciousness (fainting may occur without symptoms of tetany)

Hyperventilation causes decreased cerebral blood flow; light-headedness, convulsions, and unconsciousness may be due partly to cerebral ischemia. Symptoms of alkalotic tetany are more likely to occur if the respiratory alkalosis developed rapidly. The nurse should be alert for these symptoms in any patient having a condition likely to be associatef with respiratory alkalosis.

TABLE 14-7 *Conditions likely to be associated with respiratory alkalosis*

Anxiety

Lack of O_2

Pulmonary embolism

Pregnancy (high progesterone level sensitizes respiratory center to CO_2)

Hyperventilation resulting from inaccurately regulated mechanical respirators

Early salicylate intoxication

High fever

Neurologic conditions associated with overstimulation of respiratory center (such as meningitis or encephalitis)

High environmental temperature

CONDITIONS ASSOCIATED WITH RESPIRATORY ALKALOSIS

A number of conditions may be associated with respiratory alkalosis (Table 14-7). The most common cause is the hyperventilation that accompanies emotional upsets. Treatment consists of making the patient aware of his abnormal breathing pattern causes his symptoms. He can be shown how to relieve his symptoms by holding his breath or breathing into a large paper bag. Such measures cause an accumulation of CO_2 in the lungs and relieve the alkalosis. A sedative may be requried to relieve hyperventilation in very anxious patients. If alkalosis is severe enough to cause fainting, the increased ventilation will cease and respirations will revert to normal.

Hyperventilation can result from hypersensitivity of the respiratory center, such as occurs with meningitis and encephalitis. Respiratory alkalosis develops because excessive amounts of CO_2 are blown off by the lungs. The inaccurate regulation of a mechanical respirator, causing too deep and too rapid respiration, results in excessive CO_2 elimination—hence, respiratory alkalosis.

Overdoses of salicylates cause excessive stimulation of the respiratory center and hyperventilation. Alkalosis may occur early in salicylate intoxication; later, by the time the patient arrives at the hospital, metabolic acidosis may predominate (particularly in young children). Other causes of hyperventilation include high fever, exposure to high environmental temperatures, and O_2 lack. If the hyperventilation is prolonged, respiratory alkalosis may supervene.

ADULT RESPIRATORY DISTRESS SYNDROME

PATHOLOGIC MECHANISMS AND SYMPTOMS

The adult respiratory distress syndrome (ARDS) is a sudden, clinical, pathophysiologic syndrome characterized by severe dyspnea, hypoxemia, and diffuse bilateral pulmonary infiltrations and "stiff lungs" following massive acute lung injury in patients with no prior major lung disease.[13] The patient with ARDS is severely hypoxemic, and even high concentrations of inspired oxygen (FIO_2) may have little effect in reversing the PO_2. In addition, pulmonary edema and reduced compliance are present as a result of surfactant abnor-

malities. The shunt and deadspace volume are increased, thereby further increasing hypoxemia (PO_2 less than 60 mm Hg on 60% or greater O_2).

The lungs in the patient with ARDS have three characteristics. First, they are heavy owing to the increased lung volume associated with the pulmonary edema. Second, the air-exchanging capacity is diminished. Frequently the lungs are referred to as *airless;* this is due to alveolar collapse. Finally, owing to the altered surfactant production, the lungs are stiff.

CONDITIONS ASSOCIATED WITH ARDS

The literature has referred to ARDS by multiple names (Table 14-8), and ARDS is associated with many other conditions (Table 14-9). The important etiologic characteristics of ARDS include a catastrophic event, such as shock or sudden illness, and the absence of pulmonary or left-sided cardiac diseases. The final characteristic is respiratory distress, which is documented on assessment by tachypnea and labored breathing. Arterial blood gases demonstrate decreasing PO_2 and increasing $P(A-a)O_2$ in the presence of 60% or greater O_2 ($FIO_2 = .6$).

TREATMENT

The treatment is aimed at correcting the refractory hypoxemia. This is done with mechanical ventilation and PEEP (the artificial maintenance of positive pressure at the end of expiration). PEEP enables an increase in PO_2 levels without dangerously high FIO_2 levels. It is important to keep the FIO_2 below .6 because higher FIO_2 increases the risk of O_2 toxicity. O_2 toxicity results in stiff, noncompliant lungs; in the patient with ARDS, this can result in further clinical deterioration. In addition to increasing PO_2, PEEP de-

TABLE 14-8   *Other names for ARDS*

Adult hyaline membrane disease
Shock lung
Wet lung
White lung
DaNang lung
Congestive atelectasis

TABLE 14-9 *Conditions associated with ARDS*

Shock
    Hemorrhagic
    Cardiogenic
    Septic
    Anaphylactic

Disseminated intravascular coagulation (DIC)

Embolism
    Fat
    Air

Drug ingestion

Infections
    Pancreatitis

creases the work of breathing and improves V/P inequalities.[14]

Fluid therapy in these patients is aimed at maintaining normal physiologic parameters, as measured with the Swan-Ganz catheter.[15] It is currently unacceptable to use excessive amounts of diuretics and reduced fluid replacement, which may place the patient in a hypovolemic state. If hypovolemia does occur, the effect of PEEP is limited owing to the diminished cardiac output. In this case, the increased oxygenation at the alveoli is valueless unless it can be delivered to the tissues. The type of fluid used is not as important as the amount. Pulmonary vascular pressures need to be closely monitored to prevent hypovolemia or overzealous fluid replacement.[15]

Pharmacologic agents have also been used in the reversal of ARDS. The use of these agents is presently theoretic, and there is little experimental or clinical data to support their effectiveness. The corticosteroids may be used because of their effectiveness in reducing inflammation and decreasing capillary permeability, thereby assisting in correcting lung water imbalances. Antibiotics are used when ARDS occurs in the presence of an infectious process.

WATER BALANCE IN PROLONGED MECHANICAL VENTILATION

Mechanical ventilation provides movement of respiratory gases into and out of the pulmonary system. This mode of therapy is chosen only when ventilation can be improved or maintained through positive pressure. Mechanical ventilation can be life-supportive; however, it is not without complications.

WATER EXCESS

A study conducted at Massachusetts General Hospital reveiwed 100 patients treated with prolonged continuous mechanical ventilation.[16] Of the 100 patients, 19 developed a positive water balance, associated with weight gain and a significant drop in the plasma sodium concentration and in the hematocrit. The water retention existed primarily as pulmonary edema rather than as peripheral edema. It appeared to be in no way connected with the patient's original diagnosis and was not affected by the type of respirator used. Water overloading may be associated with the use of efficient nebulizers providing an additional 300 ml to 500 ml of water per day. Positive pressure ventilation produces changes in the dynamics of flow in the pulmonary vessels; possibly this causes excess water to move into the interstitial spaces. The water retention, with its associated weight gain and dilutional hyponatremia, may be associated with an elevated level of antidiuretic hormone (ADH).

To avoid the dangerous complications that can result from water overloading in ventilated patients, careful monitoring of intake and output is essential. The water contribution from nebulizers should be included in the intake column. Daily body weight measurement is valuable in detecting weight gain from water loading, particularly when a sensitive in-bed scale is used. The actual weight gain may appear small, but one must consider that immobilized patients with low caloric intake would normally lose about 200 g to 500 g daily. Any gain in weight, or even maintenance of a steady weight under these conditions, may be caused by water retention.

An increase in pulmonary extravascular water may be detected by radiographs. Treatment consists of water restriction and the use of diuretics. Failure to notice pulmonary water loading encourages progressive difficulty in ventilation.

FLUID DEPLETION

Patients with an infectious process who are on mechanical ventilation are at risk for fluid depletion. The febrile state increases water and electrolyte loss through the kidneys and the skin. (It should be noted that the tachypnea of fever is not present in the patient on controlled ventilation.) The use of diuretics or supplemental high-protein tube feedings leads to further fluid loss.

The nurse should recognize fluid volume deficit through identification of dry mucous membranes, decreased skin and tongue turgor, decreased body weight, and thickened pulmonary secretions. In addition, hemodynamic and serum electrolyte parameters should be assessed.

CARE OF THE NEAR-DROWNED PATIENT

The incidence of drowning has increased as water sports have become more popular. Today, accidental drowning is a common cause of death in the United States. The following factors may contribute to accidental drowning: fatigue; hyperventilation with prolonged breathholding in order to swim longer distances underwater; muscle cramps; hysteria; currents or underwater obstacles; and, intoxication.

The nurse should be acquainted with the physiologic changes occurring with drowning and with the treatment of these changes, since one may be called upon to care for near-drowned patients in the hospital or at the scene of the accident.

Most submersions are fatal after 2 to 3 minutes; respiratory arrest occurs by the third minute, and cardiac arrest occurs during the fourth minute.[17] Rescue before the third minute may result in spontaneous resuscitation. Drowning death may result from asphyxia due to laryngospasm without fluid aspiration or, frequently, from aspiration after relaxation of the larynx.[18] Salt water drowning leads to aspiration of other foreign substances that can cause chemical irritation to the respiratory tract or obstruction. In freshwater drowning, there may be some hemodilution and hemolysis due to rapid absorption of hypotonic solution into the circulation, while saltwater drowning results in hemoconcentration.[18] The clinical pictures of saltwater or freshwater drowning are fairly similar (Table 14-10). However, there are differences in electrolytes and replacement fluid therapy that must be recognized.

Immediate treatment of the near-drowned patient is crucial. The following procedures are important:

1. Immediate treatment at the scene consists of clearing the airway and giving mouth-to-mouth ventilation, cardiac massage if necessary, and 100% O_2 by mask as soon as available. Recent recommendations of the American Heart Association and the American Red Cross include pressing on the upper abdomen to expel water from the lungs, in a modification of the Heimlich maneuver.
2. If no pulse is palpable at the time of hospital admission, 1 ml of 1:1000 epinephrine solution should be given into the heart.
3. Electrical defibrillation should be performed if ventricular fibrillation is present.
4. Sodium bicarbonate may be used to rapidly correct metabolic acidosis (secondary to hypoxemia).
5. The trachea should be intubated and suctioned, then ventilation provided by a volume-cycled respirator.
6. Frequent blood gas studies are indicated to guide O_2 administration.
7. A gastric tube should be inserted to remove swallowed water.

SYNDROME OF SECRETION OF INAPPROPRIATE ANTIDIURETIC HORMONE (SIADH) ASSOCIATED WITH RESPIRATORY DISEASES

The pathophysiology of SIADH associated with respiratory diseases was initially documented in patients with bronchogenic carcinoma. Briefly, this syndrome is due to an increase in ADH secretion, which leads to increase in total body water. There is a hypo-osmolar state and abnormal dilution of the ECF. (See Chapter 2 for a more thorough discussion of the physiological changes induced by SIADH.)

SIADH in the respiratory patient can be induced by neoplasms, drugs, infections, intrathoracic processes, and mechanical ventilation. Neoplasms are believed to cause SIADH by secretion of an ADH-like substance by the tumor itself. Vincristine (Oncovin), a chemotherapeutic agent used to treat bronchogenic cancer, is believed to act on the supraoptic hypophyseal system to increase production or release of ADH.[19]

Intrathoracic processes such as tuberculosis and pneumonia are believed to cause SIADH. Mechanical ventilation (especially with PEEP) increases the risk of SIADH, but the mechanism of this is not clearly defined.

TABLE 14-10 *Clinical picture of the near-drowned patient*

	Salt water	Fresh water
Vital signs		
Heart rate	↑	↑
Respiration rate	↑	↑
Temperature	↑	↑
Cardiac arrhythmias	Present	Present
Pulmonary assessment		
Rales	Present	Present
Bronchospasm	Present	Present
Pulmonary edema	Present	Present
Cyanosis	Present	Present
Air hunger	Present	Present
Central nervous system assessment		
Agitation	Present	Present
Seizuring	Present	Present
Posturing	Present	Present
Arterial blood gases	Consistent with combined metabolic and respiratory acidemia with hypoxemia	Consistent with combined metabolic and respiratory acidemia with hypoxemia
Significant serum electrolytes		
Sodium	Elevated	Normal or decreased
Potassium	Normal or decreased	Elevated

↑, increased.

The clinical picture of SIADH and its treatment are detailed in Chapters 2 and 13.

REFERENCES

1. Stothert JC, Carrico CJ: Fluid therapy in adult respiratory distress syndrome: A pathophysiologic approach. *Semin Respir Med* 2, No. 3:123, 1981
2. Sorensen KC, Luckmann J: Basic Nursing: A Psychophysiologic Approach, p 480. Philadelphia, WB Saunders, 1979.
3. West JB: Respiratory Physiology: The Essentials, 2nd ed, p 18. Baltimore, Williams & Wilkins, 1979
4. *Ibid*, p 122
5. *Ibid*, p 123
6. Burton GG, Gee GN, Hodgkin JE (eds): Respiratory Care: A Guide to Clinical Practice, p 236. Philadelphia, JB Lippincott, 1977
7. Shapiro B et al: Clinical Application of Blood Gases, 2nd ed, p 140. Chicago, Year Book Medical Publishers, 1977
8. Wade J: Respiratory Nursing Care, 2nd ed, p 74. St. Louis, CV Mosby, 1977
9. *Ibid*, p 133
10. Burton et al, p 240
11. *Ibid*, p 242
12. *Ibid*, p 241
13. Petty TL: Adult respiratory distress syndrome: Historical perspective and definition. *Semin Respir Med* 2: No. 3:99, 1981
14. Shapiro B et al: Clinical Application of Respiratory Care, 2nd ed, p 364. Chicago, Yearbook Medical Publishers, 1979
15. Stothert & Carrico, p 126
16. Sladen A, Laver M, Pontoppidan H: Pulmonary complications and water retention in prolonged mechanical ventilation. *N Engl J Med* August 1968.
17. Graef J, Cone T: Manual of Pediatric Therapeutics, p 54. Boston, Little, Brown & Co, 1974
18. Baum GZ (ed): Textbook of Pulmonary Diseases, p 869. Boston, Little, Brown & Co, 1974
19. Kubo A, Grant P: The syndrome of inappropriate secretion of antidiuretic hormone. *Heart Lung* 7, No. 3:471, 1978

15

Water and Electrolyte Disturbances From Heat Exposure

Although heat disorders occur most often in tropical zones, the temperate climate of North America can cause heat stress. Many persons living in a temperate climate withstand heat stress poorly, hence the increased number of deaths during heat waves. The literature reports numerous deaths in the elderly during the hot summer months. In fact, two thirds of the cases of heatstroke occur in persons over the age of 50.[1]

A recent social phenomenon is the increase in the number of persons who have taken up recreational jogging; there are reportedly more than 28 million joggers in North America![2] This running fad has increased the potential for serious heat disorders to catastrophic numbers; the danger is greatest in novice runners unaccustomed to vigorous exercise in the heat. Heat disorders are also common occurrences on the football field, as well as in other strenuous sports. They are particularly common in athletes training vigorously before they have become acclimated to heat stress. Heatstroke remains the second leading cause of death among American athletes.[1]

New military recruits undergoing basic training in southern states also are at high risk for heat disorders.

Industry is often associated with artificially created hot climates, resulting from or deliberately designed for some industrial process. For example, persons working in the textile weaving and processing industry are often subjected to an artificially induced, warm, humid climate, because these conditions are best suited to textile processing. Certain segments of the glass, rubber, steel, and mining industries are also associated with high environmental temperatures . Laundry, construction, and agricultural workers and firemen are often exposed to heat stress. Because so many persons in our society may be subject to heat disorders, the nurse should become familiar with their prevention, recognition, and treatment.

A brief review of body thermoregulation mechanisms will promote a more thorough understanding of the section of this chapter concerning specific heat disorders.

THERMOREGULATION IN THE BODY

PHYSIOLOGY OF HEAT LOSS Thermal balance in the body is determined by three factors: heat exchange between the body and its surroundings, metabolic heat production, and heat loss by sweat evaporation. Heat is gained when the environmental temperature exceeds body temperature and when body energy expenditures are high, as in the performance of heavy exercise.

The body is able to maintain constant normal temperature by changing the volume of blood flow to the skin and by changing the rate of the skin's heat loss. This mechanism is initiated in the hypothalamus where local receptors sense the temperature of blood in the area. The immediate response to heat is cutaneous vasodilatation with an increase in blood flow to the skin and an increase in sweat production. Cutaneous vasodilatation permits heat loss through radiation and convection and supplies the metabolic needs for the production of sweat. Sweating is an efficient means of cooling when the humidity is low, with up to 600 K-cal dissipated for each liter of sweat that evaporates.[3] Acclimated persons can sweat more efficiently, losing less sodium, than nonacclimated persons, which explains why serious heat disorders are less common in these persons.

Sweating is initiated when the environmental temperature exceeds 82.4° to 86°F (28°–30C). The ensuing evaporation of sweat from the body surface causes cooling of peripheral blood; the cooled blood is returned to core parts of the body and serves to reduce body temperature. A decrease in the amount and rate of sweating severely interferes with the mechanism for body heat loss.

Of the body's sweat glands, approximately 2 million have thermoregulation as their chief function. Sweat is normally a hypotonic fluid containing several solutes, the chief of which are sodium chloride and potassium. Excessive sweat loss can thus lead to water deficit, salt deficit, and potassium deficit.

A normal person can sweat about 1 liter/hour for 2 hours and lose up to 5% of body weight without straining the cooling mechanism.[4] However, a loss of 7% of body weight is dangerous; at this point, the sweating mechanism may shut down to conserve water.[4]

A high relative humidity predisposes to heat stress because it interferes with the evaporation of sweat from the body. Effective sweat evaporation diminishes rapidly at 60% relative humidity and virtually ceases at 75%.[4] Body temperature begins to rise when the relative humidity is 100% and the environmental temperature is above 94°F (34.4°C). As a rule, when heatstroke occurs, the temperature is over 95°F. In most reported epidemics, more than 48 hours of environmental temperatures averaging over 90°F were recorded, and relative humidity was in the range of 50% to 75%.[5]

BODILY RESPONSES TO HEAT STRESS Exposure of the body to heat stress elicits the following responses:

1. Increased peripheral vasodilatation, to allow more blood to come to the surface for cooling
2. Increased sweating, to allow for cooling by evaporation
3. Increased cardiac output and pulse rate
4. Increased aldosterone secretion, to allow the conservation of body salt; aldosterone causes sodium retention by the kidneys and may exert a similar effect on the sweat glands. *Unfortunately, this mechanism causes potassium loss to be further increased.*

ACCLIMATIZATION TO HEAT Acclimatization to heat is a physiologic process by which a person becomes able to tolerate heat stress more safely and comfortably. It has long been known that persons accustomed to high temperature, either in a natural hot climate or in their work, tolerate heat stress much better than those accustomed to cool temperatures. Yet, the latter can gradually develop a tolerance for heat when repeatedly exposed to it. Acclimatization is accompanied by a series of physiologic changes, which serve to ameliorate the effects of heat stress.

Physiologic changes Persons exposed repeatedly to heat stress gradually experience fewer of the disagreeable sensations induced by heat, such as lassitude and general discomfort, because of the physiologic changes induced by acclimatization. The sweat contains a lower concentration of sodium chloride and a higher quantity of potassium than normal, as a result of increased production of aldosterone. (Recall that aldosterone causes sodium retention and potassium loss.) Retention of sodium causes water retention and, thus, an increased plasma volume. Other specific changes include a progressive decrease in rectal and skin temperatures, a decreased pulse rate, and increased sweating ability.

Rate of acclimatization Most of the changes brought about by acclimatization occur in the first 4 to 7 days of heat exposure; they usually attain their maximum after 2 weeks of daily heat exposure. A person does not have to be subjected to heat stress 24 hours per day in order to become acclimated.

The degree of acclimatization achieved is enhanced by previous physical conditioning and by exercise. Once achieved, acclimatization is maintained for weeks without reexposure to heat, or for longer periods with brief reexposure to heat, or for longer periods with brief reexposures. The extent of acclimatization depends on the physical conditioning of the person and the degree of heat exposure and should not be thought of as an all-or-none phenomenon.

HEAT DISORDERS

Continuation of heat stress without relief can eventually lead to heat cramps, heat exhaustion, or heatstroke. Some patients, after heat exposure, have a clinical picture that includes elements of more than one of these syndromes; usually, however, the heat disorders are seen in their "pure" form.

HEAT CRAMPS The heat cramps syndrome is the mildest of the three heat disorders. Its characteristics and treatment are summarized in Tables 15-1 and 15-2.

HEAT EXHAUSTION Next on the continuum of heat disorders is the heat exhaustion syndrome. Its characteristics and treatment are summarized in Tables 15-3 and 15-4.

HEATSTROKE Heatstroke is the most serious of the three heat disorders; it constitutes a medical emergency and has a mortality rate of up to 80%.

Types There are two major types of heatstroke: exercise-induced and non-exercise-induced. The former tends to occur in the young athlete, military recruit, or

TABLE 15-1 *Heat cramps: characteristics*

Characteristically occurs in well-conditioned athletes following intensive exercise[6]

Occurs in persons who sweat profusely; the exact mechanism of its production is not known

Believed to result from an acute deficiency of sodium in the muscles[7,8]

Rarely occurs in the elderly, since they are seldom able to perform the degree of physical activity necessary to produce such large sweating losses

Characterized by brief, intermittent, painful spasms of the large skeletal muscles that have been subjected to intense activity; pain may be mild to severe

Onset usually sudden

Body temperature normal

Thirst mechanisms intact; person often replaces fluid loss with water only

Patient is rational and alert

TABLE 15-2 *Heat cramps: treatment*

Ingestion of fluid containing salt (*e.g.,* a solution of 1 tsp NaCl in 1 qt of water)[9]

Some authorities favor adding extra salt to food in preference to the use of salt tablets[10]

Salt tablets are not as widely used today as in the past; if they are used, one tablet of salt, enteric coated, for every pound of weight lost by perspiration during the day has been suggested as the ideal amount to avoid gastrointestinal irritation[11]

Be aware that salt ingestion without adequate water intake, or vice versa, may make matters worse

Rest at least 12 hours

In the event of severe, repeated, unrelenting cramps, oral or IV salt solutions rapidly relieve all symptoms[12]

TABLE 15-3 *Heat exhaustion: characteristics*

Results from inadequate cardiovascular responsiveness to the circulatory changes brought about by heat (the diversion of blood flow to the skin is not adequately compensated for by vasoconstriction in other parts of the body or by volume expansion)[10]

Frequently seen among elderly who are at special risk because of a high incidence of cardiovascular disease, sluggish vascular autonomic responses, and frequent use of diuretic agents

Signaled by cool, pale, moist skin (Note: patient continues to sweat)

Extreme weakness and copious sweating, giddiness, and postural syncope; sometimes accompanied by nausea, vomiting, and the urge to defecate

Body temperature may be normal or elevated

Patient usually conscious but may be unconscious

Tachycardia and hypotension may be present

In young athletes, signs of heat exhaustion may mimic a viral upper respiratory infection[6]

Can present with or without heat cramps

TABLE 15-4 *Heat exhaustion: treatment*

Cool patient by swabbing (may or may not need tub cooling, depending on initial body temperature)

Patient should lie down and rest in a cool place

Oral replacement of water and electrolytes is indicated if the patient is able to tolerate oral fluids

Intravenous fluid replacement may enhance recovery but is usually unnecessary[13]

    (The type of parenteral fluid to be administered must be delineated by clinical laboratory findings, since patients may present with low, normal, or high sodium levels. For example, if the patient has predominantly a water deficit, D_5W may be used; if the primary problem is salt deficit, isotonic saline may be used.)

Hospitalization is indicated for patients over 65 years of age or those with predisposing disease, severe vomiting, diarrhea, muscle necrosis, or unstable blood pressure

Rest for 2 or 3 days

worker; the latter tends to occur in the sedentary elderly person.

Strenuous muscular work contributes to heatstroke by enhancing production of body heat. The hyperpyrexia can be associated with extensive muscle damage characterized by myoglobinuria, oliguria, and severe hyperkalemia. These complications are not often seen in the elderly, in whom severe muscular work is not ordinarily a contributory factor.

A common contributory factor to both exercise-induced and non-exercise-induced heatstroke is the use of medications interfering with thermoregulation (see the next section for a list of these medications). Contributory factors to exercise-induced heatstroke include inadequate acclimatization and obesity. Contributing to the non-exercise-induced variety of heatstroke are chronic illnesses (such as arteriosclerotic heart disease, congestive heart failure, diabetes, malnutrition, and alcoholism) and sweat gland dysfunction (as in cystic fibrosis or scleroderma).

Risk factors for heat exhaustion and heatstroke The similarities between heat exhaustion and heatstroke suggest that the pathophysiologies of the two disorders may be similar and that they may represent closely related points along a continuum of responses to heat stress.[14] At risk for heat exhaustion and heatstroke are the following:

    Those exposed to high ambient temperature and humidity without adequate acclimatization

    The elderly, particularly those with known cardiovascular disease

Those with myocardial ischemia, arteriosclerosis, and hypertension

Those performing heavy physical work in a hot environment, since they undergo the double stress of two sources of excessive heat

Those with recent illnesses, particularly when accompanied by fever

Those with a water deficiency

Diabetics, since they frequently have autonomic dysfunction and heart disease and thus may be unable to make the important hemodynamic alterations necessary to dissipate a heat load imposed by increased environmental temperatures[15]

The obese person, who has less body surface in proportion to body weight than does a person of slight build; thus, there is greater difficulty in dissipating heat

Those with impaired ability to sweat, as in dermatologic conditions affecting large areas of skin

Those with potassium deficiency

Those taking medications that interfere with thermoregulation:

Diuretics, because they promote fluid loss

Antiparkinsonian drugs ⎫
Antihistamines ⎬ because they suppress sweating
Anticholinergics ⎭

Phenothiazines, because they suppress sweating and possibly disturb hypothalamic temperature regulation

Haloperidol (Haldol), because it decreases thirst recognition and might disturb hypothalamic temperature regulation

Amphetamines ⎫
Vasodilators ⎬ because they increase metabolic heat production
Thyroid extract ⎪
Lysergic acid diethylamide (LSD) ⎭

Alcohol, because it causes an increased metabolic load, promotes diuresis, and impairs judgment and critical thinking

Propranolol, because it impairs sweating and decreases cardiac output[16]

Tricyclics, because they increase motor activity and heat production

Unalert bedridden patients who are unable to throw off bedclothes and to drink at will

Those wearing impermeable clothing and exercising strenuously to achieve weight loss

Short, stocky, heavily muscled athletes appear to be at higher risk than tall, lanky athletes

Those who falsely believe that water deprivation accelerates physical conditioning and thus avoid drinking adequate water during heavy exercise

Those who overzealously take salt supplements; ingestion of excessive sodium and insufficient water predisposes to severe water-depletion heat exhaustion that can culminate in frank heatstroke

Hemodynamic and metabolic effects Consideration of the metabolic and hemodynamic alterations of heatstroke is important for an understanding of the disorder, its complications, and treatment.

As stated earlier, increased environmental temperature causes increased heart rate, increased cardiac output, and decreased systemic vascular resistance; these are normal circulatory adjustments required to dissipate the heat load imposed by increased environmental temperatures. These changes permit maximum heat losses through radiation, convection, and evaporation.

Many elderly persons are unable to compensate adequately with tachycardia and decreased systemic resistance and therefore are predisposed to the development of heatstroke. Uptake of heat from the environment increases proportionally with age; in this context, the prevalence of heatstroke in the elderly during heat waves is easily understood.[17]

Patients with heatstroke may hyperventilate and develop respiratory alkalosis. If heatstroke is accompanied by hypotension or hypovolemia or has occurred in a heavily exercised person, lactic acidosis can result. Also, it is possible that a combination of respiratory alkalosis and lactic acidosis can be present. Before assessing arterial blood gas readings, a temperature correction of the blood gases must be made. (Recall that blood gas analyzers are controlled for normal body temperature; thus, the technician must correct the results to the patient's actual temperature by use of the appropriate nomogram if severe hyperthemia is present.)

Most cases of heatstroke are associated with mild hypernatremia, hypokalemia, and acidemia (pH values to 6.9 have been recorded.)[18] Hypokalemia occurs in approximately half of the patients with heatstroke. In patients with exercise-induced heatstroke, hypokalemia usually represents frank potassium depletion. In the presence of oliguria, the release of potassium ions from injured tissues frequently corrects the hypokalemia and leads to hyperkalemia. Knochel and associates have pointed out the role of hypokalemia in the exercise-induced type of heatstroke.[19] Studies performed on recruits undergoing physical conditioning showed that sweat losses and lack of adequate volume replacement led to secondary aldosteronism, which in turn induced potassium depletion. This predisposes to rhabdomyolysis and heatstroke by inhibiting the vasodilatation of exercising skeletal muscle.

Rhabdomyolysis (skeletal muscle necrosis) is present in nearly all cases of exertional heatstroke but is less common in nonexertional heatstroke. Dangers associated with rhabdomyolysis include hyperkalemia and its attendant danger of cardiac arrest, myoglobinuria and its attendant danger of acute renal failure, and shock as a result of sequestration of fluid into injured muscle cells.[20]

Severe heatstroke may be associated with hypophosphatemia and hypocalcemia. Hypophosphatemia is generally observed within hours after onset, but hypocalcemia is usually noted on the second or third day, after hypophosphatemia has undergone spontaneous correction.[21] Hypocalcemia in rhabdomyolysis follows sequestration of calcium, as calcium phosphate, and calcium carbonate in the dead or injured skeletal muscle.[21] Calcium administration should be avoided if possible, since it can cause further sequestration of calcium salts in skeletal muscle.

Hypophosphatemia in patients with heatstroke has been ascribed to respiratory alkalosis and an increased cellular uptake of phosphate. Other possible causes of hypophosphatemia include hyperthermia, diabetes mellitus, starvation, and hypomagnesemia.[22]

Dehydration disrupts circulation of blood and heat dissipation; it also impairs the ability to sweat. It is not obligatory that dehydration be present in heatstroke; in fact, heatstroke may set in before considerable water loss has occurred. Dehydration, although modest, may play a significant role in the development of heatstroke.

Disseminated intravascular coagulation has been observed in a large number of patients with severe heatstroke, particularly the type caused by muscular exertion. Decreased fibrinogen concentration, thrombocytopenia, and elevated fibrin split products occur in the bulk of patients with heatstroke. Purpura is likely to occur on the second or third day. In the event of gross hemorrhage, heparin therapy may be considered. Decreased production of clotting factors in the liver, due to acute liver injury, can account for some of the bleeding diathesis in heatstroke patients.

Patients with non-exercise-induced heatstroke frequently present with hyperglycemia, whereas those with exertion-induced heatstroke may manifest hypoglycemia.

Characteristics and treatment The characteristics of heatstroke and recommended treatment modalities are summarized in Tables 15-5 and 15-6. *Quick and effective reduction of the high body temperature is essential.* Even a few hours delay may leave the patient with severe neurologic deficits. The longer the temperature remains high, the greater the possibility of irreversible brain damage. When the body temperature is above 106°F (41.1°C), damage to the parenchyma of cells throughout the entire body occurs. Especially devastating is the loss of neurons, since neuronal cells, once destroyed, cannot be replaced. When the body temperature reaches 110° to 114°F, (43.3°–45.5°C), the patient can live only a few hours unless the temperature is reduced rapidly. (Once the body temperature has reached 110°F or 43.3°C, the body metabolism has doubled.) Regardless of the degree of temperature elevation, it should be reduced to 102°F (38.9°C) within the first hour of treatment. Recovery from heatstroke depends largely on reducing the degree and duration of fever.

A highly effective method of cooling is immersion in a tub of water to which ice is slowly added. This measure may seem drastic, yet the temperature must be rapidly lowered to 102°F (38.9°C). (The use of a bathtub filled with ice can usually lower body temperature to this level within one hour.) Since the patient may be comatose, or at least disoriented, he must be protected from drowning. The trunk and extremities should be massaged during the bath to bring blood to the periphery for cooling. The body temperature should be checked every 3 to 5 minutes. When it reaches approximately 102°F (38.9°C), the patient should be removed from the ice water bath since, as a rule, the body temperature continues to fall another 3° to 4°F. Reduction of the temperature to 102°F (38.9°C) causes the patient to feel better; slight paralysis is sometimes relieved by the temperature reduction. After removal from the tub, the body temperature should be measured at frequent intervals so that any rise can be noted early. (It may be necessary to repeat hypothermic treatment.) The patient should next be placed in a cool room with low humidity.

Some industries with a high risk of heat disorders are equipped with a latticelike bed to suspend the patient while he is sprayed with cold water and exposed to air movement supplied by an electric fan. A less effective cooling method includes sponging the patient with alcohol or cool water; this method is made more

TABLE 15-5 *Heatstroke: characteristics*

Sudden onset in most cases

May occur with premonitory symptoms (headache, fatigue, and disorientation)

Rectal temperature usually 105.8°F (41°C) or higher

Signaled by very hot, dry skin (Note: patient has usually ceased to sweat)

Presents with varying degrees of impairment of consciousness, such as lethargy, stupor, or coma, due to shunting of blood away from central nervous system

The classic picture is a comatose patient with hot, dry skin (this type occurs most often in the elderly, alcoholic, or diabetic person)

In young athletes, signs of heatstroke may resemble psychiatric illness: euphoria, confusion, belligerence, and assaultiveness[23]

Sinus tachycardia with heart rates of 140–150 is the rule (may approach 170 in severe heatstroke)

Can occur with or without heat exhaustion or heat cramps preceding it

In the elderly, exogenous heat can cause a quietude resembling a senile dementia

Some patients report having felt chilly or "having goose flesh" just before collapsing (piloerection of upper chest and arms)

Hypotension may be present, likely due to pooling of blood in the skin; should diminish as core temperature cools

Dehydration has been reported to be present in 10%–20% of elderly patients[24]

Hyperpnea often present

Electrolyte and acid–base disturbances may include:[25]
    Hypernatremia
    Hypokalemia
    Hyperkalemia
    Hypocalcemia
    Hypophosphatemia
    Respiratory alkalosis
    Lactic acidosis

effective when good air movement is ensured, as with an electric fan. Electrically-controlled hypothermic mattresses have been used successfully in the treatment of heatstroke. Antipyretics are too slow and do not lower the body temperature sufficiently to be of value in the initial treatment of heatstroke.

In the presence of shock, fluids must be administered carefully while the central venous pressure is being monitored. Care should be taken to avoid precipitation of pulmonary edema by too rapid fluid replacement. The initial status of body water and electrolytes depends upon the state of hydration prior to the onset of heatstroke. Some patients have no water or salt loss; others have a severe depletion of salt or water or both. Hypernatremia is present in a large number of the cases and requires treatment with dextrose and water

or hypotonic electrolyte solutions or both. Usually metabolic acidosis is present and may be treated with an isotonic sodium bicarbonate solution. Vasopressors are contraindicated initially, since they cause vasoconstriction and, therefore, interfere with heat dissipation.

Sometimes a phenothiazine derivative (chlorpromazine) is used intravenously for its hypothermic and metabolism-lowering effects and for its sedative action; however, a significant drop in blood pressure may occur with its use, so the blood pressure should be monitored closely.

Usually an indwelling catheter is inserted into the bladder to monitor hourly urine output. If the hourly output is consistently less than 20 ml/hr, mannitol or another diuretic, such as furosemide or ethacrynic acid, may be used in an effort to prevent the develop-

TABLE 15-6 *Heatstroke: treatment*

Reduce core temperature promptly to below 102.2°F (39°C), by ice baths, rubs, or packs; this treatment is crucial and should be instituted immediately

Monitor rectal temperature every few minutes, along with other vital signs

When core temperature drops to 102.°F (39°C), discontinue hypothermic measures and continue to monitor vital signs

Some authorities recommend the use of chlorpromazine (Thorazine), 50 mg, when core temperature drops; otherwise, at about 104°F (40°C) violent shaking may occur

Monitor urinary output closely; remember, oliguria does not necessarily mean that dehydration is present; it can be due to renal failure

Administer IV fluids cautiously; avoid overloading the circulatory system and causing congestive heart failure and pulmonary edema. Remember, most patients with heatstroke are not markedly dehydrated; 500 ml–1000 ml of fluid may be all that is needed.[26] The type of parenteral fluid to be administered must be delineated by clinical laboratory findings since patients may present with low, normal, or high sodium levels.

Avoid antipyretic drugs, such as aspirin or acetaminophen; they are not helpful because their action requires the presence of intact heat-losing mechanisms, and they may be harmful because of their tendency to produce bleeding[27]

Obtain electrocardiogram, particularly if patient is over 50 years of age

Obtain lab work: complete blood count, platelet count, prothrombin and fibrinogen times, blood urea nitrogen, creatinine, glucose, electrolytes, and liver function tests

Observe for possible complications:
    Spontaneous recurrence of hyperthermia
    Acute tubular necrosis
    Myocardial infarction
    Clotting abnormalities
    Hepatic necrosis

Avoid heparin and dextran solutions, since heatstroke patients harbor a risk of bleeding disorders that these agents only compound; coagulation characteristics usually revert to normal as cooling is accomplished[28]

Preventive measures are listed at the end of the chapter

---

ment of acute tubular necrosis. Acute renal failure is reported to occur in 5% of the patients with non-exercise-induced heatstroke and in 40% of those with exercise-induced heatstroke.

After 1 to 3 days of intensive treatment, the sweat glands should become functional again, although it may be as long as 6 months before they begin to secrete normally. The patient should continue to rest in bed and avoid exposure to sunlight for 1 to 2 weeks after temperature reduction.

Emergency measures before medical aid is available Because time is so vital in preventing fatalities from heatstroke, one should take every measure to reduce body temperature as soon as possible. Unfortunately, heatstroke may occur in an area some distance from medical aid; furthermore, facilities for ice water baths or even sponging may not be available. *The following points should be kept in mind to care for the heatstroke victim before medical aid is available:*

1. Move the patient out of the sun to the coolest, best ventilated spot available.
2. Remove most of the patient's clothing.
3. Summon medical aid; if necessary, move the patient to medical aid. The transporting vehicle should be air-conditioned or have all of its windows opened so that a draft can blow on the patient to promote cooling. If moving the patient entails further exposure to high heat stress, it is better to wait until a more suitable means of transportation is available. Additional heat stress could cause death.
4. Investigate surroundings for *any* immediate means of reducing the patient's temperature until more effective measures can be made available. For example, if heatstroke occurs during an outing near a

body of water, the patient may be partially immersed to promote cooling. Or, if a water hose is available, the patient can be sprayed continuously with water. If nothing but a drinking water supply is available, the patient can be sponged with it.

5. Massage the patient's skin vigorously; this maintains circulation, aids in accelerating heat loss, and stimulates the return of cool peripheral blood to the overheated brain and viscera. Body heat may be lost rapidly in this manner.

MEASURES TO PREVENT HEAT DISORDERS

Heat disorders are usually both predictable and preventable by simple measures. The nurse should become familiar with the preventive measures listed below so that she can offer sound advice about heat disorders. The following points should be remembered:

Strenuous activity should be curtailed as much as possible on hot days. Prolonged exercise such as training programs or athletic events should be held in the cooler parts of the day, such as before 8 AM or after 6 PM. Care should be taken to avoid the hours of 11 AM to 2 PM, since these are the hours of greatest solar heat. ("Mad dogs and Englishmen go out in the mid-day sun. . . . ." Noel Coward)

Persons moving from a temperate climate to a hot climate, or those subjected to heat stress in their work, should gradually build up a tolerance to heat through planned acclimatization. Sudden exposure of an unacclimated person to high heat stress invites heat disorders. (See section dealing with acclimatization.)

Persons exposed to heat stress should maintain good physical condition. Sufficient rest and proper food and fluid intake help prevent heat disorders.

Persons exposed to heat should wear loose, porous clothing. Under no circumstances should heavy exercise be performed in the heat while wearing plastic or rubberized clothing.

Persons confined to bed should be protected from excessive bedclothing and hot, poorly ventilated rooms.

The use of salt tablets without adequate water replacement increases risks. Salt supplements in some form are mandatory during the first week or so of conditioning in hot environments; however, *provision of water is much more important.* After acclimatization is achieved, salt loss in perspiration declines. The salt consumed with meals is

generally adequate; thirst is the best guide to water regulation. The use of commercial drinks containing glucose, water, salt, and other minerals is unnecessary but probably harmless.[29] One kind of commercially available salt tablet (Thermotabs, Beecham Products, Pittsburgh, PA) contains sodium chloride, 7 grains (0.45 g), and potassium chloride, 1/2 grain (30 mg); manufacturer's directions state that a full glass of water should be consumed with each tablet.

Potassium deficit has been shown to contribute to some heat disorders. Hence, persons prone to develop potassium deficit—those taking diuretics, for example—should guard against potassium deficit in hot weather. Food intake, and thus potassium intake, usually decreases in hot weather; at the same time, more potassium than normal is lost in heavy sweating. Thus, potassium supplements are indicated. A decrease in the requirement for digitalis, caused by hypokalemia, may render the management of cardiac patients more difficult during their stay in a hot environment. (Recall that hypokalemia potentiates the action of digitalis on the heart.)

Medications predisposing to heat stress should be cautiously diminished or stopped temporarily during prolonged periods of extreme heat (under the direction of a physician). These medications are discussed earlier in this chapter.

Offer regular periods of rest during work or heavy exercise so the person can cool off and drink fluids.

"Islands of coolness" should be provided for workers exposed to constant heat stress in their work. It has been pointed out that the predisposition to heat exhaustion or heatstroke may be related to lack of a recovery period, such as the availability of a cool place to rest periodically.

Water or cool drink dispensers should be placed in convenient locations in high heat stress environments.

Athletes or workers performing heavy exercise in a hot environment should be advised of the early symptoms of heat injury:

Excessive sweating

Abdominal cramps

Headache

Nausea

Dizziness

Cessation of sweating with piloerection on upper chest and arms

Gradual impairment of consciousness

Persons perspiring heavily should be advised to limit water loss to no more than 5% of body weight; a loss of 7% or more is dangerous. (The nurse should go through the calculation with the patient since cognizance of a specific number of pounds means more than a percentage.)

The following facts pertain to competitive runners:[30]

Prior training in the heat can reduce the risk of heat injury

Consumption of 500 ml of water immediately prior to the run and 250 ml at every water station during the run can help forestall the development of heat injury

Competitors should run with a partner and agree to be mutually responsible for the other's well-being during the run

Planners of competitive runs should observe the following:[30]

Plan races to avoid the hottest months of July and August

Choose a course with some shade and provide adequate supplies of fluids every 2 to 3 kilometers

Alert local hospitals of the event

Provide facilities at the race site to treat heat disorders

REFERENCES

1. Sprung C: Heatstroke: Modern approach to an ancient disease. Chest 77, No. 4:462, 1980
2. Sutton J: Letter: Heatstroke from running. JAMA, May 16, 1980, p 189
3. Wheeler M: Heatstroke in the elderly. Med Clin North Am 60, No. 6:1290, 1976
4. Burch G, Knochel J, Murphy R: Stay on guard against heat syndromes. Patient Care, June 30, 1979, p 80
5. Wheeler, p 1293
6. Burch et al, p 69
7. Knochel J: Environmental heat illness: An ecclectic review. Arch Intern Med 133:849, 1974
8. Maxwell M, Kleeman C: Clinical Disorders of Fluid and Electrolyte Metabolism, 3rd ed, p 1530, New York, McGraw Hill, 1980
9. Burch et al, p 75
10. Wheeler, p 1291
11. Haraguchi K: Nurses can take the heat off of workers. Occup Health Saf, July-Aug 1978, p 40
12. Maxwell & Kleeman, p 1531
13. Wheeler, p 1292
14. Costrini A, Pitt H, Gustafson A, Uddin D: Cardiovascular and metabolic manifestations of heatstroke and severe heat exhaustion. Am J Med, 66:301, 1979
15. Sprung C, Portocarrero C, Fernaine A, Weinberg P: The metabolic and respiratory alterations of heatstroke. Arch Intern Med 140:669, 1980
16. Burch et al, p 74
17. Sprung C: Hemodynamic alterations of heatstroke in the elderly. Chest 75, No. 3:365, 1979
18. Shibolet A, Lancaster M, Danon Y: Heatstroke: A review. Aviat Space Environ Med 47:293, 1976
19. Knochel J, Dotin L, Hamburger R: Pathophysiology of intense physical conditioning in a hot climate: Mechanism of potassium depletion. J Clin Invest 51:1750, 1972
20. Maxwell & Kleeman, p 1552
21. Knochel J, Caskey J: The mechanism of hypophosphatemia in acute heatstroke. JAMA 238, No 5:425, 1977
22. Sprung C, Portocarrero C, Fernaine A, Weinberg P: The metabolic and respiratory alterations of heatstroke. Arch Intern Med 140:669, 1980
23. Burch et al, p 73
24. Levine J: Heatstroke in the aged. Am J Med, 47:251, 1969
25. Barcena C et al: Obesity, football, dog days and siriasis: A deadly combination. Am Heart J 92, No 2:239, 1976
26. A healthy case of heatstroke. Emergency Med, July 15, 1979, p 37
27. Wheeler, p 1295
28. Burch et al, p 78
29. Hodges R: Nutrition in Medical Practice, p 181. Philadelphia, WB Saunders, 1980
30. Hughson R: Primary prevention of heatstroke in Canadian long-distance runners. Can Med Assoc J 122:1119, 1980

Among the various transformations in pregnancy, none is more important than the changes in body fluids. Even these, forthright though they appear, pose many unanswered questions. For example, in the nonpregnant woman or in the male, excessive water retention invariably is accompanied by excessive retention of sodium. Whether or not there is excessive sodium retention in pregnancy remains controversial; however, many, perhaps most, clinicians believe that pregnant women do retain excessive sodium. They therefore restrict sodium intake and may even prescribe diuretics. (Diuretic drugs appear to be overused in pregnancy. One authority reports the deaths of four patients from excessive use of diuretics; three of these died from electrolyte depletion and one from hemorrhagic pancreatitis.) Other physicians believe that pregnant women are sodium wasters; they add supplemental salt to the diet to avoid preeclampsia. Both the secretion and excretion of the sodium-conserving hormone aldosterone increase during normal pregnancy; however, the effect of this increase on sodium homeostasis is poorly understood. Were it not, perhaps we could explain the puzzling situation in preeclampsia, in which there is apparent sodium retention, even though aldosterone secretion actually decreases.

In this chapter, we first examine changes in body fluids during normal pregnancy. Knowledge of the normal pregnant state enables the nurse to recognize the borderline between a physiologic and a pathologic change. One will know, for example, what is a physiologic increase in hydration and what is a pathologic increase. Next, we look at disorders of pregnancy closely related to body fluid disturbances.

PHYSIOLOGIC CHANGES IN BODY FLUIDS DURING PREGNANCY

WATER CONTENT

At term, the fetus, placenta, and amniotic fluid contain about 3.5 liters of water. An additional 3 liters has accumulated because of increases in the mother's blood volume, in the size of her breasts, and in the mass of the uterus. The average woman, therefore, retains at least 6.5 liters of extra water in the extracellular compartment during a normal pregnancy. Such hydration of the maternal tissues is physiologic in nature, and the body's physiologic processes handle it with equanimity.

16

Fluid Balance in Pregnancy

The increase in blood volume deserves comment: it averages from 40% to 45% and results from increases in plasma volume and in red cell mass. Plasma volume increases 45% to 50%, or about 1200 ml to 1400 ml, with the maximum reached 2 to 6 weeks before term. During the last weeks of pregnancy, the rate of increase in plasma volume declines. With delivery, plasma volume rapidly diminishes, so that, by the end of the first week postpartum, it has returned to the nonpregnant value. Red blood cell volume increases during pregnancy some 20% to 40%, an addition of 300 ml to 500 ml. The proportionately greater increase in plasma than in red blood cells results in the "hemodilution of pregnancy." (The hemoglobin level may fall to 10.5 g to 12 g from the nonpregnant normal of 12 g to 15 g; the hematocrit may fall to 30% to 33% from the nonpregnant normal to 35% to 45%.) In addition, there occurs a 10% increment in heart rate and as much as a 40% increase in cardiac output, reaching its maximum at from 28 to 32 weeks, then decreasing to term. A natural accompaniment of these phenomena is a linear increase in the consumption of oxygen, peaking at term. Because of these factors, the pregnant woman with heart disease may be hard pressed to meet the strenuous demands imposed on her cardiovascular system.

FLUID RETENTION Now let us review known factors that produce fluid retention during pregnancy:

An increase in venous pressure elevates the effective intracapillary hydrostatic pressure. This results from two mechanisms: first, the pregnant uterus impinges against the inferior vena cava, thus causing increased back pressure; second, the vascular congestion of the pregnant pelvis also increases pressure on the vena cava. The increased venous pressure enhances filtration from the vascular bed and often produces physiologic dependent hydrostatic edema, to be differentiated from the edema of toxemia. When the woman lies on her side, the pressures against the inferior vena cava are relieved, and the venous pressure is not elevated. Also, when the woman lies on her side, some or all of the accumulated fluid may be mobilized and excreted, thus explaining the observation that in late pregnancy the urinary volume at night approaches that excreted during the day.

Plasma albumin decreases by about 1 g/dl of plasma. This reduction in the colloidal osmotic pressure of the plasma amounts to some 20 percent; it favors plasma-to-interstitial fluid shift.

Still controversial is the question of whether excessive sodium is retained by the pregnant woman. Many obstetricians now believe that salt and water restriction in the normal pregnancy is not only unnecessary but also harmful. They would confine such restriction to those patients with a pathologic process.

Aldosterone secretion increases markedly in pregnancy, the rise occurring as early as the 15th week.[1] According to one source, the aldosterone secretion rate is about ten times higher than in nonpregnant women.[2] Increased production of steroid hormones causes increased sodium and water retention; however, the 50% increase in the glomerular filtration rate during pregnancy tends to offset fluid retention so that the normal pregnant woman has only a moderate fluid excess.

How does one go about measuring the retention of water in pregnancy? Our most useful gauge is weight gain, but water retention is also revealed by pitting edema of the ankles and legs—especially at the day's end—due to mechanical obstruction by the enlarging uterus, elevated femoral venous pressure, and the gravity effect produced by the upright position. In fact, edema of the feet and ankles occurs in about 75% of all late pregnancies, especially in the summer. This edema usually disappears overnight and should be regarded as a physiologic rather than as a pathologic phenomenon. Generalized edema may be tested for by finger swelling and tightening of finger rings. Its presence may be significant in the development of preeclampsia. However, one retrospective study showed that 20% of otherwise normal pregnant women displayed some degree of generalized edema.

Management of ankle edema or of weight gain slightly exceeding that normally expected can usually be accomplished by lateral recumbency plus mild sodium restriction. If the physician employs severe sodium restriction, diuretics, or both, the patient is in danger of becoming sodium depleted, perhaps potassium depleted as well. Sodium depletion is particularly unfortunate. Not only is it dangerous *per se*, but its signs (including oliguria, decreased glomerular filtration rate, and increased plasma concentration of uric acid) imitate preeclampsia. This may cause further sodium restriction, leading to more severe sodium depletion.

ACID–BASE BALANCE

The pregnant woman can be considered as having a moderate respiratory alkalosis as a result of hyperventilation, presumably induced by increased progesterone levels. (Progesterone increases the respiratory center's sensitivity to carbon dioxide.) Arterial PCO_2 in normal pregnancy is approximately 30 mm Hg, compared with normal values in the nonpregnant state of 40 mm Hg. Loss of carbon dioxide from the blood is compensated for by renal loss of bicarbonate. The mild respiratory alkalosis and reduced plasma bicarbonate have little significance except in cases in which severe metabolic acidosis (such as ketoacidosis) occurs; in this situation, the normal compensatory mechanisms for correction of the metabolic acidosis may be overtaxed.

With overbreathing induced by labor, the maternal PCO_2 can fall below 17 mm Hg, resulting in a delay in the initiation of respiration in the newborn infant.

CALCIUM LEVELS

Some 30 g to 40 g of calcium are deposited in the fetus, chiefly during the last trimester of pregnancy. Maternal total serum calcium concentration decreases during pregnancy, being lowest in the eighth month of gestation. The fall in calcium may be due to dilution by increased extracellular fluid volume and to hypoalbuminemia that develops in pregnancy. The average present day pregnancy diet, containing from 1.5 g to 2.5 g of calcium, appears quite adequate to supply the needs of mother and fetus without depletion of the maternal stores. Possibly because of the high phosphate content of milk, some patients—especially heavy milk drinkers—may suffer an imbalance in the ratio of calcium to phosphate in the plasma. (Recall that calcium and phosphate are antagonistic. An increase in plasma phosphate tends to decrease plasma calcium, and vice versa.) This imbalance in the calcium-phosphate ratio may be responsible for the leg cramps that sometimes occur in pregnancy. Indeed, some clinicians have reported that the leg cramps can be relieved by reducing the milk intake and administering supplemental calcium.

SODIUM LEVEL

At term, the plasma sodium concentration is approximately 5 mEq/liter less than in the nonpregnant state. The decrease in the sodium level is due primarily to excessive water retention; this phenomenon is most pronounced in the last trimester and is thought to be related to an excessive antidiuretic hormone (ADH) level, the cause of which is not known.[3] It is known that oxytocin has an antidiuretic effect; increased secretion of this substance in late pregnancy could be related to the mild hyponatremia. (See the section on water intoxication during induction of labor with oxytocin later in this chapter.)

PREGNANCY DISORDERS CLOSELY RELATED TO BODY FLUID DISTURBANCES

TOXEMIA OF PREGNANCY

The syndrome of toxemia of pregnancy has no known origin. Some theorize that toxemia is the result of autoimmunity or allergic reaction caused by the fetal presence. Lending credence to this theory is the fact that symptoms disappear within a few days after delivery. Some feel that malnutrition, especially reduced protein intake, may play a role.[4] Toxemia can be divided into two types, depending upon the severity; preeclampsia, or toxemia without convulsions; and the extremely serious eclampsia, or toxemia with convulsions. Patients with underlying vascular or renal disease are not included under the diagnosis of toxemia of pregnancy yet they are often difficult to distinguish from patients with toxemia, especially during the last trimester.

Approximately 10% of pregnant women in the US develop preeclampsia-eclampsia.[4] Five percent of patients with preeclampsia progress to eclampsia; 10% to 15% of those who develop eclampsia die.[4] Although toxemia may occur earlier, it usually begins after the 32nd week of pregnancy. Toxemia can also occur postpartum (usually 24 to 48 hours after delivery), with hypertension and convulsions; it has been reported to occur as many as 7 days postpartum. Toxemia is most frequent in young primiparas; it also occurs in older multiparas. When toxemia occurs without proteinuria, particularly in the multiparous woman, it may be an early sign of essential hypertension. A twin pregnancy or hydramnios appears to predispose to it, as may preexisting hypertension. About one third of the women who have toxemia will develop the disease in a subsequent pregnancy.[5] It seems the disease occurs more frequently in lower socioeconomic classes in which obesity and short stature are more common.

While the cause of toxemia remains unknown, abnormal sodium retention and generalized vasoconstriction explain the signs and symptoms.

Widespread vasoconstriction of arterioles affects the placental circulation, kidneys, and eye grounds. This vasoconstriction appears to be related to the increased blood pressure and may account for the visual problems experienced in severe eclampsia. Small degenerative infarcts appear in the placenta, apparently caused by vasoconstriction; the damaged placenta may separate prematurely or may fail to nourish the fetus adequately. A slight decrease in renal blood flow and glomerular filtration rate occurs in the toxemic patient. (Recall that in normal pregnancy, both the glomerular filtration rate and the renal blood flow increase.) In addition to the above changes, the toxemia patient develops fibrinoid deposits in the basement membrane of the glomerular tufts.

It appears incorrect to think that control of weight gain and limitation of sodium intake reduce the incidence of preeclampsia; this belief confuses cause and effect. Although preeclamptic patients retain water and sodium, weight gain does not cause preeclampsia. Certainly endocrine or metabolic disorders or both may be implicated in the genesis of the disease.

SIGNS AND SYMPTOMS The signs of the onset of toxemia include edema, not only of the ankles but also of the hands and face; hypertension; and, in some patients, proteinuria. A significant rise in blood pressure consists of an increase in systolic pressure greater than 15 mm Hg and in the diastolic greater than 10 mm Hg.

Proteinuria becomes significant when it exceeds 0.3 g/liter. (Proteinuria in toxemia can range from as much as 10 g–30 g/24 hr to as little as 0.5 g/24 hr.[6]) A weight gain of more than 5 lb/week or a blood pressure higher than 140/90 should cause one to consider early toxemia. However, it is important to remember that almost half of the patients in whom toxemia develops do not display excessive weight gain. The blood pressure in preeclampsia or eclampsia rarely exceeds 190/115; in fact, the systolic pressure is below 160 mm Hg in most cases. A systolic pressure greater than 200 mm Hg usually indicates underlying essential hypertension.

More advanced symptoms include generalized edema, headache, irritability, visual disturbances, epigastric pain, oliguria, nausea, and vomiting—a cluster of symptoms warning of the approach of eclampsia. Headache and irritability are due in part to cerebral edema; visual impairment may be due to retinal edema, hemorrhage, and even detachment. Eclampsia is heralded by generalized tonic-clonic convulsions. While apprehension and hyperreflexia often precede the convulsion, they do not always do so; an aura usually does not precede the convulsion.

Laboratory findings in eclampsia may include hemoconcentration, elevated blood urea nitrogen (BUN), and elevated serum uric acid. Hemoconcentration of toxemia is apparently related to the vasoconstrictive state; administration of a vasodilator allows remobilization of sequestered fluid from the interstitial space back to the vascular space.[5] The plasma uric acid level rises in toxemia due to a decrease in urate clearance.

With eclampsia, death can occur during a convulsion, or it can result from cerebral hemorrhage, which is reported to be the most frequent cause of death in eclampsia. Autopsies on patients who die of eclampsia reveal hemorrhagic lesions in the placenta, liver, kidneys, heart, brain, spleen, adrenals, and pancreas. In addition, necrosis, tissue infarction, fibrin deposits, and evidence of disseminated intravascular coagulation (DIC) may be found. Pallor of the kidney cortex may be observed, with little blood in the glomerular capillaries.

The fetus may also be seriously threatened, either by the toxemia or by the eclamptic convulsions. Should the pregnancy be terminated either therapeutically or naturally, symptoms of toxemia promptly diminish.

TREATMENT AND NURSING INTERVENTIONS The incidence of toxemia can be reduced by proper prenatal care, with attention given to monitoring of blood pressure. As stated earlier, a rise in systolic blood pressure greater than 15 mm Hg or diastolic pressure greater than 10 mm Hg is significant.

Some physicians will manage preeclamptic women at home with bed rest, sedation, and sodium restriction, provided there is a mechanism for monitoring the blood pressure and intake and output, plus daily urine tests for proteinuria. If improvement does not occur in 48 hours, hospitalization is urgently indicated.

The presence of proteinuria is considered by many physicians to be an indication for hospitalization; during hospitalization, the course of the hypertension, proteinuria, and renal function can be closely monitored.

The hospitalized preeclamptic patient should have

a private room free of bright lights to prevent photic stimuli. No visitors should be allowed. Nursing care should be planned to provide regular uninterrupted rest periods. Maternal vital signs and fetal heart tones should be checked at least every 4 hours. An accurate intake–output record must be kept, and daily body weight measurement should be recorded. Daily weights are a guide to the extent of the diuresis induced by diuretic therapy and salt restriction. There is usually no need for an indwelling catheter in the conscious patient; however, in comatose or convulsing patients, a catheter is indicated to carefully evaluate urine output. The nurse must be especially alert to a rise in blood pressure, hyperreactive reflexes, a weight increase, or a low urinary output and should promptly report any of these.

The patient with mild toxemia (blood pressure no higher than 140/90) may be treated with bed rest, salt restriction to 0.5 g to 1.0 g/24 hours, sedation with phenobarbital or diazepam, and a diuretic (such as hydrochlorothiazide or furosemide) if edema is present.[7] Some authorities avoid the use of diuretics unless pulmonary congestion is present.[8] If therapy is successful in reducing the hypertension and proteinuria, and there is no sign of central nervous system hyperexcitability or renal dysfunction, the patient may be cautiously followed as an outpatient. In severe toxemia in which the diastolic blood pressure is higher than 110 mm Hg, parenteral antihypertensive therapy (such as intravenous hydralazine) is indicated.

Delivery of the fetus is indicated in severe toxemia after the blood pressure and convulsions have been controlled; convulsions increase both fetal and maternal mortality. Fetal bradycardia is common immediately after a convulsion, probably owing to hypoxia and acidosis induced by heavy muscular activity.[9] Edema and proteinuria usually disappear within a week after delivery.[10]

Some obstetricians prescribe magnesium sulfate primarily for its anticonvulsant effect but also for its antihypertensive effect. Magnesium sulfate controls convulsions by blocking neuromuscular transmission; it depresses the central nervous system and produces an initial hypotensive effect because of peripheral vasodilating effect. (Hypermagnesemia may cause flushing and a sensation of heat.) Some authorities feel that a plasma magnesium level of 4.2 mEq to 5.8 mEq/liter is the desired therapeutic range in the convulsion-prone toxemic patient. Deep-tendon reflexes may be depressed when the plasma level exceeds 4 mEq/liter;

the patellar reflex disappears when the plasma magnesium level reaches 8.3 mEq to 10.0 mEq/liter. Respiratory paralysis can occur when the concentration reaches 10.0 mEq to 12.5 mEq/liter, hence the importance of frequent checks on the patellar reflex. Concentrations above 13 mEq/liter can cause electrocardiogram (EKG) changes (prolonged P-R interval and widened QRS complex); heart block and death may occur.

Magnesium sulfate is given IV or IM to toxemia patients, since oral doses fail to produce satisfactory blood levels. The action of IV magnesium sulfate is immediate and lasts about 30 minutes; IM doses do not become effective until about one hour after administration and last about 3 to 4 hours.[11] Since IM doses of magnesium sulfate are painful, they should be given deep in the upper outer quadrant of the buttocks. The injection site should be massaged well to encourage absorption; large doses should be equally divided between both buttocks.

After the initial dose of magnesium sulfate, subsequent doses should be given only under the following circumstances:

> There is no respiratory depression; as a rule, the physician should be notified if the respiratory rate decreases below 14 to 16/minute.
>
> Urine flow has been 100 ml or more in the previous 4 hours; 99% of the magnesium administered parenterally is excreted by the kidneys.[12] Oliguria leads to an accumulation of magnesium in the bloodstream.
>
> The patellar reflex is present; a poor to absent patellar reflex may be indicative of hypermagnesemia and should be reported to the physician.
>
> The patient is oriented to person, place, and time; excessive central nervous system depression can cause drowsiness, lethargy, slurring of speech, and eventually coma.
>
> There has been no significant drop in blood pressure or in the maternal or fetal heart rates.

To counteract possible magnesium toxicity, an ampule of 10% calcium gluconate, with a 20-ml syringe, should be kept at the bedside. Calcium is an antidote for magnesium excess, since calcium and magnesium are mutually antagonistic. (Magnesium blocks release of acetycholine at the neural endplate; calcium counteracts this effect by increasing the release of acetycholine.) The intravenous administration of 5 mEq to 10 mEq of calcium usually reverses respiratory depression and heart block.[13] Equipment for respiratory resuscitation should be available for emergency use.

Magnesium sulfate should be used cautiously in patients with impaired renal function and in those receiving digitalis.[14] Magnesium blood levels of 6 mEq to 8 mEq/liter may reduce uterine contractions and prolong labor.

The convulsing patient should be turned on her side to avoid aspiration of vomitus and mucus. A rubber airway may be inserted between the teeth to prevent biting the tongue and to maintain respiratory exchange. Suction apparatus should be on hand to aspirate fluid from the airway if necessary. If oxygen is necessary, it should be given by nasal prongs (masks and nasal catheters may produce excessive stimulation).[4] Padded siderails should be used to prevent the convulsing or heavily sedated patient from falling or otherwise injuring herself. The duration and character of each convulsion as well as the depth of coma that follows should be noted. Fetal heart tones should be checked as often as possible. The convulsing patient should be observed for signs of rapid labor; abruptio placentae and excessive bleeding may occur. Typed and cross-matched whole blood should be available. It should be remembered that a heavily sedated patient may also have a rapid labor.

HYPEREMESIS GRAVIDARUM (PERNICIOUS VOMITING OF PREGNANCY)

Half of all pregnant women become nauseated or vomit in the first trimester of pregnancy, but true pernicious vomiting of pregnancy has decreased greatly in recent years. Approximately 0.2% of pregnant women develop hyperemesis gravidarum and require hospitalization.[15] The diagnosis of true pernicious vomiting of pregnancy applies only to those patients who have lost large amounts of weight and suffer severe body fluid disturbances, including extracellular fluid volume deficit, acid–base disturbances, and potassium deficit.

Pernicious vomiting of pregnancy frequently—perhaps usually—is brought about by psychic factors. In some instances, however, liver disease may be associated with it.

It has been suggested that the nausea may be due to the large quantity of estrogen secreted by the placenta. Some have postulated that it is related to high levels of chorionic gonadotropin; this substance peaks at the tenth week, much as does the vomiting of hyperemesis gravidarum. Others feel it may be related to

decreased gastric motility and hypochlorhydria. Characteristically, the ailment starts as simple morning sickness, in which the typical patient feels nauseated when she awakens and may be unable to eat breakfast. By noon, however, her symptoms have disappeared, and she remains symptom-free until the next morning. Morning sickness usually subsides without treatment by the 14th to 16th week but, in some patients, morning sickness may develop into nausea and vomiting lasting throughout the day. This is called pernicious vomiting or hyperemesis gravidarum, which usually appears during the fifth or sixth week of pregnancy. It can last a month or two or even longer and produce a weight loss of 10 lb to 20 lb or more. Ketonuria appears as a result of starvation. (Metabolic alkalosis rarely occurs, even when vomiting is severe, since most women in early pregnancy have hypochlorhydria or achlorhydria.)

The BUN and uric acid rise slightly. Plasma chloride and CO_2 combining power decrease. Plasma potassium is likely to decline. Urinary output falls, but the concentration of the urine rises.

An ominous complication of hyperemesis gravidarum is hemorrhagic retinitis. With its appearance, the mortality rate reaches 50%.

TREATMENT AND NURSING INTERVENTIONS The patient with pernicious vomiting should be hospitalized in a private room at complete bed rest. Rest is of primary importance; visitors should not be permitted during the first day or so. Because psychic factors may play a large role in this condition, psychologic counseling may be indicated. Opportunities should be provided for the patient to verbalize her feelings and concerns in a trusting atmosphere that offers support and reassurance.

The use of sedative and antinauseant preparations deserves consideration in managing the condition, although some authorities minimize the use of drugs because of the danger of teratogenicity. In general, it is probably best to give the smallest dose that is clinically effective. Agents sometimes used to control nausea include pyridoxine hydrochloride and meclizine.

Vomiting may be so severe that fluid volume deficit, electrolyte, and acid–base disturbances develop. Prompt correction of fluid and electrolyte disturbances usually relieves the symptoms.[16] As a rule, the patient is given nothing by mouth for 48 hours and is maintained during this period on intravenous fluids. Electrolyte and glucose solutions may be administered in amounts of 3 liters or more in 24 hours. An accurate intake and output record should be kept. If there is no improvement after 48 hours, nasogastric tube feedings may be administered by *slow* drip. The tube feeding mixture must be introduced very slowly to avoid nausea. (Nursing responsibilities in tube feedings are discussed in Chapter 5.) As soon as tolerated, small dry feedings (such as toast or crackers) are alternated with small amounts of liquids. Usually hot tea or cold gingerale is better tolerated than water; lukewarm liquids are not well accepted. Food intake is gradually increased to a full diet. Trays should be prepared carefully; portions should be small and served attractively. Unpleasant odors or sights should be avoided because the slightest stimulus can initiate nausea and vomiting.

Repeated ophthalmoscopic examinations should be performed to detect early retinitis. Should hemorrhagic retinitis appear, the pregnancy should be promptly terminated. (Retinal hemorrhages and retinal detachment are unfavorable prognostic signs.) Today, therapeutic abortion is rarely required.

WATER INTOXICATION RELATED TO INDUCTION OF LABOR WITH OXYTOCIN

Induction of labor with oxytocin, particularly when delivered in an appreciable amount of an aqueous solution (such as D_5W), can lead to serious water intoxication. The pharmacologic action of oxytocin includes a potent antidiuretic effect, causing increased water retention. Whenever 20 mU/minute or more of oxytocin is infused, free water clearance by the kidney decreases appreciably.[17] A commonly used solution used for induction of labor includes 10 units of oxytocin in 1000 ml of D_5W, supplying 10 mU of oxytocin/ml. (Some physicians prefer an even more dilute solution.) *Because of the danger of water intoxication, it is preferable to mix the oxytocin in a balanced salt solution, such as lactated Ringer's solution.* To avoid accidental infusion of too much oxytocin, it is advisable to deliver the solution through a reliable infusion pump. Rarely is it necessary to exceed a rate of 20 mU/minute.[17]

Symptoms of water intoxication include the following:

Behavior changes

Headache

Blurred vision

Nausea and vomiting

Convulsions

Coma

One case was reproted in which a primigravida developed a convulsion during delivery, after receiving 4.5 liters of 5% dextrose in water with oxytocin within a 3 1/2-hour period.

The nurse should carefully monitor the rate of infusion of oxytocin, particularly when it is administered in D_5W, and should be on the alert for signs of water intoxication; it is wise to monitor plasma sodium levels during treatment with oxytocin, particularly when it is administered in dextrose/water solutions.

REFERENCES

1. Maxwell M, Kleeman C: Clinical Disorders of Fluid & Electrolyte Metabolism, 3rd ed, p 1385. New York, McGraw-Hill, 1980
2. Watanabe M, Meeker C, Gray M, Sims A, Solomon S: Secretion rate of aldosterone in normal pregnancy. J Clin Invest 42:1619, 1963
3. Maxwell & Kleeman, p 1386
4. Krupp M, Chatton M: Current Medical Diagnosis & Treatment, p 478. Los Altos, Lange Medical Publications, 1980
5. Maxwell & Kleeman, p 1395
6. *Ibid*, p 1393
7. *Ibid*, p 1398
8. Pritchard J, MacDonald P: Williams Obstetrics, 16th ed, p 690. New York, Appleton-Century-Crofts, 1980
9. *Ibid*, p 691
10. *Ibid*, p 686
11. Butts P: Magnesium sulfate in the treatment of toxemia. Am J Nurs, Aug 1977, p 1295
12. Pritchard & MacDonald, p 688
13. Butts, p 1297
14. Gahart B: Intravenous Medications, 3rd ed, p 129. St. Louis, CV Mosby, 1981
15. Krupp & Chatton, p 475
16. Pritchard & MacDonald, p 323
17. *Ibid*, p 790

Water and electrolyte disturbances occur more frequently in children than in adults. Although one recognizes and manages fluid imbalances in children in much the same way as in adults, there are also important differences. The younger the child, the more pronounced are the differences. The nurse should understand the peculiar problems posed by the child with a body fluid disturbance.

An obvious and important difference between small children and adults is size. Yet, *children are not merely miniature adults,* for the child's body composition and homeostatic controls differ from those of the adult. It is helpful to compare the child's body composition with that of the adult and to review the salient characteristics of the child's homeostatic and metabolic functioning.

DIFFERENCES IN WATER AND ELECTROLYTE BALANCE IN INFANTS, CHILDREN, AND ADULTS

BODY WATER CONTENT The premature infant's body is approximately 90% water; the newborn infant's body, 70% to 80%; the adult's body about 60%. The infant has proportionately more water in the extracellular compartment than does the adult. For example, 40% of the newborn infant's body water may be in the extracellular compartment, as compared to less than 20% in the case of the adult.

As the infant becomes older, his total body water percentage decreases, possibly a result of progressive growth of cells at the expense of the extracellular fluid. The decrease is particularly rapid during the first few days of life, but continues throughout the first 6 months. After the first year, the total body water is about 64% (34% in the cellular compartment and 30% in the extracellular compartment). By the end of the second year, the total body water approaches the adult percentage of approximately 60% (36% in the cellular compartment and 24% in the extracellular compartment). At puberty, the adult body water composition is attained. For the first time, there is a sex differentiation: females have slightly less water because they have a higher percentage of body fat.

DAILY BODY WATER TURNOVER IN INFANTS AND ADULTS The infant's relatively greater total

17
Fluid Balance Disturbances in Infants and Children

body water content does not always protect him from excessive fluid loss. *On the contrary, the infant is more vulnerable to fluid volume deficit than is the adult* because he ingests and excretes a relatively greater daily water volume. An infant may exchange half of his extracellular fluid daily, whereas the adult may exchange only one sixth of his in the same period. Proportionately, therefore, the infant has less reserve of body fluid than does the adult.

The daily fluid exchange is relatively greater in infants, in part because their metabolic rate is two times higher per unit of weight than that of adults. Owing to the high metabolic rate, the infant has a large amount of metabolic wastes to excrete. Because water is needed by the kidneys to excrete these wastes, a large urinary volume is formed each day. Contributing to this volume is the inability of the infant's immature kidneys to concentrate urine efficiently. In addition, relatively greater fluid loss occurs through the infant's skin because of his proportionately greater body surface. The premature has approximately five times as much body surface area in relation to weight, and the newborn, three times, as do the older child and adult. Therefore, any condition causing a pronounced decrease in intake or increase in output of water and electrolytes threatens the body fluid economy of the infant. According to Gamble, an infant can live only 3 to 4 days without water, while an adult may live 10 days.

ELECTROLYTE CONCENTRATIONS AND METABOLIC ACID FORMATION Plasma electrolyte concentrations do not vary strikingly among infants, small children, and adults. The plasma sodium concentration changes little from birth to adulthood. Potassium concentration is higher in the first few months of life than at any other time, as is the plasma chloride concentration. Magnesium and calcium are both low in the first 24 hours after birth. The serum phosphate level is higher in infants and children than in adults. Inability of the premature infant to regulate his calcium ion concentration can bring on hypocalcemic tetany.

Newborn and premature infants have less homeostatic buffering capacity than do older children. They have a tendency toward metabolic acidosis, with pH averages slightly lower (7.30–7.38) than normal.[1] The mild metabolic acidosis (base bicarbonate deficit) is thought to be related to high metabolic acid production and to renal immaturity. The premature infant is even more acidotic than the newborn. Because cow's milk has higher phosphate and sulfate concentrations than breast milk, newborns fed cow's milk have a lower pH than do breast-fed babies.

KIDNEY FUNCTION The newborn's renal function is not yet completely developed. Thus, if infant and adult renal functions are compared on the basis of total body water, the infant's kidneys appear to become mature by the end of the first month of life. However, if body surface area is used as the criterion for comparison, the child's kidneys appear immature for the first 2 years of life. Since the infant's kidneys have a limited concentrating ability and require more water to excrete a given amount of solute, he has difficulty in conserving body water when it is needed. Also, he may be unable to excrete an excess fluid volume.

BODY SURFACE AREA Until the child is 2 or 3 years old, his body surface area is proportionately greater than that of the adult. The skin represents an important route of fluid loss, especially in illness. Since the gastrointestinal membranes are essentially an extension of the body surface area, their area is also relatively greater in the young infant than in the older child and the adult. Hence, relatively greater losses occur from the gastrointestinal tract in the sick infant than in the older child and adult. *In comparing fluid losses in infants with those in adults, one might regard the baby as a smaller vessel with a larger spout.*

WATER REQUIREMENTS Since infants and children have higher metabolic rates than adults, they need proportionately more water. Water needs for various age-groups are listed in Table 17-1.

ELECTROLYTE DISTURBANCES IN CHILDREN

SODIUM IMBALANCES

HYPONATREMIA Recall that the normal concentration of plasma sodium ranges from 137 mEq to 147 mEq/liter (normals may vary slightly from laboratory to laboratory). Hyponatremia may result from excessive loss of sodium, as occurs in gastroenteritis.

Cystic fibrosis patients subjected to heat stress lose excessive amounts of NaCl through sweating; hyponatremia is likely to result if salt-free liquids are used to

TABLE 17-1 *Mean ranges of daily water requirements of children at different ages under ordinary conditions*

Age	Average body weight (kg)	Total H₂O requirements per 24 hours (ml)	H₂O requirements per kg in 24 hours (ml)
3 days	3.0	250–300	80–100
10 days	3.2	400–500	125–150
3 months	5.4	750–850	140–160
6 months	7.3	950–1100	130–155
9 months	8.6	1100–1250	125–145
1 year	9.5	1150–1300	120–135
2 years	11.8	1350–1500	115–125
4 years	16.2	1600–1800	100–110
6 years	20.0	1800–2000	90–100
10 years	28.7	2000–2500	70–85
14 years	45.0	2200–2700	50–60
18 years	54.0	2200–2700	40–50

(Nelson W (ed): Textbook of Pediatrics, 11th ed, p 175. Philadelphia, WB Saunders, 1979)

replace the lost body fluid. (Recall that patients with cystic fibrosis have high [60 mEq–120 mEq/liter] concentrations of NaCl in their sweat.[2]) Provision must be made for extra salt intake for cystic fibrosis patients during periods of sustained heat stress.

Hyponatremia may also result from water overloading. This situation can be caused by overadministration of electrolyte-free parenteral fluids to the pediatric patient, particularly when impaired renal function exists.

HYPERNATREMIA Hypernatremia may result from decreased intake or increased output of water or from increased intake or decreased output of sodium. Infants, very young children, and unconscious or retarded patients are unable to communicate thirst; therefore, they may develop hypernatremia, particularly if excessive water losses are sustained (as in prolonged fever, tracheobronchitis, and diabetes insipidus). Hydrocephalus or other neurologic conditions may disrupt the thirst center in the hypothalamus, causing inadequate water intake.

The accidental substitution of salt (NaCl) for sugar in formula preparation results in a disastrously high sodium intake. (Widespread use of commercially prepared formulas has decreased the likelihood of this problem.) A grossly hypertonic formula causes the infant to cry; if his cry is interpreted as indicating hunger, more of the hypertonic formula is given and the condition worsens. Symptoms accompanying the excessive ingestion of sodium include these:

Dry, sticky mucous membranes
Avid thirst
Irritability when disturbed (otherwise lethargy)
Tremors and convulsions
Nuchal rigidity
Muscle rigidity
Elevation of protein and chloride concentrations of the spinal fluid
Expansion of the extracellular fluid
Visible edema
Brain damage in some patients

(Hypertonic dehydration [fluid volume deficit with sodium excess] caused by severe watery diarrhea is discussed later in this chapter.)

High-protein tube feedings, plus inadequate water intake, can produce hypernatremia regardless of whether or not the feeding solution contains an excessive amount of salt. High-solute feedings act as osmotic diuretics and "pull" water out of the body, particularly in infants because of their immature renal function. (See the section on tube feedings later in this chapter.)

The premature infant sometimes requires the assistance of a ventilator and an artificial heating device. It should be remembered that increased water loss can occur with the use of low-humidity assisted ventilation. Contributing to the water loss (through the skin in this instance) is the use of an artificial heating device. Excessive water loss can, of course, result in hypernatremia.

POTASSIUM IMBALANCES

HYPOKALEMIA Recall that the normal plasma potassium level ranges from 4.0 to 5.6 (normals vary slightly from laboratory to laboratory). Excessive loss of potassium-rich body fluids, such as occurs in vomiting and diarrhea, results in hypokalemia. Other causes include use of potassium-losing diuretics, adrenocorticosteroids, and excessive sweating. Potassium depletion is often associated with metabolic alkalosis.

Symptoms include apathy, abdominal distention, and muscular weakness; severe hypokalemia causes flaccid paralysis and cardiac arrhythmias.

Hypokalemia is treated by potassium replacement, either orally or intravenously. Potassium-rich fruit juices include orange, tomato, prune, grapefruit, and pineapple. (See Table 5-4 for precise potassium content of various foods.) As always, great care should be exercised when potassium is administered intravenously. The concentration of potassium should not exceed 20 mEq to 40 mEq/liter; the suitably diluted solution should be given at the precise prescribed rate. A microdrip and volume-controlled pediatric set should be used to limit the chance of accidental excessive fluid administration. The normal potassium requirements are approximately 2 mEq/kg/24 hr, or 50 mEq/m^2/24 hr. As a rule, no more than 4 mEq/kg of potassium should be given daily to correct hypokalemia, unless a severe deficit is present. Potassium solutions should not be administered to anuric or oliguric children since dangerous hyperkalemia can result. Suitably diluted potassium solutions should be administered evenly over the 24-hour period; they should never be given undiluted or by IV push. (Excessive or too rapid administration of potassium can cause cardiac arrest.)

HYPERKALEMIA Hyperkalemia may be caused by renal failure, hemolysis, Addison's disease, and excessive, or too rapid, administration of potassium solutions. Symptoms may include muscle weakness, paresthesia, and flaccid paralysis. Cardiac changes are the principal manifestations of hyperkalemia and may prove lethal. (These changes are described in Chapter 11.)

Calcium administration relieves the cardiotoxic effects of hyperkalemia, temporarily, without actually decreasing the plasma potassium level. Sodium bicarbonate alkalinizes plasma and drives potassium into the cells. Dextrose and insulin cause potassium to enter the cells by the process of glycogen formation.

CALCIUM IMBALANCES

HYPOCALCEMIA Recall that the normal concentration of calcium in plasma ranges from 4.5 mEq to 5.8 mEq/liter (normals may vary slightly from laboratory to laboratory). Hypocalcemia may occur in the pediatric patient for a variety of reasons. It is more apt to occur following abnormal pregnancies for reasons that are not clear; sometimes it occurs in the presence of other diseases or may be an isolated finding. Transient hypoparathyroidism, with hypocalcemia, may occur in the offspring of hyperparathyroid or diabetic mothers.

Prolonged hypocalcemia may produce a variety of symptoms in the young child. Symptoms include numbness, twitching, cramps, carpopedal spasm (Fig. 12-2), positive Chvostek's sign, laryngospasm, positive peroneal sign (tapping the peroneal nerve over the fibular side of the leg produces abduction and dorsiflexion of the foot), irritability, convulsions, and retarded physical and mental growth. Other symptoms of prolonged hypocalcemia may include poor dentition, photophobia, conjunctivitis, and cataracts.

A disorder called "tetany of the newborn," occurring in the third or fourth week of life, is sometimes seen in infants fed a milk formula with a high phosphate/calcium ratio (such as cow's milk).

High phosphate intake from large quantities of cow's milk tends to raise the phosphate concentration in the bloodstream and, thus, to lower the calcium level. (Recall that a reciprocal relationship exists between calcium and phosphate; a rise in one causes a decrease in the other.) It is not known why only a few babies fed cow's milk are subject to this disorder; possible reasons include immature renal function and relative hypoparathyroidism. Treatment includes lowering the solute and phosphate loads in the formula and providing extra calcium.

Babies with low serum calcium levels usually are symptom-free; however, twitching or frank convulsions may occur. Calcium can be replaced either orally, as calcium lactate or gluconate, or intravenously. The intravenous route is used only for severe hypocalcemia, since it can cause serious cardiac problems. A suitably diluted calcium preparation is infused slowly while the heart rate is being monitored; bradycardia is a sign to stop the infusion immediately. (Recall that cardiac slowing and arrest can result from rapid administration of calcium intravenously.) Calcium salts should not be added to a solution containing sodium bicarbonate, since an insoluble precipitate (calcium carbonate) will form.

HYPERCALCEMIA A brief period of hypercalcemia may cause abdominal pain, nausea, and vomiting; prolonged hypercalcemia may cause precipitation of calcium stones in the renal system.

A rare condition known as idiopathic hypercalcemia may occur in children. Although its cause is unknown, some believe it might be due to increased intake of vitamin D during pregnancy, abnormal sterol synthesis, or a defect in vitamin D metabolism. Idiopathic hypercalcemia sometimes occurs in a severe form characterized by elfin facies, prominent lips and eyes, hanging jowls, large low-set ears, motor and mental retardation, hypotonia, polyuria, polydipsia, and cardiac defects. Milder forms have a favorable prognosis, but severe forms may be fatal. Treatment consists of limiting dietary calcium and vitamin D intake; if necessary, adrenocorticosteroids may be used to interfere with calcium absorption from the gut.

MAGNESIUM IMBALANCES

HYPOMAGNESEMIA Recall that the normal plasma magnesium concentration is 1.4 mEq to 2.3 mEq/liter. Hypomagnesemia may be seen in the newborn as a familial condition or with hypoparathyroidism. It may also be a problem in artificially fed neonates. Symptoms may include positive Chvostek's sign, irritability, tremors, and especially in the newborn, convulsions.

Magnesium deficiency simulates calcium deficiency. Magnesium sulfate may be administered either orally or parenterally as indicated to correct hypomagnesemia.

HYPERMAGNESEMIA Hypermagnesemia may be seen rarely in the newborn after the mother has received magnesium sulfate. It is particularly apt to occur when the mother received frequent and repetitive doses of magnesium sulfate before delivery. It may also be seen in children in renal failure. (Recall that magnesium is normally excreted by the kidneys; renal failure causes magnesium buildup in the bloodstream.)

Symptoms of hypermagnesemia do not usually appear until the magnesium concentration in the plasma exceeds 4 mEq/liter; drowsiness and hypotension may be early signs. Loss of deep-tendon reflexes can occur when the plasma magnesium concentration is greater than 7 mEq/liter; respiratory failure and heart block can occur when the concentration exceeds 10 mEq/liter. The newborn with hypermagnesemia has a weak or absent cry. Usually, treatment of hypermagnesemia is not necessary unless the plasma level exceeds 7 mEq/liter; in this case, dialysis may be indicated.

pH DISTURBANCES

RESPIRATORY ALKALOSIS Respiratory alkalosis may occur in pediatric patients as a result of hysterical hyperventilation, salicylate intoxication, and abnormal irritability of the respiratory center caused by encephalitis or meningitis. Symptoms may include circumoral paresthesia, dizziness, palpitations, chest discomfort, and convulsions secondary to hypocalcemia (recall that calcium ionizes poorly in alkalosis). The patient's respirations are increased, causing excessive loss of carbon dioxide (the primary disturbance). In respiratory alkalosis, the increased respirations are the *cause* of the imbalance. Plasma pH is elevated (may be above 7.6) and the PCO_2 is decreased (sometimes as low as 10 mm Hg). Treatment depends on the condition causing respiratory alkalosis. The hysterical hyperventilating child may be treated by rebreathing into a bag and by the administration of tranquilizers; psychological help may also be indicated. (Salicylate intoxication treatment is described later in the chapter.)

RESPIRATORY ACIDOSIS Chronic respiratory acidosis may occur in cystic fibrosis, advanced muscular dystrophy, asthma, rickets, and bulbar poliomyelitis. The patient may be dyspneic, confused, and uncooperative. Plasma pH is slightly below normal and the PCO_2 is elevated as a result of hypoventilation. The base excess is above normal owing to renal conservation of bicarbonate to compensate for the excess carbonic acid concentration in the blood.

Acute respiratory acidosis may result from upper airway obstruction (croup or aspiration of a foreign body), hyaline membrane disease, respiratory failure, or cardiac arrest. Acute air hunger is present and the child often appears cyanotic. A severe drop in plasma pH (may be as low as 7.0) occurs along with an elevated PCO_2. Treatment depends on the underlying cause; therapy is always directed at improving ventilation. If possible, correction of an obstruction is performed; artificial ventilation is used as indicated. A sodium bicarbonate solution may be administered slowly IV as needed to correct severe acidemia.

Respiratory distress syndrome occurs in some immature infants because of a deficiency of surface-active agents, causing inadequate alveolar expansion. The decreased functional lung surface area leads to a decreased PO_2, an elevated PCO_2, and a decreased pH (respiratory acidosis). Hypoxia follows and leads to metabolic acidosis as a result of excessive lactic acid formation. Acidosis causes increased pulmonary artery resistance and diminished pulmonary blood flow. Frequent arterial blood gas studies are necessary to assess the infant's condition; blood may be obtained by means of a catheter in the umbilical artery or from puncture of the radial, temporal, or brachial arteries. Symptoms of respiratory distress syndrome include expiratory grunting, retractions of the chest wall, rapid seesaw respirations, and cyanosis. Therapy is directed at supporting ventilation until the infant matures enough to produce surface-active agents. The infant is placed in an isolette and is given enough oxygen to keep the arterial PO_2 at a level of 50 mm to 70 mm Hg; an endotracheal tube and positive pressure ventilation may be necessary if difficulty is encountered in elevating the PO_2 to the minimal level. High concentrations of oxygen must be used cautiously, since they may cause retrolental fibroplasia or pulmonary oxygen toxicity; the goal of oxygen therapy is to keep the arterial PO_2 between 50 mm to 70 mm Hg (it should never exceed 100 mm Hg). A PCO_2 higher than 70 mm Hg may necessitate use of a respirator. An arterial pH of less than 7.25 compromises pulmonary blood flow and may be treated with an alkalinizing agent, such as sodium bicarbonate.

METABOLIC ALKALOSIS Metabolic alkalosis results when the plasma bicarbonate level rises and the pH increases. A frequent cause of metabolic alkalosis is loss of gastric juice from nasogastric suction or from vomiting. This imbalance is particularly likely to occur in pyloric stenosis, since only gastric juice is lost. The lungs attempt to compensate for metabolic alkalosis by conserving carbon dioxide (hypoventilation); however, this form of compensation is limited by oxygen need. Treatment of this imbalance consists of surgical correction of the stenosed pylorus and administration of chloride-containing solutions and potassium.

METABOLIC ACIDOSIS Metabolic acidosis results when the plasma bicarbonate level falls and the pH decreases. Conditions associated with metabolic acidosis include diabetes mellitus, salicylate poisoning in children under 5 years of age, starvation, renal failure, hypoxia, and loss of alkaline intestinal secretions. Remember that infants tend toward metabolic acidosis because of their high basal metabolic rates. Hyperventilation occurs in an attempt to increase the respiratory loss of carbon dioxide, thus relieving the acidemia. Treatment is aimed at eliminating the precipitating

condition; sodium bicarbonate is given, as indicated, to alkalinize the plasma.

NURSING ASSESSMENT OF CHILDREN WITH FLUID IMBALANCES

Nursing observations can be immensely helpful to the physician or can mean nothing, depending on whether the nurse knows what to look for and takes the time to record the observations on the nursing notes. Because small children cannot describe their problems, the pediatric nurse has to be especially observant. Some of the major areas in which observations should be made are described below.

TISSUE TURGOR Tissue turgor is best palpated in the abdominal areas and on the medial aspects of the thighs. In a normal person, pinched skin will fall back to its normal configuration when released. In a patient with fluid volume deficit, the skin may remain slightly raised for a few seconds. Skin turgor begins to be lost after 3% to 5% of the body weight is lost as fluid. Severe malnutrition, particularly in infants, can cause depressed skin turgor even in the absence of fluid depletion.

Obese infants with fluid volume deficit often have skin turgor that is deceptively normal in appearance. An infant with water loss in excess of sodium loss (sodium excess), such as occurs in some types of diarrhea, has a firm thick-feeling skin. This same phenomenon is observed in the child who has sodium excess owing to an excessive sodium intake, as occurs in salt poisoning.

MUCOUS MEMBRANES Dry mouth may be due to a fluid volume deficit or to mouth breathing. When in doubt, the nurse should run her finger along the oral cavity to feel the mucous membrane where the cheek and gums meet; dryness in this area indicates a true fluid volume deficit. The tongue of the fluid-depleted child is smaller than normal. Mucous membranes in the child with sodium excess are dry and sticky.

BREATHING RATE, DEPTH, AND PATTERN The nurse should observe the rate, depth, and pattern of respiration. Hyperpnea, such as occurs in metabolic acidosis resulting from diarrhea or salicylate poisoning, can double the water loss by way of the lungs. The overbreathing of metabolic acidosis is not always as obvious in the child as it is in the adult.

Accelerated breathing should be reported so that water losses through this route can be replaced. Older children and adults with metabolic alkalosis have decreased rate and depth of respiration, with irregular rhythm. The young infant may normally have irregular respiration; thus, changes in respiratory rhythm are not dependable in detecting metabolic alkalosis.

Changes in respiratory rate and depth are significant in evaluating the child's response to therapy. For example, a change from deep, rapid respiration to slower, less deep respiration indicates improvement in the child with metabolic acidosis.

TEARING AND SALIVATION The absence of tearing and salivation is a sign of fluid volume deficit and should be noted on the chart. It becomes obvious with a fluid loss of 5% of the total body weight.

THIRST Avid thirst indicates increased tonicity of the extracellular fluid with cellular dehydration. An infant can be tested for thirst with water, although the presence of nausea may mask the symptom.

BEHAVIOR The child's general behavior is also significant in evaluating the response of fluid imbalances to therapy. When the very ill child begins to display appropriate responses to people, ceases to have irritable, purposeless movements, and is less lethargic, his condition has improved.

GENERAL APPEARANCE A child with a fluid volume deficit has a pinched, drawn facial expression (Fig. 17-

1). A fluid volume deficit of 10% of body weight causes decreased intraocular pressure and, thus, the eyes appear sunken and feel soft to the touch. Suture lines in the skull become prominent. If the anterior fontanel is still patent, it may be depressed. A grayish skin color, owing to decreased peripheral circulation, also accompanies severe fluid volume deficit. The occurrence of mottling of the skin is an ominous sign.

NATURE OF CRY The cry of an ill infant is higher pitched and less energetic than normal. With improvement in his condition, the cry becomes less high pitched and more lusty.

BODY TEMPERATURE Fluid volume deficit is often associated with a subnormal temperature because of reduced energy output. Depending on the underlying disease, however, fever can accompany fluid volume deficit. If fever is present, its height should be recorded frequently. The rate of insensible water loss is greatly increased with fever; the amount of water lost depends on the height and duration of the fever. Fever may indicate excessive water loss from the body with resultant sodium excess, or it may be caused by an infection. The extremities are cold to the touch in severe fluid volume deficit, even when fever is present; this is due to decreased peripheral blood flow.

URINE OUTPUT In addition to noting the number of voidings, the nurse should estimate how much of the diaper is saturated with urine. Occasionally one would do well to weigh a dry diaper and compare its weight with that of the same diaper after the child has voided. The nurse should also note the urine's concentration, as revealed by its color. Failure to record urinary out-

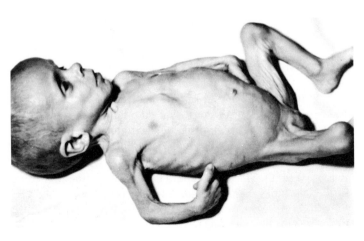

FIG. 17-1.
Severely dehydrated infant. (Waechter E, Blake F: Nursing Care of Children, 9th ed. Philadelphia, JB Lippincott, 1976)

put accurately makes treatment far more difficult. When the physician requests an accurate hourly recording of urine output, and a catheter is not inserted, the nurse must devise a method to catch all the urine passed.

Some hospitals have metabolic beds, consisting of synthetic cloth mesh suspended from the bed frame with a sloping surface beneath to catch the urine from the diaperless child and divert it into a collecting device. For male infants, a finger cot with the blind end cut off can be used to conduct urine from the penis to the drainage tube. A pediatric urine collection bag may be used; its major drawback is that it can rub the perineal skin raw very quickly. Regardless of the type of collection apparatus used, it should be checked frequently for leakage. Good skin care is necessary to prevent irritation of the genitalia.

A child with fluid volume deficit has a decreased urinary output and an increased urinary specific gravity. If the fluid deficit is severe, he may go as long as 18 to 24 hours without voiding and still not have a distended bladder. If a child with a known fluid volume deficit excretes large amounts of dilute urine, he probably has renal damage. If renal concentrating ability is impaired, or if the patient is receiving a high-solute diet, the urine volume will be somewhat above normal in order to clear all the metabolic wastes. The same is true if a hypercatabolic state (such as fever or infection or both) is present.

Urine concentration may be monitored as necessary by testing the specific gravity of each specimen. A refractometer requires only a drop of urine for the test and is thus useful to have on a pediatric nursing unit.[3]

CENTRAL VENOUS PRESSURE Central venous pressure (CVP) can be measured in the newborn by inserting an umbilical vein catheter in the superior vena cava at a level above the diaphragm. The child should be in the supine position with the zero level of the manometer at the midaxillary line. A pressure of 4 cm to 7 cm of water is within the normal range in premature and newborn infants.[4] In infants and older children, the CVP catheter may be inserted into the external jugular vein or through a peripheral vein into the superior vena cava. A pressure of 6 cm to 15 cm of water is within the normal range in children.[4] (The procedure for measuring CVP is described in Chapter 6; a rough measure of venous pressure in adults and older children is described in Chapter 4.)

Normally, the level of transition between collapse and distention of the jugular vein can be observed 1 cm to 2 cm above the suprasternal notch when the patient is at a 45 degree angle. This procedure is not very helpful in infants and young children since they have short, fat necks.

STOOLS Again, it is not enough just to chart the number of stools. The quantity of the stool should be estimated as nearly as possible; its character should be described. Thus, if a stool appears normal, it should be so described on the chart. If the stool is liquid, the degree of saturation of the diaper should be noted. Any abnormal contents, such as blood or mucus, should also be recorded.

VOMITING It is important to chart the number of times the patient has vomited, when he vomited, the quantity of vomitus (approximated if necessary), and the nature of the vomitus. Merely charting the number of times the patient vomited helps little in planning fluid replacement therapy, since the amount of fluid lost can vary widely from one attack of vomiting to another. Failure to describe the vomitus may make fluid replacement therapy more difficult. For example, if the vomitus is bile stained, one can conclude that it came from below the pylorus. Since fluids from below the pylorus are chiefly alkaline, fluid therapy must be designed to replace alkaline losses, using, for example, an intestinal replacement solution.

WEIGHT CHANGES Weight loss can be caused by loss of fluid or by catabolism of body tissues. The weight loss associated with fluid volume deficit occurs more rapidly than that caused by starvation. A mild fluid volume deficit in an infant or child entails a loss of from 3% to 5% of the normal body weight; a moderate fluid volume deficit, from 5% to 9%; a severe fluid volume deficit, 10% or more. (A loss of 15% of the body weight will likely cause hypovolemic shock.)

If possible, the child's weight before the onset of the illness should be obtained from the parents, or from the family physician, who may have a record of the normal weight from a recent office visit.

If weighing is not performed accurately, it is useless. Even a minor error is important when the patient is small. The child should be weighed at the same time each day, before he has eaten, after he has voided. The same scales should be used each time, and the child should be weighed naked.

FLUID REPLACEMENT THERAPY IN CHILDREN

DAILY REQUIREMENTS

The basic requirements of water and electrolytes must be met daily. In addition, one should supply the amounts necessary to correct preexisting deficits, as well as concurrent abnormal losses, such as those that occur from diarrhea, vomiting, suction drainage, and the like. *A major nursing function is the collection and accurate measurement of abnormal fluid losses.* The normal water requirements per kilogram of body weight at various ages are listed in Table 17-1.

Although the adult can go without food for several days without developing gross ketonuria, infants and children react quickly to the omission of calories. Ketonuria can occur within a few hours after the onset of fasting. For this reason, carbohydrate must be incorporated in fluids designed to meet daily maintenance needs.

ORAL REPLACEMENT

Water and electrolyte replacement is best accomplished by the oral route for these reasons:

Fluids taken into the gastrointestinal tract are slowly absorbed, while parenterally administered fluids pass directly into the circulation. The body is less adversely affected if excessive amounts of water or electrolytes are given by the oral than by the parenteral route.

Oral fluid replacement allows the child free movement and activity, as opposed to hours of being restrained during intravenous infusions.

It should be emphasized, however, that, even in the case of the oral route, it is relatively easy to overwhelm the body's homeostatic capabilities, particularly in the case of the infant and small child. For this reason, the dose for fluids administered orally should be calculated with the same care and precision as doses to be administered parenterally. Moreover, although the oral route is far safer than the intravenous route, potassium should not be given by mouth when oliguria or anuria is present.

Pediatric oral electrolyte solutions are available commercially. Homemade electrolyte solutions (made with measured amounts of salt, baking soda, or sugar) are risky, since they may be incorrectly prepared, lead-

ing to serious electrolyte disturbances (Tables 17-2, 17-3). For this reason, parents should not be told to prepare such oral electrolyte solutions at home. Table 17-2 lists the electrolyte contents of solutions of table salt. Table 17-3 lists the electrolyte contents of two commercially available oral electrolyte solutions.

TABLE 17-2 *Electrolyte contents of solutions of table salt*

Concentration salt/1 quart of tap water	Approximate composition	
	Na mEq/liter	Cl mEq/liter
⅛ tsp	10–15	10–15
¼ tsp	20–30	20–30
½ tsp	45–60	45–60
1 tsp	120	120
1 tbs	350	350

(Statland H: Fluid and Electrolytes in Practice, 3rd ed, p 201. Philadelphia, JB Lippincott, 1963)

TABLE 17-3 *Commercially available oral electrolyte solutions*

Pedialyte (Ross Laboratories, Columbus, OH)		
	per 8 fl oz	*per liter*
Sodium	7.1 mEq	30 mEq
Potassium	4.7 mEq	20 mEq
Calcium	0.9 mEq	4 mEq
Magnesium	0.9 mEq	4 mEq
Chloride	7.1 mEq	30 mEq
Citrate	6.6 mEq	28 mEq
Dextrose	11.8 g	50 g
Calories	48	200

Hydra-Lyte (Jayco Pharmaceuticals, Camp Hill, PA)	
	per packet
Sodium	84 mEq
Potassium	10 mEq
Chloride	59 mEq
Bicarbonate	10 mEq
Glucose	67 mM
Sucrose	35 mM

TUBE FEEDINGS

Sometimes tube feedings are used to feed unconscious children or those with swallowing difficulties. Many tube feeding mixtures contain large solute loads that, if associated with limited water intake, may increase plasma osmolarity. The nurse will do well to watch for signs of inadequate water intake, particularly when the feedings are high in protein content. Should insufficient water be provided, water will be drawn from the tissues to supply the needed volume for urinary excretion of the increased solute load. Eventually, dehydration (sodium excess) will occur, along with an accumulation of nitrogenous waste products in the bloodstream.

Hypernatremia may cause the skin to feel inelastic and "doughy" and the fatty tissue to feel unusually firm. Excessive protein in relation to water causes nausea, vomiting, diarrhea, and, eventually, ileus. If the condition is allowed to go uncorrected, it will result in high fever and disorientation. Laboratory tests will reveal an elevated blood urea nitrogen (BUN) level. Early in protein overloading, the urine volume is large even though water intake is inadequate. The large urine volume can easily lead the staff to think that water intake is adequate; in such instances, however, output actually exceeds intake—the extra fluid is being withdrawn from the body tissues. Very young children and unconscious patients should be observed carefully for inadequate water intake, since they are unable to express thirst.

PARENTERAL FLUID THERAPY

SUBCUTANEOUS ROUTE Subcutaneous fluids are easier to start than are intravenous fluids, particularly in infants and small children. For this reason, the subcutaneous route is sometimes used in pediatrics, even though it has serious hazards and limitations. Subcutaneous fluids are poorly absorbed in the child with a severe fluid volume deficit because of the frequently associated peripheral circulatory collapse. The route, therefore, is not dependable in patients with severe fluid balance disturbances. Another disadvantage of the subcutaneous route is the limitation of the types of fluids that can be administered by this route with even a modicum of safety. (The reader is referred to Chapter 6 for a discussion of which fluids can be given subcutaneously.)

Fluids may be administered to an infant using the subcutaneous tissues of the anterior or posterior axillary folds, medial aspect of the midthigh, inferior aspect of the scapula, or the lower abdominal wall. Older children may receive fluids in the outer aspect of the midthigh. It must be emphasized that the subcutaneous route is erratic and less desirable than the intravenous avenue of fluid administration. Fortunately, the use of the subcutaneous route for fluid administration has been largely abandoned in recent years.

INTRAVENOUS ROUTE *Procedure* Prior to performing the venipuncture, the child should be properly restrained according to the area involved. Restraints should not be so tight that they impede breathing and circulation. Venipuncture sites for the pediatric patient may include superficial veins of the scalp, wrist, hand, arm, leg, or foot. Since superficial veins of the scalp do not have valves, fluids may be infused in either direction.[5] Special care in cleansing the skin is mandatory before venipuncture is made in superficial scalp veins, since they communicate with the dural sinuses. A sandbag may be necessary to hold the infant's head in position. The scalp-vein infusion set (see Fig. 6-3) is attached to a syringe filled with sterile saline. After the needle enters the vein (as evidenced by aspirating blood into the syringe), the operator slowly injects 1 ml of saline while observing for swelling at the site. If no swelling is noted, the infusion set is attached to the IV tubing and taped into place.

Venipuncture may also be performed with an intracatheter (catheter threaded through a metal needle) or with a plastic needle (catheter mounted on a metal needle). A cutdown (surgical insertion of a catheter into a vein) may be performed if fluids are urgently needed and difficulty is encountered in finding a suitable vein. The internal saphenous vein can be entered at any point along its course and is easily identified.

After the venipuncture is accomplished, the designated rate of flow must be established and thereafter carefully monitored. *The danger of administering an excessive fluid volume is a real one in all age-groups.* Infants and small children are faced with special dangers simply because of their small size and the ease of supplying fluids in adult-sized bottles. The accidental administration of an extra 500 ml of fluid, such as 5% dextrose in water or isotonic solution of sodium chloride, might mean little to the adult, but it can be disastrous to the infant or small child. This group of patients has greater difficulty excreting excessive fluid volume. Moreover, they are more susceptible to pulmonary edema than are adults. In fact, pneumonia from overhydration is thought to be one of the most common treatment-related diseases of hospitalized children.

Measures to assure accurate flow rate Measures should be taken to avoid an overdose of intravenous fluids. The volume available in the bottle, which might run rapidly into the patient, should be limited. *It has been recommended that quantities exceeding 150 ml not be connected to children under 2 years of age, no more than 250 ml to children under 5, and no more than 500 ml to children under 10.*

The development of special administration sets for pediatric use has added a greater margin of safety to fluid administration. (See the section on volume controlled sets in Chapter 6.) There should never be more than enough fluid for 2 hours in any of these volumetric sets. A microdrip adaptor may be supplied with these sets; one delivering 60 drops/ml is particularly convenient in that the number of drops delivered per minute is also the number of milliliters delivered per hour. (For example, 60 drops/min delivers 60 ml/hr, 20 drops/min delivers 20 ml/hr, and so forth.)

The physician prescribes the total volume of fluid to be given over a designated period of time, as well as the desired number of milliliters per hour or drops per minute. When the nurse knows the total volume to be administered in a fixed period of time, plus the drop factor of the set to be used, the flow rate can easily be calculated. (The reader is referred to the discussion on calculation of flow rates in Chapter 6.) Great care should be taken to ensure accuracy when the flow rate is calculated.

Even though drop size adaptors and small containers are used to reduce the possibility of error, the nurse must still keep a close vigil on the flow rate, as well as on the patient's response to the fluids. The flow rate should be counted every 15 minutes and adjusted as necessary. (Factors that can alter the flow rate are discussed in Chapter 6.) *A pediatric parenteral fluid sheet should be kept at the bedside of infants and small children to record the observed flow rate, the amount of fluid absorbed each hour, and the amount of fluid left in the bottle.* Such frequent observations and notations greatly reduce the risk of excessive fluid administration. *Obviously, fluid monitoring is a major aspect of pediatric nursing!*

Occasionally, veins may be too small in some infants to allow administration of fluids by gravity (even when the container is elevated to its highest position);

in these cases, a mechanical infusion pump is indicated. Because it is often necessary to administer very small volumes to pediatric patients, it is advisable to use an intravenous infusion pump whenever such an apparatus is available. Most infusion pumps can deliver a constant, exact flow rate, a definite improvement over the erratic flow of a gravity apparatus. (Infusion pumps are described in Chapter 6.)

Tubing and dressing changes should be made daily, always using strict aseptic technique. Complications such as thrombophlebitis and infiltration should be observed for frequently. (See Chapter 6 for a discussion of the complications of intravenous therapy.)

FACTORS INFLUENCING FLUID INTAKE

The child's condition should be reassessed at frequent intervals to reevaluate his fluid requirements. The clinical status of infants can improve or deteriorate remarkably in a few hours. The high humidity in an incubator or isolette minimizes insensible water loss and, thus, decreases the total amount of fluid to be given by about a third. Fever increases fluid maintenance requirements by about 10% for each Fahrenheit degree elevation. As a general rule, the fluid intake should be adequate to keep the child's urine output in the normal range for his age. The desired urine flow for a child one year old or under is approximately 5 ml to 10 ml/hr. Children from 1 to 10 years of age should excrete approximately 10 ml to 25 ml of urine/hr. However, normal outputs are usually not achieved in the immediate post-operative period because of the influence of stress.

CLINICAL CONDITIONS COMMONLY ASSOCIATED WITH FLUID IMBALANCES IN SMALL CHILDREN

DIARRHEA

Diarrhea is a common cause of water and electrolyte disturbances in infants and small children. The large loss of liquid stools can rapidly deplete the young child's extracellular fluid volume, especially when it is combined with vomiting. Usually water and electrolytes are lost in isotonic proportions (fluid volume deficit or isotonic dehydration). However, water can be lost in excess of electrolytes (fluid volume deficit with sodium excess or hypertonic dehydration), and electrolytes can be lost in excess of water (fluid volume deficit with sodium deficit or hypotonic dehydration). Because sodium is the chief extracellular ion, its excess or deficit is of primary importance in producing symptoms.

Intestinal fluids are alkaline; therefore, large losses of fluids in diarrhea may result in metabolic acidosis. Potassium deficit is another frequent accompaniment of diarrhea.

EXTRACELLULAR FLUID VOLUME DEFICIT (ISOTONIC DEHYDRATION) *Symptoms* Approximately 70% of patients with severe diarrhea undergo a proportionate loss of water and electrolytes. Symptoms of fluid volume deficit due to infantile diarrhea include the following:

History of large quantities of liquid stools

Weight loss

Dry skin with poor tissue turgor

Diminished tearing

Soft eyeballs with a sunken appearance (resulting from decreased intraocular pressure)

Depression of anterior fontanel, if it is still present

Skin ashen or gray in color and extremities cold (owing to inadequate peripheral circulation)

Depressed body temperature, unless fever accompanies the diarrhea, as in an infection

Lethargy

Signs of hypovolemic shock

Weak rapid pulse

Decreased blood pressure

Oliguria

Metabolic acidosis usually accompanies frequent liquid stools. (Recall that the intestinal secretions are alkaline because of their high bicarbonate content. Therefore, loss of alkaline secretions in diarrheal stools results in metabolic acidosis.) Decreased dietary intake contributes to metabolic acidosis; thus, in the absence of adequate food intake, the body utilizes its own fats for energy purposes. The metabolism of these fats causes the accumulation of acidic ketone bodies in the blood, further contributing to the metabolic acidosis caused by bicarbonate loss.

A major symptom of metabolic acidosis is the increased depth of respiration, a body compensatory mechanism that blows off carbon dioxide, thus reducing the carbonic acid content of the blood and influencing the carbonic acid–base bicarbonate balance in the direction of an increased pH. If ketosis of starvation is present, an acetone odor may be noted on the breath. Symptoms of severe potassium deficit include weakness, anorexia, vomiting, excessive abdominal gas, and muscles flabby, like half-filled water bottles.

Treatment The first goal of fluid replacement therapy is to expand the extracellular fluid volume sufficiently to prevent or correct symptoms of hypovolemic shock. A restored blood volume permits adequate renal blood flow and increased urine formation. Improvement of renal function helps the body eliminate organic acids and, thus, correct acidosis. Physicians vary in the precise fluid therapy employed. All agree that if kidney function is depressed because of extracellular fluid volume deficit, a special solution should be administered to correct renal depression. Renal depression is indicated by the following:

Urinary specific gravity above 1.030

Oliguria, revealed by the history of voiding fewer than three times during the previous 24 hours

Anuria, shown by absence of urine in the bladder

Renal depression is assumed to be present when there has been a recent fluid loss of great magnitude, such as occurs with severe infectious diarrhea of explosive onset.

When renal depression is present, the physician administers an initial hydrating solution. A typical initiate solution is one third isotonic solution of NaCl in 5% dextrose. Such a solution supplies 51 mEq of sodium and 51 mEq of chloride.

With renal flow established, one can then administer a repair solution. Some physicians use lactated Ringer's solution with dextrose and added potassium. Doses of such solutions are usually based on milliliters per kilogram of body weight.

The water and electrolyte requirements for infants and children, when expressed in units/kilogram, vary considerably for children of different ages and weights. For this reason, many physicians prefer to use body surface area as the dose criterion, since this is independent of age and weight, except in the case of prematures.

FLUID VOLUME DEFICIT WITH SODIUM EXCESS (HYPERTONIC DEHYDRATION) *Symptoms* Approximately 20% of patients with severe diarrhea have suffered a relatively greater loss of water than of electrolytes. If the infant has ingested a high-solute–containing formula during his illness or has been inadvertently given an overly concentrated electrolyte mixture, the renal water loss intensifies the sodium excess already present. Because the infant cannot concentrate urine efficiently, large volumes of water are needed to excrete solutes. The infant's need for water is intensified by the fact that his insensible water loss is great because of his large body surface area.

Symptoms of fluid volume deficit and sodium excess caused by diarrhea include these:

> History of large quantities of liquid stools associated with a low water intake, high solute intake, poor renal function, or all three
>
> Weight loss
>
> Skin elasticity and turgor not lost (however, the skin has a thickened, firm feeling)
>
> Avid thirst (hypertonic extracellular fluid draws water from the cells, producing cellular dehydration)
>
> Irritability displayed when disturbed; otherwise behavior is lethargic
>
> Tremors and convulsions
>
> Muscle rigidity
>
> Nuchal rigidity
>
> Chloride and protein concentration of spinal fluid elevated
>
> Although signs of extracellular fluid volume deficit are not present for first few days, they eventually occur with symptoms of hypovolemic shock
>
> Brain injury (may be due to intracranial hemorrhage and effusion into the subdural space)

Treatment Small amounts of electrolytes are used in the repair solutions to prevent a too-rapid correction of the sodium excess, since a rapid return of the plasma sodium concentration to normal may precipitate cerebral edema.

Symptoms of sodium deficit can result from the abnormal uptake of water by the cells, secondary to the inability of the immature kidneys to maintain the normal relationship between water and solute. Convulsions can result from the too-rapid reduction of the sodium concentration in the extracellular fluid; they are less likely to occur when the correction of the sodium excess is carried out gradually.

Hypocalcemia can occur during treatment, possibly caused by the loss of calcium in the stool or by the decreased ionization of available extracellular calcium, which occurs with correction of the metabolic acidosis. Symptoms of hypocalcemia occur less frequently when calcium is included in the treatment solution. The administration of 10 ml to 30 ml of 10% calcium gluconate added daily to one of the infusions may prevent calcium deficit from developing.

FLUID VOLUME DEFICIT WITH SODIUM DEFICIT (HYPOTONIC DEHYDRATION) Approximately 10% of patients with severe diarrhea have suffered a relatively greater loss of electrolytes than of water, usually because fluid losses have been replaced with plain water in dextrose or some other electrolyte-free solution, which dilutes the electrolyte concentration of the extracellular fluid. Because sodium is the chief extracellular ion, the primary symptoms are due to its deficit. Symptoms of fluid volume deficit and sodium deficit caused by diarrheal losses include these:

> Clammy skin
>
> Lethargy
>
> Hypovolemic shock, in severe cases

MEASURES TO PREVENT OR MINIMIZE WATER AND ELECTROLYTE LOSS IN DIARRHEA The following measures should be taken to prevent or minimize imbalances caused by infant diarrhea:

1. Hospitalized infants with diarrhea should be isolated so as to prevent infecting other children in the unit. Meticulous attention should be paid to isolation technique.

2. Diarrhea can be caused by infections transmitted to the infant by contaminated formula or equipment. Unless technique is impeccable, this can occur readily in the hospital. The nurse should see that the mother knows how to prepare the baby's formula safely before she goes home.

3. Water and electrolyte losses can be minimized if diarrhea is reported immediately, so that treatment can be started. Mothers should be instructed to report diarrhea as soon as it is noticed.

4. Liquid stool losses are greater if the baby continues to take oral feedings than if his gastrointestinal tract is put to rest temporarily. Mothers should be instructed to withhold formula feedings when diarrhea occurs until they have checked with the physician. Because of the high solute content of skim milk, boiled skim milk that is undiluted should not be used in the treatment of infants with diarrhea. Because of the infant's poor renal concentrating ability, he needs large quantities of water to excrete the large renal solute load presented by undiluted

skim milk. Boiled skim milk should never be used unless diluted with at least an equal volume of water, plus added carbohydrate.

5. Mothers should be encouraged to follow the physician's instructions precisely in returning the infant to full formula feedings following a bout of diarrhea. In most cases the infant with diarrhea is given initially an oral electrolyte solution with glucose. When tolerated, a little milk is added each day until full feedings are resumed.

VOMITING CAUSED BY HYPERTROPHIC PYLORIC STENOSIS

Symptoms Hypertrophic pyloric stenosis is a common cause of vomiting in small infants, usually under 6 weeks old. Males are reportedly affected five times more often than females.

Because of the repeated vomiting, the infants are poorly nourished. While hypertrophic pyloric stenosis can be corrected surgically, preoperative correction of the water and electrolyte disturbances caused by the prolonged vomiting is mandatory.

Vomiting causes the same imbalances in children as it does in adults. These include metabolic alkalosis, potassium deficit, sodium deficit, and fluid volume deficit. Metabolic alkalosis occurs because of the excessive loss of potassium, hydrogen, and chloride in the vomitus. Loss of chloride causes a compensatory increase in the number of bicarbonate ions; this occurs because both are anions (negatively charged ions). Total cations must always equal the total anions so that electrical equality can be maintained; if the quantity of one anion is decreased, another anion must increase in compensation. The bicarbonate side of the carbonic acid–base bicarbonate ratio is increased, and the pH increases—that is, becomes more alkaline. Sodium and potassium are plentiful in gastric juice; prolonged vomiting leads to deficits of both. Since water is also lost in the vomitus, fluid volume deficit occurs. Because the losses are sustained over a relatively long period, circulatory collapse is not prominent in hypertrophic pyloric stenosis as it is in severe diarrhea.

The infant with hypertrophic pyloric stenosis has the following symptoms:

> Difficulty in retaining feedings, which becomes progressively worse during the first few weeks of life; eventually, projectile vomiting follows each feeding

Appearance of malnutrition

Symptoms of fluid volume deficit

Decreased respiration (compensatory action of lungs to retain carbon dioxide and increase the carbonic acid content of the blood)

Tetany accompanying alkalosis (owing to decreased calcium ionization in an alkaline pH)

Despite starvation, ketosis does not usually appear

Palpable pyloric tumor

Treatment *Preoperative period* To minimize fluid losses, oral feedings should be discontinued and fluids given parenterally. If gastric suction is used, the water and electrolytes lost by this procedure should be replaced. A solution of 5% dextrose in 0.45% NaCl, with added potassium, may be used as a correction fluid for metabolic alkalosis.

Postoperative period The young child does not retain sodium after a surgical operation, as do many adults. Sodium should, therefore, be included in the postoperative repair solution. There is excellent rationale for including potassium likewise.

BURNS

The treatment of children with burns is essentially the same as that described for adults in Chapter 8. The major difference lies in the calculation of the percentage of the body involved in the burn, since the child's body proportions are different from those of the adult (See Table 8-2).

Another major difference between the child and the adult lies in the need for greater accuracy in fluid administration in children, since the child has greater sensitivity to minor errors in fluid administration. Both pulmonary edema and shock develop more quickly in children than in adults.

A second degree burn of more than 10% of the body surface in a child under one year of age or of 15% in an older child represents serious injury and requires hospitalization. To evaluate the effectiveness of fluid replacement, it is necessary to measure the urinary output at regular intervals. A child under one year of age should have an hourly output of from 5 ml to 10 ml; a child between 1 and 10 years between 10 ml and 25 ml; or, 10 ml to 30 ml/m^2 of body surface-hour, regardless of age.

SALICYLATE INTOXICATION

Salicylate intoxication is commonly seen in children and accounts for 12% of all toxic ingestions. It has resulted from leaving aspirin in the reach of the small child, particularly flavored aspirin. Recent use of safety caps, however, has decreased the occurrence of this problem. Half of the cases of salicylate intoxication now are due to parent or practitioner errors; that is, some children are sensitive to theoretically permissable doses.[6] Occasionally an older child may attempt suicide by aspirin ingestion. Salicylate poisoning can result also from ingesting sodium salicylate tablets or from drinking methyl salicylate (oil of wintergreen).

Symptoms Salicylates cause the respiratory center to be more sensitive to carbon dioxide; they cause respiration to become deep and rapid. As a result, excessive amounts of carbon dioxide are eliminated from the lungs and respiratory alkalosis develops. Symptoms of a deficit of ionized calcium caused by the alkalosis appear. They include numbness and tingling of the face and extremities, positive Chvostek's sign, muscle twitching, and convulsions.

Respiratory alkalosis is the major disturbance in adults and older children. Management consists of suitable fluid and electrolyte infusions to promote excretion of salicylates.

Children under 5 years of age usually develop a more complicated acid–base disturbance. Quickly following the initial respiratory alkalosis, they develop metabolic acidosis, as a result of inadequate utilization of carbohydrate caused by the toxic doses of salicylates and the resultant increased utilization of body fat. Symptoms of severe salicylate intoxication in the child consist of severe hyperpnea, vomiting, and fluid volume deficit. Hyperthermia is common and is manifested by a flushed appearance and sweating. Bleeding may occur, owing to prolonged prothrombin time and platelet dysfunction. The serum salicylate level is over 40 mg/dl. Hypokalemia may result from excessive loss of potassium in vomitus.

The blood pH is affected by two imbalances— respiratory alkalosis and metabolic acidosis. Sometimes the two imbalances neutralize each other and the pH remains normal. If respiratory alkalosis is more severe than metabolic acidosis, as is frequently the case in older children and adults, the pH is elevated above normal. If metabolic acidosis is more severe than respiratory alkalosis, as is frequently the case in small children, the pH is decreased below normal.

Adults and older children can cope with the imparied metabolism of salicylate intoxication better than can young children.

Treatment Emergency therapy is directed at removing as much of the salicylate from the stomach as possible before it is absorbed. Syrup of ipecac may be used to induce emesis or the stomach may be lavaged with an isotonic solution of sodium chloride.

After the stomach contents have been emptied, attention is given to supplying adequate fluids to promote excretion of salicylates by the kidneys, which account for the excretion of about 80% of ingested salicylates. Carbohydrate is administered to prevent or combat ketosis. The amount of fluids required depends largely upon the length of time elapsed since poisoning occurred. Since severe hyperpnea usually accompanies metabolic acidosis, large quantities of water are lost by way of the lungs. Some patients who have not received prompt therapy suffer severe fluid volume deficit, which must be repaired. Potassium may be given as indicated to correct hypokalemia and allow alkalinization of the urine once urine flow is established.

General treatment principles and details are similar to those given under extracellular fluid volume deficit.

The urine can be alkalinized (above pH 7.5) as a means of promoting salicylate excretion. This can be accomplished by the intravenous administration of lactate containing solutions.

Calcium gluconate can be given to relieve symptoms of ionized calcium deficit, such as tetany. Vitamin K_1 can be used to prevent excessive bleeding resulting from hypoprothrombinemia. Dialysis is indicated when potentially fatal serum levels exist or when oliguria or anuria is present.

REFERENCES

1. Maxwell M, Kleeman C (eds): Clinical Disorders of Fluid and Electrolyte Metabolism, 3rd ed, p 1570. New York, McGraw-Hill, 1980
2. *Ibid*, p 1578
3. McGrath B: Fluids, electrolytes, and replacement therapy in pediatric nursing. Matern Child Nurs, Jan/Feb 1980, p 61
4. Graef J, Cone T (eds): Manual of Pediatric Therapeutics, 2nd ed, p 35. Boston, Little, Brown & Co, 1980
5. Kempe C et al: Current Pediatric Diagnosis and Treatment, 6th ed, p 1000. Los Altos, Lange Medical Publications, 1980
6. Waechter E, Blake F: Nursing Care of Children, 9th ed. Philadelphia, JB Lippincott, 1976

18

Fluid Balance in the Geriatric Patient

CHANGES IN WATER AND ELECTROLYTE HOMEOSTASIS ASSOCIATED WITH AGING

Homeostasis may be defined as the body's ability to restore equilibrium under stress. The aged are particularly vulnerable to fluid and electrolyte imbalances when under stress because their major homeostatic mechanisms have undergone numerous physiologic changes.

RENAL CHANGES A 50% reduction in kidney function may occur between early maturity and old age. With advanced age, there is reduction in blood flow to the kidney, a decrease in the number of nephrons, a decline in glomerular filtration rate (GFR), and diminished renal concentration ability. Renal blood flow is reduced owing to decreased cardiac output and increased peripheral resistance. The normal GFR of 120 ml/min at age 40 decreases to 60 ml to 70 ml/min at age 85.[1] One of the most common changes in renal function in old age is the inability to concentrate urine when fluid intake is decreased. A young person may be able to concentrate urine to a specific gravity (SG) of 1.035 to 1.040; persons in the seventh and eighth decades of life are often not able to concentrate urine beyond a SG of 1.022 to 1.026.[2] Decrease in renal concentrating ability causes mild increasing polyuria and can result in nocturia. Owing to loss of maximum concentrating ability, the definition of oliguria in the elderly should encompass higher urine volumes than for young adults. For example, a urine output of 900 ml/day may be inappropriately low for many elderly persons and reflect oliguric renal failure.[3] Because the aged patient has diminished renal concentrating ability, he needs more fluid to excrete a given amount of solute. It is interesting to note that the mean blood urea nitrogen (BUN) concentration at age 40 is 12.9 mg/dl; after age 70, it is 21.2 mg/dl.[4] (This reflects the reduction in GFR in the aged kidney.)

The elderly are more vulnerable to drug toxicity owing to decreased renal function. The usual patient over age 64 probably takes an average of three drugs.[3] Potentially toxic antibiotics that are excreted virtually unchanged by the kidney (such as streptomycin, kanamycin, tobramycin, and neomycin) should be used cautiously in aged persons. Antacids containing calcium or magnesium should be used with extreme caution in aged patients with renal insufficiency.

RESPIRATORY CHANGES Problems related to the respiratory mechanism in aged persons include increased rigidity of chest walls, weakening of respiratory muscles, decreased elasticity of lung tissue, defective alveolar ventilation, inadequate diffusion of respiratory gases, and interference with bronchial elimination of secretions. The vital capacity is reduced with age, and an increase of residual air volume occurs. There is a decrease in the normal partial pressure of oxygen (PO_2) in the elderly; a general guideline is to subtract 1 mm Hg from the minimal 80 mm Hg for every year over the age of 60 (up to age 90).[5] Recall that regulation of pH is largely controlled by the respiratory system by either elimination or retention of carbon dioxide (controlling carbonic acid content). In the aged, diminished respiratory function leads to less reserve in the face of major illness, surgery, burns, or trauma and thus interferes with acid–base regulation.

CARDIOVASCULAR CHANGES With advanced age, blood flow through the coronary arteries is decreased; also, there is diminished ability of the myocardium to use oxygen. Cardiac output and stroke volume are decreased. In general, the aged person's heart is less well equipped to deal with stressful situations.

The myocardium and circulatory system of the aged are less able to manage the hypotension associated with shock; shock becomes irreversible sooner in the aged and thus requires aggressive treatment. (See Chapter 7 for a discussion of shock.)

A moderate increase in blood pressure occurs in men and women up to the age of 65; particularly increased are systolic pressure and pulse pressure. Associated with the rise in blood pressure is a decrease in elasticity of the arterial vasculature. Systolic hypertension develops as the aorta becomes less elastic and there is increased resistance to blood flow; closely related is an increase in cholesterol and calcium deposits in the arteries. From pumping against increased resistance, and from advanced age, heart contractions weaken and cause diminished cardiac output. Buildup of pressure on the venous side of the circulatory system can raise capillary pressure and force fluid out into the tissues, producing edema. (Edema is discussed in Chapter 4.)

Orthostatic hypotension often develops in the aged owing to increased lability of the vasopressor control mechanism; changing positions more slowly helps control this problem.

SKIN CHANGES The dermis loses elasticity in the aged and becomes relatively dehydrated. Atrophying sweat glands and a diminished capillary bed in the skin reduce the ability to sweat and to control body temperature; because of this, heatstroke is a formidable problem in elderly persons. (This subject is discussed later in this chapter and in Chapter 15.)

Skin turgor is not a valid sign of fluid balance status in the aged, since skin loses elasticity with age; probably the best gauge of hydration is tongue turgor, which, fortunately, does not change appreciably with age. (Tongue turgor is described in Chapter 4.)

Aged persons (with their atrophic dermal papillae, sweat glands, and hair follicles) will develop deep burns from the same amount of heat intensity that produces only moderate second-degree burns in middle-aged adults.

GASTROINTESTINAL CHANGES Both volume and acidity of gastric juice gradually decrease with age owing to atrophy of the gastric mucosa; hypochlorhydria or achlorhydria is present in about 35% of patients over the age of 60 years. The geriatric achlorhydric patient experiencing vomiting or gastric suction is more likely to develop hypokalemia and hyponatremia than other imbalances. Iron uptake is impaired in the absence of free hydrochloric acid and bile, predisposing to iron-deficiency anemia.

Loss of muscle tone of the intestinal tract and a lessened sense of the need to eliminate can lead to chronic constipation, a problem that often becomes upper-most in the minds of many aged persons. (Constipation is discussed later in this chapter.)

NEUROLOGIC CHANGES There is a progressive loss of central nervous system cells with advancing age; the loss of neurons affects both the brain and spinal cord. A decrease in the sense of smell and in the number of taste buds perhaps contributes to decreased appetite in the elderly. The aged patient also experiences a decreased tactile sense and increased pain tolerance.

The thirst mechanism in the aged is feeble and serves as a poor guide to the need of liquids in patients who are ill. The aged patient may be too weak to verbalize his thirst or to reach for a glass of water, or he may have a clouded sensorium interfering with an appropriate response to the thirst mechanism. (Recall that the sensation of thirst is dependent on excitation of the cortical centers of consciousness.) The use of antianxiety agents, sedatives, or hypnotics in the elderly can lead to confusion and disorientation, causing the patient to forget to drink fluids. Because of this, hypernatremia is a common imbalance in the aged. One should also remember that elderly persons often deliberately restrict fluid intake in order to avoid embarrassing incontinence. The autonomic nervous system is sluggish in the aged person, causing slower response to shock and to heat stress.

ENDOCRINE CHANGES As age advances, there is a corresponding decrease in glucose tolerance until the age of 75 or 80. Thus, return to the fasting level in the glucose tolerance test is slower in the aged than in younger adults. Because of this, standards for the glucose tolerance test are adjusted for age; if "normal" standards were applied to the elderly, half of them would be classified as diabetics. When diabetes does occur in the aged, the most frequently encountered metabolic problem is nonketotic hyperosmolar coma (NKHC). (This condition is discussed later in this chapter.) Hormonal deficiencies occurring with aging are decreased estrogen at menopause and decreased testosterone during the male climacteric.

TOTAL BODY WATER CHANGES The total amount of body water decreases significantly with age. (Water comprises approximately 60% of the total body weight in young males, 52% in elderly males, 52% in young females, and 46% in elderly females.)[6] The decrease in total body water is apparently due to loss of cellular fluid; there is little change in the absolute quantity of extracellular fluid.

A decrease in the lean body mass and a relative increase in body fat occurs with aging. Recall that the total amount of body water varies indirectly with the level of obesity; the more obese the person, the less the proportion of body water.

The nurse should be aware that the rate of parenteral fluid administration for an elderly person is somewhat less than for a younger adult to prevent circulatory overload. However, it is important to remember that when the aged patient is volume depleted, he should be treated aggressively, since hypovolemia is also dangerous. (There is a tendency to undertreat aged persons who are volume depleted.) When necessary, heart function and response to fluid administration can be monitored by observing the central venous pressure (CVP). At times, a Swan-Ganz catheter is in-

dicated to monitor pressures in both the right and left sides of the heart (see Chapter 11.) With such sophisticated monitoring devices, there is no reason to avoid aggressive treatment of the elderly in need of intravenous fluid therapy.

SPECIAL PROBLEMS IN THE AGED

HYPERNATREMIA AND PROTEIN OVERLOAD WITH TUBE FEEDINGS Elderly persons are frequently given high-protein tube feedings to facilitate healing of decubitus ulcers, fractures, and the like. The nurse should keep in mind that the aged patient has a decreased renal concentrating ability, making it necessary to give him extra water to facilitate excretion of the high solute load. Hypernatremia and an elevated BUN may develop after administration of an excessive solute load with inadequate water provision and lead to osmotic dehydration. Early in protein overloading, the urine volume is large even though the water intake is inadequate; actually, fluid is being drawn from the plasma and tissue spaces for urine formation. This can easily lead the staff to think that the water intake is adequate. In such instances, however, output actually exceeds intake. To properly assess the situation, the nurse should compare the total 24-hour intake with the total 24-hour output. An output greatly exceeding the intake is an indication to increase the water intake, as is an elevated BUN. Additional water should be given as necessary to maintain a satisfactory urinary output. Schulte advocates allowing at least 0.5 ml of water for every milliliter of tube feeding.[7] (Tube feedings are discussed in greater detail in Chapter 5.)

Thirst is an early indicator of need for extra water; confused or unconscious patients should be observed carefully for inadequate water intake, since they are not aware of thirst. Clinical signs of dehydration (hypernatremia) include sticky tongue and mucous membranes, confusion, elevated plasma sodium level, thirst, and elevated body temperature.

IMBALANCES ASSOCIATED WITH USE OF DIURETICS Diuretics are frequently used in the elderly to treat hypertension or congestive heart failure. Both thiazides and furosemide (Lasix) are potassium-losing diuretics, having a greater tendency to induce hypokalemia in the aged than in younger adults. Unfortunately, the symptoms of hypokalemia, such as muscle weakness, lethargy, and leg cramps, are often dismissed as being due to "old age." It is important to remember that hypokalemia potentiates the action of digitalis (another drug commonly prescribed for elderly persons) and can precipitate toxic symptoms. The use of a potassium-sparing diuretic (such as triamterene) has been shown to produce a higher incidence of hyperkalemia in the aged than in younger adults.

In summary, the aged are more prone to develop both hypokalemia and hyperkalemia than younger adults when diuretics producing these conditions are given. It is important that the nurse instruct the patient as to the reason for the use of the prescribed diuretics and the possible side-effects. The need for adherence to the correct dosage should also be emphasized.

Excessive loss of sodium can also occur with the use of diuretics, particularly when the patient is on a sodium-restricted diet or is losing sodium by some other route (such as vomiting or diarrhea).

IMBALANCES RELATED TO CONSTIPATION AND LAXATIVE ABUSE Reduced motility of the intestinal tract and a lessened sense of the need to eliminate can lead to chronic constipation with laxative and enema dependency. Prolonged use of strong laxatives predisposes to hypokalemia and fluid volume deficit. (Electrolyte imbalances associated with prolonged use of laxatives and enemas are discussed in Chapter 9.)

It has been reported that persons over 70 years of age take laxatives twice as often as those in the 40 to 50 year age-group.[8] In addition to decreased gastrointestinal motility with aging, certain drugs, such as anticholinergics and antacids containing calcium carbonate or aluminum hydroxide, predispose to constipation. Unfortunately, the use of laxatives in the aged often becomes habit-forming, requiring larger and more frequent doses to achieve results.

Increased fluid and bulk intake, plus regular exercise, should be encouraged to correct constipation. Stool softeners are useful physiologic tools for increasing the softness of stools, and a glass of warm water or hot coffee first thing in the morning can stimulate the evacuation reflex.

Chronic constipation in the aged predisposes to ammonia formation by the putrefactive action of bacteria on nitrogenous substrates. Normally the ammonia formed by this process is metabolized in the liver; however, if liver function is impaired, the ammonia

concentration may rise to toxic levels. The feeding of high-protein diets also contributes to this problem.

IMBALANCES ASSOCIATED WITH PREPARATION FOR DIAGNOSTIC STUDIES OF THE COLON Standard colon cleansing techniques for diagnostic studies usually include several days of dietary restrictions (such as low-residue or liquid diets), purgatives (such as castor oil or magnesium citrate), and numerous cleansing enemas (either saline or tap water). Studies have indicated that the rigorous catharsis associated with this kind of preparation leads to significant shifts of fluids among body compartments; if the shifts occur rapidly in an elderly person with cardiovascular disease, the results can be dangerous.[9]

The elderly patient undergoing rigorous bowel preparation should be observed closely for adverse reactions. Care should be taken to perform the procedure correctly the first time to avoid the need for repeated radiographs, requiring more cathartics, more enemas, and more fluid restriction. The elderly patient can ill afford to undergo one test upon another without a "rest period" in between; it is frequently up to the nurse to intervene in this area on the patient's behalf.

The already reduced GFR in the elderly potentiates the hazards of any further decrease in extracellular fluid volume, as occurs in vigorous catharsis. The aged patient having gastrointestinal radiographs should probably receive intravenous fluids during the preparation period of reduced oral intake and increased fluid loss by catharsis and enema.[4]

It should also be noted that the use of radiocontrast agents in diagnostic radiology, particularly in large doses, has been incriminated as predisposing to acute renal failure, especially when the patient has a severe fluid volume deficit. (Acute renal failure is discussed in Chapter 10.)

OSTEOPOROSIS There is a progressive loss of bone after age 40; by age 80, total bone mass may be reduced to one half of what it was at age 40.[10] Usually in osteoporosis the rate of bone formation is normal, but the rate of bone resorption is increased. Bone resorption is especially increased in women. In fact, one in three postmenopausal women has osteoporosis; one in five will incur a hip or vertebral-crush fracture.[10] Postmenopausal osteoporosis is most common in small-framed white women past age 50.[11] Back pain is the most frequent symptom, and spinal deformity is the most frequent sign. "Dowager's hump" (a hunchback posture due to severe loss of anterior vertebral height) is a common finding. The patient's height may be 1 to 6 inches shorter than the arm span if multiple wedge compression fractures have occurred. Involutional bone alterations may reduce cortical bone mass of the femurs by a factor of 30% to 50%.

Serum concentrations of calcium, phosphate, and alkaline phosphatase are normal even though negative calcium, phosphate, and nitrogen balances exist.

Reduced calcium intake seems to play a role in the development of post-menopausal osteoporosis. Also implicated are diminished physical activity, impaired intestinal calcium absorption, increased renal calcium loss, increased parathyroid hormone (PTH) effect, and reduced secretion of estrogen.[12] Increased PTH secretion in postmenopausal osteoporosis may be related to a slightly decreased ionized calcium level owing to a number of hepatic, renal, and intestinal changes that accompany aging.[12]

It seems that oral calcium, vitamin D, and estrogen can reverse negative calcium balance related to menopause and probably delay or prevent the onset of clinical osteoporosis.[13] Calcium supplements help replace the gross deficiency of calcium associated with osteoporosis. Vitamin D increases calcium absorption from the intestines. Oral estrogen administration has been shown to be the most effective treatment modality; it reduces negative calcium balance and stimulates positive calcium balance. It should be remembered that large daily doses of estrogen have been shown to be associated with sodium retention, hypertension, myocardial infarction, and increased risk of endometrial carcinoma. Many physicians feel that the anticatabolic effect of estrogen, which reduces bone resorption, outweighs the risk of these untoward effects. Other therapies for osteoporosis may involve increased exposure to real or artificial sunlight (infrared light) to increase the level of 25-hydroxy vitamin D, and decreased intake of dietary phosphate. Recent studies suggest that an excess of phosphate may be an important cause of bone loss; "junk foods" tend to be high in phosphate content.

HEAT EXHAUSTION AND HEATSTROKE Two thirds of the cases of heatstroke occur in persons over the age of 50.[14] The type of heatstroke occurring in the aged is usually referred to as "non-exercise-induced" (as opposed to "exercise-induced" in the young athlete).

The elderly are particularly predisposed to heatstroke because of a high incidence of cardiovascular disease, sluggish vascular autonomic responses, and frequent use of medications interfering with thermoregulation. Many elderly persons are unable to compensate for increased environmental temperature with tachycardia, increased cardiac output, and decreased systemic vascular resistance (recall that these responses are normal and permit maximum heat losses from the body). The fact that uptake of heat from the environment increases proportionally with age makes it easier to understand the increased incidence of heatstroke in the elderly during heat waves.[15] Recall that a number of medications frequently prescribed for aged patients can interfere in some way with thermoregulation (e.g., diuretics, antiparkinsonian drugs, and propranolol; see the section dealing with risk factors for heat exhaustion and heatstroke in Chapter 15). Hemodynamic and metabolic effects of heatstroke are also discussed in Chapter 15, as are characteristics and treatment of this disorder. Heat cramps rarely occur in the elderly, since they are seldom able to perform the degree of physical activity necessary to produce large sweat losses.

MALNUTRITION Occurring in the aged are decreased taste sensations, decreased absorption of nutrients, and a greater need for vitamin supplements. In addition to these factors, the elderly are frequently malnourished becuase they typically have diminished incomes, frequently live alone, often wear (or need) dentures, tend to lack interest in food, and frequently have mobility problems that interfere with shopping and preparation of food.

Persons who are unable or unmotivated to eat a normal diet are susceptible to fluid and electrolyte imbalances. The elderly are particularly prone to imbalances owing to decreased absorption of nutrients from the gastrointestinal tract, excessive use of laxatives and antacids, and inadequate intake of nutrients. (Nutritional assessment is discussed in Chapter 5.) Malnutrition probably contributes to the loss of scalp hair and impairment in immunologic mechanisms seen in aged persons.

It should be noted that the "routine" preparations for many laboratory and radiologic tests (and some types of surgery) are often nutritionally depleting, particularly if they must be repeated. Unfortunately, it is not uncommon for errors to occur, causing tests to have to be repeated or surgery to be cancelled and then rescheduled. Rarely are the fluid, protein, and calories omitted in lost meals make up at a later time.

Prolonged use of "routine" intravenous fluids also predisposes to malnutrition. If parenteral fluids are needed longer than a week, provision for supplying nutrients such as amino acids, carbohydrates, and fats should be considered. (Intravenous fluids are discussed in Chapter 6.)

SPECIAL CONSIDERATIONS FOR THE AGED SURGICAL PATIENT Aged persons tolerate major surgery and its complications less well than younger adults; therefore, preoperatively, every effort should be made to treat conditions likely to cause postoperative difficulties. Cardiac complications are the greatest single cause of postoperative mortality; pneumonia and atelectasis are second.[16] Conscientious preoperative preparation often means the difference between success or failure of surgery in the aged, since the decreased body homeostatic adaptability of these patients predisposes to difficulty when they are exposed to stress. (Preoperative preparation is discussed in Chapter 7.)

The following facts apply to the aged:

Moderate fluid volume deficit and decreased circulating blood volume are not uncommon in the aged *before* operation. Fluid volume deficit predisposes to renal insufficiency, particularly in the aged. Administration of adequate intravenous fluids *before* surgery improves renal blood flow and renal function; in contrast, administration of the same fluids *after* induction of anesthesia will have only minimal effect in increasing renal blood flow. It is important that parenteral fluids be started *before* surgery and an adequate urine volume established *prior to induction* of anesthesia. Urine flow should be at least 50 ml/hr (preferably 75 ml to 100 ml/hr).

The plasma pH tends to remain fixed on the low side of normal in the aged, largely because of decreased pulmonary and renal function.

Since diminished respiratory function interferes with carbon dioxide elimination, many aged patients are in a state of impending respiratory acidosis. Because of this decline in pulmonary function, the nurse must help the aged patient achieve maximal ventilation. This can be accomplished by keeping the respiratory tract free of excessive secretions, providing maximal allowed activity, turning the bedfast patient from side to side at regular intervals, and avoiding restrictive clothing and chest restraints.

Because renal response to pH disturbances is not as efficient in the aged, imbalances occur faster. There is a tendency toward metabolic acidosis as a result of decreased renal function. It behooves the health care team to detect imbalances early and to intervene early.

As stated earlier, changes in pH are less well tolerated in the aged. Anemia, with its decreased hemoglobin, depletes one of the major buffer systems; emphysema is not uncommon in the aged and also disrupts pH control. Blood should be administered as needed to correct anemia, preferably several days before the operation. Measures to improve pulmonary function should also be employed preoperatively.

Hypoxia is not well tolerated in the aged; for this reason, local or spinal anesthetics are preferable to inhalation anesthetics. The operation should be completed in the shortest possible time consistent with safety.

Hypotension is poorly tolerated by the aged, and, unless corrected quickly, is frequently complicated by renal damage, stroke, or myocardial infarct. Shock becomes irreversible earlier than in younger patients. (Shock is discussed in Chapter 7.)

The aged patient will develop sodium deficit faster than younger adults; thus, the nurse should be particularly alert for this imbalance when the patient is losing body fluids containing sodium. Hyponatremia is particularly apt to occur when there is a free intake of water orally or when excessive volumes of 5% dextrose in water are administered IV.

Malnutrition is more common in the aged than in younger adults, contributing to the increased incidence of postoperative complications. Preoperative dietary management is particularly important. Optimal nutrition helps the aged patient withstand the electrolyte deficits and pH changes occurring with surgery. If the patient is unable to eat, tube feedings or parenteral nutrients are indicated to meet nutritional needs and build up operative reserve. (Tube feedings are discussed in Chapter 5, parenteral nutrition in Chapter 6.)

Ambulation and activity improve appetite and sleep and prevent the complications of bed rest. Bed rest in the preoperative period can be very damaging to the aged patient, since it predisposes to negative nitrogen balance, osteoporosis, muscle weakness, pneumonia, phlebitis, bedsores, bladder and bowel dysfunction, decrease in myocardial reserve, and diminished pulmonary ventilation and tidal volume.

A conservative dose of preoperative medication helps the aged patient avoid respiratory depression and hypoxia.

Gentle handling of tissue and careful technique are essential in geriatric surgery to avoid the development of complications; it has been recommended that only very experienced surgeons operate on elderly debilitated patients.

NONKETOTIC HYPEROSMOLAR COMA Most patients who develop nonketotic hyperosmolar coma (NKHC) are elderly persons with relatively mild (perhaps not yet detected) diabetes mellitus. Severe hyperglycemia exists; usually the history reveals a decrease in fluid intake secondary to illness, administration of potent medications, or invasive diagnostic or surgical procedures. Because the number of aged persons receiving more and more aggressive therapies is increasing, one would anticipate an increasing occurrence of NKHC. Since mortality is high in this condition, it is important that it be recognized and treated *early*. Monitoring the adequacy of fluid intake in the aged diabetic is an important nursing assessment that can lead to earlier detection of NKHC. A thorough discussion of NKHC is presented in Chapter 12; the reader is encouraged to review this section.

REFERENCES

1. Papper S: The effects of age in reducing renal function. Geriatrics, May 1973, p 84
2. Cole W: Medical differences between the young and the aged J Am Geriatr Soc, Aug 1970, p 592
3. Mitchell J: Chronic renal failure in the elderly: What to look for when it starts. Geriatrics, Nov 1980, p 28
4. Papper, p 87
5. Shapiro B, Harrison R, Walton J: Clinical Application of Blood Gases, 2nd ed, p 140. Chicago, Year Book Medical Publishers, 1977
6. Cole W, p 594
7. Condon R, Nyhus L (eds): Manual of Surgical Therapeutics, 4th ed, p 246. Boston, Little, Brown & Co, 1978
8. Berman P, Kirsner J: Recognizing and avoiding adverse gastrointestinal effects of drugs. Geriatrics, June 1974, p 60
9. Rogers B, Silvis S, Nebel O et al: Complications of flexible fiberoptic colonoscopy and polypectomy. Gastrointest Endosc 24, No 1:25, 1977
10. Skillman T: Can osteoporosis be prevented? Geriatrics, Feb 1980, p 95
11. *Ibid*, p 96
12. *Ibid*, p 98
13. *Ibid*, p 99
14. Sprung C: Heatstroke; Modern approach to an ancient disease. Chest 77, No. 4:462, 1980
15. Sprung C: Hemodynamic alterations of heatstroke in the elderly. Chest 75, No. 3:365, 1979
16. Cole, p 590

BIBLIOGRAPHY

Baker M et al: Plasma 25-hydroxy vitamin D concentrations in patients with fractures of the femoral neck. Br Med J 1 (6163):589, 1979
Libow L, Sherman F: The Core of Geriatric Medicine. St. Louis, CV Mosby, 1981

Fluid and electrolyte problems associated with the development and treatment of cancer in man are being recognized more frequently owing to increased technology and longer survival of cancer patients. These fluid and electrolyte problems may occur rapidly, and hence constitute emergency states, or they may be chronic in nature. Fluid and electrolyte problems may cause the initial symptoms exhibited by the undiagnosed cancer patient, or they may be related to the development of metastatic disease. *Fifty to seventy-five percent of all cancer patients will experience a problem with fluid and electrolyte regulation during their illness.*[1]

CAUSES OF FLUID AND ELECTROLYTE PROBLEMS IN THE CANCER PATIENT

Electrolyte imbalances in the cancer patient may be caused by the primary disease or they may be a consequence of its treatment. The underlying mechanisms for imbalances may be the direct result of increased electrolyte loss or decreased intake or the indirect result of faulty hormonal regulation of body water and electrolytes. (A summary of fluid and electrolyte imbalances in the cancer patient, and of their causes, is presented in Table 19-1.)

FAULTY REGULATION OR AVAILABILITY OF ELECTROLYTES

Fluid and electrolyte problems associated with cancer are most commonly related to faulty regulation or availability of one of the following substances:

> Calcium
> Uric acid
> Sodium
> Potassium
> Phosphate

These substances may be present in excessive or diminished amounts, accounting for the variance in clinical signs exhibited by cancer patients with fluid balance problems.

*This chapter was written by Roberta P. Scofield, R.N., M.S.N., Oncology Clinical Nurse Specialist, Veterans Administration Medical Center, St Louis, Missouri.

19

Fluid Balance in the Patient with Malignant Disease*

TABLE 19-1 *Fluid and electrolyte imbalances in cancer patients*

Electrolyte imbalance	Physiologic process	Underlying cause/type of cancer
Hypercalcemia	Increased osteolytic activity of tumor	Breast tumor Multiple myeloma Prostate tumor Cervical tumor Melanoma
	Prostaglandin activity	? Lymphoma Leukemia
	Ectopic PTH production	Epidermoid lung tumor Renal adenocarcinoma Oral cavity squamous tumor Bladder tumor Parotid tumor Esophageal tumor
	Ectopic OAF release	Lymphoma Multiple myeloma Acute and chronic leukemia
	Metabolic alterations	Treatment with: • Estrogens • Progestins • Androgens • Antiestrogens
Hypocalcemia	Increased utilization of calcium	Thyrocalcitonin release from thyroid tumors "Hungry-bone syndrome" (rapid healing of osseous lesions)
	Decreased intake of calcium	Starvation Hyperalimentation (if solution is deficient in calcium)
	Decreased absorption of calcium	Fistulas (allow loss of fluid before absorption can occur) Lymphoma of small bowel (tumor of such magnitude that it interferes with small bowel absorptive area)
	Decreased serum albumin	Starvation Transudation of protein into third-space fluid accumulation
Hyperuricemia	Increase in uric acid precursors	Polycythemia vera Chronic granulocytic leukemia
	Rapid tumor dissolution with increased nucleic acid release	Induction therapy for acute leukemia Radiosensitive/chemotherapy-sensitive lymphomas
Hyponatremia	Loss of sodium	Fistulas Prolonged nausea and vomiting associated with chemotherapy
	Ectopic ADH production, causing abnormal water retention	Oat cell lung tumor Acute nonlymphocytic leukemia Chemotherapy-induced: • High-dose cyclophosphamide (Cytoxan)—over 1.5 g • Vincristine (Oncovin)

(continued)

TABLE 19-1 *Fluid and electrolyte imbalances in cancer patients* (continued)

Electrolyte imbalance	Physiologic process	Underlying cause/type of cancer
Hypernatremia	Increased water loss	Renal damage associated with multiple myeloma
Hyperkalemia	Increased release of potassium from cellular breakdown	Induction therapy for acute leukemia (due to dissolution of large tumor masses)
Hypokalemia	Ectopic ACTH production from tumor, resulting in potassium loss	Oat cell lung tumor
		Thymoma
		Islet cell pancreatic tumor
		Bronchial carcinoid tumor
	Increased gastrointestinal losses of potassium	Prolonged vomiting
		Suction
		Diarrhea
		Fistulous drainage
	Increased renal tubular loss of potassium	Tubular damage from antibiotic therapy for fever in neutropenic leukemia patients
		Tubular damage from induction therapy for acute leukemia
Hypophosphatemia	Increased urinary excretion of phosphate	Multiple myeloma
	Respiratory alkalosis[2]	Sepsis (recall that fever causes hyperventilation, which in turn causes respiratory alkalosis)
	Decreased phosphate intake	Starvation
		Hyperalimentation (if solution is deficient in phosphate)
Hyperphosphatemia	Increased release of phosphate from cellular breakdown	Acute leukemia
		Lymphoma

ALTERED HORMONAL REGULATORY MECHANISMS

Electrolyte imbalances in the cancer patient are often due to alterations in various hormonal regulatory mechanisms caused by a variety of tumors. Various malignant tumors produce ectopic hormones or pseudohormone substances that interfere with water and electrolyte balance. These hormones (or hormonelike substances) are listed below:

Antidiuretic hormone (ADH)
Parathyroid hormone (PTH)
Osteoclastic activating factor (OAF)
Adrenocorticotropic hormone (ACTH)

The production or release of the tumor hormone is not regulated by normal suppression feedback loops; consequently, the ectopic hormone continues to be released by the tumor, often causing severe, life-threatening electrolyte imbalances.

THIRD-SPACE FLUID ACCUMULATION

Cancer patients frequently develop third-space fluid accumulations; these may be due to the following:

Malignant effusions into the peritoneal, pericardial, or pleural compartments. Direct tumor involvement of the serous surface of the cavity appears to be the most frequent cause of effusions in cancer patients.

Edema; trapping of excess fluid in the interstitial fluid space due to obstruction of lymphatic drainage or venous return secondary to tumor pressure. Protein seepage through the capillary bed in the edematous site pulls fluid with it, making the edema more severe.

Major shifts of water and electrolytes into potential fluid spaces result in an increase in total body fluid, yet a clinically defined state of fluid volume deficit is present. Recall that fluid volume deficit is manifested by poor tongue and skin turgor, decreased urinary output, increased urinary specific gravity, a drawn facial expression, sunken eyeballs, and postural hypotension. There is *not* the expected decrease in body weight with the presence of trapped fluid. Unfortunately, administration of intravenous fluids to correct the fluid volume deficit also allows an increase in the fluid volume trapped in the third space, adding to the patient's discomfort.

TREATMENT-RELATED IMBALANCES

It should be noted that the treatment of malignancies can *create* fluid and electrolyte imbalances. Examples of treatment-induced imbalances include the following:

Hypercalcemia associated with hormonal treatment (tamoxifen [Nolvadex]) for breast cancer

Hyponatremia and hypokalemia associated with nausea and vomiting frequently caused by chemotherapy

Hyponatremia (water intoxication) associated with the use of certain chemotherapeutic drugs (vincristine [Oncovin] and cyclophosphamide [Cytoxan])

FLUID AND ELECTROLYTE DISTURBANCES

Fluid balance problems common to patients with cancer are described in this section. Mechanisms responsible for the imbalances, as well as the tumors (or associated problems) causing the imbalances to occur, are described. A summary of this information is presented in Table 19-1.

Treatment to correct common electrolyte imbalances is also described in this section. It should be noted that treatment of tumor-related fluid and electrolyte problems is directed first at relieving immediate life-threatening problems and secondarily at controlling the tumor underlying the basic physiological response.

CALCIUM

HYPERCALCEMIA The most common nonendocrine cause of hypercalcemia is malignant neoplastic disease.[3] Elevated serum calcium levels are associated with a wide variety of malignant tumors and treatment programs, including the following:

Solid tumors with bony metastases (breast, prostate, cervical, and malignant melanoma) and without bony metastases (lung, head and neck, and renal tumors)

Hematologic tumors (lymphoma, acute leukemia, and myeloma)

Selected chemotherapeutic regimens (treatment with androgens, estrogens, progestins, and antiestrogens)

Hypercalcemia associated with solid tumors occurs in cancer of the breast, lung, kidney, head and neck, cervix, and prostate, in neuroblastoma, and in other miscellaneous solid tumors. The highest incidence of hypercalcemia is found in patients with breast cancer.[4] In a series of 430 hypercalcemic cancer patients, Myers found that 86% of the subjects had scintiscan-demonstrated skeletal lesions suggesting bony destruction as the cause of the hypercalcemia.[5] An additional 10% of the patients had solid tumors without scintiscan demonstration of bony lesions; these patients were also hypercalcemic. In the same population of 430 patients, Myers noted that patients with nonsolid hematologic tumors (lymphomas, leukemias, and multiple myeloma) also had elevated serum calcium levels. The following brief review suggests a variety of pathways other than osseous metastases as a basis for the hypercalcemia associated with cancer.

Mechanisms The basic mechanism responsible for an elevated serum calcium level is an increase in calcium release from bone that exceeds calcium excretion from renal tubules. The following physiologic mechanisms may be responsible for the increase in release of calcium from bone in neoplastic states:

Tumor-related:
Direct tumor invasion of bone
Prostaglandin activity
Tumor release of PTH or pseudo-PTH substance
Immune cell release of OAF
Tumor release of vitamin D-like sterols
Treatment-related:
Treatment with androgens, estrogens, or progestins
Non-tumor-related:
Coincidental primary hyperparathyroidism

The presence of malignant tumor cells in bony spaces has been demonstrated to cause an increased release of calcium from bone and a loss of skeletal integrity.[6] This direct osteolytic action is commonly the mechanism for hypercalcemia in women with breast cancer. In addition to the direct resorption of bone by tumor cells, prostaglandin action has been proposed as a mechanism for increased calcium release from bone.[7] Prostaglandins may be produced from one of the following:[8]

Tumor cells
Immune cells
Osteoclasts

Animal models have demonstrated that prostaglandins have osteolytic activity.[8] It is not known if this animal model principle can be transferred to humans.

Solid tumors without bony metastases have been associated with hypercalcemic states, as documented in patients with squamous cell carcinoma of the head and neck and carcinomas of the lung and kidney. A possible explanation for this nonosteolytic action is tumor production of either a PTH-like mediator of calcium release or a vitamin D-like sterol.[9] The tumor either elaborates ectopically the PTH-like substance or produces a substance that stimulates the parathyroids to produce excessive PTH.[10] (Parathyroid hyperplasia does not appear to be associated with the second hypothesis.) More investigation is needed to demonstrate the actual mechanisms responsible for hypercalcemic clinical syndromes in patients with radiographically negative skeletal series.

Immune cells (lymphocytes and monocytes) have also been implicated in the production of hypercalcemia. There are three possible mechanisms whereby monocytes may lead to the release of calcium from bone:[11]

Release of prostaglandins
Direct bone resorption through contact action
Release of OAF

Monocytes, through direct contact with bony tissue, cause a dissolution of bone. Kahn has demonstrated that monocytes, frequently present in metastatic bony lesions, release an enzyme that causes bone destruction.[12] Bockman has developed a model that demonstrates the synergistic action between tumor cells and immune cells in producing hypercalcemia:[13]

1. Immediate:
   Tumor cells initiate osteolytic process at metastatic site
   Host cells (monocytes) and tumor cells cause active bone resorption

2. Intermediate:
   Bony resorption stimulates an increase in migration of tumor cells and monocytic phagocytes to the lesion
   Lytic process is augmented

3. Late:
   Tumor cell resorption is primary lytic process

In addition to prostaglandin release and direct contact-mediated bone resorption, immunocytes may release an osteoclastlike activating factor (OAF) that increases the release of calcium from bone. This is a proposed mechanism for hypercalcemia in patients with lymphomas and leukemias.[14]

The only treatment-related cause of hypercalcemia is that associated with the use of androgens, estrogens, antiestrogens, and progestins. Metabolic alterations associated with the use of these agents cause calcium release from the bone, resulting in elevation of the serum calcium level. *Any patient receiving these agents should have frequent serum calcium determinations; any elevation of the serum calcium level warrants discontinuance of the drug.* No further therapy is indicated once the hormonal therapy is stopped.[15]

Clinical presentation and diagnosis The symptoms exhibited by the cancer patient with hypercalcemia are the same as those in hypercalcemic patients with primary hyperparathyroidism (see Chapter 12). In review, symptoms of hypercalcemia may include anorexia, nausea and vomiting, constipation, muscular weakness and hyporeflexia, confusion, psychosis, tremor, and lethargy. The differential diagnosis in a cancer patient with hypercalcemia should include primary hyperparathyroidism. The following factors should be considered in distinguishing a tumor-associated hypercalcemia from that seen in primary hyperparathyroidism:[16]

The type of malignant tumor. Is it one commonly associated with hypercalcemia?

The patient's history. Was there a history of hypercalcemia *prior* to the diagnosis of a malignancy?

The effect of anticancer therapy on the hypercalcemia. If hypercalcemia is tumor-induced, anticancer therapy should alleviate it.

Treatment The treatment for tumor-related hypercalcemia is similar to that of other clinical states of elevated serum calcium levels (Table 19-2). A more thorough discussion of the treatment of hypercalcemia is presented in Chapter 12 in the section dealing with

TABLE 19-2 *Management of hypercalcemia in the cancer patient*

Short-term/Acute

Decrease serum calcium level by increasing renal excretion of calcium:

Increase hydration by means of intravenous fluids, 3 to 4 liters or more per day (including saline infusions)

Increase renal calcium excretion by means of furosemide (Lasix) or ethacrynic acid (Edecrin) administration
(Recall that sodium excretion is accompanied by the excretion of calcium; therefore, inducing sodium diuresis by the above two methods is helpful in reducing the serum calcium level.)

Decrease the tumor's hypercalcemic effect by means of chemotherapy, surgery, or radiation

Decrease serum calcium level by decreasing bone resorption:

Increase patient mobilization (recall that immobilization favors hypercalcemia)

Administer phosphates and steroids

If refractory to the above two treatments, it may be necessary to administer mithramycin (Mithracin) or actinomycin D

Long-term/Chronic

Control malignant tumor

Institute hemodialysis or peritoneal dialysis

Administer phosphates (long-term)

---

hyperparathyroidism. In patients in whom the calcium level is less than 13 mg/dl and symptoms are not clinically evident, treatment with traditional anticancer therapy may be all that is needed. If mithramycin (Mithracin) is needed to treat the hypercalcemic cancer patient, one should be alert for problems associated with thrombocytopenia and bleeding diatheses. (Recall that the patient who has received previous chemotherapy may be thrombocytopenic *before* the additional platelet insult of mithramycin.)

In addition to the immediate control of symptoms, long-term therapy for the cancer patient with tumor-induced hypercalcemia must be considered. One should remember that while cancer patients with hypercalcemia may respond rapidly to emergency treatment, the serum calcium level may return to an elevated state and remain a chronic management problem unless the malignancy is controlled with antineoplastic therapy.

HYPOCALCEMIA Malnutrition with resultant decrease in serum albumin is a major cause of hypocalcemia in cancer patients. A decrease in serum albumin of 1 g/dl will result in a reduced serum calcium level.[2] Although uncommon in cancer patients, hypocalcemia may occur secondary to the following:

Severe liver disease

Malabsorption due to extensive small bowel resection or tumor replacement (tumor in small intestine of such magnitude that it interferes with bowel absorptive surface)

"Hungry bone syndrome" due to rapid bone healing after successful treatment of bony lesions

Magnesium depletion

Symptoms of hypocalcemia include weakness, fatigue, irritability, progressive paresthesia, carpopedal spasms, and tetany. Treatment of hypocalcemia consists of oral or intravenous calcium gluconate administration (utilizing serial serum calcium levels until normal states are achieved).[17]

URIC ACID

HYPERURICEMIA An elevation of the serum uric acid level in cancer patients is usually related to the following:

Rapid release of nucleic acids from tumor cell lysis after successful radiation or chemotherapy such as occurs in lymphomas, acute leukemia, or myeloma.[18] Hyperuricemia is commonly observed in patients who are receiving chemotherapy.

A tumor that overproduces uric acid precursors, such as polycythemia vera and chronic granulocytic leukemia. Certain rapidly proliferating neoplasms with a high nucleic acid turnover may present with hyperuricemia even in the absence of prior chemotherapy.

Malignant tumors with high growth rates and high cell mass release excessive amounts of nucleic acids when treated; in turn, nucleic acid metabolism leads to elevation of the serum uric acid level. It should be noted that leukemias and lymphomas traditionally respond rapidly to treatment; this response can create a potential for hyperuricemia if the patient is not treated prophylactically. Hyperuricemia should be viewed as a *preventable* complication rather than as an emergency condition inasmuch as uric acid formation can be inhibited.

Prophylactic treatment of hyperuricemia consists of adequate hydration, administration of allopurinol (Zyloprim), and alkalinization of the urine. Allopurinol inhibits uric acid formation, hydration aids in the excretion of uric acid, and alkalinization of the urine increases solubility of uric acid (Table 19-3.) Untreated serum uric acid levels may reach 20 mg/dl in cancer patients receiving induction therapy for acute leukemia.[19] The danger of uric acid nephropathy is present when the serum uric acid concentration exceeds 15 mg/dl. The patient with an elevated uric acid level may exhibit ureteral colic, oliguria, and azotemia. Serum uric acid levels greater than 80 mg/dl are lethal. Hemodialysis should be considered in patients with extremely high uric acid levels. Renal damage due to urate nephropathy should be prevented, particularly since patients who suffer from hyperuricemia may be entering a stage of remission and have a good prognosis.

HYPONATREMIA Isaacs reports that hyponatremia is the most common electrolyte imbalance seen in her patient population.[20] Reasons cited for decreased serum sodium levels in cancer patients include the following:

Ectopic release of an ADH-like substance from certain tumors, especially oat cell bronchogenic cancer

Salt loss from the gastrointestinal tract, as occurs in vomiting or fistulous drainage

Hyperproteinemia states, including multiple myeloma and hyperglobulinemia

The syndrome of secretion of inappropriate antidiuretic hormone (SIADH) may occur in cancer patients with certain tumors. These tumors produce an ectopic ADH-like substance in the face of decreasing serum sodium and serum osmolality. The serum sodium level falls dangerously low and creates a myriad of clinical manifestations (see Table 13-1). A number of pharmacologic agents can also cause SIADH (see Table 2-2). Antitumor agents associated with SIADH include cyclophosphamide (Cytoxan) and vincristine (Oncovin).

Surgical removal of the tumor is the most direct treatment. Chemotherapy and radiation therapy may be palliative and temporarily increase the sodium level by decreasing formation of the ectopic hormone. Other standard measures to control water excess and restore normal serum sodium concentrations are described in Chapter 2. On a short-term basis, fluid intake is restricted to the extent that negative water balance is induced; this may require as little as 400 ml to 700 ml/24 hr. Administration of furosemide (Lasix) plus salt replacement has accomplished an increase in the serum sodium concentration. The use of demeclocycline (Declomycin) for long-term control of hyponatremia may be indicated, particularly in patients with uncontrolled malignancies; this agent inhibits the action of ADH on the kidneys, allowing increased urinary output and thus liberalization of fluid intake. (Many patients find the long-term restriction of water extremely unpleasant.)

POTASSIUM

HYPERKALEMIA Rapid cell destruction following successful treatment of large tumors may cause a precipitous release of cellular potassium and an increase in the serum potassium level. This is most frequently

---

TABLE 19-3 *Management of hyperuricemia in the cancer patient*

I. Prevention

Administer allopurinol (Zyloprim) 100 mg three times daily (starting one day before chemotherapy) to inhibit uric acid formation

II. Emergency Treatment

A. Decrease uric acid concentration in the urine by means of adequate hydration (sufficient fluid intake to achieve an output of 4 liters/day)

B. Decrease uric acid production by administration of allopurinol 300 mg/day (1800 mg/day in severe cases)

C. Increase uric acid solubility by increasing urine *p*H to 7.0 or above by means of sodium bicarbonate administration

III. Once *nephropathy* is established, emergency hemodialysis or peritoneal dialysis may be indicated; excessive fluid administration in the presence of renal failure is dangerous

seen during induction therapy for acute leukemia. An increased serum potassium level can also be a result of acute renal failure due to untreated hyperuricemia.

A falsely high serum potassium level may be reported in the presence of high white blood cell counts. (Increased leukocyte fragility and lysis of cells after venous sampling may falsify the level in the clotted serum sample.) Validation of a supposed hyperkalemic state with expected electrocardiogram changes and other clinical signs of hyperkalemia is needed to determine if the serum potassium level is truly elevated in a patient with a high white count due to a leukemic process.[21]

HYPOKALEMIA Excessive loss of potassium due to renal tubular damage occurs in 60% of patients during induction therapy for acute nonlymphocytic leukemia.[22] The renal tubular damage is attributed to two processes:

Acute tubular necrosis due to antibiotic therapy during neutropenic states; drugs that may produce this effect include gentamycin, carbenicillin, and cephalothin

Elevated lysozymuria due to cellular breakdown

Both of these mechanisms for increased potassium loss revert to normal after the patient recovers from induction therapy.

Nonhematologic malignancies may also produce a decreased serum potassium level related to an ACTH-like substance released by tumor cells. These tumors include the following:

Oat cell carcinoma of the lung

Thymoma

Pancreatic islet cell carcinomas

Medullary carcinoma of the thyroid[23]

Imura reports that 50% of patients with bronchogenic carcinoma exhibit decreased serum potassium due to ACTH-like activity.[24]

Patients with ACTH-like tumor-related hypokalemia do not usually exhibit other clinical signs of Cushing's disease. The main manifestations of ectopic ACTH tumor release are an elevated urinary 17-hydroxycorticosteroid content, decreased serum potassium, edema, and the presence of ACTH in the tumor extract. As with ectopic ADH production, normal feedback regulatory inhibition does *not* occur; thus, testing with normal-dose dexamethasone suppression is not valid. Only high-dose dexamethasone administration will depress ACTH levels due to ectopic tumor pro-

duction.[25] Surgical removal of the tumor is warranted to control the production of the ectopic ACTH. If this is not possible, surgical adrenalectomy or treatment with aminoglutethimide may be indicated. Potassium losses may be replaced with oral supplements.

PHOSPHATE

HYPERPHOSPHATEMIA Massive tumor breakdown in acute leukemic patients undergoing chemotherapy and in lymphoma patients receiving radiation may result in an elevated serum phosphate level. Hyperphosphatemia usually occurs concomitantly with hyperuricemia and hyperkalemia. The administration of phosphate-binding gels with close monitoring of serum phosphate levels is indicated in the treatment of patients with hyperphosphatemia. (Careful observation of the serum calcium level is indicated, as concurrent hypocalcemia may ensue during the presence of an elevated serum phosphate level.[2])

HYPOPHOSPHATEMIA Hypophosphatemia may be associated with increased renal losses of phosphate due to renal tubular defects in multiple myeloma, decreased intake of phosphate due to starvation or prolonged hyperalimentation, and respiratory alkalosis associated with septicemic episodes in neutropenic patients.

Treatment consists of oral phosphate replacement by administration of potassium–phosphate solutions. Intravenous replacement may be indicated if the loss is severe.[26]

SUMMARY OF TREATMENT

Many of the fluid and electrolyte problems experienced by patients with cancer are managed traditionally with replacement or restrictive treatment modalities. However, the underlying cause of the electrolyte problem must be kept in mind or the problem will become a chronic one. Elimination or control of the tumor is a primary consideration. Careful, continued observation of the patient is indicated to effectively control the malignancy and its associated fluid balance problems. The reader is referred to Chapter 4 for a review of assessment parameters associated with fluid balance.

REFERENCES

1. Dorr R, Fritz W: *Cancer Chemotherapy Handbook*, p 180. New York, Elsevier, 1980
2. Isaacs M: Life threatening fluid and electrolyte abnormalities in patients with cancer. Curr Probl Cancer 4, No. 3:8, 1979
3. Besorab A, Caro J: Mechanisms of hypercalcemia in malignancy. Cancer 4:2276, 1978
4. Myers W: Differential diagnosis of hypercalcemia and cancer. CA 27, No. 5:258, 1977
5. *Ibid*, p 259
6. Bockman R: Hypercalcemia in malignancy. Clin Endocrinol Metab 9, No. 2:317, 1980
7. Besorab, p 2279
8. Bockman, p 320
9. Besorab, p 2280
10. Myers, p 264
11. Bockman, pp 320–324
12. Kahn A, Stewart C, Teitelbaum S: Contact mediated bone resorption by human monocytes in virto. Science 199:988, 1978
13. Bockman, p 322
14. Myers, p 266
15. Dorr & Fritz, p 153
16. Shnider B, Manalo A: Paraneoplastic syndromes: Unusual manifestations of malignant disease. DM 25, No. 5:39, 1979
17. Haskell C: Cancer Treatment, p 998. Philadelphia, WB Saunders, 1980
18. *Ibid*, p 999
19. See-Lasley K, Ignoffo R: Manual of Oncology Therapeutics, p 357. St Louis, CV Mosby, 1981
20. Isaacs, p 11
21. O'Reagan S et al: Electrolyte and acid–base disturbances in the management of leukemia. Blood 49, No. 3:348, 1977
22. See-Lasley & Ignoffo, p 314
23. Blackman M et al: Ectopic hormones. Adv Intern Med 23:93, 1978
24. Imura H: Ectopic hormone syndromes. Clin Endocrinol Metab 9, No. 2:239, 1980
25. *Ibid*, p 242
26. Isaacs, p 9

Index